29.50
JUN. 1975

Prentice-Hall Computer Applications in Electrical Engineering Series
FRANKLIN F. KUO, *Editor*

ABRAMSON AND KUO *Computer-Communication Networks*
BOWERS AND SEDORE *SCEPTRE:*
  A Computer Program for Circuit and Systems Analysis
CADZOW *Discrete-Time Systems: An Introduction with Interdisciplinary Applications*
CADZOW AND MARTENS *Discrete-Time and Computer Control Systems*
DAVIS *Computer Data Displays*
FRIEDMAN AND MENON *Fault Detection in Digital Circuits*
HUELSMAN *Basic Circuit Theory*
JENSEN AND LIEBERMAN *IBM Circuit Analysis Program:*
  Techniques and Applications
JENSEN AND WATKINS *Network Analysis: Theory and Computer Methods*
KOCHENBURGER *Computer Simulation of Dynamic Systems*
KUO AND MAGNUSON *Computer Oriented Circuit Design*
LIN *An Introduction to Error-Correcting Codes*
NAGLE, CARROLL, AND IRWIN *An Introduction to Computer Logic*
RHYNE *Fundamentals of Digital Systems Design*
SIFFERLEN AND VARTANIAN *Digital Electronics with Engineering Applications*
STAUDHAMMER *Circuit Analysis by Digital Computer*
STOUTEMYER *PL/1 Programming for Engineering and Science*

CIRCUIT ANALYSIS BY DIGITAL COMPUTER

JOHN STAUDHAMMER

North Carolina State University at Raleigh

PRENTICE-HALL, INC.

Englewood Cliffs, New Jersey

Library of Congress Cataloging in Publication Data

Staudhammer, John
   Circuit analysis by digital computer.

  Includes bibliographical references.
   1. Electronic data processing—Electric networks.
I. Title.
TK454.2.S73   621.319′2′02854   74-19080
ISBN 0-13-133967-2

© 1975 by Prentice-Hall, Inc.
Englewood Cliffs, N. J.

All rights reserved. No part of this book
may be reproduced in any form or by any means
without permission in writing from the publisher.

Printed in the United States of America

10 9 8 7 6 5 4 3 2 1

PRENTICE-HALL INTERNATIONAL, INC., *London*
PRENTICE-HALL OF AUSTRALIA, PTY. LTD., *Sydney*
PRENTICE-HALL OF CANADA, LTD., *Toronto*
PRENTICE-HALL OF INDIA PRIVATE LIMITED, *New Delhi*
PRENTICE-HALL OF JAPAN, INC., *Tokyo*

To the Memory of my Teachers

LOUIS ALBERT PIPES (1910–1971)
SAMUEL TERRIL YUSTER (1903–1958)
L. M. K. BOELTER, Dean (1898–1966)

*Professors of Engineering at the
University of California, Los Angeles*

CONTENTS

PREFACE   xi

Chapter 1. **THE CIRCUIT ANALYSIS TASK**   1

1.1	The Nature of Circuit Analysis	1
1.2	The Tools of Network Analysis	5
1.3	Computational Considerations	7
1.4	Modeling	10
1.5	Nonlinear Networks	13
1.6	Desirable Features of Programs	14
	References	16

Chapter 2. **SOME ELEMENTS OF LINEAR ALGEBRA**   18

2.1	Basic Definitions	19
2.2	Tabular Matrix Multiplication	22
2.3	Linear Simultaneous Equations	24
2.4	Gauss Elimination Program	30
2.5	Gauss-Seidel Procedure	32
2.6	Matrix Inversion	35
2.7	Numerical Considerations	37
2.8	In-Place Inversion with Pivoting	43
2.9	Summary	46
	References	46
	Problems	47

Chapter 3. **DC AND AC ANALYSIS OF NETWORKS**   50

3.1	Basic Relationships	51
3.2	Analysis of Arbitrary Networks	64
3.3	Program for DC Analysis	72
3.4	AC Analysis of Linear Networks	83
3.5	A Simple AC Analysis Program	86

3.6	Mutual Inductance Model	96
3.7	Direct Calculation of the Node Admittance Matrix	97
3.8	Sparse Matrix Analysis	100
3.9	Unaccompanied Energy Sources	105
3.10	Dependent Voltage Sources	108
3.11	Free-Form Input	113
3.12	Summary	121
	References	122
	Problems	122

Chapter 4. TRANSIENT ANALYSIS OF NETWORKS — 130

4.1	Models for Energy Storage Elements	131
4.2	A Rudimentary Transient Analysis Program	138
4.3	General Equations	141
4.4	Time-Variant Sources	147
4.5	A Basic Transient Analysis Program	150
4.6	Final-Value Solutions	165
4.7	Piecewise Linear Networks	166
4.8	Sparse Matrix Analysis	170
4.9	Iteration on Element Values	170
4.10	Summary	171
	References	172
	Problems	173

Chapter 5. STATE SPACE ANALYSIS — 176

5.1	General Relationships	177
5.2	Reduction to Repeated DC Analysis	179
5.3	Detection of Surplus Variables	185
5.4	Time-Domain Solution	190
5.5	Program for the Formulation of the State Equations	200
5.6	Poles and Zeros	220
5.7	Frequency Response	238
5.8	Stiff Differential Equations	240
5.9	Summary	247
	References	247
	Problems	248

Chapter 6. TOLERANCE ANALYSIS — 252

6.1	Derivatives and Sensitivity	254
6.2	Statistical Variations	267
6.3	Worst-Case Analysis	274
6.4	A Simple Sensitivity Program	279
6.5	Correlated Parameters	289
6.6	Monte Carlo Analysis	292
6.7	A Comprehensive Example	299

6.8	Summary	307
	References	308
	Problems	308

Chapter 7. NONLINEAR ANALYSIS   311

7.1	Basic Relationships	312
7.2	Diode Equivalent Circuits	322
7.3	Large-Signal Model for Transistors	333
7.4	Piecewise Linear Transistor Model*	342
7.5	Equation Iterations	349
7.6	Organization of A Nonlinear Program	352
7.7	Input Language Considerations	354
7.8	Illustrative Examples	360
7.9	Automatic Modeling	374
7.10	Summary	383
	References	384
	Problems	385

Appendix A. COMPUTER PROGRAM LISTINGS   387

A.1	Subroutine SOLVER	387
A.2	Subroutine INVRT	388
A.3	Code Changes for Subroutine SOLVX	389
A.4	Code Changes for Subroutine INVRX	390

Appendix B. PLOTTING SUBROUTINES   391

B.1	Instructions for Use of Plotting Routines	391
B.2	Subroutine PLOTXY	394
B.3	Subroutine PLT2F	395
B.4	Subroutine PLT4XY	397

Appendix C. SEMICONDUCTOR MODEL LIBRARY   398

C.1	Diode Parameters	398
C.2	Transistor Parameters	401

INDEX   409

PREFACE

This book is intended as an introduction to the systematic analysis of lumped-parameter networks. Throughout the book the node method of analysis is used. This basically consists of summing the currents at each node in the network with unknown voltages at the nodes and solving the resultant equations for these node voltages. Emphasis is placed upon procedures which generate the node voltage equations automatically from a knowledge of the circuit topology and the circuit components. The reason for choosing this method is that in a network the node voltages with respect to ground generally are uniquely determined and the current summations at each node yield a sufficient number of equations for the solution of the node voltages. In contrast, the independent loops in a network are not unique and care must be exercised in choosing a sufficient number of them for solving a given network. A secondary reason is that in the vast majority of networks there will result fewer node equations than loop equations. For these reasons, most automated circuit analysis programs use the node method of analysis.

The book is intended as an introduction to computer-oriented procedures for the analysis of circuits. Rather than exploring the theoretical foundations of circuit analysis, the material is concerned more with an exposition of the problems and ideas for solution for various aspects of circuit analysis. For the same reason, problems of uniqueness and existence of solutions are not discussed. The circuits for which analysis procedures are discussed will tacitly be assumed to possess solutions; if the resultant programs are then applied to circuits where solutions do not exist, the procedures will lead to unreasonable numerical results. For circuits arising from production analysis and reasonable design processes the methods presented here will result in usable programs.

It is assumed that the reader has available computing facilities in the form of time-sharing services, batch facilities or medium-scale laboratory computers. The reader is expected to be conversant with FORTRAN. Through the book, an attempt is made to use elementary FORTRAN statements in earlier chapters with more advanced FORTRAN being introduced gradually. The computer programs presented were checked out on an IBM 370/165 system using WATFIV and the FORTRAN-IV,

Level G compilers. A number of the programs were also run on a CDC 6600 computer using the Extended FORTRAN compiler, on a UNIVAC 1108 system in FORTRAN-V, on the GE Timeshare service, MARK II, and on an ADAGE AGT/30 system using FORTRAN-4. While a few differences arose due to compiler requirements, chiefly because of differences in DO-statement interpretation and due to word-length differences, the programs are checked out and should execute with only minor differences on all ASI FORTRAN-IV compilers. An attempt was made to remove all dialectical peculiarities from the program listings.

While the computer procedures are presented in FORTRAN, a number of the programs were on occasion written in BASIC and a few were implemented in PL/1. Each of these languages has its own advantages; the choice of FORTRAN for the presentation in this book was dictated by the wide use of that language in engineering practice.

The material in this book was presented in a senior-graduate course at North Carolina State University for several years. Portions of this book were used in note-form for short courses presented by the author for a period of 8 years at the University of California, Los Angeles; at Utah State University; at the University of Houston; at the University of Wisconsin and in courses at Arizona State University, as well as in several industrial and governmental organizations. In the regular courses, a secondary teaching objective was to expand the students' expertise in programming and in matrix manipulations on a computer.

The background expected of readers of this book is basically some experience in elementary linear circuit analysis, an understanding of terms in circuit theory, familiarity with a higher-level programming language, preferably FORTRAN-IV, some work in elementary matrices and numerical procedures. Normally all this background could be acquired by the end of the sophomore year in an electrical engineering curriculum. Some introduction to numerical matrix procedures is given in Chapter 2, but the serious reader will probably want a deeper background, such as a one-semester course in linear algebra. Likewise, a course in linear system analysis, normally a junior-year course, would be desirable for a deeper appreciation of the material presented here. Thus the normal background of students for this book would be achieved at the end of the junior year in a curriculum of electrical engineering, physics or systems engineering.

The material in this book deals with procedures for the automated analysis of networks, linear as well as non-linear. The bulk of the material presented deals directly with linear circuits. The extension to nonlinear networks is done later and it is shown how the analysis of such networks can be reduced to a series of analyses of linear circuits which approximate the action of the nonlinear network.

Some motivation for and uses of the procedures are discussed in Chapter 1, an introduction to the materials necessary for the background of the problem. Chapter 2 is a discussion of simple numerical procedures for matrix manipulations by computer, solution of simultaneous linear equations and matrix inversion. The emphasis is on exposing the ideas underlying these procedures, rather than the resultant programs: every computer center has fully checked-out programs for these tasks. However, since modifications in these programs are often desired, listings of the resultant programs are included in Appendix A for use in problem solutions in later chapters.

Procedures for linear circuit analysis are presented in Chapters 3 through 6. The fundamental equations of node analysis in forms suitable for DC networks and AC steady-state analysis are derived in the early sections of Chapter 3. The derivation presented there forms the basis of all analysis procedures in this book. Later sections of Chapter 3 streamline the computer implementations of the analysis equations. Some consideration is given to computer memory use and numerical difficulties and ideas for improvement of execution speed are presented. The chapter discusses implementation of a simple input language so that personnel with minimal computer training can make full use of these analysis programs.

The material is then expanded in Chapter 4 to the calculation of time-variant responses in networks, to the computation of transient responses. This chapter consists of a method of solving a DC analysis problem at each time-increment with some source voltages and currents representing the accumulated capacitor charges and stored inductive energy. Again the underlying matrix relations are mechanized in a computer program and ideas are introduced for making the process computationally more efficient and easier to use for personnel with a small amount of computer training. Similar extension of the basic material of Chapter 3 is done in Chapter 5 to the calculation of poles and zeros of an arbitrary linear network. The procedure discussed is the solution of a simple DC circuit for each energy storage element in the network and for each input to the network. The process is simple and quite accurate if done carefully; the implementation discussed gives an adequate program for all practical networks. Problems associated with the solution of the resultant equations are then discussed. A deeper understanding of linear algebra is necessary; thus a discussion of the numerical matrix eigenvalue problem is included with a number of problems calculated in detail.

The final topic for predominantly linear circuits is the discussion of analysis for tolerances and the effect of small variations in components given in Chapter 6. Here the fundamental node voltage equations are differentiated and the resultant relations are mechanized. An introduction is made to problems arising from component variations and the statistical nature of component values. Some procedures for the prediction of variations of performance functions are discussed and are summarized with an extensive numerical example.

The basic voltage equations are used in Chapter 7 as the basis for calculating the node voltages of a non-linear network. The procedure used is to approximate the nonlinear voltage-current characteristics with a linear element and a voltage source. Thus an equivalent circuit is formed whose voltages and currents are good approximations to the actual voltages and currents in the neighborhood of the point where the approximation is made. The nonlinear analysis then becomes a series of operating point assumptions followed by an analysis to verify that the assumed points are in fact the operating points. Refinements are made until a set of sufficiently close results are found. Thus the nonlinear analysis is reduced to a convergent sequence of DC analyses of the type discussed in Chapter 3. Exploration of problems of organizing a viable nonlinear analysis package and in obtaining realistic data for semiconductor devices completes this book. A semiconductor device library is appended.

The material in Section 7.4 on a simplified transistor model was contributed by Dr. C. O. Harbourt, University of Missouri, Columbia.

I am indebted to Dr. George B. Hoadley, Head of the Department of Electrical Engineering, North Carolina State University, for arranging my schedule so that I could be involved in several consecutive semesters of the course Automated Circuit Analysis. This course allowed me to collate and update sets of notes into the manuscript of this book. I want to express my thanks to the Triangle Universities Computation Center for the use of their facilities for my classes and for the development of the programs herein. The help of my wife Monique in preparing the near-endless drafts was the key element in completion of the manuscript. I wish to thank Mrs. Anna A. Stroh and Mrs. Marillyn Mulholland for the preparation of the final drafts.

J. S.

CIRCUIT ANALYSIS
BY DIGITAL COMPUTER

THE CIRCUIT ANALYSIS TASK

1

In this book we shall discuss procedures for the calculation of voltages and currents in electronic networks. These procedures will all be computer oriented so that responses of networks can be determined efficiently and inexpensively. We need to know how to formulate the network equations, how to solve them, and how to interpret the results. We will need to look closely at numerical problems that arise from the analysis task and at solution techniques for equations. The terms *circuit* and *network* will be used interchangeably throughout this book.

In this chapter we explore the context of the analysis task. It is important to get a proper perspective on the nature of the task, the uses to which the results are to be put, and the tools required for the analysis. We hope to provide a motivation for a detailed study of general network analysis procedures and for computer-oriented formulation of the network equations. We hope to furnish a framework for interrelating types of analysis programs and an appreciation of computational considerations in these procedures.

1.1 The Nature of Network Analysis

Electronic circuits consist of an aggregate of electrically and/or magnetically connected building blocks. Each block has two or more electrical connections. The junctions of two or more electrically connected components are termed *nodes* of the network. Voltages within the network will be measured as differences in voltage between these nodes. Currents leaving the nodes, the branch currents, will flow through circuit components thus generating the voltage differences between the nodes to which the components are connected. The intent of this book is to present methods and procedures for the analysis of networks made up of practical components. Such components are the familiar two-terminal devices (resistors, capacitors, inductors, diodes, voltage sources, and current sources) three-terminal devices (transistors) and multiple-terminal devices (transformers). In the absence of a power source, be it a signal input source or the familiar DC power supply, networks of these elements will have no currents flowing, thus having no voltage differences between the various nodes.

The subject of this book is the calculation of voltages and currents in electronic networks subject to input excitations and the calculation of other closely related problems. These calculations we refer to as the analysis of the network. We note that circuits need to be analyzed only in order to derive information about the network so that its predicted behavior can be verified or when changes to the network are to be made. The analysis of a network for the joy of analyzing is typically an academic exercise; the vast majority of analysis tasks arise from the need of designers to verify their conceived designs or to predict the effects of changes in a circuit. We note that the analysis task starts with a mathematical model of the network that was chosen by the analyst to represent the behavior of the actual network. In no case can the computed results give solutions that exceed the accuracy of the approximation used in choosing the mathematical model. For example, a wire-wound resistor may be treated as having only resistance, while the inductive and capacitive effects could well dominate the behavior under some operating conditions (such as at high frequencies). It is engineering practice to think of electronic circuits normally as containing lumped parameter components that obey some simple voltage-current characteristic. This characteristic is often an equation, for example, Ohm's law for resistors. These characteristics or characteristic equations are based on observed behavior in which the finite speed of propagation of the electromagnetic energy becomes negligible, i.e., the components become infinitesimally small compared with the wavelengths involved. For example, a wire leading from the left to right sides of a circuit card can well be regarded as having the same potential everywhere, but a power transmission line wire leading halfway across a continent must be treated as a transmission line whose voltages and currents are calculated from Maxwell's field equations.

The environment in which circuit analysis exists is diagrammed in Fig. 1.1, a flowchart of the circuit design task. It is assumed for now that a large number of copies of the circuit must be made, as would be the case for a consumer product such as a table radio. The need for the circuit is usually perceived as a problem for whose solution many different alternatives may exist. Since the cost, reliability, size, and speed of electronic components has improved steadily over the years with respect to other mechanizations, an increasing share of the problems is being solved by electronic means. The methods to be developed in this book are of direct use in the electronic mechanizations; however, they can also be employed in areas of mechanical, thermal, fluidic, and other analyses. Normally, an electric analog is constructed [2]* and then the analog is analyzed directly. This procedure is used in setting up analog computer

*Numbers in brackets refer to the references at the end of chapter.

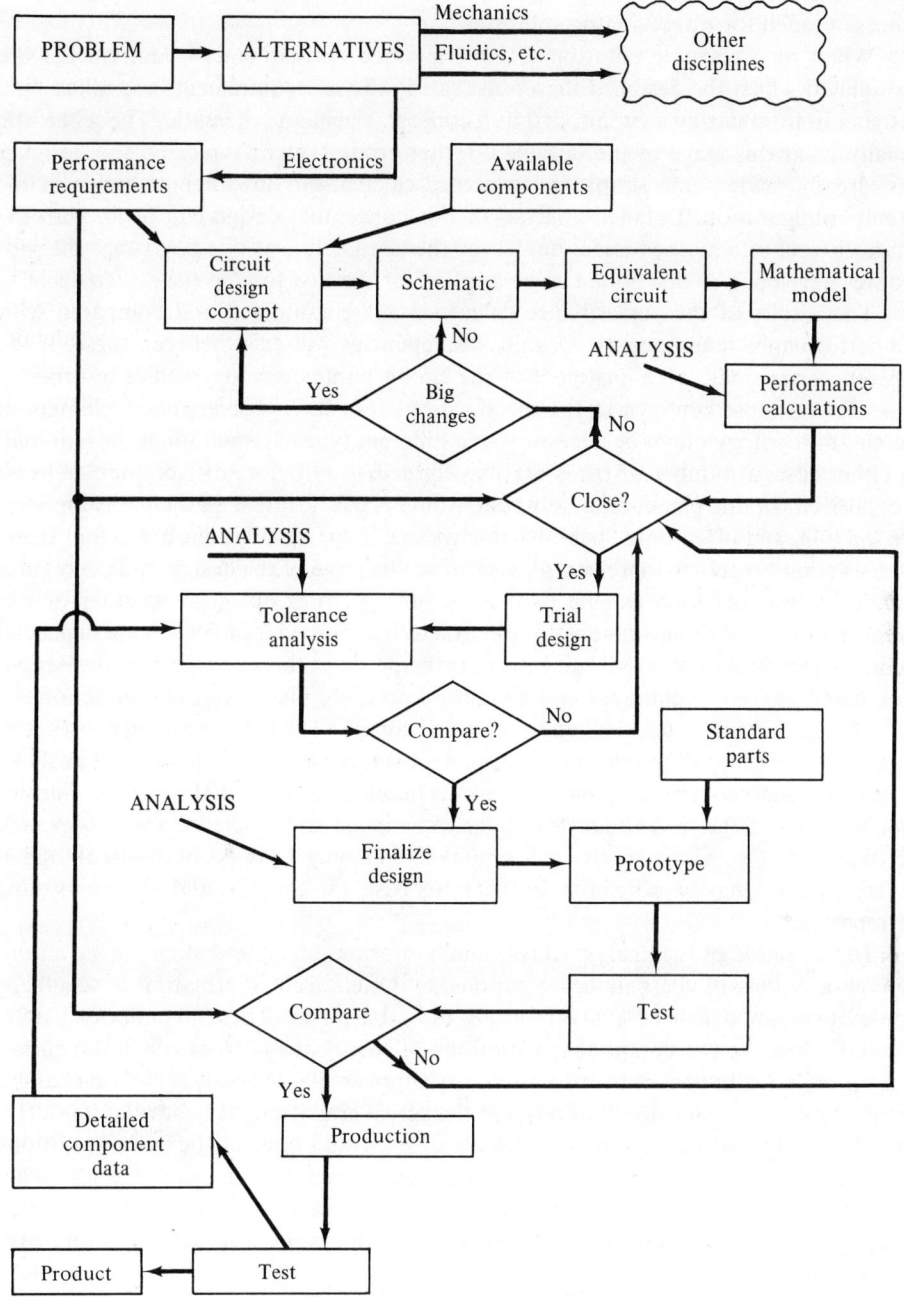

Fig. 1.1 Circuit Design Task

solutions to non-electrical problems. Here we merely point out that an electric analog can be analyzed digitally with great ease and often with greater convenience than the process needed for direct analog solutions.

When an electronic solution is contemplated, performance requirements are formulated which the designed item must satisfy. These requirements are taken by a designer in formulating a circuit, first as a concept, then as a schematic. The schematic usually is, at this stage in the design, a rather crude sketch; typically, the designer has already made some simple performance calculations in order to arrive at his circuit configuration. Detailed analysis of this conceptual design can begin with the establishment of a mathematical model for the circuit. It is at this point that the procedures developed in this book can be applied for the first time in the design cycle.

The results of the performance calculations are examined and compared with the performance requirements. Usually discrepancies will exist between these. If the differences are small, an adjustment of the circuit parameters may suffice in order to make the response conform to the specifications. For large discrepancies, changes in the circuit topology might be necessary or a different type of circuit might be required. In either case, a number of trials are usually necessary before an acceptable circuit is conceived. In this phase, relatively coarse data is used, simple mathematical models are the rule, and often only "back-of-an-envelope" type calculations are made. However, as circuits become more complex, even at this stage of the design cycle, a regular circuit analysis program can be used to advantage. It is not clear whether or not simple models that neglect much of the characteristic behavior of devices are beneficial to the understanding of the circuit action; rather, even at this early stage in the design, automated analysis techniques can give more accurate pictures of circuit actions.

Once a circuit is decided upon, detailed analysis of the circuit responses and detailed calculations of the effects of component variations must be made. The analyses may show unacceptable responses or unacceptable behavior of the circuit. Further modifications of the trial circuit may have to be made and a detailed re-analysis may be necessary. Occasionally, the conceptual circuit may have to be replaced and a different circuit may be called for. In this latter case, the conceptual design steps must be repeated.

In this phase of the analyst's task, much information is needed about the nature of the variations of the network components. Often such information is scanty or non-existent, in which case assumptions must be made. These assumptions must correctly describe the component variations of the elements from which the circuit is to be built. On many occasions a change in suppliers might result in radical changes in the component value distributions and the actual circuit might behave quite differently from what was expected. The discrepancies would typically be in the variations of the performance functions rather than in their nominal values. Good design practice dictates, however, that the circuit performance be affected only to a small degree by the parameter variations so that a good design normally will result in a very easy analysis task.

When an acceptable circuit has been achieved through these trials and analyses, the final design item is built and tested. This prototype is checked against the original performance requirements. Again, if discrepancies are discovered, earlier design steps may have to be repeated. Due consideration must be given to the actual circuit component values used in comparison with the allowable variations on the component values. An acceptable prototype will then be put into production. As the final production items are checked out, new data may become available for the component data base so that subsequent analyses may benefit from the earlier measured component variations.

For lower-volume production items, the process to be used is obviously less elaborate with many steps in Fig. 1.1 either omitted or coalesced into a few steps. However, the analysis task throughout the design remains the prediction of circuit behavior of a known set of interconnected components to specified inputs. The analyst must be prepared to cast a circuit diagram into quantified form and perform a series of mathematical operations. He must be able to relate his results to a set of performance criteria. Since the analyst and the designer usually is the same person, the design-analysis cycle can not easily be dichotomized. The most convenient arrangement is to let the designer-analyst work an interactive computer system and obtain an optimized solution by trial and refinement.

We note that although analysis *per se* does not have much application, it is, nevertheless, an integral part of design. Successful analysis procedures must therefore be judged by their applicability, ease of use, and relevance to the design cycle. The discussion presented here indicated briefly the areas of competence that an analyst should possess: firm foundation in linear circuit theory, a knowledge of modeling, optimization, statistics, and computer procedures. In this chapter we discuss further some of these tools and how they interrelate and relate to the topics in this book.

1.2 The Tools of Network Analysis

An examination of the basic design cycle outline shown in Fig. 1.1 reveals a number of places where analysis process is used. At various points different types of analyses must be made. In the early phases of the design, simple calculations of node voltages, currents, and powers are needed. Here often simple bias circuits must be considered, and step and frequency responses must be determined. The basic mathematics involved is the solution of simultaneous equations. These equations are linear for linear circuits, but they become nonlinear for any other types of circuits. Usually, the designer thinks in terms of simple relationships and the circuits that he works with are relatively small. The tools for such circuits are the basic voltage-current relationships for the circuit elements, rules for generating simultaneous equations, and solution procedures for these equations.

In order to have an efficient computer method, general procedures must be examined instead of an implementation of the various special processes used in hand calculations. Because of the high cost of hand calculations, a myriad of processes exist for solving special circuits. For example, models of low frequency, mid-range, and high-frequency amplifier circuits are used routinely; however, in computer work it is easier to use a model in which all components are present for all calculations. Similarly, although most amplifier models can be worked systematically node-by-node from input to output, a recognition of the topology of the circuit and a knowledge of the circuit action are required. Computers are capable of carrying out a vast number of detailed computations; however, they can not conveniently be made to recognize topological relationships. Thus computer methods tend to have much numerical detail but no insight into the problem. The analyst will prefer to write a computer program for a wider class of problems than what he has to solve at the moment when he designs his program. Hence, generalized methods are the rule in computer aided analysis. One can, of course, write specialized routines for special problems and reduce the cost of computation for that problem, but there will be a vastly increased price of total program design.

General analysis methods will solve simultaneously for the entire set of node voltages rather than follow manual processes of segmenting the calculations. This allows the inclusion of arbitrary node-to-node interrelations, as would exist in a circuit with many feedback paths. Normal network models do not include weak feedback effects, such as the transistor voltage feedback h_r, because of the computational difficulties associated with that term. In computer formulation of the network equations, the additional overhead incurred by such terms is negligible and therefore is normally included. In order to handle the network equations simultaneously, matrix methods must be used. Matrices formalize the handling of sets of relationships. It is expected that readers of this book will have had some familiarity with the concepts of matrix algebra and have had some experience working with simple matrices. The simultaneous node voltage equations for instantaneous voltages lead to matrices with real coefficients for most types of analyses, except for steady-state AC analysis in which matrices with complex coefficients are generated. The same rules and the same computer procedures apply in either case. Thus conceptually and computationally there is little difference in AC or DC analyses [1].

For frequency response calculations, one would normally repeat the entire analysis calculations stepping the frequency each time.

The reader is assumed to be familiar with basic circuit analysis calculations [3]. The tools required for this task include the use of linear transforms. Both entail capabilities in calculating with complex numbers to find singular points in the transform domain (poles). When a set of equations is transformed, the finding of the poles of the various network functions is equivalent to finding the eigenvalues of the network differential equations. The techniques for this task are drawn from linear algebra and the theory of differential equations. Implementation on a computer depends on judicious use of advanced numerical techniques for eigenvalue computations. Thus the calculation of poles (and zeros) will depend on combining the linear techniques used in simple DC circuits with advanced numerical techniques normally considered in numerical analysis courses.

The analyst's task extends to the calculation of variations in circuit performance functions since they depend on the probabilistic nature of the circuit parameter values. Usually, the circuit components will be manufactured separately and the variations of the individual component values will be independent of one another. For example, resistor values will not depend on capacitor values. For such circuits, methods of basic statistics can be used successfully; averages and variances can be calculated easily from a knowledge of the variations of the individual circuit elements. However, for circuit components that contribute more than one significant parameter to the circuit action, not only the individual parameter variations but also their statistical interaction must be considered. For example, diodes can be characterized by a reverse leakage current and a reverse-biased junction capacitance. The two numbers will be correlated, i.e., for a diode with lower leakage there will normally be a smaller junction capacitance also. The techniques necessary for dealing with variations of this type are drawn from more advanced statistics; hence the successful analyst will need approximately one year of college statistics to cope with these problems. Variations are, of course, relatively small for many circuit parameters; hence first-order approximations may be used to advantage. In terms of circuit equations, first-order approximations mean that derivatives must be calculated. For linear networks these derivatives are obtained from a set of linear equations and are therefore rather simple. All that is required is a way of handling sets of derivatives, each of them of very simple nature.

A further examination of the design task outlined in Fig. 1.1 indicates that the task includes not only the calculation of responses for a defined topology and defined inputs but also the optimization of the circuit. Most often this optimization must be done in an economic frame of reference. The task outline indicates several loops of decisions and calculations in which optimization can be applied. The procedures employed in this phase of the analysis are drawn from the relatively new discipline of optimization theory. In this book, this phase of the design task will not be discussed; the reader is referred to other books for further study [5].

From this very broad discussion of the analyst's task, it is obvious that he needs, in addition to the normal introduction to linear circuits, a working knowledge of linear algebra, statistics, and optimization. The material presented in this book will provide introductions to all these areas, except to optimization. It is assumed throughout that the reader has had an introduction to linear circuit analysis and has performed simple analyses. The thrust here will be to present systematic procedures applicable to any circuit.

1.3 Computational Considerations

The analysis of circuits, when done manually, usually starts with a circuit diagram. The analyst then writes a set of equations for this circuit from an inspection of the topology of the element interconnections. The equations relate the voltages and currents of the elements to other voltages and currents in the network by means of Kirchhoff's laws. The equations will be linear for linear networks.

The analyst obtains his equations by scanning the circuit diagram and noting where and how components are interconnected. For computer use, however, no such method can be used. A computer does not deal with topological drawings and cannot recognize spatial relationships. Computer information must be in quantified form. Hence circuit diagrams must be replaced with numeric information. The easiest way to do this is to number the nodes and the branches of the circuit and provide an interconnection matrix that defines how branches are connected to the nodes.

Next, a procedure must be devised for writing the network equations automatically. This procedure must be general and apply to all circuits for which analysis is to be made. In manual analysis, shortcuts are possible if the circuit is of a special kind. For example, ladder networks are easy to analyze by a concatenation scheme, say with transfer matrices, but this procedure becomes very tedious for networks with feedback. An experienced analyst will select his analysis tools by basing them on the kind of circuit he deals with. Since the cost of computation becomes very low with computers, it is desirable to devise general methods that may contain many computational steps. The methods discussed in this book aim at the establishment of such general algorithms. The success of any computer procedure depends critically on finding an algorithm for the calculations.

The algorithm is implemented in a language that the computer can understand. Internal to the computer all information must be represented as a set of YES/NO decisions, bits of information. This information is stored in the computer memory in groups of such bits. One of the characteristics of a computer system is the number of bits grouped into one unit of computer memory, the memory word size. The word size of most computers used in scientific and engineering calculations is 16, 32, 36, or 60 bits, with some other word sizes also available. Part of the cost of the computer system depends on how many bits the machine must handle simultaneously. Obviously, more parallel operations will be more expensive. The speed with which the machine operates is largely independent of the number of parallel operations required. Thus the increase in word length results in more information processed simultaneously; each computer word can therefore contain more precise information.

It is normal to think of a computer as being a nearly infinite speed calculator of basically unlimited precision. In terms of engineering measurements, the vast majority of computers used in computer centers does indeed seem excessively precise. In a computer, two kinds of numbers are used for calculations: integers and floating point. Integers are meant to be used primarily for counting, floating-point numbers for calculations. Integers are represented as integers to base 2 within the computer. The size of the integer that may be used will therefore depend on the number of bits allocated to any one integer. This unit is normally a computer word. For a 32-bit machine, the largest signed integer is therefore $2^{31} \approx 2.14 \times 10^9$ (one bit is allocated to the sign of the number). When floating-point numbers are to be represented in the machine, part of the bits available are used for a characteristic and part for a mantissa. Although all numeric values within the machine are base 2 numbers, it is convenient, and correct, to think of computer internal numbers as having a decimal mantissa and a decimal characteristic. The range of the number system that may be employed will depend on the number of bits allocated to the characteristic, and the accuracy with which each value is represented will depend on the number of bits used for the mantissa.

The accuracy of computed results will depend, in general, on the number, the sequence, and the nature of the steps required as well as on the accuracy with which each step is carried out. The accuracy of each operation is a function of the number system used on the computer if floating-point numbers are used. For example, on a computer with 24 bits for the mantissa, the accuracy with which a number may be represented is 1 part in $2^{24} \approx 1.68 \times 10^7$. This seems to be a very high accuracy, but in reality it is not so. Consider that if very many (10000) small numbers, say 1, are to be added to a large number, say 10^8, the operations, to an accuracy of 10^{-7} will be

(1) $10^8 + 1 = 10^8$ (add first number)

(2) $10^8 + 1 = 10^8$ (add second number)

.
.
.

(10000) $10^8 + 1 = 10^8$ (add 10000th number)

Thus the result will be 10^8. The probable error in each partial result is $10 = (10^8 \times 10^{-7})$. The apparent answer is therefore

$$\text{sum} = 10^8 \pm 8$$

(i.e., the error is in representing 10^8 with an accuracy of 1.6×10^{-7}).

If the operations were re-ordered as follows

(1) $0 + 1 = 1$ (add first number)

(2) $1 + 1 = 2$ (add second number)

.
.
.

(10000) $9999 + 1 = 10000$ (add 10000th number)

(10001) $10000 + 10^8 = 1.0001000 \times 10^8$ (add 10^8)

The apparent answer is

$$\text{sum} = (1.0001000 \pm 0.8 \times 10^{-7}) \times 10^8$$

In this latter case, a re-ordering of operations was used, i.e., a different algorithm. It is clear that the printed results will differ and that the actual results will be different. Thus the results of a calculational scheme may well depend on the care that was expended on the details of the process.

Had the calculations been done on a computer with more bits in each of the floating-point numbers, the two processes may well have resulted in identical numerical values. In the above example, a 27-bit mantissa would be sufficient.

The instructions used by a computer must be in the machine language for that machine. Machine language is very tedious, and writing circuit analysis algorithms would be too time consuming. There are a number of English-like languages that can be used to describe a series of steps that the computer is to take. It is desirable that these languages be as natural to the user as possible and that they define the operations to be done uniquely. The FORTRAN language is used in this book for generating the algorithms. Programs exist on various machines to translate the FORTRAN statements to machine instructions. Since FORTRAN is the most widely used language for engineering and scientific work, the programs developed here will have the widest applications.

There are, however, some differences between the various FORTRAN compilers. The reader is urged to note carefully any differences in his version of FORTRAN from the one used here. The programs are designed to run in ASI FORTRAN-IV and were checked out on the IBM 370/165 system using the FORTRAN-G compiler. Numerical discrepancies can arise from word-length differences, but if marked errors are detected, one should carefully check the coding details for dialectical differences. Readers not familiar with FORTRAN are referred to [6].

In the course of computation of network functions, large tables of information will be used. The algorithms developed will be independent of the circuit size, but data storage must be allocated at the time the program is loaded into core. This is a disadvantage with the FORTRAN programming system, but little computer overhead is used and FORTRAN program executions are typically efficient. In writing the computer programs the various table sizes must be specified. One would like to have tables of sizes large enough to analyze anything that one might encounter, but since computer costs are partially determined by memory size used, the cost of executing a large program becomes uneconomical when only limited-sized circuits are analyzed. Therefore, a trade-off must be reached between the desire to write programs capable of analyzing huge circuits and the cost considerations for the bulk of the analysis tasks. The programs presented in this book are all for moderate-sized circuits. They will handle problems too complicated to be done by hand, and they can be modified for larger circuits. The programs here are designed to run in 100 K bytes of IBM 360/370 memory.

In the developments detailed in this book, considerable attention will be paid to numerical procedures and computational constraints. The reader will find a large amount of detail about numerical procedures which have direct bearing on the analysis of electronic circuits. However, the material will be introductory in nature. The reader must keep in mind that these topics are of current research interest. Although many results are known, typically there is no known best solution for all cases. Future results will, of course, ease the analysis task; the material in this book will lay a foundation enabling the reader in the future to note and appreciate applicable research results.

1.4 Modeling

The calculation of the response of a network is accomplished by working with a set of equations. These equations describe, to some level of detail, the observed terminal behavior of circuit components. Thus the calculations reflect not the expected response of the network, but rather the response of the mathematical model. Often the detailed internal voltage and current changes are immaterial for the response being calculated. An example might be in frequency response calculations of RC coupled small-signal amplifiers where the quiescent device voltages and currents have little direct bearing. In this example one can determine the small-signal parameters from a knowledge of the biasing circuits and calculate a very accurate response from these parameters alone. If minor changes are then made to the circuit components, the small-signal parameters may not have to be redetermined for accurate calculation of the response of the revised circuit.

The small-signal h-parameters of a transistor are an example of a model used extensively for the calculation of a network function. These parameters describe the terminal voltage-current *changes* adequately for use in the analysis procedures. Note that the model must be appropriate to the use it is put to: in this instance, the absolute values of current and voltage are not described well by these parameters and are of little interest because of the capacitive coupling employed. Were the response of a resistance-coupled amplifier to be sought, the small-signal h-parameters would be inadequate for the analysis. One could, of course, devise a model for both small-parameter and absolute-value terminal voltages and currents. The Ebers-Moll transistor model is a widely used model that relates all terminal voltages and currents and therefore could be used in both RC coupled amplifier analysis and in DC coupled amplifier analysis. The model is, however, overly elaborate and entails more computational detail than is necessary for either application.

We note that the appropriateness of a model depends not only on the use to which the device is being put but also on the external circuitry and the accuracy with which the analysis is to be made. For example, an operational amplifier may well be regarded as having zero output impedance, infinite gain, finite bandwidth, and infinite input impedance if the approximate behavior of a heavily fed-back circuit is sought in which the operational amplifier performs. If, however, current-gain calculations are to be made on the amplifier itself, a different model must be sought.

One possible approach to finding the proper computational model is to start with a more elaborate one than what is really required, compute the various terms that contribute to the network function being calculated, and neglect those having less than a pre-assigned contribution to the overall function. Care must be taken to account for the interaction effects of the components and to retain only the significant terms [8]. Although this process does not circumvent the need for having to use all the details inherent in the original model, in the derived model only significant elements and significant effects remain. Hence subsequent uses of this model will result in simpler computations.

In the analysis of electronic networks, semiconductor devices have the most complicated behavior for which models must be obtained. Computational ease, particularly if non-automated procedures are to be used, requires simple models. The vast majority of device models was derived for manual calculations. Since computer costs are very low, detailed models might be quite acceptable for automated procedures. One advantage of a detailed model is that it may be used in many different circuits and applications without changes. The cost of establishing a simple model can outweigh the expected additional computer costs that the more complex model might require. It is very tempting to try to maintain a universally useful transistor model library. These models can then be used for all applications of the device. Note, however, that the model must now be capable of faithfully simulating the terminal behavior of the device for all reasonable and unreasonable uses. Hence operational characteristics and failure indications must both be in the model. Data for such complicated models are difficult to obtain with any precision. Establishment of a model for a normally connected transistor, however, is reasonable. Such a transistor model should faithfully reproduce both small-signal and large-signal behavior. The nonlinear nature of the transistor will preclude models that contain just a few parameters; the model given in Chapter 7 contains some three dozen parameters for each transistor. Considerable effort must be expended in finding these numbers, but once they are found they may be stored for later use in other applications. Since the model need not be any more accurate than the uses to which it is to be put, somewhat simpler models, containing fewer parameters, are normally usable for most applications. The model library enclosed as Appendix C lists the model parameter data for a simpler model that is adequate to get approximately 10% accurate results for normally connected transistors.

There is only one item that will faithfully re-enact the behavior of a device: an identical second copy of the device. There is no way that one can ensure that this second device may exist because there are differences in manufacture. Thus the model must serve as a representative device for a group of slightly different devices that all carry the same type designation. Hence the model serves as an approximation, within a tolerable error, for the prediction of the behavior of an actual device. The conditions of operation (frequency ranges, signal magnitude ranges, temperature, etc.) provide guidance in arriving at the key approximations which are then incorporated in the model making. Although there may normally be more than one adequate model, often the ease of determination of model parameters by terminal measurements decides which model is to be used. Programming considerations also may enter this decision. The trend in computerized analysis procedures is to use circuit element models which are more elaborate than those used for manual processes. The low cost of computation favors the expenditure of extra computing in exchange for savings in setting up models. Thus capacitors and inductors may be considered always with parallel and series resistances, as is done in Chapter 7, and transistors are modeled with a large number of parameters for use in all normal connections to the transistor.

One can establish the model configuration and model parameters from a careful study of the terminal behavior without considering anything about the internal processes in the device. This "black-box" approach is useful, but it mimics behavior only over the ranges of the measurements. Another approach is to consider the dominant physical processes in the device and the internal arrangement of the device. Each of these is described in detail and juxtaposed in the same topology as the internal device arrangement. In this way, a lumped model may be derived which reflects in a one-to-one correspondence the interrelated internal processes and closely approximates the externally observed behavior over a wide range of conditions. Thus, for example, in a semiconductor junction one may describe separately the junction behavior itself with the diode voltage-current equation, consider the bulk resistance action of the P-region and the N-region separately, include the inductive and resistive effects of leads, model the surface leakage action, consider storage effects in the junction and storage effects in the device mountings. Each of these model components is well defined and their internal interconnections can be deduced from the diode construction. Hence a fairly accurate, if somewhat elaborate, model can be established. Each of these model components will have more pronounced effects under different operating conditions. As a by-product of the model making, the conditions under which the model components can be measured will also be established.

There does not exist a universal process for finding a model for an arbitrary circuit component. In some very critical applications one might have no choice but to use the actual device in a mock-up of the circuit. Here the driving functions could be computed and applied to the device; the response of the device is then used to calculate the behavior of the circuit external to the device. This mode of circuit analysis does require an elaborate hybrid computer facility, but it allows easy changes to circuit parameters. Device parameters, however, cannot be changed and therefore circuit response for only one device can be calculated.

The need for a model lies not only in its use for predicting the terminal behavior of devices. Rather, models can be used to ascertain the effects on the overall circuit behavior of changes within the devices. It is, for example, almost impossible to study the effects of changes in gain in the transistor since there is almost no way of assuring that devices will vary *only* in their gain. Such parameter variation studies are essential in arriving at required component tolerances for acceptable production yields and are thus the responsibility of the analyst.

Adequate models are essential to successful analysis. The successful analyst will also be a successful model maker. Model making, in turn, depends heavily on a firm understanding of the physical processes used in the devices. Therefore a study of the physics of the devices is very important. In modeling electronic circuits, the study of semiconductor devices is the most essential ingredient for success. Other difficult models include the action of saturable cores; however, the bulk of the circuit analyses involve only semiconductor models. A brief discussion of semiconductor device modeling is included in Chapter 7. In the rest of the book, analysis methods are discussed which can utilize any device model made up of linear circuit elements or piece-wise linear circuit elements.

1.5 Nonlinear Networks

The reader is familiar with the elementary circuit components: resistors, capacitors, inductors, and sources. Each of these is characterized by a single number that relates terminal voltages and currents. These numbers do not limit the range of applicable voltages and/or currents. Circuits made up of such elements obey the following superposition principle.

If the voltage in a component is v_1 due to a source S_1 in the network, and it is v_2 due to another source S_2, which may be located anywhere else or at the same place as S_1 in the network, then the voltage due to the simultaneous action of S_1 and S_2 is $v_1 + v_2$.*

This principle allows the calculation of the simultaneous effects of many sources to be calculated from a series of simple one-source analyses. Also, the calculations need not consider the absolute values of the source strengths since the response to a source strength KS_1 will simply be K times the value calculated for a source strength of S_1. Note that K is unlimited; theoretically, any voltage or current level would obey the same relationships. Obviously, some limits must exist for every component in which linearity will be violated. There are, as a matter of fact, no truly linear circuit elements; some do, however, have operating ranges where linearity (and superposition) can be observed. The most familiar of these is a resistor: over its normal operating range it will exhibit a linear voltage-current characteristic. For a doubling of applied voltage the current through it will also double.

A great many electronic components do not have a linear characteristic even in their normal range of operation. The semiconductor diode is, perhaps, the simplest such device. One cannot obtain the current due to two different sources through the diode by summing the effects of each of the sources acting separately. In circuits where the circuit elements must be considered as strictly nonlinear devices, the effects must be calculated separately for each forcing function combination. There are methods for dealing with specific kinds of nonlinear elements, but there is no unifying procedure for their characterization and calculation of their responses [9]. The only process that may be applied with some degree of success is to consider the network for small changes of current and voltage as a linear network. This amounts to working with small-signal changes in the network and to replacing the network elements by the slopes of their characteristics at each calculation. Large-signal changes can be computed in the network by calculating a very large number of incremental steps in succession. Thus nonlinear circuit responses may be calculated as a series of linear responses with the circuit element values changing at each incremental step.

As a consequence, the calculation of nonlinear circuit responses is characterized by a large number of individual circuit analysis calculations. In order to accomplish this task with any reasonable computer cost, efficient methods of linear analysis must be employed. This is one of the reasons for stressing details of linear analysis throughout the bulk of this book. We will also stress the analysis of semiconductor circuits and will introduce specific simplifications (models) for diode and transistor circuits in the computer program for nonlinear circuit analysis.

1.6 Desirable Features of Programs

The theory underlying the analysis of networks is well understood. Many books and articles detail all the required background. Little original work need be done if

---

*Current may be substituted for voltage without any restriction in this statement.

one were satisfied with a direct mechanization of the matrix descriptions of linear networks. Difficulties arise when nonlinearities must be analyzed or when a topological description (a circuit diagram) is used to enter the network into a computer. The easiest way to describe a network for computer use is to devise an input language for use by the engineer for communicating with the computer. The language must be understandable in engineering terms and be flexible enough to permit some latitude in input formats. Typically, the user-engineer does not know, and does not care to know, the details of the program he is using. The input language must free the user from unnecessary formalities, yet contain, in computer recognizable terms, all information necessary to specify an unambiguous analysis task.

In designing widely usable analysis programs, the following desirable features should be considered [4]:

(1) *Simple Input*. A clerk who has no circuit analysis background and no knowledge of computing should be able to enter the required input data. Input should include

(a) topology
(b) circuit values
(c) excitations
(d) output modes

Each of the above categories should be separate without the need of re-inputting all data.

(2) *Variable Models*. In active circuit analysis one of the chief considerations is the equivalent circuit used for the active components. There should be choice to utilize a variety of equivalent circuits as the accuracy of the desired analysis dictates.

(3) *Nonlinearities*. A wide range of these must be considered. Typical ones include saturation and reverse voltage breakdown, but they could also include thermal considerations.

(4) *Selective Outputs*. Because of the low cost of computation, it is no trick at all to envelop the circuit designer in reams of output data. What is required is a set of options for significant outputs and some automation in ignoring most numbers.

(5) *Automatic Parameter Modification*. Changes in parameter values should not require complete re-inputting. Such modifications are necessary in tolerance analysis and in automated design procedures.

(6) *Error Checks*. The reliability and accuracy of the answers provided should be easy to assess. Generally, every answer, no matter how inaccurate, is printed with maximum precision in every program. No such automatic error checks are presently available in circuit analysis programs.

Usually, for automated design these programs are combined with some optimization procedures. Performance criteria are calculated and are compared against the desired values. Error measures are derived and are then used to adjust the parameters and/or the topology of the test circuit. For such use the above criteria for analysis programs must be extended to:

(7) *Optimization.* This must be done simultaneously for the various parameters in the network. The network parameters interact greatly in their contribution to any performance function and any optimization, particularly if subject to constraints, is an involved numerical task. The parameter constraints are often necessary for the validity of the circuit model.

(8) *Flexible Objective Description.* In every design effort there is some function, often only verbally circumscribed, which must be optimized. Unless the objective function can be described in quantitative terms to a computer program, no optimization can begin. Presently many objectives (simplicity, reliability, ease of trouble shooting, etc.) can be expressed only incompletely and with great difficulty in numerical form. Further developments in this area are necessary either for writing compilers, which "understand" more verbal descriptions, or in educating the designers to use fewer qualitative descriptors.

This book will be concerned with the analysis of networks instead of their conception (design). Thus items 7 and 8 will only be alluded to and not examined in any detail. In the ensuing chapters we will discuss some of the background mathematics of analysis in algorithmic terms, look at the underlying mechanizations of the general node analysis procedure for various kinds of analyses, discuss methods of analyzing the effects of parameter variations and consider problems associated with nonlinear networks. In each chapter, programs will be discussed for mechanizing the just-derived material. The programs will be amenable to further refinements toward the desirable features listed here.

REFERENCES

1. GUILLEMIN, E. A., *Introductory Circuit Theory.* John Wiley & Sons, Inc., New York, 1953.
2. PASKUSZ, G. F., and B. BUSSELL, *Linear Circuit Analysis.* Prentice-Hall, Inc., Englewood Cliffs, N.J., 1962.
3. VAN VALKENBURG, M. E., *Network Analysis,* 2nd ed. Prentice-Hall, Inc., Englewood Cliffs, N.J., 1964.
4. KUO, F. F., "Network Analysis by Digital Computer," *Proc. IEEE,* Vol. 54, No. 6 (1966).
5. WILDE, D. J., and C. S. BEIGHTLER, *Foundations of Optimization.* Prentice-Hall, Inc., Englewood Cliffs, N.J., 1967.
6. LEE, R. M., *A Short Course in Basic FORTRAN IV Programming.* McGraw-Hill Book Co., New York, 1972.
7. GUILLEMIN, E. A., *The Mathematics of Circuit Analysis.* John Wiley & Sons, Inc., New York, 1951.

8. HAPP, W. W., and J. STAUDHAMMER, "Topological Techniques for the Derivation of Models with Pre-assigned Accuracy," *Archiv der elektrischen Ubertragung*, Vol. 20, (1966).

9. CHUA, L. O., *Introduction to Nonlinear Network Theory*. McGraw-Hill Book Co., New York, 1969.

SOME ELEMENTS OF LINEAR ALGEBRA

2

A matrix is a generalized number; it is a collection of numbers and/or variables arranged in a definite pattern of rows and columns. Thus a matrix consists not only of its constituent elements but of the arrangement of its elements as well.

Matrices are exceedingly useful in modern engineering because they can be used to describe sets of relationships. Real numbers and variables are useful in describing physical phenomena because of the manipulative system of symbols (the algebra) relating to them describes observed results. Matrices are useful in describing physical problems involving a large number of elements because matrices afford a way to describe sets of closely related variables and the laws of matrix algebra define manipulations on these generalized numbers which correlate well with observed phenomena.

Just as the introduction of complex numbers necessitated the formulation of complex number operations, the use of matrices must adhere to definite rules of matrix operations. In this book the axiomatic viewpoint will be adopted in that a few definitions will be given upon which a structure of matrix algebra will be built. These definitions will not be the most fundamental ones possible, but they will suffice for a working knowledge of the methods used in this book.

This chapter will introduce basic notions and basic operations (procedures); computer programs will be written for use of these basic procedures. The programs will be used in later chapters; also extensive use will be made of the various matrix definitions and notation.

2.1 Basic Definitions

Definition 1: A matrix is an array of numbers and/or variables (called elements) arranged in a regular pattern of rows and columns. Each row (each column) will contain the same number of elements.

Definition 2: An array of N rows and M columns is called an $N \times M$ (N by M) matrix. When $N = M$, the array is called a square matrix. When $N = 1$, the array is called a row matrix; when $M = 1$, the array is called a column matrix; when $N = M = 1$, the array is a single number, called a *scalar*.

Notation: The whole array, representing the whole matrix is referred to by a single bold face symbol. This system of notation is called the *direct* notation. Any element of the matrix is referred to by a lower case symbol subscripted by indexes indicating the row and the column positions of the element. Thus the matrix $\mathbf{A}$ which is of dimensions $N \times M$ contains the elements a_{ij} with $i = 1, 2, \ldots, N$ and $j = 1, 2, \ldots, M$. To indicate this relationship between $\mathbf{A}$ and the a_{ij} the notation

$$\mathbf{A} = [a_{ij}] \tag{2.1}$$

is used. The ranges of the i and j are usually inferred from the context; however, when the order must be stated explicitly, the notation

$${}_N^M\mathbf{A} = {}_N^M[a_{ij}] \tag{2.2}$$

will be used. The notation $\mathbf{A}$ is called the *whole matrix notation* and the notation $[a_{ij}]$ is called the *kernel index notation*.

Two matrices are equal if they are both of the same dimensions (have the same number of rows and the same number of columns) and contain the same elements in identical positions.

Definition 3: Two matrices $\mathbf{A}$ and $\mathbf{B}$ are added to yield a third matrix $\mathbf{C}$ by the rule

$$c_{ij} = a_{ij} + b_{ij} \tag{2.3}$$

where the subscripts i and j range over all possible values. Hence the two matrices $\mathbf{A}$ and $\mathbf{B}$ must each contain the same number of rows and the same number of columns; by the definition the matrix $\mathbf{C}$ must therefore have the same dimensions as $\mathbf{A}$ and $\mathbf{B}$:

$$ {}_N^M\mathbf{C} = {}_N^M\mathbf{A} + {}_N^M\mathbf{B} \tag{2.4}$$

From the definition it is easy to see that

$$c_{ij} = a_{ij} + b_{ij} = b_{ij} + a_{ij} \tag{2.5}$$

and hence

$$\mathbf{C} = \mathbf{A} + \mathbf{B} = \mathbf{B} + \mathbf{A} \tag{2.6}$$

i.e., the operation of matrix addition is commutative (the order of the operands entering matrix addition is immaterial).

Definition 4: A matrix **A** is multiplied by a scalar by multiplying every element of the matrix by the scalar:

$$k\mathbf{A} = k[a_{ij}] = [ka_{ij}] \tag{2.7}$$

$$\mathbf{A}k = [a_{ij}]k = [ka_{ij}] \tag{2.8}$$

Notice that the order of the scalar and the matrix is immaterial. Also, scalar multiplication does not change the order of the matrix.

EXAMPLE 2.1

Given the two matrices **A** and **B**,

$$\mathbf{A} = \begin{bmatrix} 5 & 3 \\ 1 & 0 \end{bmatrix}; \quad \mathbf{B} = \begin{bmatrix} 2 & 1 \\ x & y \end{bmatrix}$$

calculate the sum $\mathbf{A} + 5\mathbf{B}$.

Solution.

$$\mathbf{A} + 5\mathbf{B} = \begin{bmatrix} 5 & 3 \\ 1 & 0 \end{bmatrix} + 5 \begin{bmatrix} 2 & 1 \\ x & y \end{bmatrix}$$

$$= \begin{bmatrix} 5 & 3 \\ 1 & 0 \end{bmatrix} + \begin{bmatrix} 10 & 5 \\ 5x & 5y \end{bmatrix}$$

$$= \begin{bmatrix} 15 & 8 \\ 5x+1 & 5y \end{bmatrix}$$

EXAMPLE 2.2

Given the column matrices (vectors) **C**, **D**, and **P**,

$$\mathbf{C} = \begin{bmatrix} 3 \\ 2 \\ 1 \end{bmatrix}; \quad \mathbf{D} = \begin{bmatrix} 5 \\ 0 \\ 1 \end{bmatrix}; \quad \mathbf{P} = \begin{bmatrix} 1 \\ -1 \\ 0 \end{bmatrix}$$

calculate $\mathbf{C} + \mathbf{D} - 2\mathbf{P}$.

Solution.

$$\mathbf{C} + \mathbf{D} - 2\mathbf{P} = \mathbf{C} + \mathbf{D} + (-2)\mathbf{P} = \begin{bmatrix} 3 \\ 2 \\ 1 \end{bmatrix} + \begin{bmatrix} 5 \\ 0 \\ 1 \end{bmatrix} + \begin{bmatrix} -2 \\ 2 \\ 0 \end{bmatrix} = \begin{bmatrix} 6 \\ 4 \\ 2 \end{bmatrix} = 2 \begin{bmatrix} 3 \\ 2 \\ 1 \end{bmatrix} = 2\mathbf{C}$$

Definition 5: Two matrices ${}^M_N\mathbf{A}$ and ${}^K_L\mathbf{B}$ are said to be conformable in the sequence **AB** if $M = L$; they are conformable in the sequence **BA** if $N = K$. Conformable matrices may be multiplied together. Thus if $\mathbf{A} = {}^L_N[a_{ik}]$ and if $\mathbf{B} = {}^M_L[b_{ij}]$, then the *product* (also known as the *inner product*)

$$\mathbf{C} = \mathbf{A} \cdot \mathbf{B} = (\mathbf{AB}) \tag{2.9}$$

is defined. The c_{ij} elements of **C** are obtained by the formula

$$c_{ij} = \sum_{k=1}^{L} a_{ik} b_{kj} \quad i = 1, \ldots, N; \quad j = 1, \ldots, M \tag{2.10}$$

Sec. 2.1 Basic Definitions 21

Note that the resulting matrix **C** is of order $N \times M$ (has as many rows as **A** and as many columns as **B**). Also, the product **B · A** will not exist unless $N = M$.

EXAMPLE 2.3

Using the matrices **A, B, C, D**, and **P** given in Ex. 2.1 and Ex. 2.2, we calculate a number of matrix products.

(a) Calculate $F = A \cdot B$.

Solution. The dimensions of the matrices **A** and **B** are 2 by 2 and 2 by 2 respectively (i.e., $N = 2$, $M = 2$; $L = 2$, $K = 2$). Hence the product **A·B** exists $(M = L)$. The various elements of the product are

$$f_{11} = 5 \cdot 2 + 3 \cdot x = 10 + 3x$$
$$f_{12} = 5 \cdot 1 + 3 \cdot y = 5 + 3y$$
$$f_{21} = 1 \cdot 2 + 0 \cdot x = 2$$
$$f_{22} = 1 \cdot 1 + 0 \cdot y = 1$$

The resultant matrix is therefore

$$F = \begin{bmatrix} 10 + 3x & 5 + 3y \\ 2 & 1 \end{bmatrix}$$

(b) Calculate $G = BA$.

Solution. Here again the matrices **B** and **A** are conformable in the order (sequence) **BA** since the number of columns of **B** (namely 2) is equal to the number of rows of **A**. The result of the multiplication is

$$G = \begin{bmatrix} 11 & 6 \\ 5x + y & 3x \end{bmatrix}$$

(c) Calculate $H = A \cdot C$.

Solution. In this product the number of rows in **C** is 3 and the number of columns in **A** is 2; hence the product **A · C** does not exist.

(d) Calculate $T = C \cdot D$.

Solution. Here the number of rows of **D** is 3 and the number of columns in **C** is 1; hence the product does not exist.

(e) Check the following matrix product:

$$S = \begin{bmatrix} 3 & 8 & 9 \end{bmatrix} \cdot \begin{bmatrix} x & y \\ 3 & z \\ p & q \end{bmatrix} = [3x + 24 + 9p, \quad 3y + 8z + 9q]$$

(Note that a comma was used to separate two entries on the same line since **S** is a *one*-row, *two*-column matrix.)

Terminology

(1) The *transpose* of a matrix ${}_N^M\mathbf{A}$ is obtained by interchanging its rows and columns. Formally, let **Q** denote the transpose of **A**; then

$$A^T = [{}_N^M A]^T = {}^N_M Q = Q \qquad (2.11)$$

Equivalently,
$$[q_{ij}] = [a_{ji}] \tag{2.12}$$

(2) If the elements of **Q** equal the corresponding elements of **A**, then the matrix **A** must be square and it is said to be *symmetric*. In the kernel index notation, symmetric matrices satisfy the relationship
$$a_{ji} = a_{ij} \tag{2.13}$$

(3) A matrix whose elements satisfy the relationship
$$a_{ij} = -a_{ji} \tag{2.14}$$
is said to be *skew-symmetric*. Note that such a matrix contains zeros on the main diagonal (i.e., $a_{ii} = 0$).

(4) A row matrix or a column matrix is called a *vector*.

(5) A matrix containing all zero elements is called a *null matrix*.

(6) The *main diagonal* of a matrix consists of the elements a_{ii}.

(7) A square matrix containing non-zero elements only on the main diagonal is called a *diagonal* matrix.

(8) A matrix containing all ones along the main diagonal is called the *unit matrix* or the *identity matrix*. This matrix will be denoted by the letter **U**.

2.2 Tabular Matrix Multiplication

In this section a method is presented for an orderly and systematic multiplication scheme of conformable matrices. The method is directly applicable to all kinds of matrices, but it is particularly useful for numeric matrices.

We note from Definition 5 that the product **A** · **B**, where matrix **A** has N rows and L columns and matrix **B** has L rows and M columns, is a new matrix **C** having N rows and M columns. A given element c_{ij} in the result is the sum of various products whose factors come from the ith row of **A** and the jth column of **B**. If the three matrices **A**, **B**, and **C** are arranged as shown in Table 2.1, then an element c_{ij} in **C**

Table 2.1 *Basic Arrangement of the Matrices A, B, and C for the Operation C = AB*

	[B]
[A]	[C]

results from summing the products of the numbers in **A** and **B** which lie directly to the left and directly above c_{ij}. This orderly form of the multiplication will result in much clearer work when matrices are multiplied. Note also that the number of columns in **C** is the same as the matrix at the head of the table (namely **B**) and that the number of rows is equal to the number of rows in the matrix to the left of **C**. Conformability requires that the number of rows in **B** must equal the number of columns in **A**. Hence if those two numbers don't agree, the product **A** · **B** does not exist.

EXAMPLE 2.4

Calculate the product $\mathbf{A} \cdot \mathbf{B}$ where the elements of $\mathbf{A}$ and $\mathbf{B}$ are given in Table 2.2.

Solution.

Table 2.2 *Example of Tabular Matrix Multiplication*

$$\begin{bmatrix} 5 & 3 & 2 \\ 8 & 7 & 2 \\ 1 & 2 & 0 \\ 4 & 3 & 1 \end{bmatrix} = \mathbf{B}$$

$$\mathbf{A} = \begin{bmatrix} 2 & 1 & 0 & -1 \\ 3 & 2 & -1 & 0 \end{bmatrix} \quad \begin{bmatrix} 14 & 10 & 5 \\ 30 & 21 & 10 \end{bmatrix} = \mathbf{A} \cdot \mathbf{B}$$

By concatenating two or more such tables multiple products of matrices may be calculated in tabular form. Carefully note the order of the multiplication and the relative positions that the matrices assume in the following example.

EXAMPLE 2.5

Carry out the product

$$\mathbf{Z} = \mathbf{P} \cdot \mathbf{Q} \cdot \mathbf{R} \cdot \mathbf{S}$$

where the elements of the various matrices on the right side of the equation are indicated in Table 2.3.

Table 2.3 *Example of a Continued Matrix Product*

$$\begin{bmatrix} 2 & 8 & 4 \\ 3 & 2 & -1 \\ 1 & 2 & 2 \\ -1 & 0 & 1 \end{bmatrix} = \mathbf{S} \quad \text{①}$$

$$\text{②} \quad \mathbf{R} = \begin{bmatrix} 1 & 2 & 3 & -2 \\ -1 & 4 & 0 & 1 \\ 3 & 2 & 5 & 7 \end{bmatrix} \quad \begin{bmatrix} 13 & 18 & 6 \\ 9 & 0 & -7 \\ 10 & 38 & 27 \end{bmatrix} = \mathbf{R} \cdot \mathbf{S} \quad \text{③}$$

$$\text{④} \quad \mathbf{Q} = \begin{bmatrix} 1 & 0 & -1 \\ 4 & 2 & 3 \end{bmatrix} \quad \begin{bmatrix} 3 & -20 & -21 \\ 100 & 186 & 91 \end{bmatrix} = \mathbf{Q} \cdot \mathbf{R} \cdot \mathbf{S} \quad \text{⑤}$$

$$\text{⑥} \quad \mathbf{P} = \begin{bmatrix} 1 & 0 \\ -1 & 1 \\ 9 & 2 \end{bmatrix} \quad \begin{bmatrix} 3 & -20 & -21 \\ 97 & 206 & 112 \\ 227 & 192 & -7 \end{bmatrix} \begin{matrix} = \mathbf{P} \cdot \mathbf{Q} \cdot \mathbf{R} \cdot \mathbf{S} \\ = \mathbf{Z} \end{matrix} \quad \text{⑦}$$

Solution. The sequence of entries is indicated by circled numbers. The reader is urged to follow the method outlined above and verify all entries resulting in Steps 3, 5, and 7.

Note that as we proceed down the table, the partial products are carried out right to left in the sequence indicated in the table.

A completely analogous procedure starting out with the leftmost matrix in a multiple product can also be demonstrated. For Ex. 2.5 this procedure would take the form of Table 2.4 in which the circled numbers again refer to the sequence of steps necessary to complete the table.

Table 2.4 *Alternate Arrangement for Continued Matrix Products*

	② Q	④ R	⑥ S
P ①	(P · Q) ③	(P · Q) · R ⑤	(P · Q · R) · S ⑦

The second tabular procedure is not as easy to use on paper that is naturally longer up and down than it is wide. There is, however, no difference in the two methods. This book will use the first method extensively in carrying out various calculations in the following chapters.

2.3 Linear Simultaneous Equations

A set of N simultaneous linear equations in N unknowns may be written as

$$a_{11}x_1 + a_{12}x_2 + \cdots + a_{1N}x_N = b_1$$
$$a_{21}x_1 + a_{22}x_2 + \cdots + a_{2N}x_N = b_2$$
$$\vdots \tag{2.15}$$
$$a_{N1}x_1 + a_{N2}x_2 + \cdots + a_{NN}x_N = b_N$$

Alternately, by application of the various definitions given in Sec. 2.1 the above may be cast in the matrix form

$$\mathbf{Ax} = \mathbf{b} \tag{2.16}$$

where

$$\mathbf{A} = \begin{bmatrix} a_{11} & a_{12} & \cdots & a_{1N} \\ a_{21} & \cdots & \cdots & a_{2N} \\ \vdots & & & \vdots \\ a_{N1} & \cdots & \cdots & a_{NN} \end{bmatrix}$$

$$\mathbf{x} = \begin{bmatrix} x_1 \\ x_2 \\ \vdots \\ x_N \end{bmatrix} \quad \text{and} \quad \mathbf{b} = \begin{bmatrix} b_1 \\ b_2 \\ \vdots \\ b_N \end{bmatrix}$$

Eq. (2.15) is a direct result of applying Definition 5 to the left side of Eq. (2.16).

For a given set of numbers in **A** and **b**, a unique set of numbers **x** is sought such that the above equations are satisfied. The solutions are given by *Cramer's rule*:

$$x_j = \frac{\det \mathbf{A}_j}{\det \mathbf{A}} \quad (j = 1, \ldots, N) \tag{2.17}$$

where det **A** is the determinant of the N by N array of coefficients in **A** and det $\mathbf{A}_j$ is the determinant of the coefficients of the array **A** with the jth column replaced by the vector **b**. The calculation of the determinant is of no consequence here, and the reader is referred to various texts on linear algebra for a derivation of the above relations[1].

Note that a unique set of values is obtained if

$$\det \mathbf{A} \neq 0 \tag{2.18}$$

This is the only necessary and sufficient condition for the solution vector **x** to exist.

A procedure will now be described for the calculation of the solution vector **x**. For the time being no attempt will be made to check for the condition expressed in Eq. (2.18); this check results as a by-product of the process. For illustration, a numerical example will be used instead of general algebraic terms as contained in Eq. (2.15).

The procedure relies on the fact that the row-by-row equality implied in Eq. (2.16) is not destroyed

(1) If all elements of both right and left sides of the same row are multiplied by the same constant or divided by the same constant, provided the constant is not zero;

(2) If any row is replaced by the sum of that row and another row (possibly multiplied by a constant).

Of course, whatever operations are done on one side of the equal sign must also be performed on the other side. The success of the procedure rests on choosing multipliers and divisors systematically in such a way that the structure of the equations becomes simpler. The strategy consists of reducing the square array of numbers a_{11} through a_{NN} to another array that contains ones on the main diagonal and zeros below it. The Nth row will then represent the relation

$$x_N - b'_N \tag{2.19}$$

where b'_N is whatever number results from the procedure thus described in the Nth position of **b**.

The $(N - 1)$st row is the equation

$$x_{N-1} + a'_{N-1,N} x_N = b'_{N-1} \tag{2.20}$$

which in turn may be solved for x_{N-1} by using the value of x_N (which is contained in location N of **b**). Note that normally $a'_{N-1,N} \neq a_{N-1,N}$ of Eq. (2.15).

Similarly, the values of $x_{N-2}, x_{N-3}, \ldots$, etc. are obtained; the procedure terminates with the determination of x_1.

The procedure just described is known as the *Gauss elimination method* for solving simultaneous linear equations.

A numerical example will now be worked in detail. Let it be required to solve the following set of three equations in three unknowns:

$$3x_1 + 2x_2 + x_3 = 6 \tag{2.21a}$$
$$2x_1 + 2x_2 + 2x_3 = 4 \tag{2.21b}$$
$$4x_1 - 2x_2 - 2x_3 = 2 \tag{2.21c}$$

To obtain the first main diagonal one, the first equation is divided (on both sides) by the coefficient of x_1 (namely 3):

$$x_1 + \tfrac{2}{3}x_2 + \tfrac{1}{3}x_3 = 2 \tag{2.22a}$$

To produce the a zero below the main-diagonal element, Eq. (2.22a) is multiplied by the coefficient of x_1 in Eq. (2.21b) and then the result is subtracted from Eq. (2.21b):

$$\tfrac{2}{3}x_2 + \tfrac{4}{3}x_3 = 0 \tag{2.22b}$$

Similarly, multiplying Eq. (2.22a) by 4 and subtracting the result (on both sides of the equal sign) from Eq. (2.21c) results in

$$-\tfrac{14}{3}x_2 - \tfrac{10}{3}x_3 = -6 \tag{2.22c}$$

The second main-diagonal element becomes one if Eq. (2.22b) is divided by the coefficient of x_2 (namely $\tfrac{2}{3}$):

$$x_2 + 2x_3 = 0 \tag{2.23a}$$

The subtraction of Eq. (2.23a) multiplied by the coefficient of x_2 in Eq. (2.22c) results in

$$6x_3 = -6 \tag{2.23b}$$

Finally, dividing Eq. (2.23b) by the coefficient of x_3 results in

$$x_3 = -1 \tag{2.24}$$

Note that the original set of equations (2.20) was transformed successively to the sets

$$\begin{bmatrix} 1 & \tfrac{2}{3} & \tfrac{1}{3} \\ 0 & \tfrac{2}{3} & \tfrac{4}{3} \\ 0 & -\tfrac{14}{3} & -\tfrac{10}{3} \end{bmatrix} \begin{bmatrix} x_1 \\ x_2 \\ x_3 \end{bmatrix} = \begin{bmatrix} 2 \\ 0 \\ -6 \end{bmatrix} \tag{2.25a}$$

and to

$$\begin{bmatrix} 1 & \tfrac{2}{3} & \tfrac{1}{3} \\ 0 & 1 & 2 \\ 0 & 0 & 6 \end{bmatrix} \begin{bmatrix} x_1 \\ x_2 \\ x_3 \end{bmatrix} = \begin{bmatrix} 2 \\ 0 \\ -6 \end{bmatrix} \tag{2.25b}$$

and finally to

$$\begin{bmatrix} 1 & \tfrac{2}{3} & \tfrac{1}{3} \\ 0 & 1 & 2 \\ 0 & 0 & 1 \end{bmatrix} \begin{bmatrix} x_1 \\ x_2 \\ x_3 \end{bmatrix} = \begin{bmatrix} 2 \\ 0 \\ -1 \end{bmatrix} \tag{2.25c}$$

Notice that after the first cycle the first row does not change and that after the second cycle the second row does not change, etc. Hence only rows in which changes occur need be worked with.

When every main-diagonal element has been reduced to a one, no further equation reductions are possible; this signals the end of the *forward course* of equation reduction.

The unknowns are calculated as follows:

$$x_3 = -1 \tag{2.24}$$

$$x_2 = 0 - (2)\cdot(-1) = 2 \tag{2.26a}$$

$$x_1 = 2 - (\tfrac{2}{3})\cdot(2) - (\tfrac{1}{3})\cdot(-1) = 1 \tag{2.26b}$$

With the determination of the first unknown the *return course* is complete and the values of the unknowns have been determined.

In the above procedures only the coefficients of the matrices **A** and **b** are needed; the operations indicated and partial results obtained may be carried in a tabular fashion as shown in Table 2.5.

Table 2.5 *Tabular Form of the Gauss Elimination Method*

Row	Operation	A			b
a	Given	3	2	1	6
b	Given	2	2	2	4
c	Given	4	-2	-2	2
a'	$a/3$	1	$\tfrac{2}{3}$	$\tfrac{1}{3}$	2
b'	$b - a'\cdot 2$	0	$\tfrac{2}{3}$	$\tfrac{4}{3}$	0
c'	$c - a'\cdot 4$	0	$-\tfrac{14}{3}$	$-\tfrac{10}{3}$	-6
b''	$b'/(\tfrac{2}{3})$	0	1	2	0
c''	$c' - b''\cdot -\tfrac{14}{3}$	0	0	6	-6
c'''	$c''/6$	0	0	1	-1
b'''	$b'' - 2\cdot c'''$	0	1	0	2
a'''	$a' - \tfrac{2}{3}\cdot b''' - \tfrac{1}{3}\cdot c'''$	1	0	0	1

Note that each row in the table represents the numerical coefficients of an entire equation, either Eq. (2.21), (2.22), (2.23), or (2.24). Operations are performed with all elements of a row as indicated by the entries in the *Operation* column. The rows are identified by distinct letters; various superscripts (i.e., primes) indicate transformed equations (i.e., a'' is the twice-transformed version of the original equation a).

The table is filled in a row at a time from the top down. First, the coefficients of the first row of the matrices **A** and **b** are entered in corresponding column positions (for **A**). Next, the second row is completed, then the third row, etc., until all the coefficients of the original set of equations are entered. In the table these are the rows

a, b, and c. So far no operations were performed on any of the rows. The reader is urged to develop such a table on a separate sheet of paper and calculate the proper table entries as this description is developed.

The first calculation is the computation of the coefficients of Eq. (2.22a); these become the entries in row a', obtained as indicated in the Operation column. For ease of following the sequence of operations, the sources of the constants occurring in the Operation column are indicated by the dashed arrows for rows a', b', and c'. Next, the coefficients of Eq. (2.22b) and (2.22c) are calculated by subtracting the coefficients of row a' multiplied by the constant occurring in the first column in row b to obtain b'; by that occurring in the first row of c to obtain c'. This process is continued until all coefficients below the main diagonal in the transformed equations (rows b', c', etc.) are reduced to zero; in the example two such rows must be calculated, but for N equations $(N-1)$ such rows must be obtained. The end of this first cycle is indicated by the solid line drawn below the last row in these calculations (row c').

This cycle of reducing the leading main-diagonal coefficient to one and producing zeros below the main diagonal is then repeated for the second, third, etc., main-diagonal entries. In the example, rows b'' and c'' are produced in the second reduction cycle and row c''' in the third cycle.

The forward course terminates when the last main-diagonal entry has been reduced to a one; in the example this point is indicated by the double solid line. Note that the coefficient of the three equations during the process of forward reduction is contained successively in rows a, b, c, rows a', b', c'; rows a', b'', c''; and rows a', b'', c'''.

The unknowns are solved by back substitution into the last set; x_3 is known directly from c'''; x_2 is determined from b''; and x_1 is determined from a' by the operations indicated in rows b''' and a'''. Hence the entries in column b in rows c''', b''', and a''' contain the numerical values of x_3, x_2, and x_1 respectively.

Every step in the above procedure must result in unambiguous numerical values for the process to yield valid results. The only difficulty that may be encountered occurs when a main-diagonal element needed for a division operation is zero. In some instances it may be possible to interchange two rows in the table; this has the effect of changing the sequence of equations, but it does not affect the solution. If all numbers on and below the main diagonal become zeros (i.e., when a column of all zeros is produced at any time during the forward reduction), the set of equations cannot be solved. The original set was not a set of independent equations and no unique solution exists.

As a by-product of this process the value of the determinant of the coefficients is obtained as the product of the various divisors in the process:

$$\det \mathbf{A} = (3)(\tfrac{2}{3})(6) = 12$$

If the value of the determinant becomes zero, no unique solution exists and the set of equations is said to be a dependent set.

The Gauss elimination procedure (method, algorithm) is the most economical procedure known for the solution of simultaneous equations. For a large number of

Sec. 2.3   Linear Simultaneous Equations   29

equations N, the number of multiplications and subtractions is approximately $\frac{1}{3}N^3$ and the number of divisions is approximately $\frac{1}{2}N^2$.

At this point the reader is urged to acquire a good working knowledge of this method by completing Problems 2.8 (a) and 2.8 (b).

A closely related procedure for the solution of simultaneous linear algebraic equations is the Gauss-Jordan method. In this procedure the forward and return steps of the Gauss elimination process are carried out with every main-diagonal element. The procedure consists only of a forward course which eliminates at the kth cycle the kth variable from every equation except the kth one. The original square array of numbers is simply reduced to a unit matrix; thus in place of the coefficients of the **b** vector the solution vector **x** is generated.

The last example is reworked in Table 2.6 by using the Gauss-Jordan procedure. As before, the entries in the column headed Operation indicate the step-by-step application of the method. The reader is again urged to calculate all entries for himself in order to understand the method.

Note that during the reduction process the coefficients of the original set of three equations are represented successively by the rows a, b, c; rows a', b', c'; rows a'', b'', c''; and rows a''', b''', c'''. The procedure terminates when all main-diagonal elements are reduced to ones.

Table 2.6 *Tabular Form of the Gauss-Jordan Reduction Method*

Row	Operation	A			b
a	Given	3	2	1	6
b	Given	2	2	2	4
c	Given	4	-2	-2	2
a'	$a/3$	1	$\frac{2}{3}$	$\frac{1}{3}$	2
b'	$b - a' \cdot 2$	0	$\frac{2}{3}$	$\frac{4}{3}$	0
c'	$c - a' \cdot 4$	0	$-\frac{14}{3}$	$-\frac{10}{3}$	-6
b''	$b'/\frac{2}{3}$	0	1	2	0
a''	$a' - b'' \cdot \frac{2}{3}$	1	0	-1	2
b''	copy b''	0	1	2	0
c''	$c' - b'' \cdot (-\frac{14}{3})$	0	0	6	-6
c'''	$c''/6$	0	0	1	-1
a'''	$a'' - c''' \cdot (-1)$	1	0	0	1
b'''	$b'' - c''' \cdot (2)$	0	1	0	2
c'''	copy c'''	0	0	1	-1

As before, the value of the determinant of the coefficients is obtained by a product of the various divisors used.

Note that although this method requires only one kind of process (namely the forward course), the number of operations is somewhat more. For a large number of equations the total number of multiplications and subtractions each is approximately $\frac{1}{2}N^3$.

Although the Gauss-Jordan procedure requires more multiplications and subtractions than does the Gauss elimination procedure, it is a simpler procedure and is therefore easier to program and results in a shorter computer program. Normally, we shall use the Gauss elimination procedure for solving sets of linear simultaneous equations; however, the Gauss-Jordan process will be used in Sec. 2.8 to derive a program for matrix inversion.

2.4 Gauss Elimination Program

In this section a FORTRAN subroutine is developed which will go through the steps of the Gauss elimination procedure and produce the solution vector to a set of linear algebraic equations. The routine will be called SUBROUTINE GAUSS. The input parameters needed are the coefficient matrix **A**, the vector **b**, and the number of equations N. We shall assume that the coefficients of **A** and **b** are assembled in the FORTRAN array A such that the $(N+1)$st vector in A contains the elements of **b**. Further, let the storage be allocated for NA rows and $(N+1)$ columns of variables. The location DET will be used to calculate the value of the determinant. Thus the beginning of the routine becomes

$$\begin{array}{l}\text{SUBROUTINE GAUSS (A,NA,N,DET)}\\ \text{DIMENSION A(NA,NA)}\\ \text{DET=1.}\end{array} \qquad (2.27)$$

Next, the forward course is done: For each row I, the values to the right of the main diagonal are divided by the main-diagonal entry (including the entry in column $N+1$); then the newly formed row is multiplied by the coefficient below the main diagonal for row J and subtracted term-by-term from the original coefficients in row J. The above is repeated for all rows $J = I+1, \ldots, N$; and all main diagonals $I = 1, 2, \ldots, N-1$. For the Nth main diagonal, only the division step is done. These steps are summarized by the statements

$$\begin{array}{l}\left(\left(\left(a_{ij} \longleftarrow \dfrac{a_{ij}}{a_{ii}}, \quad j = i+1, \ldots, N+1\right);\right.\right.\\ (a_{jk} \longleftarrow a_{jk} - a_{ji} \cdot a_{ik}, \quad k = i+1, \ldots, N+1),\\ j = i+1, \ldots, N); \quad i = 1, \ldots, N-1);\\ a_{N,N+1} \longleftarrow \dfrac{a_{N,N+1}}{a_{N,N}}\end{array} \qquad (2.28)$$

The FORTRAN code for the above algorithm is

```
         NN=N-1
         N1=N+1
         DO 100 I=1,N
         II=I+1
         DO 30 J=II,N1
    30   A(I,J) = A(I,J)/A(I,I)
         DET=DET*A(I,I)
         IF (I.EQ.N) GO TO 100
         DO 40 J=II,N
         DO 40 K=II,N1
    40   A(J,K) = A(J,K) -A(J,I)*A(I,K)
    100  CONTINUE
```
(2.29)

Statements involving the variable DET are needed to calculate the value of the determinant of **A**. The above completes the forward course of reduction. Location A(N, N1) contains the value of x_N.

The back substitution involves solving for the unknowns x_{N-1}, x_{N-2}, etc. The following code will accomplish this task:

```
         DO 200 I=1,NN
         II=N-I
         IJ=II+1
         DO 200 K=IJ,N
    200  A(II,N1) = A(II,N1) -A(II,K)*A(K,N1)
         RETURN
         END
```
(2.30)

Note that the solution vector will be in column $N + 1$ of the array A. Also no precautions were taken against zero main-diagonal elements; such may be accomplished by inserting a check statement before using A(I, I) or A(N, N) in the above.

Since interchanging entire equations does not change the solution in any way, instead of checking for a zero in the main diagonal, the equation with the largest coefficient on or below the diagonal could be substituted for the row of coefficients naturally occurring. This substitution has the beneficial effect of tending to reduce the numerical inaccuracies encountered (see Sec. 2.7). Code for this substitution may be

```
         II = . . . . . . . . .
         IF (I.EQ.N) GO TO 27
         AMAX=ABS(A(I,I))
         IJ=I
    C        SEARCH BELOW MAIN DIAGONAL.
         DO 20 J=II,N
         IF(AMAX.GT.ABS(A(J,I))) GO TO 20
         AMAX=ABS(A(J,I))
         IJ=J
    C        A NEW MAXIMUM WAS FOUND.
    20   CONTINUE
         IF (I.EQ.IJ) GO TO 27
    C        EXCHANGE ROWS I AND IJ
         DET = -DET
         DO 27 J=I,N1
         T=A(I,J)
         A(I,J)=A(IJ,J)
    25   A(IJ,J)=T
    27   CONTINUE
         IF(ABS(A(I,I)).LT.1.E-30) GO TO 500
         DO 30 J= . . . . . .
```
(2.31)

```
      500 DET=0.
          RETURN
          END
```

Note that this code is to be inserted just preceding the statement DO 30 J = II, N1. The check for nonvanishing A(N, N) is left as an exercise. The complete code for this subroutine is shown in Appendix A.

In order to check out the above program, a main routine will be written which will read data from a set of input cards, call the above subprogram, and print the results. The coefficients used in Table 2.5 may be used for checkout. The main program may be as shown in Listing 2.1.

```
C         ROUTINE TO CHECK SUBROUTINE GAUSS
          DIMENSION A(15,16)
    1     READ(1,10) N
          IF (N.LE.0) CALL EXIT
   10     FORMAT (I3)
          N1=N+1
          DO 15 I=1,N
   15     READ(1,20) (A(I,J),J=1,N1)
   20     FORMAT (8F10.0)
          WRITE (3,30) N
   30     FORMAT (//' SOLUTION OF',I3,' EQUATIONS.'/' THE INPUT COEFFIC
         1IENTS ARE'//)
          DO 40 I=1,N
   40     WRITE (3,45) (A(I,J),J=1,N1)
   45     FORMAT (1P8E13.5/(3X,1P8E13.5))
          CALL GAUSS (A,15,N,DET)
          WRITE (3,50) DET,(A(I,N1),I=1,N)
   50     FORMAT (//' THE VALUE OF THE DETERMINANT IS ',1PE13.5//' THE S
         1OLUTION VECTOR IS'//1P8E13.5/(3X,1P8E13.5))
          GO TO 1
C             PROGRAM WILL TERMINATE ON A READ ERROR OR A BLANK INPUT CARD.
          END
```

Listing 2.1 Main Routine for Simultaneous Equations Calculations.

The reader is urged to provide the input cards for the above program and run at least one successful solution.

2.5 Gauss-Seidel Procedure

The simultaneous equations solution procedures described in Sec. 2.3 will produce the numerical solution to a set of equations in a finite number of steps. Such procedures will always work provided a solution exists and numerical inaccuracies will not make the solution vectors of questionable numerical precision (see Sec. 2.7). Both the Gauss elimination and the Gauss-Jordan procedures start with no information about the numerical values of the solution vector and proceed a finite number of steps to the solution.

Sharply in contrast to these two discussed methods is the idea of obtaining the solution vector as the result of a sequence of iterates. The fundamental idea is to start with a solution vector which is then successively corrected to the true solution. The speed with which this process works depends on the closeness of the original assumed solution to the true ("exact") result; indeed, the very success of the entire process crucially depends on this *starting* vector. Processes based on these ideas are known as iterative methods.

Conceptually, these processes produce from a starting vector $\mathbf{x}^{(0)}$ a sequence of vectors:

$$\mathbf{x}^{(0)}, \mathbf{x}^{(1)}, \mathbf{x}^{(2)}, \ldots, \mathbf{x}^{(k)} \tag{2.32}$$

such that the sequence approaches the true solution $\mathbf{x}$. Many such methods are in use; they differ in the manner of producing a new approximation from the sequence of previous approximations.

A process merely attempts to solve the jth equation of the original set, Eq. (2.15), for the jth variable:

$$x_j^{(k)} = \frac{b_j - \sum_{i=1, i \neq j}^{N} a_{ji} x_i^{(k-1)}}{a_{jj}} \quad (j = 1, \ldots, N) \tag{2.33}$$

The various iterates are produced by successively applying the above equation for $k = 1, 2, \ldots$, etc., with $\mathbf{x}^{(0)}$ being the starting vector (assumed solution). If the method converges, then the solution $\mathbf{x}$ is approached as $k \to \infty$. This method is known as the *simple iteration* procedure.

The method just outlined will work well if a_{ii} is not too small; obviously, for $a_{ii} = 0$ the method will fail. Convergence is assured if the main-diagonal elements predominate. Such may sometimes be assured by re-arranging the rows of the input array [10]. Instead of exploring various implementations of iterative methods of solution and methods of enhancing convergence, this section will look at a practical method of solution.

Problems 2.16 and 2.17 are exercises designed to illustrate the above procedure.

A little reflection will show that once the jth unknown is calculated it might as well be used in the calculation of the subsequent elements of $\mathbf{x}^{(k)}$. This modification is known as the Gauss-Seidel procedure; the basic algorithm is

$$x_j^{(k)} = \frac{b_j - \sum_{i=1}^{j-1} a_{ji} x_i^{(k-1)} - \sum_{i=j+1}^{N} a_{ji} x_i^{(k)}}{a_{jj}} \tag{2.34}$$

All iterative procedures require some rule for termination. As the sequence of iterates $\ldots, \mathbf{x}^{(k-2)}, \mathbf{x}^{(k-1)}, \mathbf{x}^{(k)}$ approaches some limiting values, corresponding entries in the vectors will change very little. Usually, iteration is stopped when two successive vectors have entries "close enough" to each other. The term "close enough" is difficult to define precisely. For example, if the solution entries consist of relatively small numbers, say each 10 or fewer, a simple examination of the difference between successive iteration may suffice and if all such differences are smaller than z,

$$|\epsilon_j| = |x_j^{(k-1)} - x_j^{(k)}| \leq z, \quad \text{all } j \tag{2.35}$$

then the iteration may be stopped. Of course the value of z will depend, at least partially, on the use that is made of the solution. If the solution consists of large numerical values, such as 1000 or more, the convergence criterion of Eq. (2.35) may be too severe or may never be met. For example, for $z = 10^{-6}$, two successive iterations of magnitude 1000 would require 9 digit accuracy; this is attainable on wide-word machines (48 or 60 bits/word) with single precision calculations, but on 32 bits/word machines double-precision arithmetic must be used. On these machines the use of relative accuracy may be more appropriate:

$$\delta_j = \left| \frac{(x_j^{(k-1)} - x_j^{(k)})}{\max [x_j^{(k-1)}, x_j^{(k)}]} \right| \tag{2.36}$$

Again, if $\delta_j \leq z$ for all j, the iteration is stopped. For most applications a combination of Eqs. (2.35) and (2.36) is used.

If the equation coefficients of the N simultaneous equations are assembled in an array with N rows and $N + 1$ columns (as was done in Eq. 2.32) and if the starting vector $\mathbf{x}^{(0)}$ is assembled in the array X, a FORTRAN code for the Gauss-Seidel algorithm is as shown in Listing 2.2.

```
          SUBROUTINE ITERT (A,NA,N,X)
          DIMENSION A(NA,NA),X(NA)
          N1=N+1
    C     SET CONVERGENCE FLAG.
    10    KK=0
          DO 100 J=1,N
          XX=A(J,N1)
          DO 50 I=1,N
          IF (I.EQ.J) GO TO 50
          XX = XX -A(J,I)*X(I)
    50    CONTINUE
          OLD = X(J)
          X(J) = XX/A(J,J)
    C         CHECK FOR CONVERGENCE.
          IF (ABS(XX-OLD).LT.1.E-6) GO TO 70
          PIVOT=ABS(OLD)
          IF(ABS(XX).GT.PIVOT) PIVOT=ABS(XX)
          IF (ABS(XX-OLD)/PIVOT .LT. 1.E-6) GO TO 70
          GO TO 100
    70    KK=KK+1
    C         THE J-TH VARIABLE HAS MET CONVERGENCE.
    100   CONTINUE
          IF (KK.EQ.N) RETURN
    C         AT LEAST ONE VARIABLE DOES NOT MEET CONVERGENCE.
          GO TO 10
          END
```

Listing 2.2 Gauss-Seidel Iteration Routine

EXAMPLE 2.6

The following results were obtained for the augmented matrix:

$$[A \mid b] = \begin{bmatrix} 1.00 & 0.50 & 0.30 & \vdots & 1.50 \\ 0.20 & 0.70 & 0.30 & \vdots & 1.00 \\ 0.50 & 0.25 & 1.00 & \vdots & 3.00 \end{bmatrix}$$

Iteration	Simple Iteration			Gauss-Seidel		
Number	x_1	x_2	x_3	x_1	x_2	x_3
0	1.000	0.000	3.000	1.000	0.000	3.000
1	0.600	−0.143	2.500	0.600	−0.029	2.707
2	0.821	0.186	2.736	0.702	0.067	2.632
3	0.586	0.021	2.543	0.677	0.107	2.635
4	0.726	0.171	2.701	0.656	0.112	2.644
5	0.604	0.063	2.594	0.651	0.109	2.647
⋮						
10	0.664	0.119	2.659	0.652	0.108	2.647
11	0.642	0.099	2.638	0.652	0.108	2.647

Simple iteration took 22 cycles for an accuracy of 10^{-3} between successive iterations; Gauss-Seidel needed only 7 cycles.

2.6 Matrix Inversion

An N by N matrix $\mathbf{Q}$ is said to have the inverse $\mathbf{P} = \mathbf{Q}^{-1}$ if

$$\mathbf{P} \cdot \mathbf{Q} = \mathbf{Q} \cdot \mathbf{P} = \mathbf{U} \qquad (2.37)$$

where $\mathbf{U}$ is the unit matrix.

From Definition 5, Sec. 2.1, the dimensions of the inverse and of $\mathbf{U}$ are found to be N by N. Notice that since the order of multiplication is immaterial in Eq. (2.37), it follows that only square matrices possess an inverse.

Now let us assume that the N by N square matrix $\mathbf{A}$ has an inverse $\mathbf{A}^{-1}$. Also suppose that the array $\mathbf{A}$ is used to transform a set of unknowns $\mathbf{x}$ into another set $\mathbf{y}$:

$$\mathbf{A}\mathbf{x} = \mathbf{y} = \mathbf{U}\mathbf{y} \qquad (2.38)$$

The actual values of $\mathbf{x}$ and $\mathbf{y}$ are of no importance here, as long as $\mathbf{x}$ is not a vector of zeros. Now, the equality in Eq. (2.38) expresses N equations. For the jth row

$$a_{j1}x_1 + a_{j2}x_2 + \cdots a_{jN}x_N$$
$$= 0 \cdot y_1 + 0 \cdot y_2 + \cdots + 0 \cdot y_{j-1} + y_j + 0 \cdot y_{j+1} \cdots + 0 \cdot y_N \qquad (2.39)$$

If the left side of the equality (**A**) is reduced to the unit matrix by row operations only, performing on the right side the same operations that are being done on the left, the original unit matrix $\mathbf{U}$ is reduced to some other N by N matrix $\mathbf{B}$:

$$\mathbf{U}\mathbf{x} = \mathbf{B}\mathbf{y} \qquad (2.40)$$

Another way of reducing Eq. (2.38) to Eq. (2.40) is to pre-multiply the former by $\mathbf{A}^{-1}$:

$$\mathbf{A}^{-1} \cdot \mathbf{A} \cdot \mathbf{x} = \mathbf{U}\mathbf{x} = \mathbf{A}^{-1} \cdot \mathbf{U} \cdot \mathbf{y} = \mathbf{A}^{-1}\mathbf{y} \qquad (2.41)$$

Comparison of Eqs. (2.40) and (2.41) shows that **B** is the inverse of **A**. Hence a way to calculate the inverse is to apply the Gauss-Jordan procedure to the matrix and a unit matrix simultaneously. The reduction of **A** to a unit matrix may be carried out provided the determinant of **A** does not vanish (see Sec. 2.3).

The subroutine shown in Listing 2.3 implements the inversion process just described. Steps are included in the routine to prevent division by zero. This is accomplished by a search for the largest number in a given column below the main diagonal and an exchange of rows which assures that the largest number in a column is used for division. The value of this determinant is also calculated.

```
      SUBROUTINE INVRT (A,NA,N,DET)
      DIMENSION A(NA,1)
C         PERFORMS GAUSS-JORDAN REDUCTION ON INPUT ARRAY AND THE
C         APPENDED UNIT MATRIX, KEEPING ROW-WISE EQUALITY.
      DET=1.
      NN=N+N
C         UNIT MATRIX IS SET IN COLUMNS N+1 THROUGH N*2
      DO 20 I=1,N
      NI=N+I
      DO 20 J=1,N
      NJ=N+J
   10 A(I,NJ)=0.
   20 A(I,NI)=1.
      DO 90 I=1,N
      II=I+1
      IF (I.EQ.N) GO TO 50
      AMAX=ABS(A(I,I))
      KJ=I
      DO 30 K=II,N
      IF (AMAX.GT.ABS(A(K,I))) GO TO 30
      AMAX=ABS(A(K,I))
      KJ=K
   30 CONTINUE
      IF (KJ.EQ.I) GO TO 50
C         NEED EXCHANGE ROWS I AND KJ.
      DO 40 K=I,NN
      T=A(I,K)
      A(I,K)=A(KJ,K)
   40 A(KJ,K)=T
C         EXCHANGE OF ROWS COMPLETE.
   50 IF (ABS(A(I,I)).LT.1.E-30) GO TO 500
      DET=DET*A(I,I)
C         DIVIDE BY MAIN-DIAGONAL ELEMENT.
      DO 60 J=II,NN
   60 A(I,J)=A(I,J)/A(I,I)
      IF (I.EQ.N) GO TO 90
      DO 80 J=1,N
      IF (I.EQ.J) GO TO 80
      DO 70 K=II,NN
   70 A(J,K)=A(J,K)-A(J,I)*A(I,K)
   80 CONTINUE
C         REDUCTION COMPLETE.
   90 CONTINUE
      RETURN
  500 DET=0.
C         INVERSION FAILED.
      RETURN
      END
```

Listing 2.3 Inversion Routine

Sec. 2.7 Numerical Considerations 37

EXAMPLE 2.7

The following data may be used to check the operation of the matrix inversion program:

$$\begin{bmatrix} 1.00 & 0.50 & 0.30 \\ 0.20 & 0.70 & 0.30 \\ 0.50 & 0.25 & 1.00 \end{bmatrix}^{-1} = \begin{bmatrix} 1.226 & -0.833 & -0.118 \\ -0.098 & 1.667 & -0.471 \\ -0.588 & 0.000 & 1.177 \end{bmatrix}$$

$$\begin{bmatrix} 2.5 & 3.5 & 4.0 \\ 1.0 & 0.4 & 0.5 \\ 1.5 & 5.0 & 1.5 \end{bmatrix}^{-1} = \begin{bmatrix} -0.185 & 1.443 & 0.015 \\ -0.073 & -0.220 & 0.269 \\ 0.430 & -0.709 & -0.245 \end{bmatrix}$$

An excellent way to verify the results of an inversion program is to multiply the output of the program (the inverse) by the original matrix. The result should be the unit matrix within numerical accuracy of the computer. The reader is urged to write such a main program and experiment with his own data. A random number generator may be used to advantage to generate data for large trial matrices. Large matrices can be used to establish realistic overall running time of the routine.

2.7 Numerical Considerations

Examples presented in most textbooks contain only relatively small sets of integers. Some of the reasons for this are a desire to economize in the presentation of the material and a wish to show concise examples. Additionally, methods presented (such as the Gauss elimination scheme) using integers and carrying out integer operations often show far more clearly the sequence of operations than decimal numbers tend to do.

Integers are presumed to have infinite precision, i.e., when we write 2 we do not mean 1.9999 ... and we do not mean 2.0000 ... 1. However, practically all problems involving measurements contain data of finite accuracy. Typical voltage and current measurements have no more than four significant digits, although some measurements are more precise. In a computer such data is represented by floating-point numbers of relatively many digits, but finite precision. For example, in an IBM 360 or 370 system the precision of floating-point numbers is about six decimal places. In all modern computers, floating-point numbers are represented relatively precisely, the precision being fixed in the hardware of the machine. Moreover, the basic operations (addition, subtraction, multiplication, and division) are also carried out with the same fixed precision. This section discusses some problems that are created by numbers of finite precision being used to implement mathematical procedures based on assumed infinite precision of the variables.

Although a detailed discussion of errors in computation is far beyond the scope of our discussion here, we shall investigate a few ideas relating to machine calculations. In our discussion we assume that in digital computers, numbers are represented as decimal digits. (But they are not. Our discussion should really be concerned with bits. Bits are representable, however, as fractional decimal digits. Our discussion would still be valid, but it would lead to slightly different numerical conclusions without altering the ideas presented here.)

Let us consider the addition of two numbers, a and b:

$$c = a + b \tag{2.42}$$

We shall represent the inherent inaccuracies of each of the numbers as differentials; hence, differentiating the above gives

$$\delta c = \delta a + \delta b \tag{2.43}$$

Therefore, in addition, the total inaccuracy of the result (the total error) is the sum of the inaccuracies of the individual summands. The relative inaccuracies (the relative error) are

$$\frac{\delta c}{a+b} = \frac{\delta a}{a+b} + \frac{\delta b}{a+b} \tag{2.44}$$

or

$$\frac{\delta c}{c} = \frac{\delta a}{a} \frac{1}{1+\frac{b}{a}} + \frac{\delta b}{b} \frac{1}{1+\frac{a}{b}} \tag{2.45}$$

If the relative inaccuracies of a and b are approximately the same, then the relative inaccuracy of c is

$$\frac{\delta c}{c} = \frac{\delta a}{a}\left(\frac{a+b}{a+b}\right) = \frac{\delta a}{a} \tag{2.46}$$

that is, there is no loss in relative accuracy. On the other hand, the absolute error will increase:

$$\delta c = \frac{c}{a}\delta a \tag{2.47}$$

For example, using eight-place arithmetic (accuracy of 10^{-8}), if $a = 1$ and $b = 10^{-7}$, then Eq. (2.42) gives

$$c = 1.0000000 + .0000001 = 1.0000001 \tag{2.48}$$

The result is good to eight places. The absolute error is

$$\delta c = 1 \times 10^{-8} + 10^{-7} \times 10^{-8} = 10^{-8} \tag{2.49}$$

The relative error is

$$\frac{\delta c}{c} = 10^{-8} \tag{2.50}$$

Similarly, if we use Eq. (2.47), we get (since $\delta b/b = \delta a/a$)

$$\delta c = \frac{c}{b}\delta b = \frac{1.0000001}{.0000001} \times 10^{-7} \times 10^{-8} = 10^{-8} \tag{2.51}$$

If, however, $a = 1$ and $b = 10^{-9}$,

$$c = 1.000\,000\,0 + .000\,000\,01 = 1.000\,000\,0$$

The relative accuracy of c is still 10^{-8}, although the contribution of b is completely masked. To calculate the errors incurred in subtraction, the method of Eq. (2.43) is

Sec. 2.7   Numerical Considerations   39

used with one of the summands (say *b*) becoming a subtrahend. Thus

$$d = a - b \tag{2.52}$$

gives formally

$$\delta d = \delta a - \delta b \tag{2.53}$$

and the relative error is

$$\frac{\delta d}{d} = \frac{\delta a}{a}\frac{1}{1-\frac{b}{a}} - \frac{\delta b}{b}\frac{1}{1-\frac{a}{b}} \tag{2.54}$$

Superficially, it might appear that the numerical example shown is of little applicability in practical machine computations. Nevertheless, one should keep in mind that the procedures discussed in subsequent chapters consist of thousands of steps and sometimes of millions of steps (as with transient analysis of networks, integration of differential equations, and nonlinear analysis which relies heavily on iterative schemes).

Consider, for example, the problem of calculating the sum of reciprocals:

$$S = \sum_{k=1}^{N} \frac{1}{k} \tag{2.55}$$

This sum is known to be bounded by

$$1 + \ln N \geq S \geq \ln(N+1) \tag{2.56}$$

The above equation shows that S increases without bound as N approaches infinity. On a modern digital computer it requires but a few seconds to carry out Eq. (2.55) for N in the order of a million. On a computer having a single precision accuracy of about six decimal places no increase in S will be found for N larger than some 100,000; in every computer a point will be reached when no further terms can be accounted for by a direct application of Eq. (2.55) using the normal floating-point system of the machine.

Similar problems occur in integration of network equations as in transient analysis and state-space analysis (see Chapters 4 and 5); also tolerance analysis and nonlinear analysis may lead to such problems (see Chapters 6 and 7).

Before proceeding further we observe that the magnitude of error in a (and b) is usually deducible from the length of the mantissa of the floating-point number of the machine being used. Since numbers within the computer are usually represented by binary information, the conversion factor between decimal digits and binary digits (bits) is derived from

$$2^{3.32} = 10^1 \tag{2.57}$$

Hence one decimal digit requires 3.32 bits. For example, on a machine of 36-bit word length with 27 bits for a mantissa, 7 bits for a characteristic, and 2 bits for signs (one for the sign of the mantissa, the other for the sign of the characteristic), the inherent numerical accuracy of normal floating-point numbers is $26.5/3.32 = 8.00$ decimal digits. Note that the inherent inaccuracy of representing a number with a set of digits is half a digit, on the average. Of course, schemes exist for extended precision calculations on all computers, but such schemes are expensive in execution time as well as code storage (core) requirements.

The reader is urged to find out from his computer installation what the floating-point number mantissa is for his machine both for single precision and multiple precision (usually double precision is available, but sometimes higher precision is also implemented). The number of bits in the mantissa has a direct bearing on the accuracy of many of the examples. The reader is reminded that examples presented in this book were calculated with an IBM 370/165 system which uses six hexadecimal (4 bits each) digits for the mantissa. Since $16^{0.830} = 10$, the inherent accuracy of the single-precision numerical examples given in this book is $5.5/0.830 = 6.62$ decimal digits. This does not mean that the numbers given in the examples are that accurate; instead, it means that no numerical example (calculated in single precision) exceeds this accuracy.

Note that the range of numbers that may be represented on a computer is independent of the accuracy. The size of the numbers depends on the numbers of bits of the mantissa (on a binary machine). In the example cited (7 bits) the size of the numbers can be up to $2^7 = 128$ bits; the corresponding decimal number range is $128/3.32 = 38.5$. Thus numbers in the range 10^{-38} to 10^{+38} may be represented with an accuracy of eight places, i.e., 27 bits; the relative accuracy is 10^{-8}. In the IBM 370/165 system the characteristic is represented by 6 bits for a hexadecimal characteristic range of $2^6 = 64$; the decimal characteristic size is $64/0.83 = 77$ for a range of numbers 10^{-38} to 10^{+38} approximately.

In representing numbers to the accuracy determined by the number of bits in the mantissa, only the size and not the sign of the relative error is known. Hence all deviations in the developments so far should be understood to represent magnitude only. Hence Eq. (2.54) should read

$$\left|\frac{\delta d}{d}\right| = \left|\frac{\delta a}{a}\right|\left|\frac{1}{1-\frac{b}{a}}\right| + \left|\frac{\delta b}{b}\right|\left|\frac{1}{1-\frac{a}{b}}\right| \tag{2.58}$$

Next we observe that when $a \approx b$ (i.e., when a and b are close together), even when their relative errors are the same, the relative error in the result may be very large:

$$\frac{\delta c}{c} = \frac{\delta a}{a}\frac{a+b}{a-b} \tag{2.59}$$

To emphasize the nature of the inherent errors we write the above as

$$\left|\frac{\delta c}{c}\right| = \left|\frac{\delta a}{a}\right|\left|\frac{a+b}{a-b}\right| \tag{2.60}$$

As $(a - b) \to 0$, the inherent relative error in the difference may be very large.

For example, let us calculate $d = a + b - c = (a + b) - c$ with $a = 1.0000000\text{E}+01$; $b = 1.5234760\text{E}-08$; $c = 1.0000057\text{E}+01$ whenever normal FORTRAN convention is used for representing the decimal equivalents of floating-point numbers. The calculations are carried out with a presumed eight-place accuracy:

$$a = 1.0000000 \quad\quad E+01$$
$$b = .000000015234760E+01$$
$$\overline{a+b = 1.0000000E+01}$$
$$c = 1.0000057E+01$$
$$\overline{(a+b) - c = -0.0000057E+01 = -5.7000000E-06}$$

The computer assumed will carry out the indicated operations from left to right in eight-place arithmetic and truncate the result after the eighth place. If the result has leading zeros, the result is shifted left until a non-zero leading digit is encountered, the rightmost digits being filled in with zeros. The result, when printed in format 1PE20.7 (as above), will give the appearance of great precision, when in fact it is not precise. Note that

$$\frac{\delta(a+b)}{a+b} \cong \frac{\delta a}{a} = 10^{-8} \tag{2.61}$$

and

$$\frac{\delta d}{d} = \frac{\delta a}{a} \cdot \frac{a+b+c}{a+b-c} = 10^{-8} \frac{2.0000057}{.0000057} \tag{2.62}$$

Hence the relative error is calculated as

$$\frac{\delta d}{d} \cong 10^{-8} \times \frac{2}{.57 \times 10^{-5}} \cong 3.5 \times 10^{-3} \tag{2.63}$$

The absolute error is approximately

$$\delta d = d \cdot 3.5 \times 10^{-3} = 5.7 \times 10^{-6} \cdot 3.5 \times 10^{-3} = 2 \times 10^{-8} \tag{2.64}$$

Hence the result is

$$d = (-5.70 \pm .02) \times 10^{-6} \tag{2.65}$$

and there is an error in the second place after the decimal point. As can easily be seen, the error is in rounding off the value of $(a+b)$. In effect, b is neglected.

By re-arrangement of the operations it is sometimes possible to refine the results substantially. In the above example, writing

$$d = (a - c) + b \tag{2.66}$$

gives the calculation sequence

$$a = 1.000\ 000\ 0E+01$$
$$c = 1.000\ 0057\ E+01$$
$$\overline{c_1 = (a-c) = \quad -.0000057\ E+01}$$
$$c_1 = -5.7000000\ E-06$$
$$b = \quad 0.0152347\ E-06$$
$$\overline{d = b + c_1 = -5.6847653\ E-06}$$

which gives the appearance of much greater precision. However the result is not any better than before because the precision of c_1 is only two decimal places.

When, however, there are many small numbers to be added to a large number, it is imperative that the order of addition be such that at each point in the addition process the precision of the partial sums is optimized. This requires the summation of the smallest numbers first.

When forming differences it is highly desirable not to subtract two nearly equal numbers from one another because of the relatively inaccurate result. Since addition of a negative number is the same as the subtraction of a positive one, and since there is no good way of establishing *a priori* whether subtraction or addition will take place in a given computational sequence, the best that can be done is to minimize the number of arithmetic operations involved in a process in order to increase the resultant probable accuracy. Another rule is to minimize in further calculations the use of results from the subtraction of two numbers of nearly equal magnitude.

The errors in multiplication/division can be discussed by the use of differentials much the same as was done with addition/subtraction. Here again

$$c = a \cdot b \quad \text{and} \quad d = \frac{a}{b} \tag{2.67}$$

lead to the absolute errors

$$\delta c = a \cdot \delta b + \delta b \cdot a \tag{2.68}$$

and the relative errors

$$\frac{\delta c}{c} = \frac{\delta a}{a} + \frac{\delta b}{b} \tag{2.69}$$

i.e., the relative errors sum for multiplication. Similarly, since only the magnitude of δb is known

$$\frac{\delta d}{d} = \frac{\delta a}{a} + \frac{\delta b}{b} \tag{2.70}$$

i.e., relative errors also sum in division.

Since most matrix operations involve linear operations only, the above discussion of addition/subtraction and multiplication/division errors forms the basis of calculation of errors in matrix manipulation routines.

An excellent discussion of the growth of error in these operations will be found in [6]. In order to minimize the effects of possible round-off errors, two rules can be formulated which are easily rationalized even without an exhaustive analysis.

(1) Minimize the number of arithmetic operations.

(2) As much as possible avoid using the results of subtraction of two nearly equal numbers in further calculations.

The first rule is self-evident: The fewer the operations, the less the chances of round-off accumulation. Fewer operations are also desirable from the standpoint of machine costs.

The second rule is easily deduced from the analysis and numerical examples given. How one would implement the rule is not nearly so apparent. At any given stage in a calculation the "results of subtraction of two nearly equal numbers" may be difficult to identify. However, such numbers have to be relatively small. Thus if the largest numbers are dealt with in preference to small ones, the probability of using numbers prone to be subject to large round-off errors is minimized.

This precaution leads to the idea of pivots. Two equations in an array during the course of the Gauss elimination or Gauss-Jordan procedure may be exchanged without altering the solution vector in any way. (Note, however, that the sign of the determinant will change.) Since every computation cycle, i.e., the elimination of the next equation, starts with the division by a main-diagonal element, we will exchange with the one normally occurring that equation which will give the largest-sized element on the main diagonal.

In practice this means that at the Ith cycle the Ith column is searched on and below the main diagonal for the largest numerical value. Let this value be located in row IJ. If I $\neq$ IJ, rows I and IJ are exchanged and the sign of the determinant is inverted. The procedure, either Gauss-Jordan or Gauss elimination, is then continued with the re-arranged array. The code implementing these steps is given in Eq. (2.31).

The method just described is known as the column-wise pivot search. Pivot search procedures may be implemented also row-wise and column-and-row wise. Both of these latter have applications in numerical analysis, but usually they yield no further appreciable reduction in computational accuracy with the procedures discussed in this book.

2.8 In-Place Inversion with Pivoting

The matrix inversion procedure discussed in Sec. 2.6 can be used for the calculation of inverses, but it requires two large arrays, one for the input matrix and the other for calculating the inverse. As long as the matrix to be inverted is small, no difficulties arise, but when the matrix is large there may be an advantage if the input matrix could be inverted without having to use an appreciable amount of additional storage. A program that inverts the matrix in its original storage locations is known as an *in-place inversion routine*.

In order to derive the algorithms for such a program, the Gauss-Jordan inversion procedure of Sec. 2.6 is applied to an example in Table 2.7. Note that as a column is reduced to all zeros except the main-diagonal entry, which is made one, a new column is added to the right of the original matrix (other than remnants of the original appended unit matrix). Hence only N columns contain at any time the matrix and its partial inverse. Since the other N columns remain a unit matrix, their elements are known and need not be stored if the need arises for their use.

The basic idea in the in-place inversion algorithm is to produce the columns of the inverse in the same locations as the columns of the original matrix, as the original entries are reduced to a column containing a single one. Thus the first column of the

44 Some Elements of Linear Algebra   Ch. 2

Table 2.7 *Example of Gauss-Jordan Inversion*

Input Data | Unit Matrix

$$\begin{bmatrix} 4.000 & 2.000 & 3.000 & | & 1.000 & 0.000 & 0.000 \\ 2.000 & 3.000 & 3.000 & | & 0.000 & 1.000 & 0.000 \\ 1.000 & 2.000 & 1.000 & | & 0.000 & 0.000 & 1.000 \end{bmatrix}$$

Augmented matrix after elimination of Column No. 1

$$\begin{bmatrix} 1.000 & 0.500 & 0.750 & | & 0.250 & 0.000 & 0.000 \\ 0.000 & 2.000 & 1.500 & | & -0.500 & 1.000 & 0.000 \\ 0.000 & 1.500 & 0.250 & | & -0.250 & 0.000 & 1.000 \end{bmatrix}$$

Augmented matrix after elimination of Column No. 2

$$\begin{bmatrix} 1.000 & 0.000 & 0.375 & | & 0.375 & -0.250 & 0.000 \\ 0.000 & 1.000 & 0.750 & | & -0.250 & 0.500 & 0.000 \\ 0.000 & 0.000 & -0.875 & | & 0.125 & -0.750 & 1.000 \end{bmatrix}$$

Augmented matrix after elimination of Column No. 3

$$\begin{bmatrix} 1.000 & 0.000 & 0.000 & | & 0.429 & -0.571 & 0.429 \\ 0.000 & 1.000 & 0.000 & | & -0.143 & -0.143 & 0.857 \\ 0.000 & 0.000 & 1.000 & | & -0.143 & 0.857 & -1.143 \end{bmatrix}$$

Inverse

inverse is produced in the first column and replaces the entries of the first column. Care must be taken to retain the entries of the original matrix long enough to accomplish the reduction of the rest of the matrix.

Also, numerical considerations dictate the use of the largest element in a given column below the main diagonal and exchanging rows if necessary. We shall omit the pivot searching and exchange sequence and assume that the main-diagonal element is the largest element; however, this omission must be rectified in order to obtain a viable program.

Note that when two rows are exchanged in the original input matrix [and the corresponding rows in the inverse are switched to maintain the equality demanded by Eq. (2.38)], the matrix to be inverted is no longer the original one. The rows are permuted. In order to produce the correct inverse, the columns in the inverse must be re-ordered properly. If the original rows are numbered consecutively in a pointer vector and the pointer vector entries are exchanged along with the corresponding rows, at the end of the elimination cycle the pointer vector entries will contain the proper column subscripts. All that remains is to re-order the columns of the inverse back to the original order.

The steps of the reduction are:

(1) Divide each element A(I, J) of row I by A(I, I). This produces a one in location A(I, I); however, do not divide the element A(I, I) since this will produce erroneous results for J > I.

(2) The first entry of the new column of the inverse is then calculated in place of the original A(I, I). The value is 1./A(I, I), the result of dividing the one in location I, I of the inverse by the original A(I, I).

Steps 1 and 2 complete the division of the Ith row of **A** and **A**$^{-1}$ by the pivot element and place the elements of the inverse in place of the original matrix entries.

(3) The zeros in column I of the original matrix are produced next.

(3a) For row J and column K,

```
A(J,K)=A(J,K)-A(J,I)*A(I,K)
```

This produces a zero in column I since in the original matrix A(I, I) = 1 (but is not so stored in A(I, I)).

(3b) K = 1, N; K ≠ I since in column I a zero is being created.

(3c) J = 1, N; J ≠ I since the Ith row was reduced already in Steps 1 and 2.

(4) To produce entries in the Ith column of the inverse:

(4a) ```A(J,I)=-A(J,I)*A(I,I)```

(4b) J = 1, N; J ≠ I since A(I, I) already contains the proper value for the inverse.

(5) Repeat Steps 1 through 4 for I - 1, N.

(6) If the determinant is desired, the value A(I, I) is multiplied into the running product DET (initialized with DET = 1.) either before or after Step 1.

Coding of the above steps is left as an exercise (see Problem 2.11). The numerical examples of Table 2.7 and Ex. 2.7 may be used to check for the correctness of the code.

The pivoting procedure must be carefully done also: When two entire rows (including the Ith column entries) are exchanged, note must be made of the original order of those rows. In the beginning the rows will be in numerical order, but during the inversion process the order may be permuted repeatedly. A good procedure is to carry along a pointer vector (initialized with the natural order of integers) and each time rows are exchanged, the corresponding entries in the pointer are also exchanged.

When the inversion process is complete, the pointer vector will indicate the original *row* indices of the rows as they appear in the result. To obtain the correct inverse, the *columns* of the scrambled inverse must be re-ordered such that the pointer will again contain the natural order.

Coding is left as an exercise (see Problem 2.11). A version of the above process appears in SUBROUTINE INVERT in Appendix A.

2.9 Summary

This chapter introduced matrix notation, defined basic matrix operations, and extended these basic ideas to the solution of simultaneous equations and the calculation of the matrix inverse.

Throughout the chapter procedures for computer coding were stressed. Several programs were developed; programs for the solution of simultaneous linear equations and for the inversion of a matrix were derived and are listed in Appendix A. These programs are used extensively in subsequent chapters.

The problems contain a set of exercises for acquiring skill in these basic matrix procedures and it is strongly suggested to undertake a good portion of these exercises.

REFERENCES

1. NERING, E. D., *Linear Algebra and Matrix Theory*. John Wiley & Sons, Inc., New York, 1964.
2. HOHN, F., *Elementary Matrix Algebra*. The Macmillan Co., New York, 1961.
3a. FADDEEV, D. K., and V. N. FADDEEVA, *Computational Methods of Linear Algebra*. Freeman Press, San Francisco, 1963.
3b. V. N. FADDEEVA, *Computational Methods of Linear Algebra*. Dover Publications, Inc., New York, 1959.
4. BRAAE, R., *Matrix Algebra for Electrical Engineers*. Sir Isaac Pitman & Son, Ltd., London, 1963.
5. PIPES, L. A., and S. A. HOVANESSIAN, *Matrix-Computer Methods in Engineering*. John Wiley & Sons, Inc., New York, 1969.
6. MCCRACKEN, D. D., and W. S. DORN, *Numerical Methods and FORTRAN Programming*. John Wiley & Sons, Inc., New York, 1964.
7. HAMMING, R. W., *Numerical Methods for Scientists and Engineers*, 2nd ed. McGraw-Hill, Inc., New York, 1973.
8. RALSTON, A., *A First Course in Numerical Analysis*. McGraw-Hill, Inc., New York, 1965.
9. TROPPER, A. M., *Matrix Theory for Electrical Engineers*. Addison-Wesley Publishing Co., Reading Mass., 1962.

10. RALSTON, A., and H. S. WILF, *Mathematical Methods for Digital Computers*. John Wiley & Sons, Inc., New York, 1964.
11. WILKINSON, J. H., *Rounding Errors in Algebraic Processes*. Prentice-Hall, Inc., Englewood Cliffs, N.J., 1963.
12. HUELSMAN, L. P., *Digital Computations in Basic Circuit Theory*. McGraw-Hill, Inc., New York, 1968.

PROBLEMS

2.1 Write a FORTRAN subroutine which will add two matrices **A** and **B** of sizes *NA* rows and *MA* columns and *NB* rows and *MB* columns respectively. Matrices **A**, **B**, and **C** are defined in the main routine by the following statement

DIMENSION A(10, 15), B(20, 10), C(30, 30)

Note that this statement defines only the amount of memory allocated to the three matrices and not the numbers of rows and columns actually occupied by the matrix entries. This subroutine should check compatibility of matrices **A** and **B**, return a zero in a variable (KFLAG) if compatibility exists, and calculate the matrix sum if it exists; otherwise KFLAG should be set to 1.

2.2 Write a FORTRAN subroutine for the multiplication of two matrices **A** and **B** with the result returned in **C**. The matrices have the same properties as defined in Problem 2.1. Here again a flag should be set if the multiplication has been carried out properly.

2.3 Modify the subroutine developed in Problem 2.2 to allow for the case of $\mathbf{A} \neq \mathbf{B}$ but either $\mathbf{A} = \mathbf{C}$ or $\mathbf{B} = \mathbf{C}$. It is possible to write such a program without having to store the entire input matrices separately in temporary arrays within the subroutine.

2.4 Modify the subroutines developed in Problems 2.2 and/or 2.3 to allow calculation of the square of an input matrix **A**. Let the square occupy a different storage area, ASQ, in the main program.

2.5 Using the programs developed in Problems 2.1, 2.2, and 2.3, verify the entries in Table 2.3 and the indicated matrix multiplications in Table 3.3.

2.6 Derive a flowchart for the procedure outlined in Table 2.5, the Gauss elimination method for solving linear simultaneous equations. Make your program for *N* equations. Note that in the computer program the ones on the main diagonal and the zeros below it need not be calculated as long as the program regards those values as ones and zeros respectively. Assume that the necessary divisions can always be carried out. Code and check out this subroutine. You may want to use the data contained in Table 2.5.

2.7 To establish a formula for the number of operations (division, multiplication, addition, or subtraction) required to solve *N* equations, a set of counters can be added to your program and the actual number of operations may be counted. Data may be taken for various values of *N* and a formula derived. Obtain from your computer service center the basic add, multiply, and divide time for your computer and calculate the amount of time spent on arithmetic in your program. (The subroutine execution time may be checked independently with various timing routines available on your computer; the arithmetic time should be about $\frac{1}{4}$ to $\frac{1}{2}$ of the total execution time.)

2.8 Solve the following simultaneous equations by using the routine developed in Problem 2.6.

(a) $5x_1 + 6x_2 + 7x_3 = 5$
$3x_1 + 4x_2 + 2x_3 = 8$
$2x_1 + 8x_2 + 3x_3 = 7$

(b) $8a + 9b - 3c + d = 0$
$4a - 9b - 2c + d = 3$
$2a + 3b + 5c - 3d = 4$
$2a + 3b - 2c - d = 2$
$6a - 6b - 4c = 5$

2.9 The code assembled in Sec. 2.4 looks for the largest number below the main diagonal to locate the pivot element for starting the reduction to zeros below the main diagonal. The actual rows are exchanged. It is not necessary to do this: one could merely use an indicator array to keep track of where the actual pivot is located and exchange the FORTRAN subscript values to effect the exchange. Since no row exchanges will occur, a faster subroutine will result. Care must be taken in re-assigning the proper subscripts during the back substitution steps. Write and check out such an improved routine. You may prefer to use it in place of subroutine SOLVER shown in Appendix A.

2.10 Apply the ideas described in Problem 2.9 to the construction of a fast subroutine for Gauss-Jordan elimination (see Table 2.6) without the necessity of performing row exchanges after pivot searches.

2.11 Complete the coding of the in-place inversion routine outlined in Sec. 2.8. Subroutine INVERT listed in Appendix A may be used as a guide for coding details.

2.12 The ideas described in Prob. 2.9 may also be applied to the in-place inversion of a matrix. Develop the algorithms necessary and complete such a subroutine. You may prefer to use it in place of subroutine INVRT shown in Appendix A.

2.13 The in-place inversion routine described in Sec. 2.8 is based on the Gauss-Jordan solution algorithm. A closely related algorithm may be derived based on the Gauss elimination procedure. Derive the necessary formulas and implement such an inversion procedure.

2.14 The solutions x_1, x_2 of the quadratic equation
$$ax^2 + bx + c = 0$$
are given by
$$x_{1,2} = \frac{-b \pm \sqrt{b^2 - 4ac}}{2a}$$
When $4ac \ll b^2$, severe round-off errors occur in the calculation of the smaller of the two roots. In order to minimize such errors, an alternate procedure is required for the calculation of the smaller solution such that subtraction of nearly equal values will be avoided. Devise such a procedure. The algorithm is useful not only on digital computers but also with slide-rule calculations [6].

2.15 Sometimes it is desirable to perform matrix inversion in double precision (approximately 17 decimal places on IBM 360–370 systems) even though the input array is single precision. Make a special subroutine, starting with subroutine INVRT listed in Appendix A, which accepts a single-precision matrix and produces a single-precision inverse in its own place, but performs all calculations in double precision. Assume that the input array contains enough storage for holding the double-precision version of the matrix, i.e. the storage is twice as large as required for the largest matrix to be inverted.

2.16 In Sec. 2.5 the Gauss-Seidel procedure for solving a set of simultaneous equations by iteration is discussed and a program is given in Listing 2.2. The routine will solve equations satisfactorily as long as the off-diagonal elements of the coefficient matrix are small enough compared with the main-diagonal terms. Note that this routine does not destroy the original equation coefficients and does not create any additional entries in the storage assigned to the original coefficients. Check out this routine with several nontrivial sets of equations of your own.

2.17 Solve the following set of simultaneous equations by the Gauss-Seidel procedure.

$$x_1 + \alpha x_2 = 1$$
$$\alpha x_1 + x_2 = 0$$

(a) Let α range from 0.5 to 2 and plot x_1 vs. x_2 for each iteration, starting with $x_1 = x_2 = 0$. For what values of α does this process converge?

(b) Solve for $x_1(n)$ algebraically. n is the iteration number. How does convergence of this series of values compare with your result for part (a) of this problem?

DC AND AC ANALYSIS OF NETWORKS

3

An electric network consists of M number of *elements* (resistors, capacitors, inductors, sources) each possessing two *terminals*; these elements are connected at *nodes*. Between the nodes elements constitute *paths*. Each element in a given path constitutes a *link*. A set of links leading from a node back to that node is termed a *loop*. The network formed by two or more elements may consist of one or more electrical *parts*; coupling between such parts may exist through magnetic means (transformers) or through dependent sources.

In order to analyze a network, the interconnection of elements, the voltage-current relationship of each element, and the past history of the elements (embodied in a set of initial conditions) must be known. Voltages are computed with reference to a *reference node*, usually the *ground-node*; currents are calculated with reference to a set of assumed current directions through the various elements.

The basic relations used for analysis are the first two circuit laws of Kirchhoff:

(1) The sum of all currents entering a node must equal the sum of all currents leaving it (current law).

(2) The sum of all voltages in a given loop must sum to zero (voltage law).

In analyzing a network, usually one or the other of the above laws is applied to every independent node or independent loop of the network. Typical engineering practice is to write enough loop equations and thus effect an analysis.

For a network of P electrical parts consisting of M elements and N_T nodes, the number of independent loops L is

$$L = -N_T + M + P \quad (\text{or } L + N_T = M + P) \qquad (3.1)$$

The number of independent nodes N is

$$N = N_T - P \quad (\text{or } N + P = N_T) \tag{3.2}$$

One important observation about practical networks is that the number of independent nodes is usually less than the number of independent loops (i.e., usually there are more parallel paths between two nodes than there are series elements in loops). Thus there are typically fewer node equations than loop equations. Circuit analysis programs usually use the node-voltage formulation as the basis for analysis.

In this chapter a series of computer programs is discussed which write and solve the node-voltage equations for direct current and steady-state alternating current networks from a basic description of the contents of the branches and the interconnections between branches. The basic operations, relations, and programs are developed in detail; later sections discuss extensions of the basic programs, but the codes themselves are given only partially. These extensions will make the basic programs into shorter codes, faster in execution, conserve computer storage, and allow more flexible input formats without changing the basic program capabilities.

A careful study of Secs. 3.3 and 3.4 is essential to the development of material in the later chapters.

3.1 Basic Relationships

Let the branches (elements) of a network be labeled consecutively $(1, 2, \ldots, M)$ and let the nodes be labeled also $(0, 1, 2, \ldots, N$, with 0 being the reference node). Then the circuit interconnections of the branches may be characterized by the set of equations

$$\mathbf{A_a I_b} = \mathbf{0} \tag{3.3}$$

where $\mathbf{I_b}$ is the set of branch currents (the current i_j is the current flowing in branch j); elements of the matrix $\mathbf{A_a}$ are given by

$$a_{kj} = \begin{cases} 1 & \text{if current } i_j \text{ leaves node } k \\ -1 & \text{if current } i_j \text{ enters node } k \\ 0 & \text{if } i_j \text{ neither enters nor leaves node } k \end{cases} \tag{3.4}$$

These relations will hold for all N_T nodes in the network; for $N(= N_T - P)$ nodes, these relations represent a set of independent equations. Each equation is an application of Kirchhoff's current law (KCL) for the kth node.

For the N independent nodes, Eq. (3.3) represents the set

$$\mathbf{A I_b} = \mathbf{0} \tag{3.5}$$

where $\mathbf{A}$ has the elements given in Eq. (3.4) and contains M columns (number of branches) and N rows (number of independent nodes).

This chapter deals with circuits whose behavior is independent of time. Circuit analyses of this type are DC analysis and AC steady-state analysis. Transients in a circuit are dependent on initial conditions; this latter type of analysis will be dealt with in the next chapter. Also, the treatment is restricted here to linear circuits; nonlinear circuits will be discussed in Chapter 7. Voltage-current relations for passive elements in linear circuits are characterized by a single (complex) number at each frequency:

$$i = yv \qquad (3.6)$$

with $y = 1/R, 1/j\omega L, j\omega C$ for a resistor, an inductor, or a capacitor respectively; i and v are the complex-valued phasor representations for the current and voltage.

Eq. (3.5) represents the application of Kirchhoff's current law (KCL) to the various independent nodes (N in number) to the network. Note that the coefficient matrix **A** represents a matrix of N rows and M columns (M is the number of branches). Any given network is the result of a definite interconnection of these M branches between the N nodes. As the result of the branch currents' flowing through the various branches, a voltage is developed across the respective branches. The set of branch voltages is denoted by the vector $\mathbf{V_b}$. The polarity of these voltages with respect to the branch currents is such that a positive current in each branch is assumed to be flowing from the positive to the negative terminal of the branch.

The branch voltages in a network will be linear combinations of the various node voltages.

$$\mathbf{V_b} = \mathbf{B}\mathbf{V_n} \qquad (3.7)$$

where $\mathbf{V_n}$ represents the set of node voltages. Node voltages in an electrically connected network are measured with respect of a datum (reference) potential usually referred to as *ground*; hence this node is taken as potential zero and all other voltages in the network are measured with respect to this point. Therefore, the set of $\mathbf{V_n}$ voltages will contain only N number of elements (rather than N_T). The elements of the matrix **B** are given by

$$b_{kj} = \begin{cases} +1 & \text{if branch } k \text{ leaves node } j \\ -1 & \text{if branch } k \text{ enters node } j \\ 0 & \text{if branch } k \text{ neither enters nor leaves node } j \end{cases} \qquad (3.8)$$

The coefficient matrix **B** contains M rows and N columns.

It is most important to note the similarities between Eqs. (3.8) and (3.4). As a direct consequence of these two definitions

$$\mathbf{B} = \mathbf{A}^T \qquad (3.9)$$

i.e., the coefficient matrix relating node voltages and branch voltages is the transpose of the coefficient matrix resulting from the application of Kirchhoff's current law to the independent nodes of the network. There are three implied points in the above:

(1) The direction of current in each branch is in the direction of positive voltage to negative voltage in each branch.

(2) The same reference voltage node(s) is (are) used in measuring the node voltages as is (are) excluded from the independent node set of the network of one (of P) electrical part(s).

(3) The order in the set $\mathbf{I_b}$ and $\mathbf{V_b}$ are the same (i.e., i_3 and v_3 refer to the current in branch 3 and the voltage across branch 3); hence the same branch numbering and the same node numbering apply to both Eqs. (3.5) and (3.7).

In the above discussion no mention was made of the make-up of a branch. In fact, the only restriction is that every "branch" must have two terminals; hence a "branch" could be a collection of passive elements and sources. If the current leaving each branch is the difference of the currents caused by independent sources (the set of which is denoted by $\mathbf{I_g}$) and all other currents (the set being denoted by $\mathbf{I_b^*}$), then

$$\mathbf{I_b} = \mathbf{I_b^*} - \mathbf{I_g} = \mathbf{Y_b V_b} - \mathbf{I_g} \qquad (3.10)$$

where $\mathbf{Y_b}$ is referred to as the branch-admittance matrix.

Combining Eqs. (3.10), (3.7), and (3.5) results in the fundamental relation for all linear networks

$$\mathbf{A Y_b A^T V_n} = \mathbf{A I_g} \qquad (3.11)$$

The quantity $\mathbf{Y_n} = \mathbf{A Y_b A^T}$ is termed the *node-admittance matrix*.

The discussion presented above is merely intended as a brief introduction to the formulation of general network equations. A more rigorous and thorough presentation of this material is given in the next section. The intent here is merely to explain terms and to provide an informal framework for the following section.

At this point an explanation of terminology to be used in this chapter is in order. Most circuit problems here, and elsewhere, are usually stated in terms of a *circuit diagram* (such as the one shown in Fig. 3.1); circuit component values may or may not be indicated on the circuit diagram.

To solve the network equations the nodes and branches of the network need to be numbered uniquely. This numbering will be done sequentially; the branches will be numbered starting with 1; the nodes will be numbered starting with 0 (zero). The node given the number zero will usually be the ground-node, which serves as the reference point from which all voltages in the circuit will be measured.

The branch numbers will be denoted by the label *Bi* (with *i* being the branch number, such that *B5* would indicate branch number five). Similarly, the node number will be denoted by *Nj* (with *j* being the node number). Both of these labels will be placed on the circuit diagram in close proximity of the items to which they refer.

A useful abstraction of the circuit diagram is the *circuit graph* or *network graph*, such as the one shown in Fig. 3.2. This diagram indicates the node-and-branch interconnection. The components that make up the various branches are not usually indicated on the graph. The assumed direction of current in each of the branches is indicated by an arrowhead. The actual current direction is of little importance at this stage: If the actual current flows in the assumed direction, a positive value for the current will be calculated; otherwise a negative value will result from the computations.

The current in branch k will be denoted by I_{b_k} (or sometimes by i_{b_k}); the voltage of node j with respect to node zero ($N0$) will be denoted by V_j (or sometimes by v_j). Similarly, the voltage across branch m will be denoted by V_{b_m} (or v_{b_m}). The set of all branch currents will be denoted by $\mathbf{I}_b$; the set of branch voltages by $\mathbf{V}_b$; and the set of node voltages by $\mathbf{V}_n$. For a network containing N nodes and M branches, $\mathbf{V}_b$ and $\mathbf{I}_b$ will be column vectors containing M entries; $\mathbf{V}_n$ will be also a column vector, but it will contain N entries. The order of the terms in $\mathbf{I}_b$, $\mathbf{V}_b$, and $\mathbf{V}_n$ will follow the branch-and-node numbering, for example, the node voltage for node 4 will be the fourth entry in the column vector $\mathbf{V}_n$.

In most of the examples that follow, the network graph will be omitted since it carries no information not contained in a circuit diagram on which the branch numbers, the node numbers, and the branch current directions are indicated (see, for example, Fig. 3.3).

Finally, the concept of *branch* must be clarified. In all discussions in this book a *branch* denotes an electrical entity possessing two terminals which in turn are connected to other such branches. Into one of the terminals current will enter; at the same time a like amount of current will leave at the other terminal. Between the two terminals a voltage will be sensed. Although the majority of circuits discussed in this book will contain a single electrical *element*, such as a resistor, a capacitor, a voltage source, a current source, or an inductor between the two terminals, it is not mandatory that a branch be so simple. All that is required is that every branch possess two terminals, and for the purposes of this chapter, the voltages across the branches and currents through the branches must be related linearly. Note that this does not demand that a branch voltage must be some multiple of the current flowing through that branch; instead, the general matrix linearity relationships must hold. For virtually all network analyses, however, the branches are restricted to very simple combinations of elements, hence the linearity relationships will be easily discernible.

To illustrate the discussion given above a set of simple examples will now be worked through.

EXAMPLE 3.1

For the circuit in Fig. 3.1 write the matrix node equations.

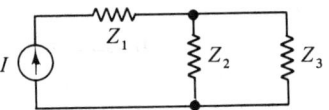

Fig. 3.1 Circuit for Example 3.1

Solution.

(a) The circuit contains four elements, three nodes, and is of one part. There are two independent nodes (V_1 and V_2 in this analysis); there are two loops.

(b) The nodes and branches are numbered and current reference directions are assigned. The *graph* in Fig. 3.2 describes the network.

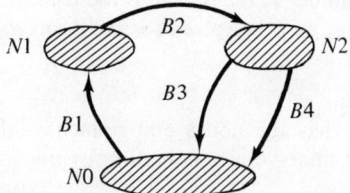

Fig. 3.2 Network Graph for Example 3.1

(c) The incidence matrix is

$$\begin{bmatrix} -1 & 1 & 0 & 0 \\ 0 & -1 & 1 & 1 \end{bmatrix} \begin{bmatrix} I_{b_1} \\ I_{b_2} \\ I_{b_3} \\ I_{b_4} \end{bmatrix} = \begin{bmatrix} 0 \\ 0 \end{bmatrix} \quad (3.12)$$

(d) The branch-admittance matrix is

$$\mathbf{Y_b} = \begin{bmatrix} 0 & 0 & 0 & 0 \\ 0 & 1/Z_1 & 0 & 0 \\ 0 & 0 & 1/Z_2 & 0 \\ 0 & 0 & 0 & 1/Z_3 \end{bmatrix} \quad (3.13)$$

(e) The current due to sources:

$$\mathbf{I_g} = \begin{bmatrix} -I \\ 0 \\ 0 \\ 0 \end{bmatrix} \quad (3.14)$$

[Note the negative sign on the first entry, a consequence of Eq. (3.10).]

(f) Hence the node-voltage equations become

$$\begin{bmatrix} -1 & 1 & 0 & 0 \\ 0 & -1 & 1 & 1 \end{bmatrix} \begin{bmatrix} 0 & 0 & 0 & 0 \\ 0 & 1/Z_1 & 0 & 0 \\ 0 & 0 & 1/Z_2 & 0 \\ 0 & 0 & 0 & 1/Z_3 \end{bmatrix} \begin{bmatrix} -1 & 0 \\ 1 & -1 \\ 0 & 1 \\ 0 & 1 \end{bmatrix} \begin{bmatrix} V_1 \\ V_2 \end{bmatrix}$$

$$= \begin{bmatrix} -1 & 1 & 0 & 0 \\ 0 & -1 & 1 & 1 \end{bmatrix} \begin{bmatrix} -I \\ 0 \\ 0 \\ 0 \end{bmatrix} \quad (3.15)$$

$$\begin{bmatrix} 1/Z_1 & -1/Z_1 \\ -1/Z_1 & 1/Z_1 + 1/Z_2 + 1/Z_3 \end{bmatrix} \begin{bmatrix} V_1 \\ V_2 \end{bmatrix} = \begin{bmatrix} I \\ 0 \end{bmatrix} \quad (3.16)$$

The first of these equations is

$$\frac{V_1 - V_2}{Z_1} = I$$

which is KCL for node number 1; the second is the KCL for node number 2, as may be verified. The equations may now be solved for the node voltages.

EXAMPLE 3.2

The network in Fig. 3.3 has the nodes and branches labeled and reference current directions indicated. Find the node-voltage equation for node 1.

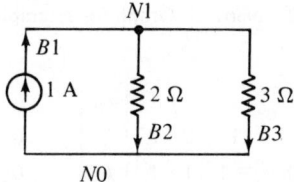

Fig. 3.3 Circuit for Example 3.2

Solution.

$$\mathbf{A} = [-1 \quad 1 \quad 1] \tag{3.17}$$

$$\mathbf{Y_b} = \begin{bmatrix} 0 & 0 & 0 \\ 0 & \tfrac{1}{2} & 0 \\ 0 & 0 & \tfrac{1}{3} \end{bmatrix} \tag{3.18}$$

$$\mathbf{I_g} = \begin{bmatrix} -1 \\ 0 \\ 0 \end{bmatrix} \tag{3.19}$$

Hence

$$\mathbf{A Y_b A^T V_n} = (\tfrac{1}{2} + \tfrac{1}{3})V_1 = [-1 \quad 1 \quad 1]\begin{bmatrix} -1 \\ 0 \\ 0 \end{bmatrix} \tag{3.20}$$

which is the node-voltage equation for node 1.

EXAMPLE 3.3

Consider the network to be made up of two branches as indicated in Fig. 3.4. Write the matrices $\mathbf{A}$, $\mathbf{Y_b}$, $\mathbf{I_i}$; solve for the node equation for node 1.

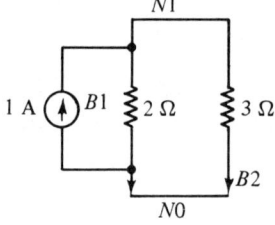

Fig. 3.4 Circuit for Example 3.3

Sec. 3.1   Basic Relationships

Solution.

The incidence matrix (using the reference currents in the passive elements) is

$$\mathbf{A} = [1 \quad 1] \tag{3.21}$$

The branch-admittance matrix is

$$\mathbf{Y}_b = \begin{bmatrix} \tfrac{1}{2} & 0 \\ 0 & \tfrac{1}{3} \end{bmatrix} \tag{3.22}$$

Now the left side of Eq. (3.8) becomes

$$\mathbf{A}\mathbf{Y}_b\mathbf{A}^T V_1 = (\tfrac{1}{2} + \tfrac{1}{3})V_1 \tag{3.23}$$

which is the left side of Eq. (3.20). The right side must involve the current flowing into node 1, which is equal to $I = 1$ ampere.

Notice that in Ex. 3.2 a current source was lumped into a branch; hence branches need not be passive circuit elements. If every branch is allowed an independent current source, whose value may be zero, in parallel with the passive element, then the currents introduced into each node are given by

$$\mathbf{I}_i = \mathbf{A}\mathbf{I}_g \tag{3.24}$$

where $\mathbf{I}_g$ is the set of branch current generators. In this latter case the node-voltage expressions become

$$\mathbf{A}\mathbf{Y}_b\mathbf{A}^T\mathbf{V}_n = \mathbf{A}\mathbf{I}_g \tag{3.25}$$

Note that the direction of the independent current sources is as indicated in Fig. 3.4 (i.e., opposite the direction of the current in the passive element). The last expression is identical to Eq. (3.11), but it was derived in a different manner.

If independent voltage sources are introduced into the various branches, the voltage-current relationships in the network become

$$\mathbf{I}_b = \mathbf{Y}_b(\mathbf{V}_b + \mathbf{V}_g) - \mathbf{I}_g \tag{3.26}$$

where the vector $\mathbf{V}_g$ represents the set of all independent voltage sources. Hence each branch becomes as shown in Fig. 3.5. Pre-multiplication of both sides of Eq. (3.26) by $\mathbf{A}$ and application of Eq. (3.3) gives

$$\mathbf{A}\mathbf{I}_b = 0 = \mathbf{A}\mathbf{Y}_b\mathbf{V}_b + \mathbf{A}\mathbf{Y}_b\mathbf{V}_g - \mathbf{A}\mathbf{I}_g \tag{3.27}$$

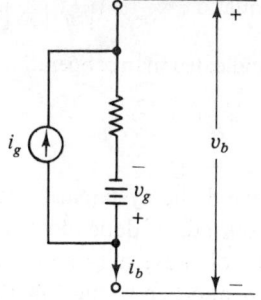

Fig. 3.5   Branch with Independent Sources

By Eq. (3.5) the above becomes

$$\mathbf{AY_b A^T V_n} = \mathbf{A(I_g - Y_b V_g)} \tag{3.28}$$

EXAMPLE 3.4

Write the matrix relationships for the network in Fig. 3.6 [see Eq. (3.28)]. Note that the branch $B1$ contains the 1-Ω resistor and the 5-V source; $B3$ contains the 3-Ω resistor and the 7-A current source.

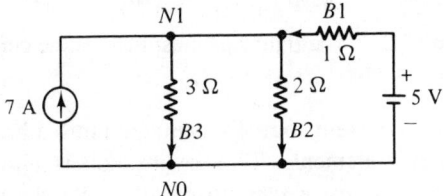

Fig. 3.6 Circuit for Example 3.4

Solution.

$$\mathbf{A} = [-1 \quad 1 \quad 1] \tag{3.29a}$$

$$\mathbf{Y_b} = \begin{bmatrix} 1 & 0 & 0 \\ 0 & \tfrac{1}{2} & 0 \\ 0 & 0 & \tfrac{1}{3} \end{bmatrix} \tag{3.29b}$$

$$\mathbf{I_g} = \begin{bmatrix} 0 \\ 0 \\ 7 \end{bmatrix} \tag{3.29c}$$

$$\mathbf{V_g} = \begin{bmatrix} 5 \\ 0 \\ 0 \end{bmatrix} \tag{3.29d}$$

The node equation becomes

$$[-1 \quad 1 \quad 1] \begin{bmatrix} 1 & 0 & 0 \\ 0 & \tfrac{1}{2} & 0 \\ 0 & 0 & \tfrac{1}{3} \end{bmatrix} \begin{bmatrix} -1 \\ 1 \\ 1 \end{bmatrix} V_1 = [-1 \quad 1 \quad 1] \left(\begin{bmatrix} 0 \\ 0 \\ 7 \end{bmatrix} - \begin{bmatrix} 1 & 0 & 0 \\ 0 & \tfrac{1}{2} & 0 \\ 0 & 0 & \tfrac{1}{3} \end{bmatrix} \begin{bmatrix} 5 \\ 0 \\ 0 \end{bmatrix} \right) \tag{3.30}$$

which becomes, upon carrying out the indicated matrix operations, the node equation for V_1

$$\tfrac{11}{6} V_1 = 12 \tag{3.31}$$

i.e., $V_1 = 6.545$ volts.

The above considerations neglected the systematic inclusion of any dependent sources in the network. There are two kinds of dependent sources: voltage generators and current generators. The strength of these may be dependent on the value of a current or a voltage associated with another branch in the network.

The four types of sources are:

Current dependent current source (CCS) $i_j = \beta_{jk} i_k$ (3.32)

Current dependent voltage source (CVS) $v_j = r_{jk} i_k$ (3.33)

Voltage dependent voltage source (VVS) $v_j = \mu_{jk} v_k$ (3.34)

Voltage dependent current source (VCS) $i_j = g_{jk} v_k$ (3.35)

where the current (voltage) associated with branch j is dependent on the current (voltage) associated with branch k.

The inclusion of the first type of source into Eq. (3.26) is

$$\mathbf{I_b} - \mathbf{I_d} = \mathbf{Y_b}(\mathbf{V_b} + \mathbf{V_g}) - \mathbf{I_g} \quad (3.36)$$

where $\mathbf{I_d}$ represents the dependent current sources. The branch now becomes as shown in Fig. 3.7.

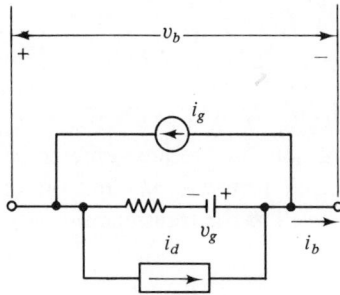

Fig. 3.7 Branch with Dependent Current Source

The dependent current sources are expressible by the relationships

$$\mathbf{I_d} = [\beta_{jk}]\mathbf{I_b} = [\beta_{jk}]\mathbf{Y_b}(\mathbf{V_b} + \mathbf{V_g}) \quad (3.37)$$

The terms $[\beta_{jk}]$ represent the strengths of the current sources in branch j as a function of the current in branch k. For one such source the matrix is zero except for the element j, k; the product $[\beta_{jk}]\mathbf{Y_b}$ gives the matrix

$$\mathbf{Y}^* = [\beta_{jk}]\mathbf{Y_b} = [\beta_{jk} y_{kk}] \quad (3.38)$$

i.e., it is a matrix of zeros except the jkth element, whose value is $\beta_{jk} y_{kk}$.

Re-arrangement of Eq. (3.37) and application of relations expressed in Eqs. (3.6) and (3.8) give

$$\mathbf{A}(\mathbf{Y_b} + \mathbf{Y}^*)\mathbf{A}^T \mathbf{V_n} = \mathbf{A}[\mathbf{I_g} - (\mathbf{Y_b} + \mathbf{Y}^*)\mathbf{V_g}] \quad (3.39)$$

The matrix $(\mathbf{Y_b} + \mathbf{Y}^*)$ contains the branch admittances along the main diagonal (because of $\mathbf{Y_b}$) and off-diagonal elements in the jkth position because of the sources β_{jk}. The value of these elements is $\beta_{jk} y_{kk}$, implying that the values y_{kk} must be known before the entry y_{jk} can be calculated. The matrix $(\mathbf{Y_b} + \mathbf{Y}^*)$ is usually referred to as $\mathbf{Y_b}$; that convention is used from here on.

Hence if only current controlled current sources are to be considered for the circuit, the complete circuit description and set-up of matrices can be accomplished by the following steps:

(1) The circuit is numbered for nodes, starting with zero for the reference node; the branches are labeled in sequence.

(2) The description of each branch (passive element value H, independent voltage value V, and independent current I_g) is entered. (Usually, the branches are entered in order.) For each branch (k) the "from-node" (i) and the "to-node" (j) are entered also. From this information two matrices ($\mathbf{A}$ and $\mathbf{Y_b}$) and two vectors ($\mathbf{V_g}$ and $\mathbf{I_g}$) are established.

$$\mathbf{A}: \quad a_{ik} = 1 \quad i \neq 0 \\ \phantom{\mathbf{A}:} \quad a_{jk} = -1 \quad j \neq 0 \tag{3.40}$$

$$\mathbf{Y_b}: \quad y_{kk} = \frac{1}{H} \tag{3.41}$$

$$\mathbf{V_g}: \quad v_k = V \tag{3.42}$$

$$\mathbf{I_g}: \quad i_k = I_g \tag{3.43}$$

(3) The dependent current sources are next entered. For each source the controlling branch (the "from-branch") number (k) and the branch number where the current source is located (j) as well as the transfer value (β_{jk}) must be known. This is used to modify the matrix $\mathbf{Y_b}$

$$y_{jk} = \beta_{jk} y_{kk} \tag{3.44}$$

The above three steps are sufficient to establish the node equations for the network as given in Eq. (3.39) [with ($\mathbf{Y_b} + \mathbf{Y}^*$) called $\mathbf{Y_b}$ in the following.] Note that β refers to the transfer ratio of a current in the passive element in a branch to the current introduced in another branch.

The inclusion of other types of sources in the network is somewhat more complicated. For a voltage controlled current source (i.e., a current source in branch j whose value is g times the voltage across a passive element in branch k),

$$\mathbf{I_d} = \mathbf{G}(\mathbf{V_b} + \mathbf{V_g}) \tag{3.45}$$

where $\mathbf{G}$ is a matrix of zeros except $g_{jk} \neq 0$. Starting with Eq. (3.36), using the above, we obtain the following:

$$\mathbf{A}(\mathbf{Y_b} + \mathbf{G})\mathbf{A}^T \mathbf{V_n} = \mathbf{A}(\mathbf{I_g} - (\mathbf{Y_b} + \mathbf{G})\mathbf{V_g}) \tag{3.46}$$

Here again a modification to the $\mathbf{Y_b}$ matrix is necessary; however, there is only an addition to the elements of $\mathbf{Y_b}$, which can be done at any time.

If, however, it is desired to have a voltage controlled current source in branch j whose strength is g_{jk} times the voltage across the kth branch, Eq. (3.45) must be replaced by

$$\mathbf{I_d} = \mathbf{G_1} \mathbf{V_b} \tag{3.47}$$

where $\mathbf{G_1}$ is the matrix of transconductances.

Sec. 3.1   Basic Relationships   61

The expression corresponding to Eq. (3.46) is

$$A(Y_b + G_1)A^T V_n = A(I_g - Y_b V_g) \qquad (3.48)$$

and the admittance matrices on the two sides of Eq. (3.46) are unlike (i.e., $Y_b + G_1$ and Y_b).

The addition of dependent voltage sources in series with the independent voltage source in the branches would merely replace V_g with

$V_g + [K]V_b$	if the source is controlled by the voltage across another branch;
$V_g + [K](V_b + V_g)$	if the source is controlled by the voltage across the passive element in another branch;
$V_g + [r]I_b$	if the source is controlled by the current in another branch;
$V_g + [r](I_b + I_g)$	if the source is controlled by the current in another branch, excluding the independent currents;
$V_g + [r](I_b + I_g - I_d)$	if the source is controlled by the current through the passive element in another branch;

where $[r]$ is the matrix of transfer resistances (mutual resistances), $[K]$ is the matrix of voltage transfers, and I_d is the set of dependent current generators.

Expressions similar to Eq. (3.46) may be written for each of the above five cases. However, because of the confusion due to the various definitions of the controlling quantity usually only one type of control, *element* voltage or current, is implemented in circuit analysis programs.

The above completes the setting up of the node equations. These equations are solved by "standard" means (usually by use of Gauss elimination using pivoting). The branch voltages are then calculated from Eq. (3.7) and the currents through the branches are obtained from

$$I_b = Y_b(V_b + V_g) \qquad (3.49)$$

Several procedures for solving the simultaneous equations were given in Chapter 2.

EXAMPLE 3.5

Solve for the node voltages, branch voltages, and currents in the network shown in Fig. 3.8.

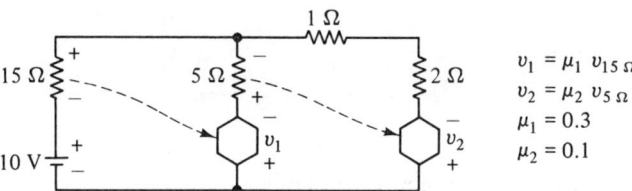

Fig. 3.8 Circuit for Example 3.5

Solution.

(1) The graph in Fig. 3.9 describes the network.

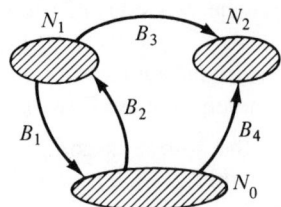

Fig. 3.9 Network Graph for Example 3.5

(2) The incidence matrix is

$$\mathbf{A} = \begin{bmatrix} 1 & -1 & 1 & 0 \\ 0 & 0 & -1 & -1 \end{bmatrix}$$

(3) The independent voltage and current source vectors are

$$\mathbf{I_g} = \begin{bmatrix} 0 \\ 0 \\ 0 \\ 0 \end{bmatrix}; \quad \mathbf{V_g} = \begin{bmatrix} -10 \\ 0 \\ 0 \\ 0 \end{bmatrix}$$

(4) The branch-admittance matrix may be established from the relation

$$\mathbf{I_b} = \mathbf{Y_b V_b} \quad \text{(without regard to } \mathbf{I_g} \text{ and } \mathbf{V_g})$$

Hence the ith row of $\mathbf{Y_b}$ will represent the coefficients entering the linear combination of the elements of $\mathbf{V_b}$ to give $\mathbf{I_b}$. Note that this relationship will hold regardless of the values (or even the presence) of the independent sources; hence one may establish $\mathbf{Y_b}$ disregarding all independent sources. The various $\mathbf{I_b}$ elements are

$$i_{b_1} = \tfrac{1}{15} v_{b_1}$$
$$i_{b_2} = \tfrac{1}{5}(v_{b_2} - v_1) = \tfrac{1}{5} v_{b_2} - \tfrac{1}{5}\mu_1 v_{15\Omega} = \tfrac{1}{5} v_{b_2} - \tfrac{1}{5}\mu_1 v_{b_1}$$
$$i_{b_3} = \tfrac{1}{1} v_{b_3}$$
$$i_{b_4} = \tfrac{1}{2}(v_{b_4} - v_2) = \tfrac{1}{2} v_{b_4} - \tfrac{1}{2}\mu_2 v_{5\Omega}$$
$$= \tfrac{1}{2} v_{b_4} - \tfrac{1}{2}\mu_2(v_{b_2} - v_1)$$
$$= \tfrac{1}{2} v_{b_4} - \tfrac{1}{2}\mu_2 v_{b_2} + \tfrac{1}{2}\mu_1\mu_2 v_{b_1}$$

The above relations yield the branch-admittance matrix

$$\mathbf{Y_b} = \begin{bmatrix} \tfrac{1}{15} & 0 & 0 & 0 \\ -\tfrac{1}{5}\mu_1 & \tfrac{1}{5} & 0 & 0 \\ 0 & 0 & 1 & 0 \\ \tfrac{1}{2}\mu_1\mu_2 & -\tfrac{1}{2}\mu_2 & 0 & \tfrac{1}{2} \end{bmatrix}$$

The node-voltage equations can now be established.

$$\mathbf{Y_n V_n} = \mathbf{I_s}$$

with $\mathbf{Y_n} = \mathbf{A Y_b A}^T$ and $\mathbf{I_s} = \mathbf{A}(\mathbf{I_g} - \mathbf{Y_b V_g})$.

Sec. 3.1   Basic Relationships   63

Y_n and I_s are then combined in the augmented matrix $[Y_n \mid I_s]$, which represents all the coefficients needed to solve for V_n by a simultaneous equations solution procedure such as the one given in Sec. 2.4. The numerical work is shown in Table 3.1; slide-rule accuracy was used. The numerical triple matrix products AY_bA^T and AY_bV_g were carried out by means of the tabular matrix multiplication technique given in Sec. 2.2.

Table 3.1 Calculation of the Network Node Equations for Ex. 3.5

	A^T	V_g		A^T		V_g
$Y_b \to$		Y_bV_g		1	0	-10
				-1	0	0
A	Y_n	$-I_s$		1	-1	0
				0	-1	0

(following Y_b block)

$\frac{1}{15}$	0	0	0	$\frac{1}{15}$	0	$-\frac{10}{15}$
$-\frac{1}{5}\mu_1$	$\frac{1}{5}$	0	0	$-\frac{1}{5}\mu_1 - \frac{1}{5}$	0	$\frac{10}{5}\mu_1$
0	0	1	0	1	-1	0
$\frac{1}{2}\mu_1\mu_2$	$-\frac{1}{2}\mu_2$	0	$\frac{1}{2}$	$\frac{1}{2}\mu_1\mu_2 + \frac{1}{2}\mu_2$	$-\frac{1}{2}$	$-\frac{10}{2}\mu_1\mu_2$

| 1 | -1 | 1 | 0 | $(\frac{1}{15} + \frac{1}{5} + \frac{1}{5}\mu_1 + 1)$ | (-1) | $-\frac{10}{15} - \frac{10}{5}\mu_1$ |
| 0 | 0 | -1 | -1 | $(-1 - \frac{1}{2}\mu_1\mu_2 - \frac{1}{2}\mu_2)$ | $(1 + \frac{1}{2})$ | $\frac{10}{2}\mu_1\mu_2$ |

$$[Y_n \mid I_s] = \begin{bmatrix} \frac{1}{15} + \frac{1}{5} + \frac{1}{5}\mu_1 + 1 & -1 & \mid & \frac{10}{15} + \frac{10}{5}\mu_1 \\ -1 - \frac{1}{2}\mu_1\mu_2 - \frac{1}{2}\mu_2 & 1 + \frac{1}{2} & \mid & -\frac{10}{2}\mu_1\mu_2 \end{bmatrix}$$

For $\mu_1 = 0.3$ and $\mu_2 = 0.1$.

$$[Y_n \mid I_s] = \begin{bmatrix} 1.327 & -1.000 & \mid & 1.267 \\ -1.065 & 1.500 & \mid & -0.150 \end{bmatrix}$$

This is now solved for v_1 and v_2 by the methods of Sec. 2.3. (Carrying fractions in the calculations in Table 3.2 yields $v_2 = \frac{46}{37}$ volts and $v_1 = \frac{70}{37}$ volts.)

Table 3.2 Reduction of the Simultaneous Equations for Ex. 3.5

$[Y_n \mid I_s] =$	Row 1 Row 2	1.327 -1.065	-1.000 1.500	1.267 -0.150	
	Row 3 = Row 1/1.327 Row 4 = Row 2 $-$ (-1.065)·Row 3	1.000 0	-0.754 0.697	0.954 0.867	
	Row 5 = Row 4/0.697 Row 6 = Row 3 $-$ (-0.754)·Row 4	0 1.000	1.000 0	1.248 1.898	$= v_2$ $= v_1$

The branch voltages are

$$\mathbf{V_b} = \mathbf{A}^T \mathbf{V_n} = \mathbf{A}^T \begin{bmatrix} v_1 \\ v_2 \end{bmatrix}$$

$$\begin{bmatrix} v_{b_1} \\ v_{b_2} \\ v_{b_3} \\ v_{b_4} \end{bmatrix} = \begin{bmatrix} 1 & 0 \\ -1 & 0 \\ 1 & -1 \\ 0 & -1 \end{bmatrix} \begin{bmatrix} 1.898 \\ 1.248 \end{bmatrix} = \begin{bmatrix} 1.898 \\ -1.898 \\ 0.650 \\ -1.248 \end{bmatrix}$$

These are the voltages across the branches including the dependent and independent voltage generators.

The branch currents are

$$\mathbf{I_b} = \mathbf{Y_b}(\mathbf{V_b} + \mathbf{V_g}) - \mathbf{I_g}$$

$$\begin{bmatrix} i_{b_1} \\ i_{b_2} \\ i_{b_3} \\ i_{b_4} \end{bmatrix} = \begin{bmatrix} 0.0667 & 0 & 0 & 0 \\ -0.0600 & 0.200 & 0 & 0 \\ 0 & 0 & 1.000 & 0 \\ 0.0150 & -0.050 & 0 & 0.500 \end{bmatrix} \begin{bmatrix} -8.102 \\ -1.898 \\ 0.650 \\ -1.245 \end{bmatrix} = \begin{bmatrix} -0.540 \\ 0.109 \\ 0.650 \\ -0.652 \end{bmatrix} \text{ (amps)}$$

The power dissipation in each branch is the branch current multiplied by the corresponding branch voltage. In this case the branch power dissipations are

$$\mathbf{P} = \begin{bmatrix} (-0.540)\cdot(1.898) \\ (0.109)\cdot(-1.898) \\ (0.650)\cdot(0.650) \\ (-0.652)\cdot(-1.248) \end{bmatrix} = \begin{bmatrix} -1.03 \\ -0.207 \\ 0.423 \\ 0.814 \end{bmatrix} \text{ (watts)}$$

It should be pointed out again that the element power dissipations are not calculated above; instead, the powers dissipated or produced in the various branches are obtained.

In this section, terminology relating to node-admittance analysis of linear networks was introduced. The basic relationships were explored and numerical examples were used to demonstrate these relationships. A more systematic treatment is given in the next section.

3.2 Analysis of Arbitrary Networks

Let a general network made up of M branches of the type shown in Fig. 3.10 be interconnected at N nodes plus at the ground node. The voltages and currents in each branch of the network are related as follows:

$$i_b = i_e + i_d - i_g \tag{3.50a}$$

$$v_b = v_e + v_d - v_g \tag{3.50b}$$

Additionally, the element voltage and current satisfy the relationship

$$i_e = y_e \cdot v_e \tag{3.50c}$$

where y_e is the conductance of the passive element in the branch. The above three

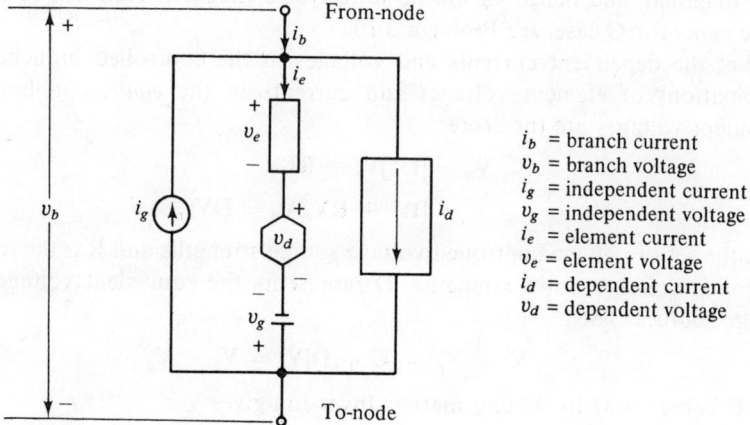

Fig. 3.10 Complete Network Branch

equations will hold for each of the M branches in the network; the branches are numbered sequentially starting with one. Normally, a second numeric subscript is appended to the subscripts used in Eqs. (3.50a, b, c) to indicate the branch number of the particular branch for which the equations are written.

The set of all branch currents $i_{b_1}, i_{b_2}, \ldots, i_{b_M}$ is arranged in the M-vector

$$\mathbf{I_b} = \begin{bmatrix} i_{b_1} \\ i_{b_2} \\ \vdots \\ i_{b_M} \end{bmatrix} \qquad (3.51)$$

Similarly, the M-vectors $\mathbf{I_e}, \mathbf{I_d}, \mathbf{I_g}, \mathbf{V_b}, \mathbf{V_e}, \mathbf{V_d}$, and $\mathbf{V_g}$ are defined. Hence the sets of currents and voltages for all the network branches satisfy the following vector equations:

$$\mathbf{I_b} = \mathbf{I_e} + \mathbf{I_d} - \mathbf{I_g} \qquad (3.52a)$$
$$\mathbf{V_b} = \mathbf{V_e} + \mathbf{V_d} - \mathbf{V_g} \qquad (3.52b)$$

The voltage-current relationships for the set of passive elements lead to the matrix equation

$$\mathbf{I_e} = \mathbf{Y_e} \cdot \mathbf{V_e} \qquad (3.52c)$$

Note that $\mathbf{Y_e}$ is a diagonal matrix of M rows and M columns if the network consists solely of resistances/conductances, capacitances, and inductances. These are elements whose voltage-current characteristic is expressible by a single number, the element admittance (self-admittance). In general DC analysis these are, of course, the only possible circuit elements. For AC analysis, however, mutual inductances will complicate the terms appearing in the above equation: there will be terms also off the main diagonal, and hence $\mathbf{Y}_e$ will no longer be a diagonal matrix. For a discussion of the general AC case, see Problem 3.10.

Let the dependent currents and voltages in the controlled branches be linear combinations of element voltages and currents in the *controlling* branches. The dependent voltages are therefore

$$\begin{aligned}\mathbf{V}_d &= [\mathbf{D}^*]\mathbf{V}_e + \mathbf{R}\mathbf{I}_e \\ &= [\mathbf{D}^* + \mathbf{R}\mathbf{Y}_e]\mathbf{V}_e = \mathbf{D}\mathbf{V}_e\end{aligned} \quad (3.53)$$

$\mathbf{D}^*$ is the set of voltage controlled voltage source strengths and $\mathbf{R}$ is the set of current controlled voltage source strengths. $\mathbf{D}$ represents the equivalent voltage controlled voltage sources. Then

$$\mathbf{V}_e + \mathbf{V}_d = [\mathbf{U} + \mathbf{D}]\mathbf{V}_e = \mathbf{V}_b + \mathbf{V}_g \quad (3.54)$$

with $\mathbf{U}$ being an M by M unit matrix. Inversion gives $\mathbf{V}_e$:

$$\mathbf{V}_e = [\mathbf{U} + \mathbf{D}]^{-1}[\mathbf{V}_b + \mathbf{V}_g] \quad (3.55)$$

Similarly, let the dependent currents be

$$\begin{aligned}\mathbf{I}_d &= \mathbf{G}\mathbf{V}_e + \mathbf{B}^*\mathbf{I}_e \\ &= [\mathbf{G}\mathbf{Y}_e^{-1} + \mathbf{B}^*]\mathbf{I}_e = \mathbf{B}\mathbf{I}_e\end{aligned} \quad (3.56)$$

Note that since $\mathbf{Y}_e$ is a diagonal matrix its inverse consists of a diagonal matrix whose elements are the reciprocals of the original $\mathbf{Y}_e$ main-diagonal elements:

$$[\mathbf{Y}_e]^{-1} = [1/y_{e_{ii}}] \quad (3.57)$$

Hence the element currents are

$$\mathbf{I}_e = \mathbf{Y}_e\mathbf{V}_e = \mathbf{Y}_e[\mathbf{U} + \mathbf{D}]^{-1}(\mathbf{V}_b + \mathbf{V}_g) \quad (3.58)$$

The branch currents are

$$\begin{aligned}\mathbf{I}_b &= \mathbf{I}_e + \mathbf{I}_d - \mathbf{I}_g \\ &= [\mathbf{U} + \mathbf{B}]\mathbf{I}_e - \mathbf{I}_g\end{aligned} \quad (3.59)$$

Using $\mathbf{I}_e$ from Eq. (3.58), we get,

$$\mathbf{I}_b = [\mathbf{U} + \mathbf{B}]\mathbf{Y}_e[\mathbf{U} + \mathbf{D}]^{-1}(\mathbf{V}_b + \mathbf{V}_g) - \mathbf{I}_g \quad (3.60)$$

The M by M matrix occurring above is termed the branch-admittance matrix $\mathbf{Y}_b$:

$$\mathbf{Y}_b = [\mathbf{U} + \mathbf{B}]\mathbf{Y}_e[\mathbf{U} + \mathbf{D}]^{-1} \quad (3.61)$$

By Kirchhoff's current law, the summation of currents at each node is zero. There are N independent nodes in the network; thus (see Eq. 3.4)

$$\mathbf{A}\mathbf{I}_b = 0 \quad (3.62)$$

where $\mathbf{A}$ is the M by N incidence matrix of the network. Hence

$$\mathbf{A}[\mathbf{Y}_b(\mathbf{V}_b + \mathbf{V}_g) - \mathbf{I}_g] = 0 \quad (3.63)$$

Finally, the node and branch voltages are related by [see Eq. 3.7]

$$\mathbf{A}^T\mathbf{V}_n = \mathbf{V}_b \quad (3.64)$$

Thus the node-voltage expression becomes

$$\mathbf{A}\mathbf{Y}_b\mathbf{A}^T\mathbf{V}_n + \mathbf{A}\mathbf{Y}_b\mathbf{V}_g - \mathbf{A}\mathbf{I}_g = 0 \qquad (3.65)$$

In the above set the node voltages are to be solved for, knowing the independent voltages, currents, interconnection (incidence matrix), and the branch admittances. The terms involving $\mathbf{V}_g$ and $\mathbf{I}_g$ are lumped together and are called the equivalent currents $\mathbf{I}_s$:

$$\mathbf{I}_s = \mathbf{A}(\mathbf{I}_g - \mathbf{Y}_b\mathbf{V}_g) \qquad (3.66)$$

The N by N matrix pre-multiplying $\mathbf{V}_n$ in Eq. (3.65) is known as the node-admittance matrix $\mathbf{Y}_n$:

$$\mathbf{Y}_n = \mathbf{A}\mathbf{Y}_b\mathbf{A}^T \qquad (3.67)$$

Hence the following augmented matrix

$$\mathbf{Y}_{na} = [\mathbf{Y}_n \mid \mathbf{I}_s] \qquad (3.68)$$

must be solved for the node voltages $\mathbf{V}_n$. This is accomplished by using a simultaneous equation solving procedure, such as discussed in Sec. 2.4.

If the node voltages are known, the branch voltages are solved from Eq. (3.64) and the branch currents are solved from Eq. (3.60):

$$\mathbf{V}_b = \mathbf{A}^T\mathbf{V}_n \qquad (3.64)$$

$$\mathbf{I}_b = \mathbf{Y}_b(\mathbf{V}_b + \mathbf{V}_g) - \mathbf{I}_g \qquad (3.60)$$

To calculate the element voltages and currents one can solve for $\mathbf{V}_e$ from Eq. (3.54)

$$\mathbf{V}_e = (\mathbf{U} + \mathbf{D})^{-1}[\mathbf{V}_b + \mathbf{V}_g] \qquad (3.69)$$

and then obtain

$$\mathbf{I}_e = \mathbf{Y}_e\mathbf{V}_e \qquad (3.70)$$

These last two relations have the advantage of using $(\mathbf{U} + \mathbf{D})^{-1}$ which had to be calculated for Eq. (3.61).

Alternately, one could calculate $\mathbf{I}_e$:

$$[\mathbf{U} + \mathbf{B}]\mathbf{I}_e = \mathbf{I}_b + \mathbf{I}_g \qquad (3.71)$$

by using a simultaneous equation solution routine followed by

$$\mathbf{V}_e = [\mathbf{Y}_e]^{-1}\mathbf{I}_e \qquad (3.72)$$

The above analysis applies to DC as well as AC circuits; the major differences are that all voltage, currents, and admittances must be carried as complex quantities and in changing the $\mathbf{Y}_b$ matrix at each frequency.

Note that in order to obtain the element voltages and currents, either the matrix $(\mathbf{U} + \mathbf{B})$ or the matrix $(\mathbf{U} + \mathbf{D})^{-1}$ must be used. If no controlled voltage sources are allowed in the network, then $\mathbf{D} = 0$ and the element voltages can easily be obtained $(\mathbf{V}_e = \mathbf{V}_b + \mathbf{V}_g)$; the element currents then follow from Eq. (3.70).

Also, because of the special nature of the $\mathbf{U}$ and $\mathbf{Y}_e$ matrices occurring in Eq. (3.61) the product $(\mathbf{U} + \mathbf{B})\mathbf{Y}_e$ can be set up directly; this procedure will be outlined and used in the next section.

EXAMPLE 3.6

Find all voltages and currents in the network of Fig. 3.11.

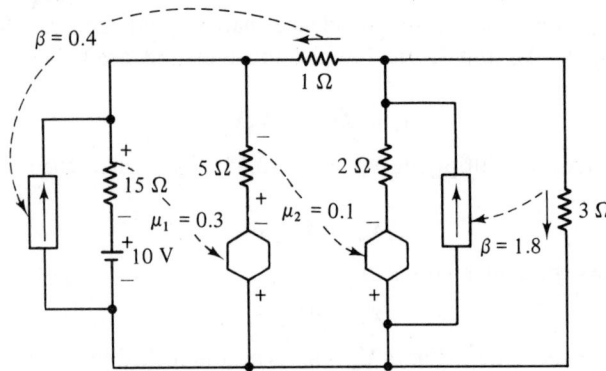

Fig. 3.11 Circuit for Example 3.6

Solution.

(1) The network graph is shown in Fig. 3.12.

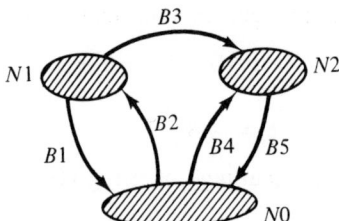

Fig. 3.12 Network Graph for Example 3.6

(2) The matrix **A**, the vectors $\mathbf{V_g}$ and $\mathbf{I_g}$ are:

$$\mathbf{A} = \begin{bmatrix} +1 & -1 & +1 & 0 & 0 \\ 0 & 0 & -1 & -1 & +1 \end{bmatrix}$$

$$[\mathbf{V_g}]^T = \begin{bmatrix} -10 & 0 & 0 & 0 & 0 \end{bmatrix}$$

$$[\mathbf{I_g}]^T = \begin{bmatrix} 0 & 0 & 0 & 0 & 0 \end{bmatrix}$$

(3) The dependent voltage sources are related to the element voltages:

$$\mathbf{D} = \mathbf{D}^* = \begin{bmatrix} 0 & 0 & 0 & 0 & 0 \\ 0.3 & 0 & 0 & 0 & 0 \\ 0 & 0 & 0 & 0 & 0 \\ 0 & 0.1 & 0 & 0 & 0 \\ 0 & 0 & 0 & 0 & 0 \end{bmatrix}$$

Sec. 3.2   Analysis of Arbitrary Networks   69

$$[U + D] = \begin{bmatrix} 1.0 & 0 & 0 & 0 & 0 \\ 0.3 & 1.0 & 0 & 0 & 0 \\ 0 & 0 & 1.0 & 0 & 0 \\ 0 & 0.1 & 0 & 1.0 & 0 \\ 0 & 0 & 0 & 0 & 1.0 \end{bmatrix}$$

(4) $[U + D]$ is now inverted by using the procedure outlined in Sec. 2.6. The result is

$$[U + D]^{-1} = \begin{bmatrix} 1.0 & 0 & 0 & 0 & 0 \\ -0.3 & 1.0 & 0 & 0 & 0 \\ 0 & 0 & 1.0 & 0 & 0 \\ 0.3 \times 0.1 & -0.1 & 0 & 1.0 & 0 \\ 0 & 0 & 0 & 0 & 1.0 \end{bmatrix}$$

(5) The Y_e matrix is

$$Y_e = \begin{bmatrix} \frac{1}{15} & 0 & 0 & 0 & 0 \\ 0 & \frac{1}{5} & 0 & 0 & 0 \\ 0 & 0 & 1 & 0 & 0 \\ 0 & 0 & 0 & \frac{1}{2} & 0 \\ 0 & 0 & 0 & 0 & \frac{1}{3} \end{bmatrix}$$

(6) The dependent currents are entered in the **B** matrix. The result is

$$[U + B] = \begin{bmatrix} 1 & 0 & 0.4 & 0 & 0 \\ 0 & 1 & 0 & 0 & 0 \\ 0 & 0 & 1 & 0 & 0 \\ 0 & 0 & 0 & 1 & 1.8 \\ 0 & 0 & 0 & 0 & 1 \end{bmatrix}$$

(7) The branch-admittance matrix is now formed by using the tabular matrix multiplication scheme introduced in Sec. 2.1. The calculations are indicated in Table 3.3.

Alternately, Y_b could be established from direct examination of the current-voltage dependencies of the branches, as was done in the previous example. The "double dependency" of the voltage source in branch 4, however, needs to be considered carefully. (The various fractions in Y_b were left in the above form to facilitate tracing such a development of Y_b.)

The node voltages, branch voltages, and currents are calculated in Table 3.4. The calculation sequence is indicated by circled numbers.

To begin with, the data necessary for setting up the network node voltage equations are systematically entered in the table. The data needed are elements of A^T, Y_b, A, V_g; they are entered in positions of the calculation table as indicated in the upper left-hand diagram accompanying the table. The first phase of the computation is the calculation of Y_n; this is done in two parts: the computation of $Y_b A^T$, followed by pre-multiplication with A (Steps 6 and 7). The second phase is to calculate I_s; this is done in three parts; the calculation of $Y_b V_g$, followed by subtraction of this product from I_g, and finally the pre-multiplication by A (Steps 8, 9, and 10). Note that I_s is put into a column immediately to the right of Y_n. Hence between Steps 6 and 10 the entire augmented matrix $[Y_n \mid I_s]$ was produced; thus all the

Table 3.3 *Tabular Multiplication of the Elements of Y_b for Ex. 3.6*

$$Y_b = [U + B]Y_e[U + D]^{-1}$$

						1	0	0	0	0	
						−0.3	1	0	0	0	
						0	0	1	0	0	$= [U + D]^{-1}$
						0.3×0.1	−0.1	0	1	0	
						0	0	0	0	1	
	$\frac{1}{15}$	0	0	0	0	$\frac{1}{15}$	0	0	0	0	
	0	$\frac{1}{5}$	0	0	0	$-\frac{0.3}{5}$	$\frac{1}{5}$	0	0	0	
$Y_e =$	0	0	1	0	0	0	0	1	0	0	
	0	0	0	$\frac{1}{2}$	0	$\frac{0.3 \times 0.1}{2}$	$-\frac{0.1}{2}$	0	$\frac{1}{2}$	0	
	0	0	0	0	$\frac{1}{3}$	0	0	0	0	$\frac{1}{3}$	
	1	0	0.4	0	0	$\frac{1}{15}$	0	$\frac{0.4}{1}$	0	0	
	0	1	0	0	0	$-\frac{0.3}{5}$	$\frac{1}{5}$	0	0	0	
$[U + B] =$	0	0	1	0	0	0	0	1	0	0	$= Y_b$
	0	0	0	1	1.8	$\frac{0.3 \times 0.1}{2}$	$-\frac{0.1}{2}$	0	$\frac{1}{2}$	$\frac{1.8}{3}$	
	0	0	0	0	1	0	0	0	0	$\frac{1}{3}$	

coefficients of the matrix equation $Y_n V_n = I_s$ are in place for application of the tabular simultaneous equation solution scheme outlined in Sec. 2.3.

The reduction of the equations (phase 3) is indicated in Step 11; the result is the node-voltage vector V_n shown in the lower right corner of Table 3.4. Thus the first part of the network analysis procedure is completed. Often this work is the only required part of the analysis.

To obtain the branch voltages (phase 4), the newly calculated values of V_n are transposed just above A^T; the calculation of $V_b = A^T V_n$ is then very easily accomplished (Step 13).

The fifth phase is to calculate I_b. This is accomplished by the addition of V_g (from Step 4) to V_b and is Step 14 of this procedure. Pre-multiplication of the resultant column by Y_b is easily carried out (Step 15) and results in the vector I_b. Thus Table 3.4 contains all the results for the node voltages, the branch voltages, and the branch currents.

The five phases of the calculation are indicated in the lower left corner of Table 3.4.

All work shown in the table was carried out using a slide rule. Independent checks can now be made to test the validity of the results. For example, summations of currents at all nodes should result in zeros and summations of voltages around each loop should also result in zeros. The reader is urged to make such checks.

(8) A separate calculation is made to obtain the element voltages and currents by using Eqs. (3.69) and (3.70). This calculation is shown at the bottom of page 71.

Sec. 3.2   Analysis of Arbitrary Networks   71

Table 3.4 Calculation of Voltages and Currents for Ex. 3.6

V_n^T

⑫

| A^T | V_g | I_g | V_b | $V_b + Y_g$ |
| ① | ④ | ⑤ | ⑬ | ⑭ |

| Y_b | Y_bA^T | Y_bV_g | ⑨ $I_g -$ | I_b |
| ② | ⑥ | ⑧ | Y_bV_g | ⑮ |

A ③	Y_n ⑦	I_s ⑩
Reduce	⑪	
	V_n	

2.105	1.69		V_b

+1	0	−10.	0.	2.105	−7.895
−1	0	0.	0.	−2.105	−2.105
+1	−1	0.	0.	0.415	0.415
0	−1	0.	0.	−1.69	−1.69
0	+1	0.	0.	1.69	1.69

0.0667	0.	0.40	0.	0.	0.4667	−0.40	−0.667	0.667	−0.360
−0.060	0.20	0.	0.	0.	−0.26	0.	0.600	−0.60	0.053
0.	0.	1.00	0.	0.	1.000	−1.000	0.	0.	0.410
0.015	−0.050	0.	0.500	0.600	0.065	0.100	−0.150	0.15	0.156
0.	0.	0.	0.	0.333	0.	0.300	0.	0.	0.563

| +1 | −1 | +1 | 0 | 0 | 1.727 | −1.400 | 1.267 |
| 0 | 0 | −1 | −1 | +1 | −1.065 | 1.233 | −0.150 |

Phases of the calculation:
(1) $Y_n = AY_bA^T$
(2) $I_s = A(I_g - Y_bV_g)$
(3) $[Y_n \vdots I_s] \rightarrow V_n$
(4) $V_b = A^TV_n$
(5) $I_b = Y_b[V_b + V_g]$

| 1.000 | −0.810 | 0.731 |
| 0. | 0.348 | −0.141 |

| 1.00 | 0. | 2.105 | $\Big\} = V_n$
| 0. | 1.00 | 1.69 |

$$[U + D]^{-1} \qquad V_b + V_g \qquad V_e$$

$$\begin{bmatrix} 1.0 & 0 & 0 & 0 & 0 \\ -0.3 & 1.0 & 0 & 0 & 0 \\ 0 & 0 & 1.0 & 0 & 0 \\ 0.03 & -0.1 & 0 & 1.0 & 0 \\ 0 & 0 & 0 & 0 & 1.0 \end{bmatrix} \begin{bmatrix} -7.895 \\ -2.105 \\ 0.415 \\ -1.69 \\ 1.69 \end{bmatrix} = \begin{bmatrix} -7.895 \\ 0.254 \\ 0.415 \\ -1.68 \\ 1.69 \end{bmatrix}$$

$$I_e = [(Y_b)_{ii}]V_e = \begin{bmatrix} -0.526 \\ 0.053 \\ 0.415 \\ -0.840 \\ 0.563 \end{bmatrix} ; \quad P_e = [(V_e)_i \cdot (I_e)_i] = \begin{bmatrix} 5.16 \\ 0.0135 \\ 0.172 \\ 1.41 \\ 0.95 \end{bmatrix}$$

(9) The voltages and currents in this network are summarized in Fig. 3.13.

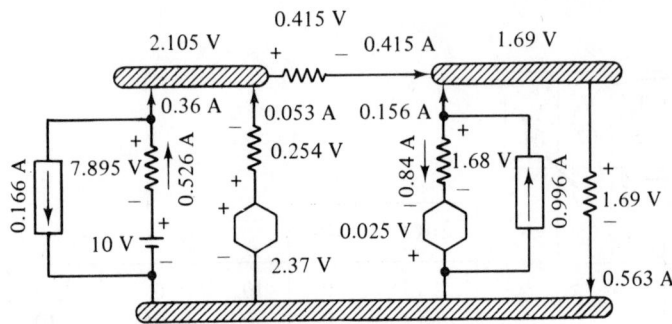

Fig. 3.13 Solution of Example 3.6

3.3 Program for DC Analysis

In the analysis of linear networks, various element types must be considered. Each element will be assigned a number for ease of computational identification. The following is a list of network elements used in this book:

(I) Passive Elements
    (a) Conductance (Type 0)
    (b) Resistance   (Type 1)
    (c) Capacitance (Type 2)
    (d) Inductance   (Type 3)

(II) Active Elements (Sources)
    (e) Independent voltage source (Type 4)
    (f) Independent current source (Type 5)
    (g) Current controlled current source (Type 6)
    (h) Voltage controlled current source (Type 7)
    (i) Current controlled voltage source (Type 8)
    (j) Voltage controlled voltage source (Type 9)

(III) Mutual Element
    (k) Mutual inductance (Type 10)

The above elements will now be defined.

(a) An element is a conductance if it is characterized by a single real number G such that the current through the element is G times the voltage across the element. G is measured in mhos.

(b) An element is a resistance if it is characterized by a single real number R such that the voltage across the element is R times the current through that element. Note that a given branch (if applicable) may equally well be characterized either as a resistance or as a conductance with the value of R being the reciprocal of the value of G. R is measured in ohms.

(c) An element is a capacitance if it is characterized by a single real number C such that the current through the element is C times the rate of change of the voltage across that element. C is measured in farads.

(d) An element is an inductance if it is characterized by a single real number L such that the voltage across the element is L times the time rate of change of the current through the element. L is measured in henries.

(e) An element is an independent voltage source if the voltage across it is independent of the current through it. The voltage value is measured in volts.

(f) An element is an independent current source if the current flowing through it is independent of the voltage across it. The current is measured in amperes.

(g) An element is a current controlled current source if it is characterized by a single number β such that the current flowing through the element is β times the current flowing through some passive element in the network. Usually, β is restricted to real numbers (always real when direct currents are involved) and dependency from the passive element located in the same branch as the controlled source is not allowed, i.e., the controlling element must be located in a branch other than the one in which the controlled source is located. β is dimensionless.

(h) An element is a voltage controlled current source if it is characterized by a single number g_m such that the current flowing through it is g_m times the voltage across some passive element in the network. The same restrictions apply to g_m as applied to β; g_m is measured in mhos.

(i) An element is a current controlled voltage source if it is characterized by a single number r_t such that the voltage across it is r_t times the current flowing in some passive element in the network. The same restrictions apply to r_t as applied to β; r_t is measured in ohms.

(j) An element is a voltage controlled voltage source if it is characterized by a single number μ such that the voltage across that element is μ times the voltage across a passive element in the network. The same restrictions apply to μ as applied to β; μ is dimensionless.

(k) Two inductors may be mutually coupled if the time rate of change of current in one coil induces a proportional voltage in the other coil in addition to the voltage caused by the change of current through the other coil. The constant of proportionality M is termed the *mutual inductance*; it is a real number measured in henries. Mutual inductances will be discussed in Sec. 3.4.

The above element definitions are consistent with the universally used definitions of the basic network elements and are valid for DC, AC, and transient analyses. The various voltage-current relationships defined above are summarized in Table 3.5.

Table 3.5 Voltage-Current Relationships

Type Number	Symbol	Name	Unit	Circuit Diagram	Equation
0	G	Conductance	mho		$i = Ge$
1	R	Resistance	ohm		$e = Ri$
2	C	Capacitance	farad		$i = C\dfrac{de}{dt}$
3	L	Inductance	henry		$e = L\dfrac{di}{dt}$
4	E	Voltage source	volt		$e = E$
5	I	Current source	ampere		$i = I$
6	β	CCCS*	—		$i = \beta i'$
7	g_m	VCCS†	mho		$i = g_m e'$
8	r_t	CCVS‡	ohm		$e = r_t i'$
9	μ	VCVS§	—		$e = \mu e'$
10	M	Mutual inductance	henry		$e = L_1 \dfrac{di}{dt} + M \dfrac{di'}{dt}$

*CCCS = Current controlled current source.
†VCCS = Voltage controlled current source.
‡CCVS = Current controlled voltage source.
§VCVS = Voltage controlled voltage source.

In this section a very simple circuit analysis program is discussed which will calculate voltages and currents in a DC network. The network will be limited to the following components:

(1) Resistors (Type 1) or conductors (Type 0)
(2) Independent voltage sources (Type 4)
(3) Independent current sources (Type 5)
(4) Dependent current sources:
 (a) Current controlled (Type 6)
 (b) Voltage controlled (Type 7)

Capacitors (Type 2) and inductors (Type 3) will also be allowed; for this DC analysis such elements will be replaced by resistors, of 10 megohms and 0.10 ohm value respectively.

In order to bypass the complications introduced by dependent voltage sources, dependent voltage sources will not be allowed in this analysis program. Dependent voltage sources will be discussed in Sec. 3.10; the present program may be reworked to include such sources for DC analysis (see Problem 3.5).

If a dependent voltage source must be considered in a circuit, in using this program for analysis, such a source may be replaced by an appropriate circuit branch containing a dependent current source. Let the original circuit dependency be as shown in Fig. 3.14 in which an equivalent circuit is also shown. In the equivalent circuit the dependent source is replaced by a 1 Ω resistor and a voltage dependent current source (Type 7); if the value of g_m is picked such that numerically $g_m = \mu$, then the voltage across the 1 Ω resistor is

$$g_m \cdot e_1 \cdot 1 = \mu e_1 \tag{3.73}$$

provided *none* of the current in R_2 goes through the 1 Ω resistor. This latter condition is ensured by putting a second dependent current source in parallel with the 1 Ω resistor; the current through this source is the current through R_2 (i.e., current is $I_2 = 1 \cdot I_{R_2}$). Careful attention must be paid to the polarities of the two dependent current sources and to the reference directions of I_{R_1} and I_{R_2}.

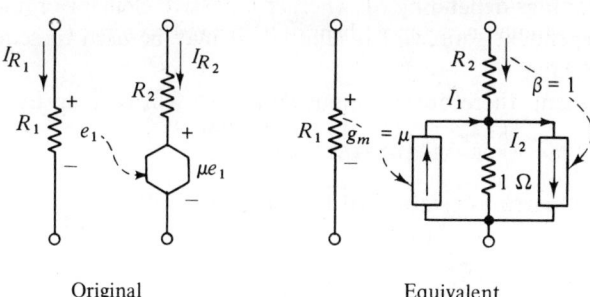

Original   Equivalent

Fig. 3.14 Voltage Controlled Voltage Source

A similar procedure may be used to replace current controlled voltage sources by current controlled current sources (Type 6).

The equations relating to the circuit analysis were developed in the previous section and are repeated here:

$\mathbf{A}$ = Incidence matrix—has as many rows as there are independent nodes in the network; has as many columns as there are elements in the network.

$\mathbf{Y_b}$ = Branch-admittance matrix of the network—has as many rows and columns as there are branches.

$\mathbf{V_n}$ = Node-voltage vector—as many rows as there are independent nodes.

$\mathbf{I_g}$ = Indpendent current generators feeding into the nodes.

$\mathbf{V_g}$ = Independent voltage generators in series with the various branches.

The node-voltage equations become

$$\mathbf{A Y_b A^T V_n} = \mathbf{A(I_g - Y_b V_b)} \tag{3.65}$$

This must be solved for $\mathbf{V_n}$. The branch voltages are

$$\mathbf{V_b} = \mathbf{A^T V_n} \tag{3.64}$$

and the branch currents become

$$\mathbf{I_b} = \mathbf{Y_b(V_b + V_g)} \tag{3.49}$$

where $\mathbf{V_g}$ = the independent voltages in the branches. For each circuit element (there may be several in a branch, see Fig. 3.10), at most five items of information are needed.

For a passive component the branch number (I), the from-node number (J), the to-node number (K), the component type (IX = 0, 1, 2, or 3), and the component value (VAL) are needed. Since independent sources will always occur in conjunction with a passive component, only the branch number (I), the source type (IX = 4 or 5), and the component value (VAL) are needed. For dependent sources, only the from-branch number (J), the to-branch number (K), the element type (IX = 6 or 7), and the element value VAL are needed. In the above description the variables J and K have different meanings depending on whether a passive element or a source is being described. For dependent sources, I is ignored; it may be used to sequence number parts of the input data.

For each element, therefore, an input data card will be read by the statements

```
          READ (1,10) I,J,K,IX,VAL
    10    FORMAT (4I3,F10.0)
```
(3.74)

The incidence matrix is established by using Eq. (3.4):

```
          IF (J.LE.0) GO TO 25
          A(J,I)=1.
    25    IF (K.LE.0) GO TO 35
          A(K,I)=-1.
    35    CONTINUE
```
(3.75)

Sec. 3.3   Program for DC Analysis

Note that node 0 is taken as the reference node; all voltages will be calculated with respect to this node.

The matrix $(\mathbf{U} + \mathbf{B})\mathbf{Y}_e = \mathbf{Y}_b$ occurring in the analysis need not be calculated separately from $(\mathbf{U} + \mathbf{B})$ and $\mathbf{Y}_e$. The main-diagonal elements of $\mathbf{Y}_b$ are

$$(\mathbf{Y}_b)_{kk} = (\mathbf{Y}_e)_{kk} \tag{3.76}$$

i.e., the kth main-diagonal entries in $\mathbf{Y}_b$ is the kth branch admittance. The off-diagonal entries come from $\mathbf{B}$, containing the dependent current sources. For a current being injected into the kth branch of the network from a controlling quantity in the jth branch, $\mathbf{B}$ contains an entry in the kjth position: b_{kj}. This element leads to one entry in $\mathbf{Y}_b$

$$(\mathbf{Y}_b)_{kj} = b_{kj}(\mathbf{Y}_b)_{jj} \tag{3.77}$$

If the source is current controlled (Type 7), then

$$b_{kj} = \beta \tag{3.78}$$

where β is the current transfer ratio to branch k from the current in the passive element in branch j. If the source is voltage controlled, i.e., the current injected in branch k is a constant (g_m) times the current in branch j, then the entry b_{kj} is

$$b_{kj} = \frac{g_m}{(\mathbf{Y}_b)_{jj}} \tag{3.79}$$

Hence if the injected current into branch k is β times the current in the passive element in branch j, the off-diagonal entry in $\mathbf{Y}_b$ will be

$$(\mathbf{Y}_b)_{kj} = \beta \cdot (\mathbf{Y}_b)_{jj} \tag{3.80}$$

and if the current injected into branch k is g_m times the voltage across the passive element in branch j,

$$(\mathbf{Y}_b)_{kj} = g_m \tag{3.81}$$

The FORTRAN statements for establishing the $\mathbf{Y}_b$ matrix (called YB in the program) are as follows:

The branch admittance matrix elements become

```
YB(I,I)=1./VAL
```
(3.82a)

for the resistors and for conductances.

```
YB(I,I)=VAL
```
(3.82b)

For independent currents (the FORTRAN variable CG will be used), the entry in the $\mathbf{I}_g$ vector will be

```
CG(I)=VAL
```
(3.83)

For independent voltages,

```
VG(I)=VAL
```
(3.84)

For each current controlled current source,

```
YB(K,J)=VAL*YB(J,J)
```
(3.85)

For each voltage controlled current source,

$$YB(K,J)=VAL \tag{3.86}$$

For a capacitor in the ith branch,

$$YB(I,I)=1.E-7 \tag{3.87}$$

and for an inductor in the ith branch,

$$YB(I,I)=10. \tag{3.88}$$

Once the A, YB, VG, CG arrays are completed, straightforward matrix operations are used to establish $[\mathbf{Y_n} \mid \mathbf{I_s}]$, $\mathbf{V_n}$, $\mathbf{V_b}$, $\mathbf{I_b}$, and $\mathbf{P_b}$ (where $\mathbf{P_b}$ is the power dissipated in the various branches of the network). Routine SOLVER, discussed in Sec. 2.4, is used to solve for $\mathbf{V_n}$ from the augmented matrix $[\mathbf{Y_n} \mid \mathbf{I_s}]$.

Use of the READ statement in Eq. (3.74) requires that each element in the network be inputted on a separate card. Circuit analysis programs in wide use replace this cumbersome procedure with a much more flexible input language (consisting of a relatively long and involved set of subroutines). In this discussion no attempt is made to streamline the input form; the emphasis here is to demonstrate viable matrix procedures for network analysis.

The completion of the network information will be indicated by a blank card. The output information will consist of node voltages, branch voltages, branch currents, branch powers, passive element voltages, currents, and powers (a total of seven columns). A flowchart of the program is shown in Fig. 3.15. The number of branches (M) and the number of independent nodes (N) is input first; a blank card here will stop the program. The rest of the program follows the steps already outlined. Note that since no dependent voltages will be allowed, the element voltages are

$$\mathbf{V_e} = \mathbf{V_b} + \mathbf{V_g} \tag{3.89}$$

and the ith element current is

$$(\mathbf{I_e})_i = (\mathbf{V_e})_i \cdot (\mathbf{Y_b})_{ii} \tag{3.90}$$

Consequently, the ith element power is

$$(\mathbf{P_e})_i = (\mathbf{V_e})_i \cdot (\mathbf{I_e})_i \tag{3.91}$$

In the program itself no attempt is made to economize either on computer storage or on execution time. Some techniques for saving storage space (such as dispensing with the **A** matrix) are given in Sec. 3.6. Note that since the elements of **A** are 0, +1, −1, any multiplication with the elements of **A** could have been replaced by an addition, a subtraction, or by elimination of that operation.

A listing of this program is given in Listing 3.1. The program was run on an IBM 360/75 and an IBM 370/165 computer system using FORTRAN-G and FORTRAN-H compilers. The program will handle 25 branches and 20 independent nodes, but by changing the DIMENSION statements, much larger circuits may be handled with ease.

For a listing of SUBROUTINE SOLVER, see Appendix A.

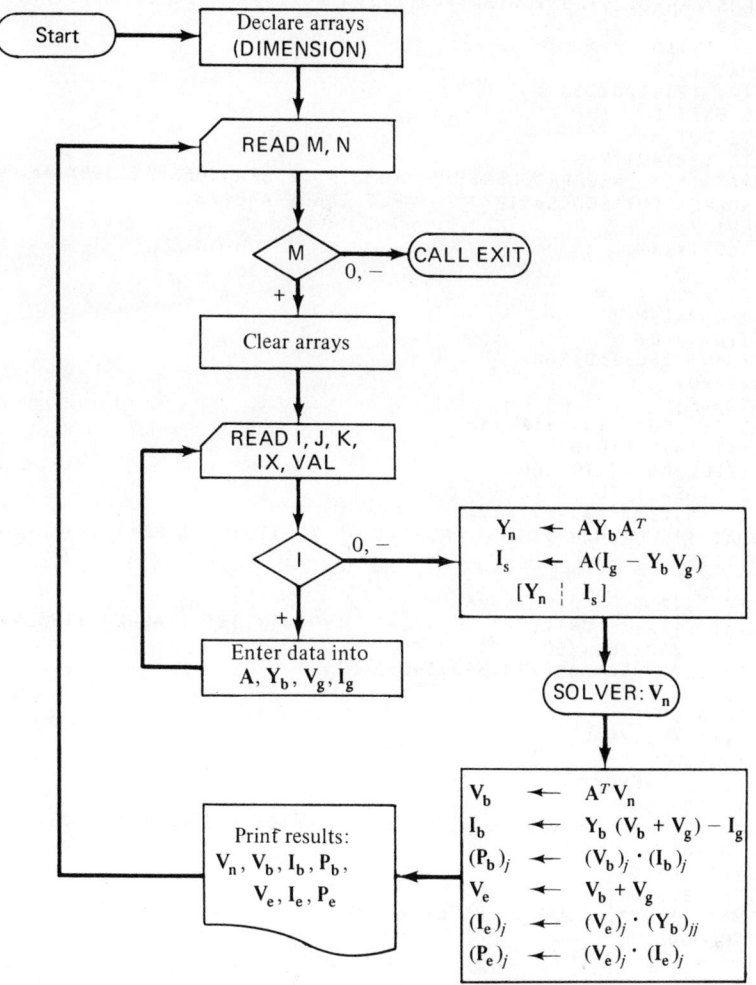

Fig. 3.15 Flowchart of DC Analysis Program

EXAMPLE 3.7

Example 3.6 is used again to demonstrate the use of the program just developed. Since the example circuit has dependent voltage sources, these sources must be changed to current dependent sources. The replacement technique shown in Fig. 3.14 is used. The resulting equivalent circuit is shown in Fig. 3.16 with the various transfer quantities indicated sequentially as $T1$, $T2$, etc.

Note the creation of two new branches (6 and 7) and two nodes (3 and 4). The input cards to the program and the final results are shown in Listing 3.2. The rather voluminous intermediate printouts are not reproduced.

```
C     THIS IS PROGRAM DCAN2
      DIMENSION YB(25,25),CG(25),VG(25),A(20,25),YN(20,21),T(25),BC(25)
      NNK=20
  100 READ (1,110) M,N
  110 FORMAT (2I3)
      IF (M) 120,120,130
  120 CALL EXIT
C     WRITE OUT M,N
  130 WRITE (3,140) M,N
  140 FORMAT ('1      NEW PROBLEM.'/'   NUMBER OF BRANCHES=',I3,5X,'NUMBER O
     1F INDEPENDENT NODES=',I3//'   INPUT CARDS ARE'//)
      N2=N+1
      DO 160 I=1,M
      CG(I) = 0.
      VG(I) = 0.
      DO 160 J=1,M
      YB(I,J) = 0.
      IF (J-N) 150,150,160
  150 A(J,I)=0.
  160 CONTINUE
  170 READ (1,180) I,J,K,IX,VAL
  180 FORMAT (4I3,F10.0)
      IF (I.LE.0) GO TO 400
      IF (IX.GE.6) GO TO 220
      WRITE (3,210) I,J,K,IX,VAL
  210 FORMAT (' BRANCH',I3,' FROM',I3,' TO',I3,'  TYPE=',I3,'   VALUE='
     1,1PE15.5)
      GO TO 240
  220 WRITE (3,230)   J,K,IX,VAL
  230 FORMAT (13X,'FROM',I3,' TO',I3,'  TYPE=',I3,'   VALUE=',1PE15.5)
  240 IF (IX) 260,260,250
  250 GO TO (270,280,300,320,330,340,350),IX
  260 YB(I,I) = VAL
      GO TO 360
  270 YB(I,I) = 1./VAL
      GO TO 360
  280 YB(I,I)= 1.E-07
      WRITE (3,290)
  290 FORMAT ('  ** CAPACITOR REPLACED BY A 10. MEGOHM RESISTOR.')
      GO TO 360
  300 YB(I,I) = 10.
      WRITE (3,310)
  310 FORMAT ('  ** INDUCTOR REPLACED BY A  0.10 OHM RESISTOR.')
      GO TO 360
  320 VG(I) = VAL
      GO TO 170
  330 CG(I) = VAL
      GO TO 170
  340 YB(K,J)=YB(K,J) + YB(J,J)*VAL
C         CURRENT CONTROLLED CURRENT SOURCE.
      GO TO 170
  350 YB(K,J) = YB(K,J) + VAL
C         VOLTAGE CONTROLLED CURRENT SOURCE.
      GO TO 170
  360 IF(J.GT.0) A(J,I) = 1.
      IF (K.GT.0) A(K,I) = -1.
      GO TO 170
C         M=NUMBER OF BRANCHES
C         N= NUMBER OF INDEPENDENT NODES.
C         READY FOR SETTING UP CIRCUIT MATRICES. A,Y,CG,VG  ARE COMPLETE.
  400 DO 410 I=1,M
      T(I)= CG(I)
```

Listing 3.1   DC Analysis Program

```
              DO 410 J=1,M
  410 T(I)= T(I) - YB(I,J)*VG(J)
C             FORM    CG - YB*VG
      DO 420 I=1,N
      YN(I,N2) = 0.
      DO 420 J=1,M
  420 YN(I,N2) = YN(I,N2)+A(I,J)*T(J)
C             CALCULATE COEFFICIENTS OF YN.
      DO 440 K=1,N
      DO 430  I=1,M
      T(I)=0.
      DO 430  J=1,M
  430 T(I) = T(I) + YB(I,J)* A(K,J)
      DO 440 I=1,N
      YN(I,K)=0.
      DO 440 J=1,M
  440 YN(I,K)=YN(I,K) + A(I,J)*T(J)
C             PRINT OUT A,YB,YN,CG,VG,T
C             THE NODE ADMITTANCE MATRIX IS COMPLETE.
      WRITE (3,450)
  450 FORMAT (//'  THE INCIDENCE MATRIX (A-MATRIX) IS'/)
      DO 460 I=1,N
  460 WRITE (3,470) (A(I,J),J=1,M)
  470 FORMAT (1X, 20F4.0/(2X,20F4.0/))
      WRITE  (3,480)
  480 FORMAT (//'  THE BRANCH-ADMITTANCE MATRIX (YB-MATRIX) IS'/)
      DO 490  I=1,M
  490 WRITE (3,500)  (YB(I,J),J=1,M)
  500 FORMAT (1X,1P8E14.5/(2X,1P8E14.5/))
      WRITE (3,510)
  510 FORMAT (//'  THE VECTOR OF INDEPENDENT CURRENTS (CG-VECTOR) IS'/)
      WRITE (3,500)  (CG(I),I=1,M)
      WRITE (3,520)
  520 FORMAT (//'  THE VECTOR OF INDEPENDENT VOLTAGES (VG-VECTOR) IS'/)
      WRITE (3,500) (VG(I),I=1,M)
      WRITE (3,530)
  530 FORMAT(//'  THE AUGMENTED NODE-ADMITTANCE MATRIX (YN-MATRIX) IS'/)
      DO 540 I=1,N
  540 WRITE (3,500) (YN(I,J),J=1,N2)
C             ALL MATRICES ARE PRINTED NOW.
C             THE AUGMENTED MATRIX (YN,A*(CG-YB*VG) IS COMPLETE.
      CALL SOLVER (YN,NNK,N,DET)
C             NODE VOLTAGES ARE NOW IN COL. N+1 OF YN
C             PRINT NODE VOLTAGES.
C             CALCULATE BRANCH VOLTAGES,CURRENTS,POWERS.
C             BV = (A-TRANSPOSE)*VN
      DO 550 I=1,M
      T(I)=0.
      DO 550 J=1,N
  550 T(I)=T(I)+A(J,I)*YN(J,N2)
C             T-VECTOR CONTAINS THE BRANCH VOLTAGES.
      DO 560 I=1,M
      BC(I)= -CG(I)
      DO 560 J=1,M
  560 BC(I)=BC(I)  +YB(I,J)*(VG(J)+T(J))
C             VECTOR BC CONTAINS THE BRANCH CURRENTS.
      DO 570 I=1,M
C             SAVE ELEMENT ADMITTANCES IN COL. 1 OF YB
  570 YB(I,1)=YB(I,I)
      DO 580 I=1,M
      YB(I,2)=VG(I)+T(I)
C             ELEMENT VOLTAGES ARE NOW IN COL. 2 OF YB. (NOT VALID IF
```

Listing 3.1—*Cont.*

```
C            DEPENDENT VOLTAGES ARE ALLOWED.)
C        PUT ELEMENT CURRENTS INTO COL. 3
      YB(I,3)=YB(I,2)*YB(I,1)
C        PUT ELEMENT POWERS INTO COL. 4
      YB(I,4)=YB(I,2)*YB(I,3)
C        PUT BRANCH POWERS INTO COL. 5
  580 YB(I,5)=T(I)*BC(I)
C        THE YB-MATRIX IS DESTROYED.
      WRITE (3,590)
  590 FORMAT ('1'//10X,'NODE',18X,3('BRANCH',9X),3X,2('ELEMENT',8X),
     1'ELEMENT'/ 9X,'VOLTS',5X,'NUMBER',5X,2(3X,'VOLTS',8X,'CURRENT',
     210X,'POWER',11X)/)
      DO 640 I=1,M
      IF (I-N) 600,600,620
  600 WRITE (3,610)YN(I,N2),I,T(I),BC(I),YB(I,5),YB(I,2),YB(I,3),YB(I,4)
  610 FORMAT (1PE15.5,I9,3E15.5,4X,1P3E15.5)
      GO TO 640
  620 WRITE(3,630)           I,T(I),BC(I),YB(I,5),YB(I,2),YB(I,3),YB(I,4)
  630 FORMAT (I24,1P3E15.5,4X,1P3E15.5)
  640 CONTINUE
C        ALL RESULTS ARE PRINTED.
      GO TO 100
      END
```

Listing 3.1—*Cont.*

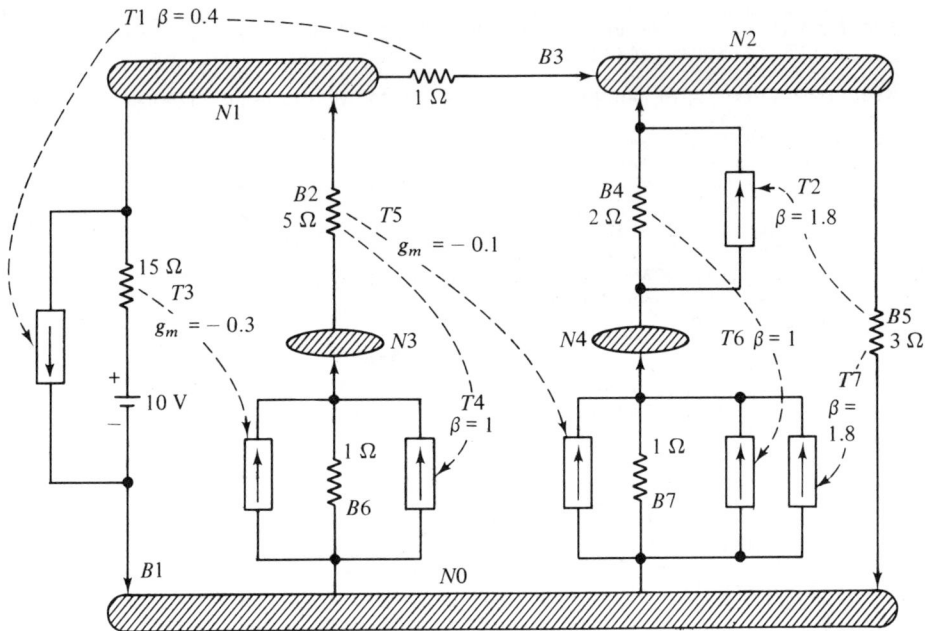

Fig. 3.16 Circuit for Example 3.7

Sec. 3.4 AC Analysis of Linear Networks 83

```
NEW PROBLEM
NUMBER OF BRANCHES=  7      NUMBER OF INDEPENDENT NODES=  4

INPUT CARDS ARE

BRANCH  1   FROM  1  TO  0   TYPE=  1   VALUE=   1.50000E 01
BRANCH  2   FROM  3  TO  1   TYPE=  1   VALUE=   5.00000E 00
BRANCH  3   FROM  1  TO  2   TYPE=  1   VALUE=   1.00000E 00
BRANCH  4   FROM  4  TO  2   TYPE=  1   VALUE=   2.00000E 00
BRANCH  5   FROM  2  TO  0   TYPE=  1   VALUE=   3.00000E 00
BRANCH  6   FROM  0  TO  3   TYPE=  1   VALUE=   1.00000E 00
BRANCH  7   FROM  0  TO  4   TYPE=  1   VALUE=   1.00000E 00
BRANCH  1   FROM  0  TO  0   TYPE=  4   VALUE=  -1.00000E 01
            FROM  3  TO  1   TYPE=  6   VALUE=   4.00000E-01
            FROM  5  TO  4   TYPE=  6   VALUE=   1.80000E 00
            FROM  1  TO  6   TYPE=  7   VALUE=  -3.00000E-01
            FROM  2  TO  6   TYPE=  6   VALUE=   1.00000E 00
            FROM  2  TO  7   TYPE=  7   VALUE=  -1.00000E-01
            FROM  4  TO  7   TYPE=  6   VALUE=   1.00000E 00
            FROM  5  TO  7   TYPE=  6   VALUE=   1.80000E 00
```

THE AUGMENTED NODE ADMITTANCE MATRIX (YN-MATRIX) IS

```
 1.66667E 00  -1.40000E 00  -2.00000E-01   0.0           6.66667E-01
-1.00000E 00   1.23333E 00   0.0          -5.00000E-01   0.0
 3.00000E-01   0.0           1.00000E 00   0.0           3.00000E 00
-1.00000E-01   0.0           1.00000E 00   1.00000E 00   0.0
```

NODE VOLTS	NUMBER	BRANCH VOLTS	BRANCH CURRENT	BRANCH POWER
2.11762E 00	1	2.11762E 00	-3.61232E-01	-7.64954E-01
1.70697E 00	2	2.47089E-01	4.94179E-02	1.22106E-02
2.36471E 00	3	4.10648E-01	4.10648E-01	1.68632E-01
-2.47087E-02	4	-1.73168E 00	1.58343E-01	-2.74200E-01
	5	1.70697E 00	5.68992E-02	9.71254E-01
	6	-2.36471E 00	4.94165E-02	-1.16856E-01
	7	2.47087E-02	1.58343E-01	3.91244E-03

		ELEMENT VOLTS	ELEMENT CURRENT	ELEMENT POWER
		-7.88238E 00	-5.25492E-01	4.14212E 00
		2.47089E-01	4.94179E-02	1.22106E-02
		4.10648E-01	4.10648E-01	1.68632E-01
		-1.73168E 00	-8.65841E-01	1.49936E 00
		1.70697E 00	5.68992E-01	9.71254E-01
		-2.36471E 00	-2.36471E 00	5.59187E 00
		2.47087E-02	2.47087E-02	6.10520E-04

Listing 3.2 Output from Program DCAN2 for Analysis of the Circuit Shown in Fig. 3.16

3.4 AC Analysis of Linear Networks

The node-voltage formulation of the network equations given in Sec. 3.2 can be used without change in AC steady-state analysis also. Care must be taken to set up the branch admittances for each frequency of interest, ω; the node equations may be solved directly. The formal circuit equations are then as follows:

$$\mathbf{A}\mathbf{Y}_b(\omega)\mathbf{A}^T\mathbf{V}_n = \mathbf{A}[\mathbf{I}_g(\omega) - \mathbf{Y}_b\mathbf{V}_g(\omega)] \tag{3.92}$$

where $\mathbf{Y}_b(\omega)$ is the frequency dependent branch-admittance matrix and includes the effect of dependent current sources (as in most "usual" circuit analysis programs) and could include other sources as indicated in Sec. 3.2. The major difference in DC and AC steady-state analysis is the fact that $\mathbf{Y}_b$ is complex and all numbers in the AC analysis are complex quantities. In addition, a series of frequency increments usually is taken and a complete analysis is performed at each frequency. Hence for any given frequency the same steps are taken for AC analysis as were taken for DC analysis with each quantity being complex. The main effect is that the calculations are algebraically more involved and that a different analysis must be done for each frequency.

The setting up of the required incidence matrix ($\mathbf{A}$), current generators ($\mathbf{I}_g$), and voltage generators ($\mathbf{V}_g$) proceeds exactly as in the DC case; the establishment of the dependent voltage matrix $\mathbf{D}$ is as outlined in Sec. 3.2. Differences between the DC and the AC analysis program so far concern the fact that $\mathbf{V}_g$ and $\mathbf{I}_g$ could be complex sources (sources having different phase angles).

A major change is needed for setting up the $\mathbf{Y}_b$ matrix. This matrix is a set of complex numbers containing admittances calculated from the various circuit elements and the frequency $j\omega$. Hence before $\mathbf{Y}_b$ can be established a frequency must be picked; every time the frequency is changed, as in a frequency response calculation, a new $\mathbf{Y}_b$ matrix must be established. Therefore, as the branch elements are entered into the program in the initialization phase, the values of the elements are stored in an array. Subsequent to storing the whole circuit description these numbers are used in establishing $\mathbf{Y}_b(\omega)$.

In AC steady-state analysis, three circuit components not considered in DC analysis must be included: inductances, capacitances, and mutual inductances. The admittance of an inductor L is $-j/\omega L$; an inductor of value VAL in branch I will produce the $\mathbf{Y}_e$ matrix entry

$$(\mathbf{Y}_e)_{ii} \leftarrow \mathtt{CMPLX(0.,-1./(VAL*W))} \tag{3.93}$$

where $W = 2\pi f$ (with f the frequency in Hz) for which the analysis is effected; and CMPLX is the FORTRAN library function that forms a complex quantity from two REAL arguments.

The corresponding entry for a capacitor is $j\omega C$, which is accomplished by the quasi-FORTRAN statement

$$(\mathbf{Y}_e)_{ii} \leftarrow \mathtt{CMPLX(0.,VAL*W)} \tag{3.94}$$

where VAL is the value of the capacitor in branch I.

Resistors and sources, both controlled and independent, produce the same entries as they did in the DC analysis [see Eqs. (3.82) through (3.84)].

An additional circuit component to be accounted for in AC analysis is the mutual inductance. In general, mutual inductances may couple any branch to any number of other branches. The computational complications introduced by such general mutual dependence are considerable; it is left as an exercise to develop the program for this. Here we shall limit ourselves to the case of a single mutual inductance coupling two branches (primary and secondary; p and s) of a circuit, each of which contains an inductor. Note that the mutual inductance is not an element in the circuit; instead, it is the equivalent of two controlled generators. The basic steady-state relationships for two coupled inductors (L_p in branch p and L_s in branch s coupled by $M_{ps} = M_{sp} = M$) are

$$E_p = j\omega L_p I_p + j\omega M_{ps} I_s \qquad (3.95)$$

$$E_s = j\omega L_s I_s + j\omega M_{sp} I_p \qquad (3.96)$$

where the currents and voltages are those of the passive elements L_s and L_p.

Solving these equations for I_p and I_s, we get

$$I_p = \frac{E_p}{j\omega L_p} \cdot \frac{L_p L_s}{L_p L_s - M^2} - \frac{E_s}{j\omega L_s} \cdot \frac{M L_s}{L_p L_s - M^2} \qquad (3.97)$$

$$I_s = \frac{E_s}{j\omega L_s} \cdot \frac{L_s L_p}{L_p L_s - M^2} - \frac{E_p}{j\omega L_p} \cdot \frac{M L_p}{L_p L_s - M^2} \qquad (3.98)$$

The coefficient of coupling (k) between two coils is defined as

$$k = \frac{M}{\sqrt{L_s L_p}} \qquad (3.99)$$

Note that as $k \to 1$, Eqs. (3.97) and (3.98) become indeterminate. Real coupled coils have a coefficient of coupling less than .995 even for devices such as good transformers; however, "ideal transformers" used in theoretical circuit analysis have a coupling coefficient $k = 1$. We shall not allow ideal transformers in our analysis program.

The entries produced in the Y_b matrix will now be examined. If the admittances of the branches p and s have been entered in the Y_b matrix, then those admittances will be the main-diagonal entries $(Y_b)_{pp}$ and $(Y_b)_{ss}$. The cross-coupling terms representing Eqs. (3.97) and (3.98) will be given by the FORTRAN statements (P and S represent the branch numbers, L is an array of inductors, and MX is the mutual inductance).

```
T1=YB(P,P)
T2=YB(S,S)
T3=L(S)*L(P)-M*M
YB(P,P)=T1*L(P)*L(S)/T3
YB(P,S)=-T2*M*L(S)/T3
YB(S,S)=T2*L(P)*L(S)/T3
YB(S,P)=-T1*M*L(P)/T3
```
(3.100)

A check could be made after T3 is calculated to see if it is acceptably large. Mutual inductance was assigned the type number 10 (see Table 3.4).

The case of mutual inductance between more than two branches is discussed in Problem 3.11.

The remaining steps of the analysis for AC steady-state conditions are identical to the DC analysis. The difficulties represented by controlled voltage sources are precisely the same as in the DC analysis (i.e., the **D** matrix); disallowing such sources will result in an AC analysis program quite analogous to the DCAN2 program developed in the previous section.

In programming, allowance must be made for complex quantities being produced in the vectors $\mathbf{I_g}$, $\mathbf{V_g}$, $\mathbf{V_n}$, $\mathbf{V_b}$, $\mathbf{I_b}$, and $\mathbf{P_b}$, the matrices $\mathbf{Y_b}$ and $\mathbf{Y_n}$, and all intermediate storage quantities as well.

The next section outlines a minimum-effort conversion of DCAN2 to an AC analysis program.

3.5 A Simple AC Analysis Program

The various steps outlined in the previous section are incorporated in the flowchart shown in Fig. 3.17. Note that since the $\mathbf{Y_b}$ matrix must be set up at each frequency used in the calculations, the input branch element values are stored instead of entered directly into the $\mathbf{Y_b}$ matrix. Hence several storage arrays are set up which are not found in the DC analysis program discussed in Sec. 3.3.

After all input data are read in, a complete node-voltage analysis (paralleling the previously discussed DC analysis) is carried out for a sequence of frequencies from FL to FH in multiplicative increments of FM. At the end of each analysis all results are printed; hence the program will produce voluminous outputs.

A comparison of Figs. 3.17 and 3.15 shows that major portions of the DC analysis program can also be used for AC analysis but the arrays and variables in the calculations must be declared COMPLEX. Similarly, SUBROUTINE SOLVER must be modified to allow complex numbers in its input and output arrays.

The DC analysis program revised for AC analysis is listed in Listing 3.3; the flowchart is shown in Fig. 3.17. These statements constitute a workable AC analysis program which does not, however, allow dependent voltage sources. Extension to this latter case is considered in Sec. 3.10.

A careful review of the actual listing is interesting. A number of "tricks" were used which are worth remembering. Also note that although program DCAN2 was written in FORTRAN II (i.e., IBM System/360 FORTRAN-E), this AC analysis program is a FORTRAN IV (i.e., FORTRAN-G) code.

The arrays RR, LL, CC, CG, VG are used to hold the value of conductance, inverse inductance, capacitance, independent current value, and independent voltage value associated with each branch of the network. Note that LL is declared explicitly to be an array of REAL numbers. Since the actual $\mathbf{Y_b}$ entry is made up in statement number 152, each of the branches will allow a parallel combination of conductance, inductance, and capacitance as the passive element; if any of the passive components is missing, the program uses a zero value for that missing component. Thus the program handles missing components as "open circuits" (i.e., zero conductance) by default. This default is forced in the FORTRAN DO-loop involving statement number 160.

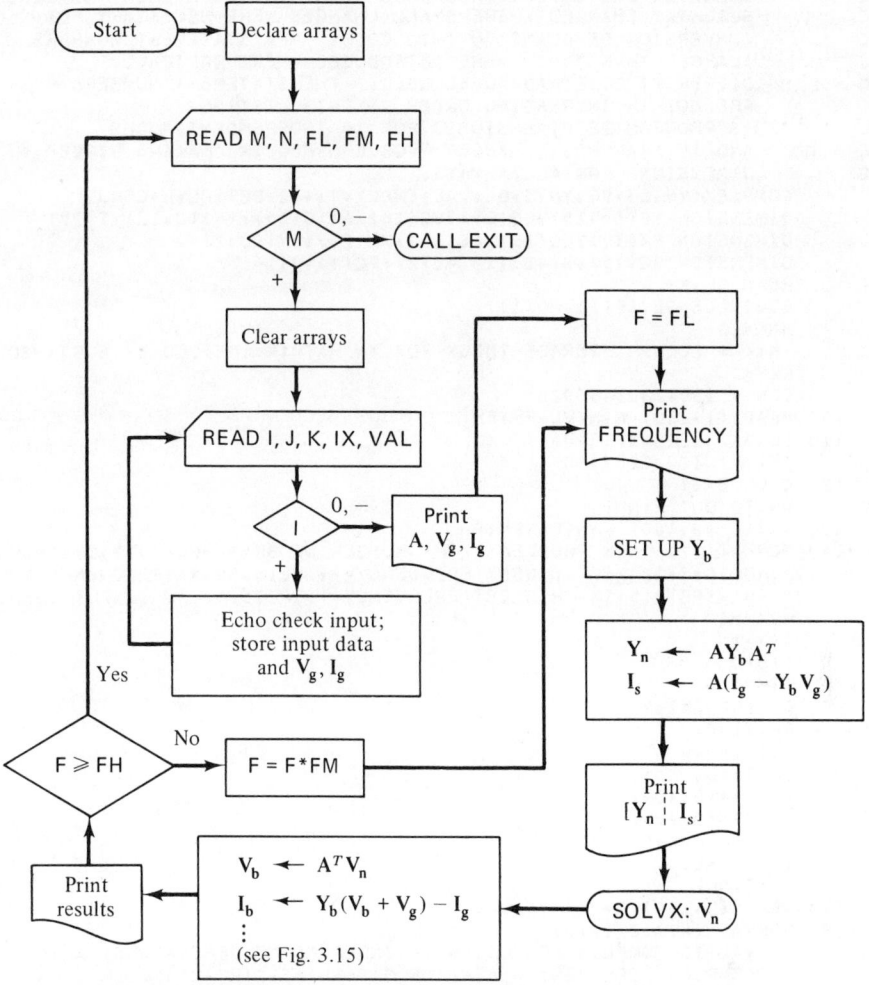

Fig. 3.17 Flowchart of AC Analysis Program

The input card formats are similar to the DC program. The two exceptions are:

(1) On the first card the frequency range and the frequency step must be indicated in addition to the number of branches and independent nodes.

(2) The element value VAL is a complex quantity using 2F10.0 format on the input card.

The value read for VAL is complex, and only the real part is retained for most elements (such as RR, LL, CC), but the entire complex number is used for dependent sources and independent generator values. In these cases the input value is the real part followed by the imaginary part of the number (*not* magnitude and phase angle).

```
C         THIS IS PROGRAM ACAN2, A CONVERSION OF DCAN2
C            THE STATEMENT NUMBERS OF PROGRAM DCAN2 HAVE BEEN RETAINED
C            WHENEVER THE CODE CORRESPONDS CLOSELY.  STATEMENT NUMBERS WERE
C            SLIGHTLY CHANGED WHERE SMALL CHANGES WERE NECESSARY FOR
C            CONVERSION OF DCAN2 TO THIS CODE.  NEW STATEMENT NUMBERS
C            (LARGER THAN 700) WERE INTRODUCED WHERE RADICALLY
C            DIFFERENT CODE HAD TO BE USED.  THE STATEMENT NUMBERS
C            ARE NOT IN INCREASING ORDER IN THIS LISTING.
C         THIS PROGRAM IS DIMENSIONED FOR 10 INDEPENDENT NODES
C            AND 15 BRANCHES.  LARGER PROBLEMS REQUIRE MAKING BIGGER
C            DIMENSIONS FOR ALL ARRAYS.
          COMPLEX YB,CG,VG,YN,T,BC,VAL,CMPLX,T1,T2,DET,PC,B,CONJG
          DIMENSION YB(15,15),CG(15),VG(15),A(10,15),YN(10,11),T(15)
          DIMENSION RR(15),LL(15),CC(15),MM(20),IM(20,2)
          DIMENSION BC(15),B(40),IB(40,2),PC(7),P(14)
          REAL LL,MM
          EQUIVALENCE (P(1),PC(1))
          NNK=10
C         NNK = COLUMN STORAGE INDEX FOR YN MATRIX (NEEDED BY SUBR. SOLVER)
          KPR=1
          CON = 180./3.1415926
     100  READ (1,110) M,N,FL,FM,FH
     110  FORMAT (2I3,3F10.0)
          IF (M) 120,120,130
     120  CALL EXIT
C         WRITE OUT M,N
     130  WRITE (3,140) M,N,FL,FM,FH
     140  FORMAT ('1 NEW PROBLEM'//'    NUMBER OF BRANCHES=',I3,3X,'NUMBER O
         1F NODES=',I3//'    LOWEST FREQUENCY=',1PE15.5,3X,'FREQUENCY MULTIPL
         2IER=',1PE15.5,3X,'HIGHEST FREQUENCY=',1PE15.5//'    INPUT CARDS'//)
          N2=N+1
          KPR=1
          IIB=0
          IIM=0
          DO 160 I=1,M
          RR(I)=0.
          LL(I)=0.
          CC(I)=0.
          CG(I)=(0.,0.)
          VG(I)=(0.,0.)
          DO 160 J=1,N
          A(J,I)=0.
     160  CONTINUE
     170  READ (1,180) I,J,K,IX,VAL
     180  FORMAT (4I3,2F10.0)
C         VAL IS COMPLEX FOR USE WITH INDEPENDENT GENERATORS, ALSO FOR
C         COMPLEX DEPENDENT GENERATOR STRENGTHS. (REAL AND IMAGINARY
C         PARTS MUST BE ENTERED.)
          IF (I.LE.0) GO TO 451
          IF (IX.GE.6) GO TO 220
     200  WRITE (3,210) I,J,K,IX,VAL
     210  FORMAT (' BRANCH',I3,'  FROM',I3,' TO',I3,'  TYPE=',I3,'  VALUE='
         1,1P2E15.5)
          GO TO 240
     220  WRITE (3,230) J,K,IX,VAL
     230  FORMAT (13X,'FROM',I3,' TO',I3,'  TYPE=',I3,'  VALUE=',1P2E15.5)
     240  IF (IX) 260,260,250
     250  GO TO (270,280,300,320,330,351,341,355,357,359),IX
     260  RR(I) = REAL(VAL)
          GO TO 360
     270  RR(I)=1./REAL(VAL)
          GO TO 360
     280  CC(I)=REAL(VAL)
          GO TO 360
```

Listing 3.3 AC Analysis Program

```
  300 LL(I)=1./REAL(VAL)
      GO TO 360
  320 VG(I) = VAL
      GO TO 170
  330 CG(I) = VAL
      GO TO 170
  341 K = -K
C         VOLTAGE CONTROLLED CURRENT SOURCE.
  351 IIB = IIB+1
C         CURRENT CONTROLLED CURRENT SOURCE.
      IB(IIB,2) = -K
      IB(IIB,1) = J
      B(IIB) = VAL
      GO TO 170
  355 WRITE (3,356)
  356 FORMAT ('  ** CURRENT CONTROLLED VOLTAGE SOURCE IS NOT IMPLEMENTED
     1. CARD IGNORED.')
      GO TO 170
  357 WRITE (3,358)
  358 FORMAT ('  ** VOLTAGE CONTROLLED VOLTAGE SOURCE IS NOT IMPLEMENTED
     1. CARD IGNORED.')
      GO TO 170
  359 IIM = IIM+1
      IM(IIM,1) = J
      IM(IIM,2) = K
      MM(IIM)=REAL(VAL)
      GO TO 170
  360 IF (J) 380,380,370
  370 A(J,I) = 1.
  380 IF (K) 170,170,390
  390 A(K,I) =-1.
      GO TO 170
C         M=NUMBER OF BRANCHES
C         N= NUMBER OF INDEPENDENT NODES.
C         READY FOR SETTING UP CIRCUIT MATRICES. A,YB,CG,VG ARE DONE.
  451 WRITE (3,450)
  450 FORMAT (//'   THE INCIDENCE MATRIX (A-MATRIX) IS'/)
      DO 460 I=1,N
  460 WRITE (3,470) (A(I,J),J=1,M)
  470 FORMAT (1X, 20F4.0/(2X,20F4.0/))
      WRITE (3,510)
  510 FORMAT (//'   THE VECTOR OF INDEPENDENT CURRENTS (CG-VECTOR) IS'/)
      WRITE (3,500)  (CG(I),I=1,M)
      WRITE (3,520)
  520 FORMAT (//'   THE VECTOR OF INDEPENDENT VOLTAGES (VG-VECTOR) IS'/)
      WRITE (3,500) (VG(I),I=1,M)
      F=FL
  701 W=6.2831852*F
      WRITE (3,711) F
  711 FORMAT ('1'//'    FREQUENCY =',1PE15.5//)
C         SET UP YB-MATRIX FOR PRESENT FREQUENCY.
      DO 731 I=1,M
      DO 721 J=1,M
  721 YB(I,J)=(0.,0.)
  731 YB(I,I)=CMPLX(RR(I),W*CC(I)-LL(I)/W)
      IF (IIM) 761,761,741
  741 DO 751 I=1,IIM
      I2=IM(I,1)
      I3=IM(I,2)
      T1=YB(I3,I3)
      T2=YB(I2,I2)
      T3 = 1./(LL(I2)*LL(I3))- MM(I)**2
      YB(I3,I3) = T1/(T3*LL(I2)*LL(I3))
      YB(I2,I2) = T2/(T3*LL(I2)*LL(I3))
      YB(I3,I2) = -T2*MM(I)/(LL(I2)*T3)
```

Listing 3.3—*Cont.*

```
      751 YB(I2,I3) = -T1*MM(I)/(LL(I3)*T3)
      761 IF (IIB) 400,400,771
      771 DO 801 I=1,IIB
          I2=IB(I,1)
C             I2 CONTAINS THE CONTROLLING BRANCH NUMBER
          I3=IB(I,2)
C             I3 CONTAINS THE CONTROLLED BRANCH NUMBER. IT IS NEGATIVE
C             FOR CURRENT CONTROL, POSITIVE FOR VOLTAGE CONTROL.
          IF (I3) 781,791,791
      781 I3 = -I3
          YB(I3,I2)=YB(I3,I2) +YB(I2,I2)*B(I)
          GO TO 801
      791 YB(I3,I2) = YB(I3,I2) + B(I)
      801 CONTINUE
      400 DO 410 I=1,M
          T(I)= CG(I)
          DO 410 J=1,M
      410 T(I)= T(I) - YB(I,J)*VG(J)
C             FORM    CG - YB*VG
          DO 420 I=1,N
          YN(I,N2) = 0.
          DO 420 J=1,M
      420 YN(I,N2) = YN(I,N2)+A(I,J)*T(J)
C             CALCULATE COEFFICIENTS OF YN.
          DO 440 K=1,N
          DO 430  I=1,M
          T(I)=0.
          DO 430  J=1,M
      430 T(I) = T(I) + YB(I,J)* A(K,J)
          DO 440 I=1,N
          YN(I,K)=0.
          DO 440 J=1,M
      440 YN(I,K)=YN(I,K) + A(I,J)*T(J)
C             THE NODE ADMITTANCE MATRIX IS COMPLETE.
          GO TO(478,542),KPR
      478 KPR=2
          WRITE (3,480)
      480 FORMAT (//' THE BRANCH-ADMITTANCE MATRIX (YB-MATRIX) IS'/)
          DO 490  I=1,M
      490 WRITE (3,500)   (YB(I,J),J=1,M)
      500 FORMAT (/1X,4(1P2E15.5,2X)/(2X,4(1P2E15.5,2X)))
C             THIS FORMAT IS SET UP FOR COMPLEX MATRIX ELEMENTS.
          WRITE (3,530)
      530 FORMAT(//' THE AUGMENTED NODE-ADMITTANCE MATRIX (YN-MATRIX) IS'/)
          DO 540 I=1,N
      540 WRITE (3,500) (YN(I,J),J=1,N2)
C             ALL MATRICES ARE PRINTED NOW.
C             THE AUGMENTED MATRIX (YN,A*(CG-YB*VG) IS COMPLETE.
      542 CONTINUE
          CALL SOLVX  (YN,NNK,N,DET)
C   SUBR. SOLVER IS MODIFIED FOR COMPLEX ARRAY IN YN.
C             NODE VOLTAGES ARE NOW IN COL. N+1 OF YN
C             PRINT  NODE   VOLTAGES.
C             CALCULATE BRANCH VOLTAGES,CURRENTS,POWERS.
C             BV = (A-TRANSPOSE)*VN
          DO 550 I=1,M
          T(I)=0.
          DO 550 J=1,N


      550 T(I)=T(I)+A(J,I)*YN(J,N2)
C             T-VECTOR CONTAINS THE BRANCH VOLTAGES.
          DO 560 I=1,M
          BC(I)= -CG(I)
          DO 560 J=1,M
      560 BC(I)=BC(I)   +YB(I,J)*(VG(J)+T(J))
```

Listing 3.3—*Cont.*

```
C           VECTOR BC CONTAINS THE BRANCH CURRENTS.
      DO 570 I=1,M
C           SAVE ELEMENT ADMITTANCES IN COL. 1 OF YB
  570 YB(I,1)=YB(I,I)
      DO 580 I=1,M
      YB(I,2)=VG(I)+T(I)
C           ELEMENT VOLTAGES ARE NOW IN COL. 2 OF YB. (NOT VALID IF
C             DEPENDENT VOLTAGES ARE ALLOWED.)
C           PUT ELEMENT CURRENTS INTO COL. 3
      YB(I,3)=YB(I,2)*YB(I,1)
C           PUT ELEMENT POWERS INTO COL. 4
      YB(I,4)=YB(I,2)*CONJG(YB(I,3))
C           PUT BRANCH POWERS INTO COL. 5
  580 YB(I,5)=T(I)*CONJG(BC(I))
C           THE YB-MATRIX IS DESTROYED.
      WRITE (3,590)
  590 FORMAT        (10X,'NODE',18X,3('BRANCH',9X),3X,2('ELEMENT',8X),
     1'ELEMENT'/ 9X,'VOLTS',5X,'NUMBER',5X,2(3X,'VOLTS',8X,'CURRENT',
     210X,'POWER',11X)/)
      JJ=1
      DO 640 I=1,M
C           CONSTRUCT OUTPUT LINE IN ARRAY PC.
      IF (I-N) 592,592,594
  592 PC(1)=YN(I,N2)
  594 PC(2)= T(I)
      PC(3)= BC(I)
      PC(4)= YB(I,5)
      PC(5)= YB(I,2)
      PC(6)= YB(I,3)
      PC(7)= YB(I,4)
C           CONVERT TO MAGNITUDES AND ANGLES.
      IF (I-N) 596,596,595
  595 JJ=3
  596 DO 598 J=JJ,13,2
      T3=P(J)
      P(J)= SQRT (T3**2 + P(J+1)**2)
  598 P(J+1) = CON*ATAN2 (P(J+1),T3)
C           CONVERSION COMPLETED.
      IF (I-N) 601,601,621
  601 WRITE (3,611) P(1),I,(P(J),J=3,13,2)
  611 FORMAT (/' MAG',1PE15.5,I5,1P3E15.5,4X,1P3E15.5)
      WRITE (3,612) (P(J),J=2,14,2)
  612 FORMAT(' PHA',1PE15.5,5X,1P3E15.5,4X,1P3E15.5)
      GO TO 640
  621 WRITE (3,631) I,(P(J),J=3,13,2)
  631 FORMAT (/' MAG',I20,1P3E15.5,4X,1P3E15.5)
      WRITE (3,632) (P(J),J=4,14,2)
  632 FORMAT(' PHA',20X,1P3E15.5,4X,1P3E15.5)
  640 CONTINUE
      IF (F-FH) 811,100,100
  811 F=F*FM
      GO TO 701
      END


      SUBROUTINE SOLVX (A,NA,N,DET)
      IMPLICIT COMPLEX (A-H,O-Z)
      REAL CABS,AMAX
C     . . . .
C     . . . .
      AMAX=CABS(A(I,I))
C     . . . .
      IF (AMAX-CABS(A(I,J))) 15,20,20
   15 AMAX=CABS(A(I,J))
      . . . .
C           REST OF SOLVER UNCHANGED
```

Listing 3.3—*Cont*.

The passive element values are stored in one of three vectors (RR, LL, CC) in a location whose subscript is the branch number where the element is located. As a consequence of the way the entries in RR, LL, and CC are stored (statements 260 to 300 as well as statements 360 and following), the last value read for a given branch element replaces any previous one input while every interconnection datum is entered in the **A** matrix. The **A** matrix entries may replace previously input data or may create new ones; hence great care must be exercised in ensuring that all input data are consistent.

The controlled sources and the mutual inductances are assembled in two "open-ended" arrays. Each time a new dependent source (either current controlled or voltage controlled) is encountered, an additional row is added to the arrays IB and B. The array B will contain the value of the source; IB will contain the from-branch, to-branch pair of numbers. Since the branch numbers are positive integers, setting the to-branch number negative indicates the presence of a current controlled current source. The counter IIB is used to keep count of the number of current sources encountered in the input data. Theoretically, it is possible to have an arbitrary number of dependent current sources in a branch; the program will handle as many as the DIMENSION statement allows.

Mutual inductances are entered, as they are encountered, in the arrays MM (declared REAL) and IM quite similarly to the data in the arrays IB and B discussed in the preceding paragraph. The counter IIM is used to keep a count of the number of mutual inductances. Here again the number of mutual inductances is limited only by the DIMENSION statement for IM and MM.

As in the DC analysis program the end of the input sequence is indicated by a zero value for the first integer (the branch number); this is most easily done by using a blank card. The program then transfers to statement number 451, prints the **A**, $\mathbf{V_g}$, and $\mathbf{I_g}$ arrays, and sets the frequency for analysis to FL.

Printing of the frequency for analysis is the first step in a sequence of operations that is repeated for every frequency. The first block of calculations is for setting up the $\mathbf{Y}_b$ matrix. This has three steps:

(1) The entries caused by the passive elements (main-diagonal entries) are constructed with statement 731. As discussed above, a parallel combination of conductance, inverse inductance, and capacitance is allowed for the "passive element" of every branch.

(2) The mutual inductances modify the main diagonal entries and create a symmetric pair of additional entries in the $\mathbf{Y}_b$ matrix. The code implementing Eq. (3.100) is in statements 741 through 751. Note that this part of the code is entered only if IMM is greater than zero, i.e., if at least one mutual inductance is used in the circuit. In addition, the primary and secondary side of the mutual inductance must have inductance values in the LL array.

(3) The dependent current sources are examined one by one. Each will produce

an entry in the Y_b matrix according to Eq. (3.85) or (3.86), depending on the sign of the first entry in the appropriate row of the IB matrix (statements 771 through 801).

The above three steps are followed for setting up the augmented matrix $[Y_n | I_s]$, completely duplicating the statements in the DC analysis program (statements 400 through 440). By declaring YB, YN, and T as COMPLEX arrays the entire set of statements for this phase of the calculations can be lifted from the DC program and will be carried out with the proper complex quantities during execution of the program.

Upon setting up the augmented matrix YN, the entries in YB and YN are printed, if this is the first frequency for the analysis.

The remainder of the program differs from the DC program in only two respects. All output quantities (such as node voltage, branch current, etc.) are complex and therefore have two numbers associated with them. In this program the output quantities are expressed as magnitudes and phase angle just prior to printing (see statements following number 596). The conversion is accomplished by using the EQUIVALENCEd arrays PC and P; note that PC is COMPLEX and P is REAL with twice as many entries as PC. The EQUIVALENCE statement forces the two arrays to occupy the same storage locations and makes the real part of PC(K) be the location P(2K−1) and the imaginary part is in location P(2K).

The indexing of the DO-loop in statement 596 is designed to take advantage of this; the magnitude is put into P(J) and the angle (in degrees) into P(J+1).

After a two-line printout (for magnitudes and angles) has been made for each node and branch, a check is made if the frequency should be increased by FM and analysis should be repeated. If the value of FH has been reached, the program will try to read a new set of input cards.

Note also that SUBROUTINE SOLVER must be modified to accept complex input coefficients and that the pivot searching in the subroutine (see Sec. 2.4) must be done on the magnitude of a complex number. These changes are indicated in Listing 3.3.

Here again no attempt is made to economize the program; the sole purpose is to show a practical working code. The problems indicate several ways of streamlining and extending this code; later sections will indicate major improvements in this code.

EXAMPLE 3.8

To illustrate the use of the program just developed, let it be required to perform a frequency analysis of the network shown in Fig. 3.18.

Solution.

The input data for the above circuit is listed in Listing 3.4; part of the output for $f = 100$ Hz is reproduced also. The computer-produced output sheets were cut and re-arranged to fit the page.

94 DC and AC Analysis of Networks   Ch. 3

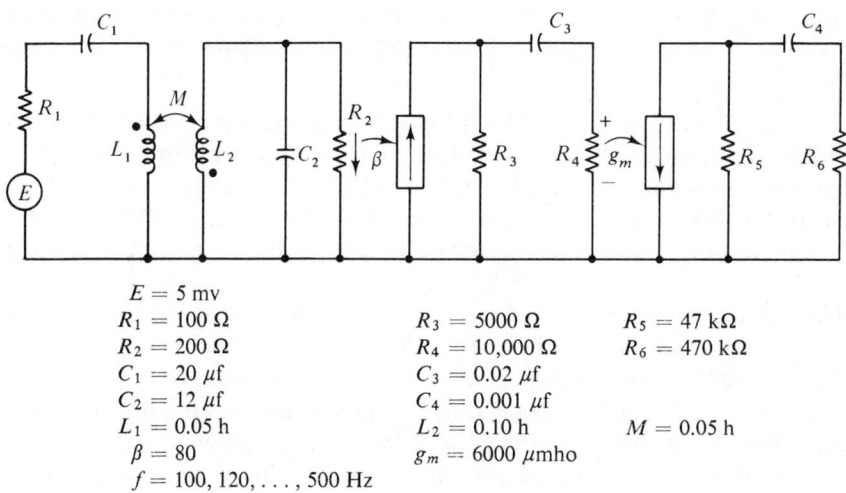

$E = 5$ mv
$R_1 = 100 \, \Omega$
$R_2 = 200 \, \Omega$
$C_1 = 20 \, \mu f$
$C_2 = 12 \, \mu f$
$L_1 = 0.05$ h
$\beta = 80$
$f = 100, 120, \ldots, 500$ Hz

$R_3 = 5000 \, \Omega$
$R_4 = 10{,}000 \, \Omega$
$C_3 = 0.02 \, \mu f$
$C_4 = 0.001 \, \mu f$
$L_2 = 0.10$ h
$g_m = 6000 \, \mu$mho

$R_5 = 47 \, k\Omega$
$R_6 = 470 \, k\Omega$

$M = 0.05$ h

Fig. 3.18 Circuit for Example 3.8

```
NEW PROBLEM

NUMBER OF BRANCHES= 12    NUMBER OF NODES=  7
LOWEST FREQUENCY= 1.00000E 02   FREQUENCY MULTIPLIER= 1.20000E 00
                HIGHEST FREQUENCY=    5.00000E 02

INPUT CARDS ARE

BRANCH  1   FROM  0  TO  1   TYPE=  1   VALUE=  1.00000E 02   0.0
BRANCH  1   FROM  0  TO  1   TYPE=  4   VALUE=  1.00000E-03   0.0
BRANCH  2   FROM  1  TO  2   TYPE=  2   VALUE=  5.00000E-05   0.0
BRANCH  3   FROM  2  TO  0   TYPE=  3   VALUE=  5.00000E-02   0.0
BRANCH  4   FROM  0  TO  3   TYPE=  3   VALUE=  1.00000E-01   0.0
BRANCH  5   FROM  3  TO  0   TYPE=  2   VALUE=  1.20000E-05   0.0
BRANCH  6   FROM  3  TO  0   TYPE=  1   VALUE=  2.00000E 02   0.0
BRANCH  7   FROM  0  TO  4   TYPE=  1   VALUE=  5.00000E 03   0.0
BRANCH  8   FROM  4  TO  5   TYPE=  2   VALUE=  2.00000E-08   0.0
BRANCH  9   FROM  5  TO  0   TYPE=  1   VALUE=  1.00000E 04   0.0
BRANCH 10   FROM  0  TO  6   TYPE=  1   VALUE=  4.70000E 04   0.0
BRANCH 11   FROM  6  TO  7   TYPE=  2   VALUE=  1.00000E-09   0.0
BRANCH 12   FROM  7  TO  0   TYPE=  1   VALUE=  4.70000E 05   0.0
            FROM  3  TO  4   TYPE= 10   VALUE=  5.00000E-02   0.0
            FROM  6  TO  7   TYPE=  6   VALUE=  8.00000E 01   0.0
            FROM  9  TO 10   TYPE=  7   VALUE=  6.00000E-03   0.0
FREQUENCY=   1.00000E 02

        NODE                BRANCH        BRANCH         BRANCH
        VOLTS     NUMBER    VOLTS         CURRENT        POWER

   MAG  1.81858E-03   1    1.81858E-03   4.14478E-05   7.53758E-08
   PHA -5.22723E 01        1.27728E 02   2.03058E 01   1.07422E 02

   MAG  1.65528E-03   2    3.29831E-03   4.14478E-05   1.36708E-07
   PHA  9.11013E 01       -6.96942E 01   2.03058E 01  -9.00000E 02

   MAG  2.12452E-03   3    1.65528E-03   4.14477E-05   1.36708E-07
   PHA -1.00530E 02        9.11013E 01   2.03058E 01   7.17955E 01
```

Listing 3.4 Results for Example 3.8

MAG	4.20835E 00	4	2.12452E-03	1.92206E-05	4.08346E-08
PHA	-1.04042E 02		7.94701E 01	-4.40800E 01	1.23550E 02
MAG	5.24710E-01	5	2.12452E-03	1.60185E-05	3.40317E-08
PHA	-2.12046E 01		-1.00530E 02	-1.05298E 01	-9.00001E 01
MAG	1.46738E 02	6	2.12452E-03	1.06226E-05	2.25679E-08
PHA	-2.27481E 01		-1.00530E 02	-1.00530E 02	0.0
MAG	4.15588E 01	7	4.20835E 00	5.24710E-05	2.20816E-04
PHA	5.07995E 01		7.59578E 01	-2.12043E 01	9.71621E 02
MAG		8	4.17551E 00	5.24710E-05	2.19093E-04
PHA			-1.11205E 02	-2.12046E 01	-9.00002E 02
MAG		9	5.24710E-01	5.24710E-05	2.75321E-05
PHA			-2.12046E 01	-2.12046E 01	0.0
MAG		10	1.46738E 02	8.84227E-05	1.29749E-02
PHA			1.57252E 02	5.08000E 01	1.06452E 02
MAG		11	1.40730E 02	8.84229E-05	1.24437E-02
PHA			-3.92005E 01	5.07995E 01	-9.00000E 02
MAG		12	4.15588E 01	8.84229E-05	3.67475E-03
PHA			5.07995E 01	5.07995E 01	0.0

ELEMENT VOLTS	ELEMENT CURRENT	ELEMENT POWER
4.14478E-03	4.14478E-05	1.71792E-07
2.03058E 01	2.03058E 01	0.0
3.29831E-03	4.14478E-05	1.36708E-07
-6.96942E 01	2.03058E 01	-9.00000E 02
1.65528E-03	1.05378E-04	1.74430E-07
9.11013E 01	1.10120E 00	9.00001E 02
2.12452E-03	6.76257E-05	1.43672E-07
7.94701E 01	-1.05298E 01	8.99999E 02
2.12452E-03	1.60185E-05	3.40317E-08
-1.00530E 02	-1.05298E 01	-9.00001E 02
2.12452E-03	1.06226E-05	2.25679E-08
-1.00530E 02	-1.00530E 02	0.0
4.20835E 00	8.41670E-04	3.54204E-03
7.59578E 01	7.59578E 01	0.0
4.17551E 00	5.24710E-05	2.19093E-04
-1.11205E 02	-2.12046E 01	-9.00002E 02
5.24710E-01	5.24710E-05	2.75321E-05
-2.12046E 01	-2.12046E 01	0.0
1.46738E 02	3.12208E-03	4.58126E-01
1.57252E 02	1.57252E 02	0.0
1.40730E 02	8.84229E-05	1.24437E-02
-3.92005E 01	5.07995E 01	-9.00000E 02
4.15588E 01	8.84229E-05	3.67475E-03
5.07995E 01	5.07995E 01	0.0

Listing 3.4—*Cont.*

3.6 Mutual Inductance Model

The program given in the previous section applies the mutual inductance equations, Eqs. (3.95) through (3.100), directly. The computation of the effects of mutual inductance can also be done quite differently.

Conceptually, a set of two coupled coils is equivalent to two coupled generators: One couples unilaterally from the primary side to the secondary, and the other couples from the secondary side to the primary. This section derives the equations for such a model, thus reducing the calculation of networks having coupled coils to those being free of mutual inductance. The equations relating to coupled coils are repeated:

$$I_p = \frac{E_p}{j\omega L_p} \cdot \frac{L_p L_s}{L_p L_s - M^2} - \frac{E_s}{j\omega L_s} \cdot \frac{M L_s}{L_p L_s - M^2} \tag{3.97}$$

$$I_s = \frac{E_s}{j\omega L_s} \cdot \frac{L_p L_s}{L_p L_s - M^2} - \frac{E_p}{j\omega L_p} \cdot \frac{M L_p}{L_p L_s - M^2} \tag{3.98}$$

If the coefficient of coupling $k = M/\sqrt{L_p L_s}$ is used, the above may be rewritten as

$$I_p = \left[\frac{1}{j\omega L_p(1 - k^2)}\right] E_p - \frac{M}{L_p} \cdot \left[\frac{1}{j\omega L_s(1 - k^2)}\right] E_s \tag{3.101}$$

$$I_s = \left[\frac{1}{j\omega L_s(1 - k^2)}\right] E_s - \frac{M}{L_s} \cdot \left[\frac{1}{j\omega L_p(1 - k^2)}\right] E_p \tag{3.102}$$

These equations are mechanized by the circuit given in Fig. 3.19. Note that the terms bracketed in Eq. (3.102) are the admittances of a primary-side inductor of value $L_1 = L_p(1 - k^2)$ and of a secondary-side inductor of value $L_2 = L_s(1 - k^2)$.

The advantage of this formulation over the method used in Sec. 3.4 is that this model uses frequency independent terms and eliminates some coding (statements relating to entries in Y_b caused by mutual inductance) from an AC analysis program. Eq. (3.100) must be re-applied at every new frequency of analysis because the terms YB(P, P), YB(S, S), etc. are frequency sensitive. Thus one could streamline the AC analysis program by changing the primary and secondary inductances from L_p

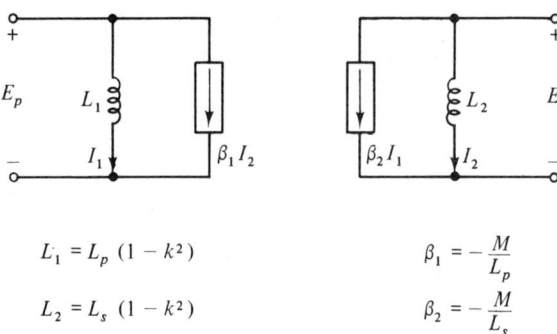

$L_1 = L_p (1 - k^2)$ $\quad\quad\quad\quad \beta_1 = -\dfrac{M}{L_p}$

$L_2 = L_s (1 - k^2)$ $\quad\quad\quad\quad \beta_2 = -\dfrac{M}{L_s}$

Fig. 3.19 Mutual Inductance Model

and L_s to $L_p(1 - k^2)$ and $L_s(1 - k^2)$ respectively, and adding two new dependent sources $\beta_1 = -M/L_p$ and $\beta_2 = -M/L_s$. The source β_1 is controlled from the secondary branch and injects current into the primary branch; the roles of the primary and secondary branch are reversed for β_2. These additions and changes need to be made just once for each circuit.

The remainder of the analysis program is unchanged. The voltages will be correctly calculated and the branch currents will also be correct; the element currents, however, will be the currents through the transformed inductances and will not be equal to the original inductor currents. Appropriate changes must be made in the calculation of the element currents to obtain the true values.

Details of this coding are left as an exercise (see Problem 3.15).

3.7 Direct Calculation of the Node Admittance Matrix

The programs presented in the previous sections are good, viable analysis programs which can be used, and have been used, for design-analysis of circuits, but they are not efficient in storage requirements. In calculating the node-admittance matrix Y_n and the equivalent current vector I_s, the branch-admittance matrix Y_b is first set up. In a normal large network there are many more branches than independent nodes; hence the maximum storage required is dictated by the number of branches (M) even though the set of simultaneous equations to be solved is only as large as there are independent nodes (N) in the network. Since in a typical large network there might be two to three times as many branches as nodes, considerable savings in memory space required can be affected by dispensing with the Y_b matrix entirely.

Another large waste in memory is the incidence matrix (A matrix). Here in every one of the M columns there is merely a $+1$ or a -1 at most. For the ith branch starting at node j and connected to node k, the entries are all zero except that $a_{ji} = +1$ and $a_{ki} = -1$ provided j and k are not zero. For j or k equal to zero there is no entry in the A matrix. The A matrix may be collapsed into a two-column incidence list IA, each column containing M entries. Hence the above incidence matrix entries are replaced by the FORTRAN statements

$$IA(I,1) = J$$
$$IA(I,2) = K \tag{3.103}$$

Note that if J or K is equal to zero, the value zero is entered in the appropriate position in IA.

This section gives the algorithms, written as FORTRAN statements, for establishing the augmented node-admittance matrix using the incidence list IA.

The input data will be recorded in the following lists:

(1) The branch-node incidence list IA (*M* rows and two columns),

(2) The branch-admittance list G (*M* entries long).

The above two lists are established as follows: For branch I containing a resistance R connected from node J to node K, the entries are as in Eq. (3.103) and an entry is made in the ith position of the branch conductance vector G

$$G(I)=1./VAL \qquad (3.104)$$

where VAL is the resistance of the ith branch.

Thus lists G and IA contain the passive element values and interconnection data. The independent voltages and currents are entered as before in the lists CG and VG:

(3) The value VAL of an independent current in branch I gives the entry

$$CG(I)=VAL \qquad (3.105)$$

(4) The value VAL of an independent voltage in branch I gives the entry

$$VG(I)=VAL \qquad (3.106)$$

The transfer quantities β and g_m will be entered in the vector B and the branch-to-branch incidence list IB similar to the IA list for the passive elements. The lengths of the lists B and IB are identical; each time a new transfer quantity is added to the circuit, new entries are added to B and IB. Each time a new transfer quantity is encountered, the counter IIB is stepped by one; thus after the last transfer quantity is entered, the variable IIB will contain the count of the transfer quantities.

(5) If the transfer quantity is a current controlled current source of strength BETA, then the entry in B is

$$B(IIB)=BETA*G(J) \qquad (3.107)$$

(6) If the transfer quantity is a voltage controlled current source of strength GM, then the entry in B is

$$B(IIB)=GM \qquad (3.108)$$

Note that the entries in B are values of mutual conductance.

(7) The entries in IB are

$$\begin{aligned}IB(IIB,1)&=K\\ IB(IIB,2)&=J\end{aligned} \qquad (3.109)$$

where J is the controlling branch and K is the controlled branch.

With the above lists the augmented node-admittance matrix $[\mathbf{Y}_n \mid \mathbf{I}_s]$ is established by using the following steps:

Let the fundamental node-voltage relation

$$\mathbf{Y}_n\mathbf{V}_n = A\mathbf{Y}_b A^T \mathbf{V}_n = A(\mathbf{I}_g - \mathbf{Y}_b \mathbf{V}_g) \qquad (3.26)$$

be rewritten as

$$A(\mathbf{Y}_c + \mathbf{Y}_t)A^T \mathbf{V}_n = A(\mathbf{I}_g - \mathbf{Y}_c \mathbf{V}_g) - A\mathbf{Y}_t \mathbf{V}_g \qquad (3.110)$$

where $\mathbf{Y}_c$ contains the self-conductances and $\mathbf{Y}_t$ contains the mutual conductances. The matrix $\mathbf{Y}_c$ is a diagonal matrix; for branch I the entry is G(I) in position (I, I) of $\mathbf{Y}_c$. Since neither the matrices $\mathbf{Y}_c$, $\mathbf{Y}_t$ nor A are actually written out, one is concerned only with how the various entries in them will show up in the augmented node-admittance matrix. The matrix $\mathbf{Y}_n$ will be represented by the FORTRAN dimensioned variable YN; for compactness let N2=N+1.

From Eq. (3.110) the augmented node-admittance matrix $[\mathbf{Y}_n \mid \mathbf{I}_s]$ is rewritten as follows:

$$\mathbf{Y}_n = \mathbf{Y}_{n1} + \mathbf{Y}_{n2} \qquad (3.111)$$

$$\mathbf{I}_s = \mathbf{I}_{s1} + \mathbf{I}_{s2} \qquad (3.112)$$

$$\mathbf{Y}_{n1} = \mathbf{A}\mathbf{Y}_c\mathbf{A}^T$$
$$\mathbf{Y}_{n2} = \mathbf{A}\mathbf{Y}_t\mathbf{A}^T \qquad (3.113)$$

$$\mathbf{I}_{s1} = \mathbf{A}(\mathbf{I}_g - \mathbf{Y}_c\mathbf{V}_g)$$
$$\mathbf{I}_{s2} = -\mathbf{A}\mathbf{Y}_t\mathbf{V}_g \qquad (3.114)$$

The above four parts of the augmented node-admittance matrix are calculated separately and superposed by the following algorithms:

(1) The entries in $\mathbf{Y}_{n1}$ and the entries in $\mathbf{I}_{s1}$: For branch l extending from node J to node K, the following entries are produced:

```
         J=IA(I,1)
         K=IA(I,2)
         IF (J.EQ.0) GO TO 11
         YN(J,J)=YN(J,J)+G(I)
         YN(J,N2)=YN(J,N2)+CG(I)-G(I)*VG(I)
   11    IF (K.EQ.0) GO TO 22
         YN(K,K)=YN(K,K)+G(I)
         YN(K,N2)=YN(K,N2)-CG(I)+G(I)*VG(I)
   22    IF (J*K.EQ.0) GO TO 33
         YN(J,K)=YN(J,K)-G(I)
         YN(K,J)=YN(K,J)-G(I)
   33    CONTINUE
```
(3.115)

(2) The above is repeated, stepping l from 1 to M for the various branches of the network.

(3) The entries in $\mathbf{Y}_{n2}$ and $\mathbf{I}_{s2}$: For the lth dependent source, whose transconductance is stored in B(I), located in branch K = IB(I, 2), and whose controlling branch is branch J = IB(I, 1), the interconnection data for branches J and K must be examined. The following algorithm establishes the entries in YN caused by such a source:

```
         J=IB(I,1)
         K=IB(I,2)
         JF=IA(J,1)
         JT=IA(J,2)
         KF=IA(K,1)
         KT=IA(K,2)
         YN(KF,JF)=YN(KF,JF)+B(I)
         YN(KF,JT)=YN(KF,JT)-B(I)
         YN(KT,JF)=YN(KT,JF)-B(I)
         YN(KT,JT)=YN(KT,JT)+B(I)
         YN(KF,N2)=YN(KF,N2)-B(I)*VG(J)
         YN(KT,N2)=YN(KT,N2)+B(I)*VG(J)
```
(3.116)

If any of the above entries were to have a zero subscript, that entry in the augmented node-admittance matrix would be skipped, as was illustrated in Eq. (3.115).

(4) The above step is repeated stepping I from 1 to IIB for the various dependent sources in the network.

The four steps outlined above will establish the augmented node-admittance matrix which then has to be solved for the node-voltage vector V_n. Upon execution, the simultaneous linear equation routine SOLVER (also used in Sec. 3.3) will replace the entries in column $(N + 1)$ of $[Y_n \vert I_s]$ with the vector V_n.

Next, the branch voltages VB must be established. Using the IA matrix, Eq. 3.7 is implemented with the statements

$$\begin{aligned}&\text{DO 220 I=1,M}\\&\quad\text{J=IA(I,1)}\\&\quad\text{K=IA(I,2)}\\&\text{220 VB(I)=YN(J,N2)-YN(K,N2)}\end{aligned} \qquad (3.117)$$

Here again modifications must be made for J or K equal to zero.

The branch element current for branch I is then obtained from

$$\text{EC(I)=G(I)*(VG(I)+VB(I))} \qquad (3.118)$$

The branch current then is calculated by subtracting from the element current the independent current and adding the dependent currents as calculated by using the IB and B arrays. Details for writing this code are left as an exercise. Test cases for this program may be run through the code presented in Sec. 3.3 for checking.

Note that the program presented in this section differs only in detail from the code of Sec. 3.3. However, savings in storage required are effected by this scheme. Basically, rather than the Y_b and Y_n matrices shown in Sec. 3.3, the procedures discussed here are employed for almost all circuit analysis programs used widely. A set of suggested exercises at the end of this chapter goes through steps that trace an orderly transition from the relatively simple analysis program of Sec. 3.3 to a much more efficient and accurate program whose basic algorithms were presented in this section.

An AC analysis version of the above program may be written with relatively minor modifications. The principal changes necessary, aside from using complex arithmetic, may be found by comparing the flowcharts, Figs. 3.15 and Fig. 3.17; writing of this version of AC analysis program is also left as an exercise (see Problem 3.18).

At this point the similarities between the code in Eqs. (3.115) and (3.116) should be noted. Advantage can be taken of this in compacting the required code somewhat more (see Problem 3.18). Such refinements and other shortcuts, which are yet to be discussed in this chapter, will be used in the following chapters to derive a more compact and more flexible set of circuit analysis programs.

3.8 Sparse Matrix Analysis

The program presented in the previous section has marked advantages over the earlier discussed program even though the structures of the two DC programs differ but very slightly. The advantages are in using less memory space (no Y_b and A matrices) and in not having to go through actual matrix multiplications for such products as AY_bA^T. In generating the Y_n matrix, extensive use was made of the sparseness of A (only two non-zero entries in each column) and simple integer values in it by using only few additions or subtractions in place of the many multiplications used in Sec. 3.3. Hence, not only is the program in Sec. 3.5 more economical of storage, it also is considerably faster in execution. This latter advantage is particularly striking in the AC analysis version alluded to at the end of the section: The additional overhead in setting up the B and IB arrays is more than offset by the speedier establishment of $[Y_n \mid I_s]$ because the zero elements in A never need to be retrieved from memory.

In a typical amplifier circuit there are about half as many nodes as branches; hence the storage savings using a Y_n matrix instead of a Y_b matrix is 75%.

The solution of the resultant $[Y_n \mid I_s]$ matrix is usually accomplished by some variation of the Gauss elimination technique as outlined in Chapter 2. This procedure is the fastest known direct solution technique for the general case of simultaneous linear equations; hence its great popularity.

For a very large circuit (of order 200 nodes), the Y_n matrix will typically contain no more than about 10% non-zero elements in any row. Hence about 90% of the number of elements will be zero; it is these "superfluous" elements that require the bulk of the storage. Their primary function is to space the non-zero elements so that those elements are in proper position for the matrix multiplication represented by $Y_nV_n = I_s$.

Another reason the zero elements are needed is that during the process of equation solution by the Gauss elimination process many new non-zero terms may be generated in positions originally occupied by zeros. Hence for the general case (random elements being non-zero), the memory locations of the zero elements will be used during the process of equation solution to store intermediate results. Therefore the N by N storage area for the Y_n matrix is required not so much because of the matrix itself but rather because of the method of solution used.

In order to have to store only the non-zero entries in the node-admittance matrix and then to use no further appreciable amounts of computer storage space, another method of solution of the augmented node-admittance matrix must be used. Considerable work has been done in this area; this section gives details of a circuit analysis program that takes full advantage of the sparseness of the Y_n matrix and, in addition, requires less time for execution in some important cases.

To circumvent the need for storing new non-zero intermediate results, the Gauss-Seidel iteration technique originally discussed in Chapter 2 is employed. It is suggested that at this point the material of Sec. 2.5 be reviewed.

The important points about the Gauss-Seidel procedure are:

(1) No new terms are generated in a sparse coefficient matrix during the course of the iteration procedure; only the original terms are used.

(2) The number of iterations required for a given accuracy depends on the closeness of the starting approximation.

(3) For a close enough starting solution, the process converges provided the determinant of the coefficients is not zero. Because of inevitable round-off errors, in practice the determinant may not be very close to zero, a condition usually satisfied by all stable networks.

Let the input data describing the network be stored as indicated in the previous section in the arrays IA, G, VG, CG, IB, B and the constants M, N, and IIB. From this description the node admittance matrix may be established.

Instead of a two-dimensional array for the matrix Y_n, the non-zero entries will be packed into a vector Y and two pointer vectors will be established to locate a given Y_n entry in Y. Suppose that the non-zero entries for a given 5 by 5 Y_n matrix are in locations indicated by an x in Fig. 3.20(a). The non-zero elements are sequence numbered as shown in Fig. 3.20(b); it is in this order that the elements are packed in the vector Y whose length will be the number of non-zero elements in Y_n. In this example the length of Y is 11 and the content of location Y(7) is the value of $Y_n(4, 1)$.

In order to pinpoint a given non-zero element, two pointers will be used. The pointer IG will contain the column subscripts of the corresponding elements in Y. Here the contents of IG will be

$$IG = (1,4,2,5,3,4,1,2,4,3,5) \qquad (3.119)$$

Note that the length of IG is the same as the length of Y. The second pointer vector, IR, of length N, will contain the largest subscript for each row in Y_n. In this example

$$IR = (2,4,6,9,11) \qquad (3.120)$$

These two pointers are set up in such a way that non-zero elements of successive rows of the original Y_n matrix may be obtained together with the column subscript for that entry. The following code will accomplish this task:

```
      KHI=0
      DO 55 J=1,N
      KLO=KHI+1
      KHI=IR(J)
      DO 25 K=KLO,KHI
      I=IY(K)
C     Y(K) IS NOW YN(J,I)
25    CONTINUE
55    CONTINUE
```
(3.121)

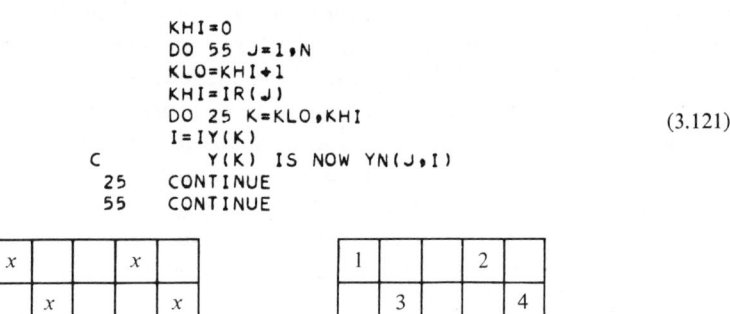

(a)   (b)

Fig. 3.20 Sparse Matrix Entries

Thus the various non-zero elements of $\mathbf{Y_n}$ are retrieved from Y in row order; each time J is incremented a new row of $\mathbf{Y_n}$ is started. Notice that for each entry in $\mathbf{Y_n}$ retrieved two memory fetches are made: one fetch for IY(K) and one to Y(K). But the fetching of a two-subscript FORTRAN variable such as YN entails the calculation of a single memory address (an integer multiplication, a subtraction, and an addition) and a single fetch. Hence the code presented here will actually retrieve the sequence of non-zero elements as fast as a single reference to an element YN(I, J) would. The exact differences depend on the actual FORTRAN compiler and computer used, but no large variations have been noted in a sampling of medium and large computers. In addition, no references are made to zero entries; hence the total time required for accessing all usable elements of a sparse matrix is only a fraction of the time required for a "standard" program. Assuming that 100 nodes and 10% fill in the various nodes, a standard Gauss elimination routine will make approximately $1/3N^3 \approx 330,000$ memory calls; the sparse matrix code above makes $0.10 \times 100 = 1000$ calls for each iteration. Thus, approximately 300 iterations could be made for the same running time for each program. Generally, up to K iterations may be done in a matrix having a P fractional average non-zero fill per row in an N by N matrix where

$$K \leq \frac{1}{3} \frac{N}{P} \qquad (3.122)$$

There remains the task of completing the particular code for the Gauss-Seidel procedure to be used here and to establish the packed Y vector in the first place. The overall strategy is as follows:

(1) The network interconnection description arrays IA and IB are scanned to establish the row and column subscripts (JR and JC) of the various entries in the $\mathbf{Y_n}$ matrix. Slight modifications of the algorithms presented in Eqs. (3.115) and (3.116) may be used to find these two subscripts which are then converted to a sequence number JS:

$$JS = JR*KLEN+JC \qquad (3.123)$$

where KLEN is the size of the row to be used (this number must be equal to or larger than N, the number of independent nodes in the network).

(?) If JR equals zero or JC equals zero, the step above is skipped.

(3) The sequence number is added to the array IG if it is not already in IG. Since the arrays IA and IB are scanned in the order of their subscript sequence, the sequence numbers IG will be in a random order.

(4) Upon completion of all entries in the IA and IB arrays, the resultant sequence numbers in IG are ordered in increasing magnitudes. Note that no duplicates will exist in IG. Let the length of the IG array be IK (IK = 11 in the example cited here).

(5) The rank-ordered entries in IG are converted to pointer-vector entries IR

and IG with the following code:

```
      DO 95 I=1,IK
      II=IG(I)/KLEN
      IR(II)=I
   95 IG(I)=IG(I)-II*KLEN
```
(3.124)

Note the extensive use of FORTRAN integer arithmetic in the above code. The array IG is made to hold both the sequence numbers in Steps 3 and 4 as well as the column pointers of this step; this is merely a memory savings trick.

(6) The entries in the Y vector are now established exactly as the YN entries

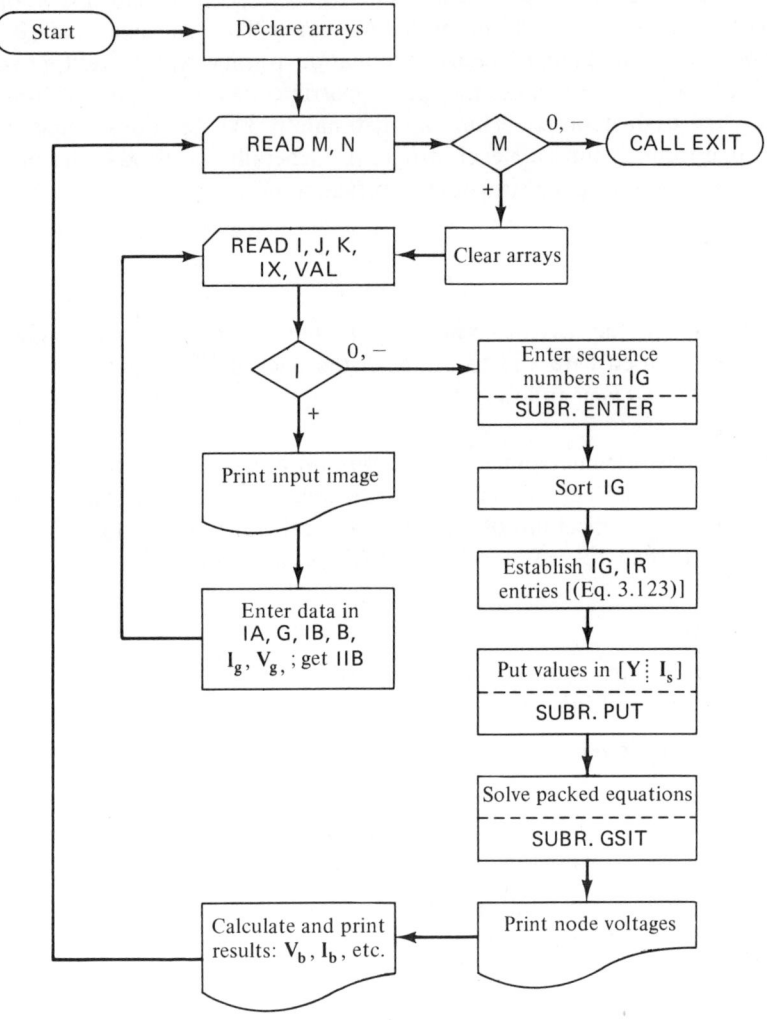

Fig. 3.21 Flowchart of Sparse Matrix Analysis Program (DC case)

were in the previous section [see Eqs. (3.115) and (3.116)]; entries in the equivalent current vector $\mathbf{I}_s$ are not put into col. N+1 of YN but are put into a separate vector CS (of length N).

A flowchart of an entire DC program based on the algorithms of this section is given in Fig. 3.21; it is left as an exercise to write the code (approximately 230 FORTRAN statements).

Although the above discussion pertained to a description of an implied DC analysis program, the advantages of this code become more apparent in an AC analysis program (or a transient analysis program as will be discussed in the next chapter). There are two major reasons for this:

(1) The calculation of the IR and IG entries needs to be done just once for each problem.

(2) At any frequency other than the first one, the previously calculated node-voltage vector can be used as a starting vector for the new node-voltage vector. Since the frequency increments are usually not very large, only relatively minor adjustments must be made for every node-voltage entry and hence only a few iterations may be required. Difficulties will be encountered in sharply tuned circuits; however, such circuits are difficult to analyze by any program.

For transient analysis (calculation of voltages and currents as a function of time), the above arguments will hold by replacing the word "frequency" with the word "time" whenever appropriate.

The ideas developed in this section do not solve all possible combinations of dependent voltage sources. If a network contains dependent voltage sources which in some way are controlled from quantities located in the same branch, then Step 6 of the algorithm will be caught in a loop without an end. The only recourse is then to set up the general equations directly (see Sec. 3.2). To detect such a condition a set of checks must be incorporated in the procedure.

3.9 Unaccompanied Energy Sources

The analysis procedures discussed up to this point require that every branch of the network contain a passive element of non-zero finite value. Often it is desirable to include pure voltage and/or current sources in a network; these sources have no accompanying series or parallel passive components. The name "unaccompanied source" will be applied to such energy sources.

Since the analysis procedures are based on Kirchhoff's current law, the inclusion of a pure current source in the analysis poses no difficulties, as demonstrated by Ex. 3.1 in Sec. 3.1. For a pure current source of strength VAL in branch K, which extends from node I to node J, the branch conductance entry is (see Sec. 3.3):

$$YB(K,K) = 0. \tag{3.125}$$

and the independent current vector entry is

$$\text{CG(K)=VAL} \qquad (3.126)$$

while the **A** matrix entries are unchanged [see Eq. (3.75)].

If the input format established in Sec. 3.3 is used, the required information for such a current source can be inputted by supplying only one card, which defines the current source strength. The interconnection data for the branch must be input separately from another card listing a conductance (Type number 0) of value zero.

The program in Sec. 3.6 (AC analysis) will also work correctly with the above data.

With unaccompanied voltage sources (i.e., ideal voltage sources), there are grave difficulties. The series resistance of such a source is zero. Application of KCL requires that every voltage source be converted to a current source; the term $\mathbf{Y}_b \mathbf{V}_g$ occurring in the expression for $\mathbf{I}_s$ is indeed no more than the generalized Norton equivalent current source for the $\mathbf{V}_g$ generators. With a term in $\mathbf{Y}_b$ being infinite (because of a zero branch resistance), the current through an unaccompanied source becomes indeterminate. A way must be found to calculate the current through such sources.

In general, let a network contain q such independent voltage sources. Let the pth such source have the strength VAL and let it be located in branch K which extends from node I to node J. By the convention adopted in Sec. 3.2, the independent voltage source rise from node I to node J may be written as

$$\mathbf{V}_{n_j} - \mathbf{V}_{n_i} = \text{VAL} \qquad (3.127)$$

Nodal analysis of the network requires the summation of currents at the various nodes. At nodes I and J the current through the pth voltage source must be included. The set of unknown node-voltage equations contains the N node voltages; to this set will be added another set of q unknowns representing the currents through the q ideal voltage sources:

$$\mathbf{V}_n^* = \begin{bmatrix} \mathbf{V}_n \\ \cdots \\ \mathbf{I}_v \end{bmatrix} \qquad (3.128)$$

where $\mathbf{I}_v$ is a vector of q entries. Therefore the current balance equation for the Ith node will contain the term $(+I_p)$; similarly, the Jth equation will contain $(-I_p)$. If the reference node is node I or J, the corresponding current balance equation will not appear explicitly.

The pth current will become the $(N+p)$th variable and Eq. (3.126) becomes the pth equation appended to the N node-voltage expressions. Note that the conductance associated with this pth unaccompanied source is zero. A diagram depicting these matrix entries is shown in Fig. 3.22.

Thus the original N unknowns are expanded to $(N+q)$ unknowns; solution of the resultant equations will yield the N node voltages and the q source currents. The only minor modifications of the programs given in earlier sections are the following:

Sec. 3.9   Unaccompanied Energy Sources   107

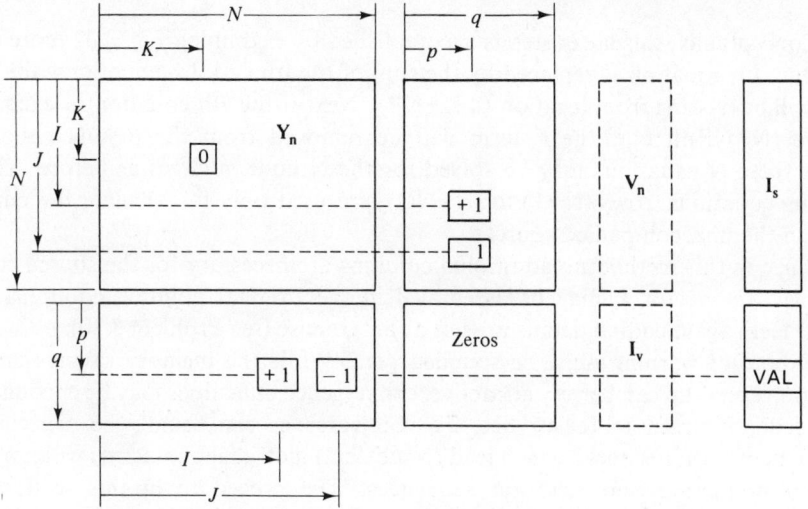

Fig. 3.22 Diagram Showing the Modified $Y_n V_n = I_s$ Relations for the pth Voltage Source of Strength VAL in Branch K which Extends from Node I to Node J (see Section 3.9)

(1) The modified I_s vector is assembled in column $N + q + 1$ of YN since N must be replaced by $N + q$ in the programs given in Secs. 3.3 and 3.5.

(2) Additional entries in the modified YN matrix are

$(p = \text{NP}; q = \text{NQ})$

```
YN(I,N+NP)=1.
YN(N+NP,I)=1.
YN(J,N+NP)=-1.                          (3.129)
YN(N+NP,J)=-1.
YN(NP,N+NQ+1)=VAL
```

If either I or J is zero, the corresponding entry in YN is not made.

(3) Step 2 is carried out for all unaccompanied sources ($p = 1, \ldots, q$).

(4) SUBROUTINE SOLVER will return the N node voltages and the q source currents in column $N+q+1$ of the array YN.

Implementation of the above steps is left as an exercise (see Problem 3.29).
Although the above procedure will work correctly, it is intuitively clear that an excess number of variables will be involved in the solution. The problem formulation is based on the node-voltage expression; hence the procedure should not require additional unknowns when relations between some of the node voltages are known. Indeed, the symmetry of the terms in Eq. (3.129) can be used to:

(1) Conserve storage: The terms due to source currents need not be stored separately in YN since the voltage source connections contain the same information.

(2) Reduce the equation set to N variables (for the N node voltages).

Conceptually, this latter step is accomplished by examining Fig. 3.22 more closely. If the Jth equation is replaced by the sum of the Ith and Jth equations, the -1.0 term will be missing from location (J,N+NP). Next if the Ith equation is exchanged for the (N+NP)th one, the I_p term will be removed from the first N equations. Hence these N equations may be solved for the N node voltages as before. The remaining equations, rows (N+1) to (N+NQ), are used only to calculate the currents through the unaccompanied sources.

Since in this method no additional columns are necessary for the source current variables, the vector I_s may be assembled in the original column, column (N+1) of YN. Here again coding details are left as an exercise (see Problem 3.30).

A note of warning must be sounded here. While the memory savings and the execution speed-up can be very attractive, convergence difficulties may be encountered. The reader is urged to review Sec. 2.5. Convergence can usually be expected for passive networks; networks which lead to substantial off-diagonal terms will normally lead to non-convergent iteration sequences. The procedure of this section has, however, been used for calculation of responses of very large (up to 300 node) stable filter networks with almost all elements being passive components.

3.10 Dependent Voltage Sources

The majority of electronic circuit analysis programs do not allow for dependent voltage sources whether they are voltage or current controlled. The main reason for this fact is readily apparent from an examination of the basic relation for the branch-admittance matrix given in Eq. (3.61):

$$\mathbf{Y}_b = [\mathbf{U} + \mathbf{B}]\mathbf{Y}_e[\mathbf{U} + \mathbf{D}]^{-1} \qquad (3.61)$$

where **B** contains the dependent current source strengths, $\mathbf{Y}_e$ the passive element admittances, and **D** embodies the dependent voltage terms. The detailed example in Sec. 3.2 readily demonstrated that the occurrence of a dependent voltage source in a branch that controls another dependent voltage source leads to more terms in the $[\mathbf{U} + \mathbf{D}]^{-1}$ matrix than there were original number of dependencies. In the example in Sec. 3.2, two dependent voltage sources led to three terms.

The generation of these terms is readily accomplished by the matrix inversion procedure, but more code and more temporary storage for the $\mathbf{U} + \mathbf{D}$ matrix and its inverse are required. An in-place inversion routine, such as the one developed in Sec. 2.9, will be useful in conserving some of the storage needed. Although constructing a program to implement directly the various equations developed in Sec. 3.2 is relatively easy, discussed here is a more sophisticated approach that can readily be applied to the methods developed in the previous section. Direct application of the relations derived in Sec. 3.2 does not readily lend itself to sparse matrix techniques.

The method basically is a computerized application of the source transformation technique developed in Sec. 3.3, Fig. 3.9, and Eq. (3.73). Essentially, every dependent voltage source will be replaced by a new branch whose passive element is a 1-ohm resistor and has at least two dependent current sources paralleling the resistor. One of the sources will represent the action of the dependent voltage source and another having a transfer quantity $\beta = 1$ will bypass the original passive element current from the 1-Ω resistor; if there were other dependent and independent current sources in the original branch, all such sources must be repeated in the newly created branch. Thus for every dependent voltage source, a new branch is created; if the circuit under consideration contains relatively few dependent voltage sources, the resultant set of node voltage equations will not be materially larger than the equation set for that circuit without the dependent voltages. If, however, the circuit contains very many such dependencies, direct calculation of the $[\mathbf{U} + \mathbf{D}]^{-1}$ will preserve the size of the $\mathbf{Y}_n$ matrix as N by N.

When branch data are entered at the outset of the program, data on the dependent voltage sources are stored in separate arrays in the same manner that data on the dependent current sources were stored in the arrays IB and B. For the IIDth dependent voltage source, the from-branch, to-branch pair of integers (J and K) will be stored in the doubly dimensioned array ID:

$$\begin{aligned} &\text{IF (IX.EQ.8) K=-K} \\ &\text{ID(IID,1)=K} \\ &\text{ID(IID,2)=J} \end{aligned} \qquad (3.130)$$

The strength of the dependent source VAL will be stored directly in the single vector D:

$$\text{D(IID)=VAL} \qquad (3.131)$$

Each time a new dependent voltage source is encountered (IX = 8 or 9), IID is incremented by one; when the circuit input is completed, the location IID will contain a count of the number of dependent voltage sources.

The following steps will be necessary in order to convert every dependent voltage source to a proper new circuit branch containing a 1-ohm resistor and the appropriate dependent current sources.

(1) If there are several dependent voltage sources in a branch, every source subsequent to the first one will be converted to a dependent current source in parallel with the first current source. No new entries will be made in the G vector and only one entry in the B and the corresponding IB vectors. The details of this code will be discussed in Step 7.

(2) For the Ith of the IID voltage sources, a new branch must be created. The from-node of the original branch is replaced by a new node number and a new branch is added to the G and IA vectors. The value of the new branch conductance is 1., the from-node is the newly created node number, the to-node the original branch to-node number.

(3) The negative of the value of the dependent voltage source D(I) is appended as the next entry at the end of the B vector; the corresponding to-branch is the new branch number, the from-branch of the D(I) source, found in ID(I, 2).

(4) A second dependent current source is added to B: The transconductance is the conductance of the original branch, the to-branch is the new branch, the from-branch is the original branch.

(5) If the original branch contains an independent current source, the same current source is inserted in the independent current vector **CG** in the new branch position.

(6) If the original branch had any dependent current sources, each such source is duplicated in the newly created branch.

(7) The check for a second voltage source in any branch is simple; if the to-node of the original branch is larger than N, the original number of nodes, a source already has been transformed and only Step 1 need be made.

```
C         CONVERSION OF ALL DEPENDENT VOLTAGE SOURCES
C         MUST BE DONE AT EACH FREQUENCY.
      IIBO=IIB
      NOLD=N
      MOLD=M
      MOLD1=M+1
      IF (IID.EQ.0) GO TO 860
      DO 850 I=1,IID
      IFLG=0
      ITO=ID(I,2)
      IF (ITO.GT.0) GO TO 805
      ITO=-ITO
      IFLG=1
  805 IFR=ID(I,1)
      KT=IA(ITO,2)
C         CHECK FOR SECOND DEPENDENT VOLTAGE SOURCE.
C         (STEP 1)
      IF (KT.LE.NOLD) GO TO 830
C         HAVE A CURRENT SOURCE ALREADY.
      DO 810 J=MOLD1,M
      IF (IA(J,1).NE.KT) GO TO 810
C         J IS THE NEW BRANCH NUMBER.
      IIB=IIB+1
      IB(IIB,1) = J
      IB(IIB,2) = IFR
      B(IIB) = -D(I)
      IF (IFLG.EQ.1) B(IIB) = -D(I)*CMPLX(RR(I),W*CC(I)-LL(I)/W)
      GO TO 850
  810 CONTINUE
      WRITE (3,820) KT
  820 FORMAT (/' NO BRANCH NUMBER ',I4,' FOUND.')
C         STOP THIS CALCULATION.
      GO TO 100
C
  830 N=N+1
      M=M+1
C         NEW DEPENDENT VOLTAGE SOURCE. CREATE NEW BRANCH, NEW NODE.
C         (STEP 2)
      IA(ITO,2) = N
      IA(M,1) =N
      IA(M,2) = KT
      RR(M) = 1.
      CC(M) =0.
      LL(M) =0.
```

Listing 3.5

Sec. 3.10 Dependent Voltage Sources

The detailed coding for the above steps is presented in the listing given in Listing 3.5. This code is to be inserted in ACAN2 after the branch admittances have been established and before the circuit node-admittance elements are calculated. In using the sparse matrix method of Sec. 3.7 for AC analysis, the IR and IG arrays remain unchanged at the various frequencies.

```
C            (STEP 3)
        IIB =IIB+1
        B(IIB) = -D(I)
        IF (IFLG.EQ.1) B(IIB) = -D(I)*CMPLX(RR(I),W*CC(I)-LL(I)/W)
        IB(IIB,1) = M
        IB(IIB,2) = IFR
C            (STEP 4)
        IIB =IIB+1
        B(IIB) = CMPLX(RR(ITO),W*CC(ITO)-LL(ITO)/W)
        IB(IIB,1) = M
        IB(IIB,2) = ITO
C       TRANSFER INDENDENT CURRENT SOURCE (STEP 5)
        CG(M) = CG(ITO)
C       CHECK FOR OTHER DEPENDENT CURRENTS (STEP 6)
        IF (IIBO.EQ.0) GO TO 850
        DO 840 J=1,IIBO
        IF (IB(J,1).NE.ITO) GO TO 840
        IIB = IIB+1
        B(IIB) = B(J)
        IB(IIB,1) = M
        IB(IIB,2) = IB(J,2)
  840   CONTINUE
  850   CONTINUE
  860   CONTINUE
C
C       N IS THE NEW NUMBER OF INDEPENDENT NODES.
C       M IS TN
C
C       M IS THE NEW NUMBER OF BRANCHES.
C       IIB IS THE NEW NUMBER OF DEPENDENT CURRENT SOURCES.
C       EQUIVALENT CIRCUIT CONTAINS DEPENDENT CURRENTS ONLY.
C       NOLD IS THE NUMBER OF ORIGINAL INDEPENDENT NODES.
C       MOLD IS THE NUMBER OF ORIGINAL BRANCHES.


C       THESE ARE ADDITIONS TO ACAN2
        DIMENSION ID(20,2),D(20)
        IID=0
  355   I=-I
C       VOLTAGE CONTROLLED VOLTAGE SOURCE.
  357   IID=IID+1
C       CURRENT CONTROLLED VOLTAGE SOURCE
        ID(IID,2) = -I
        ID(IID,1) = IBB
        D(IID) = VAL
```

Listing 3.5—*Cont.*

A flowchart of the completed AC sparse matrix program appears in Fig. 3.23. Complete coding of the program is left as an exercise (see Problem 3.31). Note that the first **NOLD** node voltages and **MOLD** branch currents will be the node voltages and branch currents of the original network.

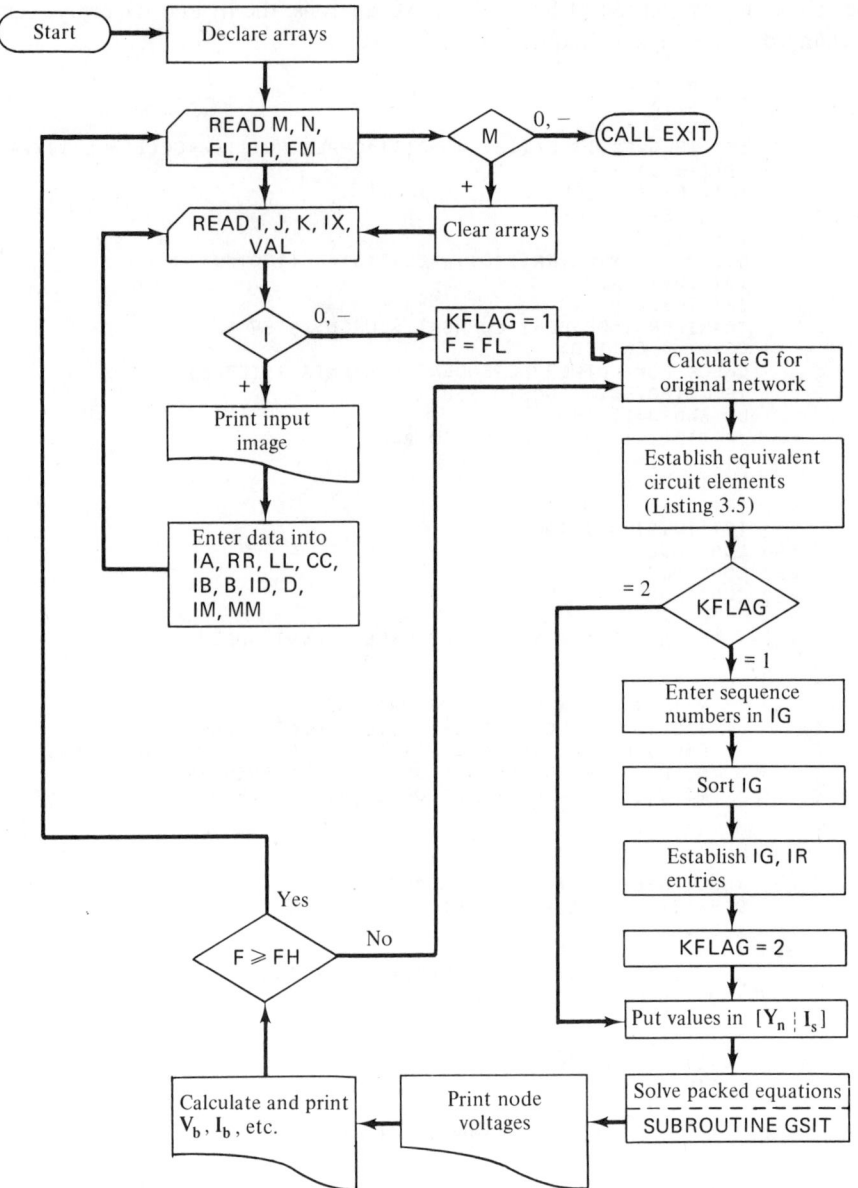

Fig. 3.23 Flowchart of Sparse-Matrix AC Analysis Program (compare with Figures 3.17 and 3.21)

3.11 Free-Form Input

One of the drawbacks of FORTRAN is the relatively rigid FORMAT required for all data. Column counts must be observed and data types must be identified. Additionally, in many small computers only numerical data (integers and floating-point numbers) may be manipulated, although alphabetic data may be read and printed.

To make a circuit analysis program useful to a wide variety of potential users, the input to such a program must be made relatively easy to use and it must be nearly a natural language for the user. Although numbers must be used within a program for branch numbers and node numbers, ideally any name (such as 'GROUND'), in addition to numbers, should be usable for labeling nodes and branches. The input formats prescribed for the various analysis programs in this chapter demand another "unnatural" number, the component type number IX. A more natural way for identifying the type number is to use alphabetic designators such as R, C, L, E, I, BETA, GM, MU, RT, M. The value of the component should be expressed in signed floating-point numbers, such as $-3.42E-05$, or integers, without regard of position on the input card. To specify a *branch* completely, the branch number, the from-node number, the to-node number, the component designator, and the component value, followed optionally by one or more component designator-component value groups, should be the only required information. Comments should be allowed in the input stream; frequency data (FL, FM, FH of Sec. 3.5) should be supplied in free-form format (i.e., three numbers, separated by any alphabetic character, usually a blank or a comma).

In general, only the input data and the final results are desired; the various matrices such as $[Y_n | I_s]$, are desirable only in special instances. Hence, normally, the user should not get these matrices in his output. Occasionally the input data may lead to circuit matrices which result in numerically near-zero determinants. Such ill-conditioned matrices will result in highly inaccurate solutions. In these cases the user should have an option in printing out the matrices used in the program.

The use of a blank card for indicating the end of input data for a circuit or end of a program run is not very natural; more descriptive would be text such as EXECUTE and END on a separate card.

In order to circumvent the need of explicitly having to supply M and N, the integers I, J, and K are put through the code

$$M=MAXO(M,I)$$
$$N=MAXO(N,J,K)$$
(3.132)

To obtain the number of branches and independent nodes at the end of the input data, i.e., upon encountering an EXECUTE card, the values of M and N are known (if they were initialized to zero before inputting any cards). Note that the above process does not require any particular order for the input data.

The above considerations are taken into account by the free-form input routine outlined in Fig. 3.24. The program operation and allowable inputs are as follows:

(1) The cards are read in alphabetic format

$$\begin{array}{ll} & \text{READ (1,10) KT} \\ 10 & \text{FORMAT (80A1)} \end{array} \qquad (3.133)$$

where KT is a singly subscripted array

$$\text{DIMENSION KT(80)}$$

Hence the ith entry in KT will contain the machine-code equivalent of the alphabetic information in column I of the input card. One by one these characters must be examined to determine their significance for the input.

(2) The card image is printed in the same format in which it was read, possibly centered on the page.

(3) A card starting with a C in column 1 will be treated as a comment; program returns to Step 1.

(4) A * in column 1 is taken as a flag to print internal tables (all internal matrices); print flags are set and the program returns to Step 1. (This option is not implemented in ACAN 2; but see Problem 3.33.)

In the above steps, as well as in some of the following steps, specific characters must be identified. In the FORTRAN language this identification may be accomplished by using ordinary IF statements, provided an array is set up whose elements contain the machine-code representation of the various characters. Using DATA statements, the code for finding a decimal integer of one or more digits starting with column IC is shown in Listing 3.6. Such a subroutine will be needed to find values of I, J, and K on the input data cards, as well as the integer following the E in a floating-point number.

In this code, liberal use is made of integer equality and non-equality tests in the various IF statements. In using FORTRAN-II, the logical IF statements must be replaced by arithmetic IFs and the DATA statements are best replaced by a READ statement which places into the array INT the internal code for the corresponding characters. This task is accomplished by using a FORMAT of A1 for each of the key characters to be searched for. These are input on the first card of the input data deck (preceding any circuit data cards).

(5) A frequency card must contain the letter F as its first non-blank character. The first value (FL) is picked up, as are all element values in this routine, starting in the column after the first equal sign ('=') that is encountered. Here again a normal IF statement may be used to find that = sign.

(6) If the first character is not an 'F' indicating a frequency card, a check is made if the first two characters are 'EN', indicating an END card. For 'EN' in the first two columns, a CALL EXIT is executed.

(7) Next, a check is made to determine if the first two characters are 'EX'

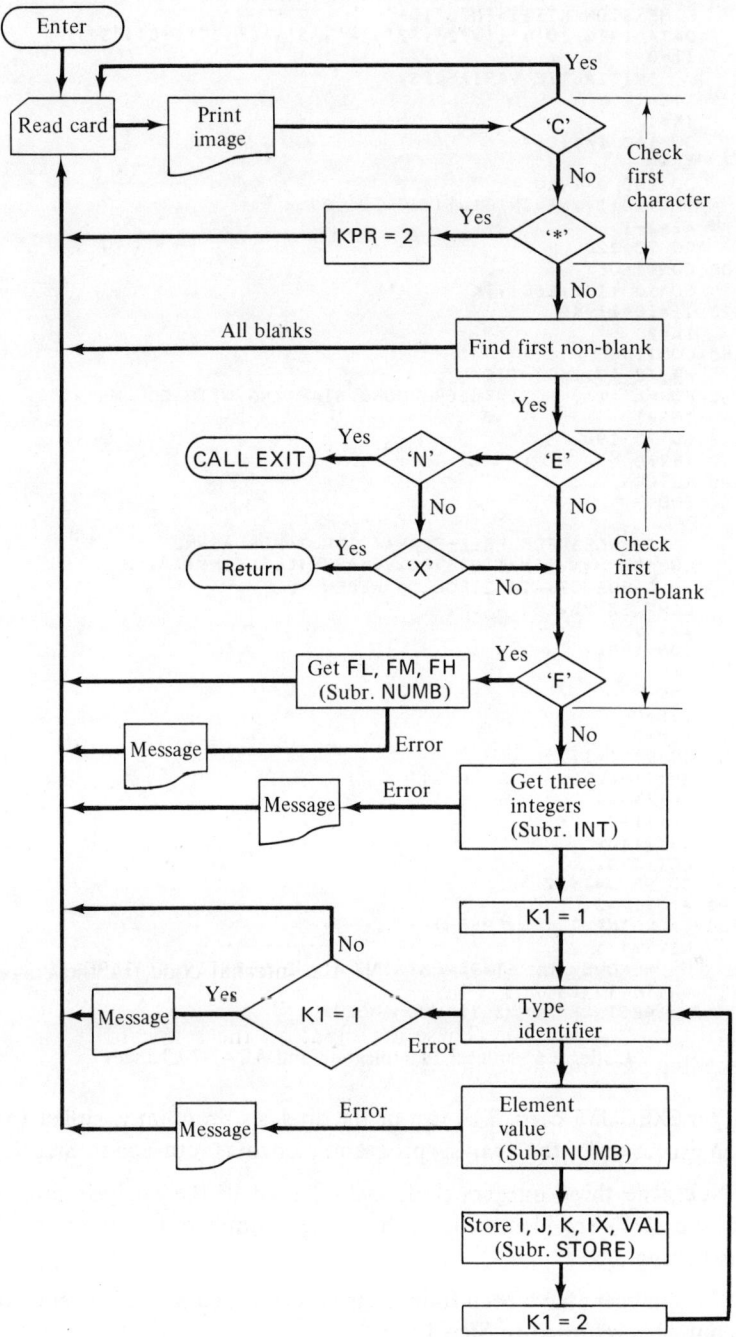

Fig. 3.24 Flowchart of Free-Form Input Routine

```
          SUBROUTINE INT (KT,IC,II,IER)
          DIMENSION KT(1),INTG(10)
          DATA INTG/'0','1','2','3','4','5','6','7','8','9'/
          II=0
C         INITIALIZE VARIABLES.
          IIC=IC
          IK=1
          DO 150 I=IIC,73
          IC=I
          DO 100 J=1,10
          IF (KT(I).NE.INTG(J)) GO TO 100
          K2=J-1
          GO TO 120
  100     CONTINUE
          GO TO (150,180),IK
  120     II=10*II+K2
          IK=2
  150     CONTINUE
          WRITE (3,160) IIC
  160     FORMAT (' NO INTEGER FOUND STARTING WITH COLUMN',I5)
          IER=1
          GO TO 190
  180     IER=0
  190     RETURN
          END
C
C         CHANGES FOR FREE-FORMAT INPUT FOR ACAN2
          COMMON M,N,IIM,IIB,B,CG,VG,A,RR,LL,CC,MM,IM,IB
C         CARDS OF ACAN2 FOLLOW HERE.
C         COMPLEX YB,CG, . . .
C         . . .
C         CON=180./ . . . .
          N=0
          M=0
          IIB=0
          IIM=0
          DO 90 I=1,15
          RR(I)=0.
          CC(I)=0.
          LL(I)=0.
          VG(I)=0.
          CG(I)=0.
          DO 90 J=1,10
   90     A(J,I)=0.
  100     CALL INPUT (FL,FM,FH)
          N2=N+1
C         REMOVE ALL STATEMENTS IN ACAN2 BETWEEN STATEMENTS 100 AND 451
  451     WRITE (3,450)
C         REST OF ACAN2 IS UNCHANGED.
```

Listing 3.6 Integer Identification and ACAN2 Changes

indicating an EXECUTE card. The rest of the analysis program is called for by this card. Upon execution of the analysis program, control is returned to Step 1.

(8) Next, the three integers (I, J, and K) used in the analysis programs (see Figs. 3.15 and 3.17) are identified. Each of these numbers is one or two integers; the code is Listing 3.6.

(9) If no integers have been found, the card is taken as a comment card and is ignored; control is returned to Step 1.

(10) After the first three integers an equal sign is searched for; it is found in

```
      SUBROUTINE INPUT (FL,FM,FH)
C         THIS TAKES THE PLACE OF STATEMENTS 100 THROUGH 451 IN ACAN2
C         INPUT ALLOWS ONLY REAL NUMBERS NOW.
      DIMENSION KT(80),IDENT(10)
      COMMON/TRACER/KPRINT
      INTEGER KT,IDENT,BLANK,C,STAR,E,EQL,N,X,F
      DATA IDENT /'G','R','C','L','E','I','A','M','U','T'/
      DATA BLANK/' '/,C/'C'/,STAR/'*'/,E/'E'/,EQL/'='/,N/'N'/,X/'X'/
      DATA F/'F'/
      KPRINT=1
C         WILL BE SET TO 2 IF 'TRACING' IS REQUIRED.  (NOT NOW IN ACAN2)
    5 READ (1,10) KT
   10 FORMAT (80A1)
      WRITE (3,20) KT
   20 FORMAT (1X,80A1)
      KT(73) = BLANK
C         THIS FORCES 72 VALID CHARACTERS IN DATA.
      IF (KT(1).EQ.C) GO TO 5
      IF (KT(1).NE.STAR) GO TO 30
      KPRINT = 2
C         SETS PRINTING FLAG.  (SEE PROBLEM      )
      GO TO 5
   30 DO 40 I=1,73
      IF (KT(I).EQ.BLANK) GO TO 40
      ICOL = I
C         ICOL DENOTES FIRST NON-BLANK CHARACTER.
      GO TO 50
   40 CONTINUE
C         A BLANK CARD WAS FOUND.  IGNORE IT.
      GO TO 5
   50 IF (KT(ICOL).NE.E) GO TO 70
C         CHECKING FOR END AND EXECUTE CARDS.
      IF (KT(ICOL+1).NE.N) GO TO 60
C         FOUND 'EN' , TAKE AS END   CARD. STOP EVERYTHING.
      CALL EXIT
   60 IF (KT(ICOL+1).NE.X) GO TO 70
C         FOUND  EX  CARD, STOP INPUT DATA.
      RETURN
   70 IF (KT(ICOL).NE.F) GO TO 120
C         FOUND A FREQUENCY CARD, LOOK FOR  =  SIGN.
      ICOL =ICOL+1
      DO 80  I=ICOL,73
      IF (KT(I).NE.EQL) GO TO 80
      ICOL = I+1
C         ICOL NOW POINTS TO ONE PAST  =  SIGN.
      GO TO 90
   80 CONTINUE
C         NO  =  SIGN FOUND, IGNORE THIS CARD.
      GO TO 5
   90 CALL NUMB(KT,ICOL,FL,IER)
      IF (IER.NE.0) GO TO 100
      ICOL=ICOL+1
      CALL NUMB(KT,ICOL,FM,IER)
      IF (IER.NE.0) GO TO 100
      ICOL=ICOL+1
      CALL NUMB(KT,ICOL,FH,IER)
      IF (IER.EQ.0) GO TO 5
  100 WRITE (3,110)
  110 FORMAT (' FREQUENCY VALUES NOT FOUND.')
      GO TO 5
  120 CONTINUE
C         A NORMAL DATA CARD WAS FOUND.
C         LOOK FOR THREE INTEGERS SEPARATED BY ANY NON-NUMERICS.
      CALL INT (KT,ICOL,I,IER)
      IF (IER.NE.0) GO TO 130
      CALL INT (KT,ICOL,J,IER)
      IF (IER.NE.0) GO TO 130
      CALL INT (KT,ICOL,K,IER)
```

Listing 3.7   Free-Form Input Routine

```
      IF (IER.EQ.0) GO TO 150
  130 WRITE (3,140)
  140 FORMAT (' FIRST THREE INTEGERS NOT FOUND.')
      GO TO 5
  150 K1=1
  160 DO 170 L=ICOL,73
C         LOOK FOR = SIGN.
      IF (KT(L).NE.EQL)  GO TO 170
      ICOL = L
      GO TO 190
  170 CONTINUE
      IF (K1.GT.1) GO TO 5
      WRITE (3,180)
  180 FORMAT (' NO = SIGN FOUND.')
      GO TO 5
  190 DO 200 L=2,ICOL
      LL=ICOL-L+1
C         LOOK FOR IDENTIFIER BEFORE = SIGN, IGNORE BLANKS.
      IF (KT(LL).EQ.BLANK) GO TO 200
      GO TO 230
  200 CONTINUE
  210 WRITE (3,220)
  220 FORMAT (' NO ELEMENT TYPE IDENTIFIER FOUND.')
      GO TO 5
  230 DO 240 L=1,10
      IF (KT(LL).NE.IDENT(L)) GO TO 240
      IX = L-1
      IF ((L.EQ.8).AND.(KT(LL-1).NE.IDENT(1))) IX=10
C         FOUND TYPE NUMBER
      GO TO 250
  240 CONTINUE
      GO TO 210
  250 ICOL = ICOL+1
C         ICOL POINTS TO ONE PAST = SIGN
      CALL NUMB (KT,ICOL,VALUE,IER)
      IF (IER.EQ.0)  GO TO 280
      WRITE (3,270)
  270 FORMAT (' NO VALUE FOUND BEYOND THE = SIGN.')
      GO TO 5
  280 CALL STORE (I,J,K,IX,VALUE)
      K1 = 2
      GO TO 160
      END
      SUBROUTINE STORE (I,J,K,IX,VAL)
      COMMON M,N,IIM,IIB,B,CG,VG,A,RR,LL,CC,MM,IM,IB
      DIMENSION CG(15),VG(15),A(10,15),RR(15),LL(15),CC(15)
      DIMENSION MM(20),IM(20,2),B(40),IB(40,2)
      REAL MM,LL
C         FOR DEPENDENT CURRENT SOURCES
C             I = A SEQUENCE NUMBER OF THE SOURCE,
C             J = FROM-BRANCH NUMBER
C             K = TO-BRANCH NUMBER.
C         FOR MUTUAL INDUCTANCES
C             I = A SEQUENCE NUMBER OF THE MUTUAL VALUE,
C             J = PRIMARY BRANCH NUMBER,
C             K = SECONDARY BRANCH NUMBER.
C         NO DEPENDENT VOLTAGE SOURCES IN THIS CODE, BUT NEED
C         ONLY TO ADD D AND ID ARRAYS AND SOME CODE TO ENTER THEM.
      COMPLEX CG,VG,B
      IF (IX.NE.0) GO TO 10
      RR(I) = VAL
      GO TO 40
   10 IF (IX.NE.1) GO TO 20
      RR(I) = 1./VAL
      GO TO 40
   20 IF (IX.NE.2) GO TO 30
      CC(I) = VAL
```

Listing 3.7—*Cont.*

```
      GO TO 40
   30 IF (IX.NE.3 ) GO TO 60
      LL(I) = 1./VAL
   40 IF (J.EQ.0) GO TO 50
      A(J,I) = 1.
   50 IF (K.EQ.0) GO TO 200
      A(K,I) = -1.
      GO TO 200
   60 IF (IX.NE.4) GO TO 70
      VG(I) = VAL
      GO TO 200
   70 IF (IX.NE.5) GO TO 80
      CG(I) = VAL
      GO TO 200
   80 IF (IX.NE.7) GO TO 90
      K=-K
      GO TO 95
   90 IF (IX.NE.6) GO TO 100
   95 IIB=MAX0(IIB,I)
      B(I) = VAL
      IB(I,1) = J
      IB(I,2) = -K
      RETURN
  100 IF (IX.NE.10) GO TO 150
      IIM=MAX0(IIM,I)
      MM(I)=VAL
      IM(I,1)=J
      IM(I,2)=K
      RETURN
  150 WRITE (3,180)
  180 FORMAT (' DEPENDENT VOLTAGE SOURCES NOT ALLOWED. CARD IGNORED.')
      RETURN
  200 M=MAX0(I,M)
      N=MAX0(N,J,K)
      RETURN
      END
      SUBROUTINE NUMB (KT,ICOL,VAL,IER)
      DIMENSION KT(1),NUM(14)
      DATA NUM/'0','1','2','3','4','5','6','7','8','9','+','-','E','.'/
C        ROUTINE PICKS UP A GENERAL FLOATING POINT NUMBER
C           STARTING IN COL. ICOL.
C        INITIALIZE VARIABLES.
      DIV=1.
      ICC=ICOL
      IDEC=1
      VAL=0.
      SIGN=1.
      KS=1
      DO 200  I=ICC,73
      ICOL=I
      DO 50 J=1,14
      IF (KT(I).NE.NUM(J)) GO TO 50
C        A NUMBER HAS BEEN FOUND.
      JJ=J
      KS=2
      GO TO 60
   50 CONTINUE
      GO TO (200,135),KS
   60 IF (JJ.LE.10) GO TO 70
      IF (JJ.EQ.11) GO TO 100
      IF (JJ.EQ.12) GO TO 140
      IF (JJ.EQ.13)  GO TO 150
C        DECIMAL POINT FOUND.
      IF (IDEC.NE.1) GO TO 180
C        CHECK FOR PREVIOUS DEC. POINT.
      IDEC=2
      GO TO 200
```

Listing 3.7—*Cont.*

```
       70 GO TO (80,90),IDEC
       80 VAL=VAL*10. + FLOAT(JJ-1)
    C        ADD NEW INTEGER BEFORE DECIMAL POINT.
          GO TO 200
       90 DIV=DIV/10.
          VAL=VAL + FLOAT(JJ-1)*DIV
    C        ADD A NEW INTEGER AFTER THE DECIMAL POINT.
          GO TO 200
      100 IF (VAL.NE.0) GO TO 110
    C        PLUS SIGN FOUND.
          SIGN = 1.
          GO TO 200
      110 NX=1
      115 ICOL=ICOL+1
      116 CALL INT(KT,ICOL,INTG,IER)
          IF (IER.EQ.0) GO TO 130
          WRITE (3,120)
      120 FORMAT (' NO PROPER EXPONENT FOUND.')
          GO TO 300
      130 VAL = VAL*SIGN*(10.**(NX*INTG))
      135 IER=0
          GO TO 300
      140 IF (VAL.NE.0.) GO TO 145
          SIGN=-1.0
          GO TO 200
      145 NX=-1
          GO TO 115
      150 NX=1
    C        E WAS FOUND.
          IF (KT(ICOL+1).NE.NUM(12)) GO TO 160
          ICOL=ICOL+2
    C        PLUS SIGN AFTER THE E .
          GO TO 116
      160 IF (KT(ICOL+1).NE.NUM(11)) GO TO 115
    C        MINUS SIGN AFTER THE E .
          ICOL=ICOL+2
          NX=-1
          GO TO 116
      180 WRITE (3,190)
      190 FORMAT (' TWO DECIMAL POINTS FOUND.')
          GO TO 290
      200 CONTINUE
          WRITE (3,280)
      280 FORMAT (' NO INTEGER FOUND.')
      290 IER=1
      300 RETURN
          END
```

Listing 3.7—*Cont.*

column i. The element designator must be in col. i-1. Here again a dimensioned array is used in much the same way that the INTEG array was used in Listing 3.6.

(11) Starting in column i+1 the element value is identified; here the ten numeric characters, and the special symbols plus (+), minus (−), decimal point (.), and the character 'E' must be identified. Consequently, a 14-character search is made of every column starting with i+1. The numeric value is identified similar to the procedure used in Step 9 and is in SUBROUTINE NUMB.

(12) Steps 11 and 12 are repeated to identify all elements in the branch, the independent voltage source, and the independent current source.

(13) If in any of the above steps the end of the card is reached, the numeric value of the element being examined is stored properly; if at least one element was identified, control is returned to Step 1. Otherwise, an error message is printed.

(14) All branch information should be input from one card; no continuation cards will be allowed.

A program embodying the above steps is flowcharted in Fig. 3.24. The code necessary to convert ACAN 2 to this mode of input is in Listing 3.7. For the other programs in this chapter, similar conversions are left as exercises.

In this program no provision is made for specifically picking up a control branch integer IBB. Since of the three integers I, J, and K only I is used, a new definition for controlled branches is introduced that does not require an IB. For each transfer quantity (BETA, GM, MU, and RT), the three first integers shall have the meanings

$I =$ number of transfer quantity, i.e., the row index for the IB and B, or ID and D, or IM and MM arrays.

$J =$ the "from-branch" number.

$K =$ the "to-branch" number.

For transfer quantities only one element value shall be searched for.

The free-form input section described here will make the use of the developed analysis programs quite usable for many purposes. The utility of such an input language is best demonstrated by carrying out the actual coding exercise indicated.

3.12 Summary

This chapter outlined all equations necessary for a complete time-independent analysis package for arbitrary linear lumped parameter networks. Coding details were presented for introductory DC and AC analysis packages. Program details were examined and markedly improved programs were discussed. Direct calculation of the node-admittance matrix was shown to result in large memory savings. Iterative solution using a sparse matrix technique was detailed for speedy frequency analysis (except for first frequency point). Finally, the salient points of free-form input programs were presented.

These techniques form the basis of the subsequent chapters. Transient analysis will be shown to be equivalent to a series of DC analyses with the VG, CG, and YB elements variables; state-space analysis will reduce to a number of simultaneous DC analyses; tolerance analysis will use repeated DC analysis; and nonlinear analysis will be reduced to repeated DC analyses with the augmented node-admittance elements functions of the voltages and currents present in those elements. A good understanding of the program details in this chapter will prove to be most beneficial for thorough understanding of all subsequent material in this book.

REFERENCES

1. SESHU, S., and N. BALABANIAN, *Linear Network Analysis*. John Wiley & Sons, Inc., New York, 1963.
2. VAN VALKENBURG, M. E., *Network Analysis*, 2nd ed. Prentice-Hall, Inc., Englewood Cliffs, N.J., 1964.
3. KUO, F. F., and J. F. KAISER, eds., *System Analysis by Digital Computer*. John Wiley & Sons, Inc., New York, 1966.
4. RAMEY, R. L., and E. J. WHITE, *Matrices and Computers in Electronic Circuit Analysis*. McGraw-Hill, Inc., New York, 1971.
5. JENSEN, R. W., and M. D. LIEBERMAN, *IBM Electronic Circuit Analysis Program*. Prentice-Hall, Inc., Englewood Cliffs, N.J., 1968.
6. WILLOUGHBY, R. A., ed., "Sparse Matrix Proceedings," RA1 (#11707), IBM Corporation, Yorktown Heights, N. Y., 1969.
7. CALAHAN, D. D., *Computer Aided Network Design*. McGraw-Hill, Inc., New York, 1972.
8. KUO, F. F., and W. G. MAGNUSON, JR., *Computer Oriented Circuit Design*. Prentice-Hall, Inc., Englewood Cliffs, N.J., 1969.
9. LEY, B. J., *Computer Aided Analysis and Design for Electrical Engineers*. Holt, Rinehart and Winston, Inc., New York, 1970.
10. JENSEN, R. W., and B. J. WATKINS, *Network Analysis: Theory and Computer Methods*. Prentice-Hall, Inc., Englewood Cliffs, N.J., 1974.
11. TINNEY, W. F., and J. W. WALKER, "Direct Solutions of Sparse Matrix Equations by Optimally Ordered Triangular Factorization," *Proc. IEEE*, Vol. 55, No. 11, (November 1969).
12. BERRY, R. D., "An Optimal Ordering of Electronic Circuit Equations for a Sparse Matrix Solution," *IEEE Trans. on Circuit Theory*, Vol. CT-18 (January 1971).

PROBLEMS

3.1 Write the matrix equations for the node voltages for the following network. It is desired to find E_{out}.

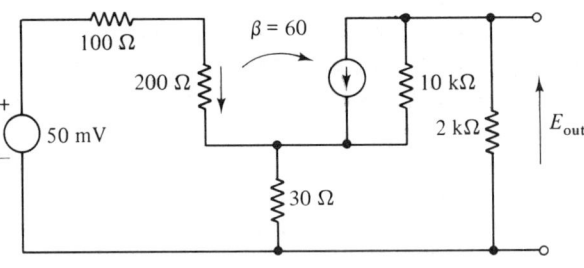

Fig. P3.1

3.2 Following the examples given in Sec. 3.1, write the node-voltage equations for the following network operating at a frequency of 10,000 Hz. Note that five simultaneous equations will result; solution by slide rule is tedious.

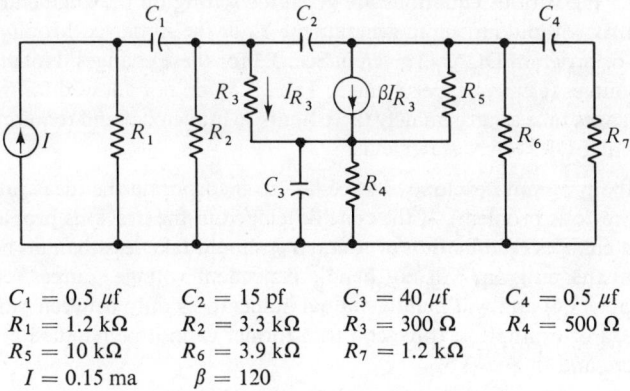

$C_1 = 0.5\ \mu f$ $C_2 = 15$ pf $C_3 = 40\ \mu f$ $C_4 = 0.5\ \mu f$
$R_1 = 1.2$ kΩ $R_2 = 3.3$ kΩ $R_3 = 300\ \Omega$ $R_4 = 500\ \Omega$
$R_5 = 10$ kΩ $R_6 = 3.9$ kΩ $R_7 = 1.2$ kΩ
$I = 0.15$ ma $\beta = 120$

Fig. P3.2

3.3 Using the equations developed in Sec. 3.2, write a computer program for the analysis of DC circuits. Call your program DCAN1.

Assume that the input to the program consists of a series of cards describing an element in a fixed FORTRAN input for the following input numbers:

(1) Branch number (I3 format)
(2) From-node number (I3)
(3) To-node number (I3)
(4) Element type number (I3) (see Table 3.5)
(5) Element value (F10.0)
(6) Control branch number (I3) (for dependent sources only)

A blank card will initiate calculation of the node voltages. A blank card, if not preceded by an element description card, will terminate the program. Allow only resistances/conductances, independent sources, and dependent current sources. Use program SOLVER (Sec. 2.4) for solving the resultant simultaneous equations.

3.4 Discuss the difficulties that may be encountered in setting up and solving the general node-voltage equations [Eq. 3.65] as the element values are varied over the range $(0, \infty)$. What added difficulties may result when elements in the range $(-\infty, 0)$ must also be considered?

3.5 Modify the program from Problem 3.3 or the program in Sec. 3.3 to include dependent voltage sources also. Subroutine INVRT (Sec. 2.9 or Appendix A) could be used for the inversion of the $(\mathbf{U} + \mathbf{D})$ matrix.

3.6 Derive the matrix node-voltage equations, similar to Eq. (3.65), for the analysis of networks where the transfer quantities β and g_m depend not on element currents and element voltages respectively, but on branch currents and branch voltages. Let the standard branch be defined as in Fig. 3.10 and disallow dependent voltage sources. The object of this problem is to rederive the matrix equations for a network; in this book only the dependencies on element voltages and currents are considered in detail.

3.7 Solve Problem 3.1; use your program DCAN1 (Problem 3.3) and check with DCAN2 discussed in Sec. 3.3.

3.8 Verify the solution of Ex. 3-6 (Sec. 3.2) by using your program from Problem 3.5.

3.9 In Sec. 3.7 the various equations are given for setting up the augmented node-admittance matrix without having to generate the Y_b or the A matrix. Modify your program DCAN1 or program DCAN2 given in Sec. 3.3 for these changes. Note that dependent voltage sources (element types 8 and 9, Table 3.5) are not allowed for in this program. These changes take approximately three hours to implement and require approximately 50 different FORTRAN statements.

3.10 Modify the program developed in Sec. 3.5 to incorporate the ideas presented in Sec. 3.7 (see previous problem). If the code developed in the previous problem is used, the suggested changes (elimination of A and Y_b) should take less than an hour. Note here again that the program will not handle dependent voltage sources (element types 8 and 9, Table 3.5) and will handle mutual inductances only between pair-wise coupled primaries. For example, a three-coil transformer cannot be handled by the code suggested here and in Sec. 3.5.

3.11 (General AC analysis). In Sec. 3.2 the equations for the analysis of arbitrary networks are discussed. The equations developed do not account fully for mutual inductances which may couple various inductors. For AC (steady-state) analysis, the voltage-current relationships between the various elements, including mutual inductances, are (terms are defined in Sec. 3.2)

$$\mathbf{V_e} = \mathbf{ZI_e}$$

where $\mathbf{Z}$ is the element impedance matrix of M rows and M columns and has the following non-zero entries:

$z_{ii} = R_i$ if branch i contains a resistor of value R_i

$\phantom{z_{ii}} = 1/(j\omega C_i)$ if branch i contains a capacitor of value C_i

$\phantom{z_{ii}} = j\omega L_i$ if branch i contains an inductor of value L_i

$z_{ik} = z_{ki} = j\omega M_{ik}$ for all mutual inductance which couple branches i and k

The element admittance matrix $\mathbf{Y_e}$ may now be obtained by direct inversion of $\mathbf{Z}$:

$$\mathbf{Y_e} = \mathbf{Z}^{-1}$$

Show that the remainder of the relationships in Sec. 3.2 are unaffected and that the final relationships for the node voltages are

$$\mathbf{AY_bA^T V_n} = \mathbf{A(I_g - Y_b V_g)}$$

with

$$\mathbf{Y_b} = \mathbf{(U + B)(Z)^{-1}(U + D)^{-1}}$$
$$= \mathbf{(U + B)[(U + D)Z]^{-1}}$$

3.12 Derive a flowchart for a general AC analysis computer program based on the equations developed in the previous problem and similar to the flowchart shown in Fig. 3.17.

3.13 Code the general AC analysis program flowcharted in the previous problem. Use a full YB matrix (M by M) and a full A matrix (N by M) similar to the one given in Listing 3.3. Note that inversion must be performed on a full M by M matrix with complex coefficients. A modification of program INVRT (see Appendix B and Sec. 2.6) must be made. Include dependent voltage sources in this program. With the simple input format used in Sec. 3.5 this problem takes about 300 FORTRAN-IV statements and requires approximately two days of effort, including simple debug time.

3.14 The program developed in the previous problem uses the A matrix directly. Note that since the inversion of $(\mathbf{U} + \mathbf{D})\mathbf{Z}$ can produce non-zero entries in the inverse where there is a zero in the original matrix, the matrix YB should be calculated directly. The condensation of YB and A into the smaller YN augmented matrix (N rows and $N + 1$ columns) can be accomplished without having to use the A matrix. Instead, a pointer array, identical to IA in Sec. 3.7, should be used. This results not only in a savings in memory required, but it also speeds up the problem solution. (Why?) Code the details of this version of the general AC analysis program.

3.15 In Sec. 3.6 a frequency-independent model for pairwise mutual inductors was derived. Implement the required code for this model. One way of implementation is to use the code of Listing 3.3 and to add two dependent current sources to the array B for each mutual inductance encountered. Corresponding entries must be made in the IB arrays and the self-inductance values in the two coupled branches must also be changed. Then the YN matrix is assembled as in Listing 3.3 without any further entries being made for the actual mutual inductances. Alternately, the code developed in Problem 3.9 may be used.

When the element currents are calculated after the node-voltage solution has been obtained, the currents in the dependent sources representing the mutual inductances must be added to the element currents otherwise calculated to yield correct results.

The above procedures must be repeated at each frequency for which the analysis is to be done. Coding time for this problem is approximately one hour.

3.16 Derive the general matrix node-voltage equation for networks in which the transfer quantities β, g_m, μ, and r_t (element types 6, 7, 8, 9 in Table 3.5) depend on branch voltages and currents instead of on element voltages and currents. Let the standard branch be defined as in Fig. 3.10. Compare results with Problem 3.6. As in Problem 3.6, the purpose of this problem is to re-examine the basic procedure for the derivation of the matrix node-voltage equations; in this book no further use is made of the relationships developed in this problem.

3.17 Derive the general matrix node-voltage equation for networks in which the transfer quantities β and g_m are dependent on the branch current and voltage of the controlling branch and the quantities μ and r_t depend on the controlling branch element voltage and current. The purpose of this problem is the same as the purpose of Problem 3.16.

3.18 In Sec. 3.7 a scheme is presented for calculating the $\mathbf{Y_n}$ matrix without recourse to the A matrix or the $\mathbf{Y_b}$ matrix explicitly. The code shown in Eq. (3.115) details the assembly of the $\mathbf{Y_{n1}}$ matrix and Eq. (3.116) shows the assembly of $\mathbf{Y_{n2}}$. The former matrix accounts for the admittances of the various branches; the latter contains the mutual conductances (more properly, mutual admittances).

A passive element admittance can be thought of as a mutual admittance when the controlling branch is the same as the controlled branch. Hence the code contained in Eq. (3.115) is superfluous if the G-vector entries (self-admittances) can be modified for use in Eq. (3.116). Make such a modification. This will result in less code for the analysis program. This modification is identical for DC and AC analysis; it is suggested that the modifications be tried on a simple DC analysis program such as the one given in Listing 3.1.

3.19 Derive the basic loop-node-parts-branches relationship in Eq. (3.1). Note that this relationship is very basic in determining the number of loop equations and node equations in a given network.

3.20 How can Eq. (3.1) be used to determine the number of independent node-voltage equations and number of independent loop-current equations for a general network containing N_v voltage sources and N_c current sources among its branches?

3.21 How must Eq. (3.1) be modified if each branch of a network is made up of "basic branches" as defined in Fig. 3.10?

3.22 In Sec. 3.3 a procedure is shown for replacing a voltage dependent voltage source (element type 9, Table 3.5) by a dependent current source and associated other sources and resistances (see Fig. 3.14). When the controlled branch contains additional independent and/or dependent current sources, the equivalent circuit shown must be modified to prevent the currents in those sources from passing through the 1 Ω resistor of the equivalent circuit. (Why?)

Devise an algorithm for this (more general) case. (An algorithm is a procedure; it need not be computer code, but it often is. In this problem a word description would suffice.) You may check your procedure by comparing Figs. 3.11 and 3.16.

3.23 Devise an algorithm similar to the one in Problem 3.19 for current dependent voltage sources (element type 8, Table 3.5).

3.24 Modify the algorithms in Problems 3.19 and 3.20 to account for branches that contain several dependent voltage sources (in series with one another and a passive element).

3.25 The procedure described in Sec. 3.3 and discussed in the previous two problems replaces controlled voltage sources by controlled current sources *and* creates additional branches *and* nodes in an equivalent network. The equivalent network may be solved without recourse to inversion; hence an increase in independent nodes can be traded for not having to invert a matrix such as [U + D]. In addition, arbitrary mutual inductances can be treated as a set of current controlled voltage sources; these in turn may be replaced by equivalent circuits containing additional nodes, branches, and controlled current sources (see also Problem 3.24). Thus the entire DC or AC analysis procedure can be reduced to an inversion-less program containing only self-admittances, controlled currents, and independent sources. These networks can be analyzed by using no **A** matrix and without having to generate Y_b directly, thus saving computer memory and increasing execution speed (since the many zero terms in **A** need not be retrieved).

In a "normal" network usually there are only a few dependent voltage sources (if any) and only a few inductances are coupled to another. Thus the increase in number of nodes (the size of the Y_n matrix) tends to be modest. Therefore the procedure outlined here is usually attractive from the standpoint of execution speed and memory storage requirements (while still retaining only the simplest solution techniques).

Devise an AC analysis program to handle all the types of elements listed in Table 3.5 without generating Y_b or **A** and without using matrix inversion. The program should be capable of handling arbitrarily coupled mutual inductances (more than two coupled coils), but of a modest number. When this program is coupled with free-format input routines (see Sec. 3.11), it represents an analysis program of considerable sophistication (see also Sec. 3.10). This problem might make a good term project for a small group of students. Typical implementation time for this problem has been aproximately two weeks including "normal" debugging.

3.26 In Sec. 3.8 a procedure is outlined for setting up the node-voltage equations without having to store its zero entries. This procedure is flowcharted in Fig. 3.21. Code this program. Several subroutines are indicated in Fig. 3.21 (ENTER, PUT, GSIT) which could ease the construction of the program. Subroutine GSIT is the Gauss-Seidel iteration procedure discussed in Sec. 2.5; see also Problem 2.15.

3.27 Modify the program developed in Problem 3.23 for AC analysis. To speed up the solution process, the last known set of node voltages should be used to start the Gauss-Seidel iteration as the frequency is being stepped.

3.28 The procedures outlined in Sec. 3.8 will become unstable (i.e., not converge to a solution) when off-diagonal elements become large enough. (The question of convergence is discussed in [3] of Chapter 2.) To minimize the storage required for any given problem it is desired to retain the essential features of the sparse matrix process discussed but also to revert to the Gauss elimination procedure since this latter does not have convergence difficulties (if a solution exists and/or the various terms do not suffer unduly from round-off errors). Since the Gauss elimination procedure will generate additional terms in place of the original zero entries, some concession must be made for an increase in the number of non-zero terms in the Y_n matrix. The forms of the Gauss elimination procedure given in Secs. 2.4 and 2.7 and in Appendix A try to minimize the round-off errors by using the largest element in the remaining subdeterminant. Another scheme is to examine each equation to be used such that the number of new non-zero terms generated is minimized. If only the non-zero terms are stored, during the process of equation solution the entries in Y_n may have to be re-arranged and new entries created a few times only. Thus some of the advantages of sparse storage are retained, although the iterative convenience is lost. Details of these ideas are discussed in [7 and 8].

Devise such an elimination strategy and code the resulting Gauss elimination subroutine. Use this subroutine in place of GSIT in Problem 3.23 or 3.24.

3.29 Sec. 3.9 outlines a procedure for including a voltage source without a series impedance in the analysis programs. The modified $[Y_u \mid I_s]$ matrix which results from this process is shown in Fig. 3.22. Use the procedure outlined around Eq. (3.128). Modify one of the analysis programs created so far (for example, Listing 3.1) to allow voltage sources of this nature.

3.30 Revise the program steps devised in Problem 3.29 to conserve storage space as indicated at the end of Sec. 3.9. Careful examination of Fig. 3.22 will be helpful.

3.31 Sec. 3.10 details the procedure and some of the code necessary for converting dependent voltage sources to dependent current sources and new branches (see also Problems 3.22, 3.23, and 3.24). Complete the indicated coding (and check for correctness) for converting Listing 3.3 to allow for the analysis of dependent voltage sources by this means. Compare also with the program developed in Problem 3.25.

3.32 Listings 3.6 and 3.7 contain the major part necessary to provide the various analysis programs with a free-format input capability. Complete the code and insert it into the

DC and the AC programs developed so far. Choose the most advanced of the programs; if none were developed so far, use Listings 3.1 and 3.3 for testing the modified codes.

3.33 In checking on the operation of various programs (which contain logic and keypunch errors during development and may contain erroneous data after debugging) it is convenient to provide many printouts for checking. For the program in Listing 3.3 (AC analysis), the various component arrays (RR, CC, LL, MM, B, etc.) and interconnection arrays (IA, IB, etc.) should be printed prior to forming Y_n. Then Y_n should be printed to check on the correctness of forming the node equations. Other intermediate results may also be useful at various stages in the development and use of this program. These print statements can produce much output and could clutter the results unduly.

One way to put such printouts under the user's option is to define a print flag which must be set specifically through a special code to produce outputs. In the free-format input a star in the first column of a card may force such a setting. If this flag is called KPRFLG in the program, then the statements

```
      IF (KPRFLG) WRITE (3,333) (PR(I),I=1,MRES)
  333 FORMAT (//' RR-ARRAY IS'/(1P8E15.5))
```

will print the resistor array if KPRFLG is set to TRUE. By default, KPRFLG = .FALSE., and its value is set to .TRUE. only if a card with a star in column 1 is encountered.

Implement such a printout feature for all intermediate arrays in the free-form version of the programs developed, particularly the ones in Problem 3.32.

3.34 The solutions produced in AC analysis might become very bulky and difficult to interpret. A simple set of graphs may be preferable. A plotting routine for printing a graph of 100 printer positions and 50 lines is given in Appendix B. This program accepts a vector of magnitude values, a vector of phase angle values (degrees), and a vector of associated frequencies and produces a db-magnitude and a phase angle vs. log frequency plots (Bode plots). The plot routine adjusts its own scale so that usable graphs result.

Add this plotting feature to the AC analysis programs discussed.

3.35 (Ideal Transformer Coupling) In analyzing networks containing large power transformers one must consider multiwinding coupled coils with very low leakage inductance and minimal losses; the coils are very close to being ideal transformers. The inductance matrix associated with such networks is very close to being singular and thus not invertible. The admittance matrix may not be numerically correct, even if inversion is carried out with no apparent difficulties.

Such networks may be analyzed by considering each transformer separately as a combination of ideally coupled coils. In *series* with each coil a small resistance is put to model coil losses and a small inductance is put to model the leakage inductance. The magnetizing losses may be put in *parallel* with each coil as a parallel combination of an inductor and a resistor; we neglect these here since they can be added external to the transformer model.

Let a network contain NUM number of transformers; let the Kth transformer contain up to NN(K) number of coils. Let the Ith coil be connected from node IM(1, I) to node IM(2, I); let this coil have MM(K) turns.

One way to analyze such networks is to append to the N node equations the MK coil currents, where MK is the total number of coupled coils. For each coil a new node (not actually carried in the analysis explicitly) is created between the series impedance (small resistor and inductor series combination) and the ideal coil. The node voltages are set up as before without considering the coupled coils. The effect of the coil is accounted for by the following steps.

1. Add or subtract the coil current to/from the appropriate current summation expressions, similar to the procedure used in Sec. 3.10.
2. For each transformer the sum of the ampere-turns is set to zero. This is a separate equation for each transformer.
3. The voltage ratios between the terminals of the ideal coils are set to the turns ratios. Each coil is thus related to another coil on the same transformer.

Between Steps 2 and 3 there will be created just as many equations as there are coils. Derive the necessary equations and implement them in your AC analysis program.

3.36 Write a user manual for your version of the most advanced DC and AC analysis programs developed in these problems. The manual should explain the permissible input forms, the print options (if any), and the plotting features and should list examples of use. If the program produces any error diagnostics, they must also be explained.

An acid test for this manual will be the comments that you will receive from other engineers familiar with circuit analysis, but who have no idea of what your programs contain. Your program should be usable by them and satisfy their needs of analysis.

TRANSIENT ANALYSIS OF NETWORKS

4

An important problem in automated circuit analysis is the calculation of the voltages and currents in a network as a function of time. In contrast to the considerations that underlay the development of the material in Chapter 3 due attention must be paid to the differential voltage-current relationships in inductors and capacitors.

Two major avenues of approach to this problem have been developed in detail. One is to write and solve the simultaneous differential equations that describe the network behavior; this method will be taken up in Chapter 5. The other method is to replace the inductors and the capacitors by approximate conductance-and-source combinations and then solve the resultant simultaneous algebraic equations. The equivalent of the capacitors and inductors will be good approximations for small changes in current and voltage only; hence the parameters of the equivalent DC network will have to be adjusted frequently. Thus it is necessary to solve a series of DC problems; this latter approach is developed in this chapter.

Since the development of the DC analysis programs was based on the node-admittance concept, methods must be found to approximate the behavior of inductors and capacitors in terms easily adapted to node-admittance equations. Thus relationships of this kind

$$i = yv \qquad (4.1)$$

where i and v are changes in voltages and currents, will be sought.

4.1 Models for Energy Storage Elements

4.1a The Inductor Model

The voltage-current relationship for an inductor of value L is

$$i(t) = \frac{1}{L} \int_{t'=0}^{t} v(t') \, dt' + i(0) \qquad (4.2)$$

where $i(0)$ is the current at time $t = 0$ in the inductor. The voltage-time history for the inductor is shown in Fig. 4-1.

In computing the inductor-voltage values a series of values will be calculated at equally spaced time intervals $t_0, t_1, \ldots, t_{k-1}, t_k, \ldots$. The current in the inductor at the end of the kth interval is

$$i_k = \frac{1}{L} \int_{t_{k-1}}^{t_k} v \, dt + i_{k-1} \qquad (4.3)$$

where $i_N = i(t) \big|_{t=t_N}$.

The integral appearing in Eq. (4.3) is replaced by trapezoid approximation

$$\int = \frac{1}{2} \Delta t (v_{k-1} + v_k) \qquad (4.4)$$

with Δt being the time step $\Delta t = t_k - t_{k-1}$ and the subscript on v is similar to the subscript on i as defined above. Now a linear relationship results

$$\begin{aligned} i_k &= i_{k-1} + \frac{1}{2} \frac{\Delta t}{L} (v_{k-1} + v_k) \\ &= \left[i_{k-1} + \frac{\Delta t}{2L} v_{k-1} \right] + \frac{\Delta t}{2L} v_k \end{aligned} \qquad (4.5)$$

This last expression is of the form

$$i = I + gv \qquad (4.6)$$

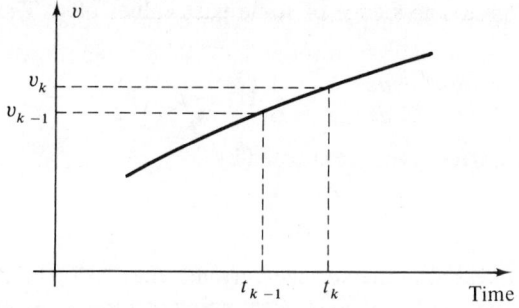

Fig. 4.1 Inductor Voltage

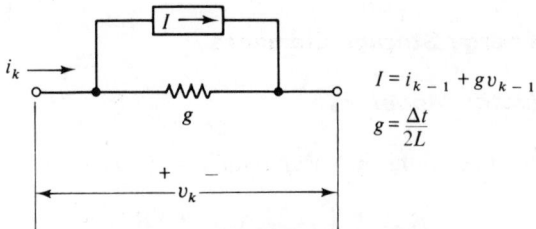

Fig. 4.2 Model of an Inductor

and thus represents a parallel combination of a current source and a conductance as shown in Fig. 4.2. Notice the directions of the various currents and polarities of the various voltages.

Since the intent is to calculate the voltage values at equally spaced instants in time, the value of g will not change during the entire calculation as time is stepped; the value of I, however, must be re-adjusted at every new time value. Also note that v_k and v_{k-1} are voltages across the inductor element, not branch voltages that must also account for independent voltage sources. Similarly, i_k and i_{k-1} are the currents through the inductor element.

This model will serve as a building block in a computer program for transient analysis to be assembled in Sec. 4.5.

4.1b The Capacitor Model

The voltage-current relationship for a capacitor of value C is

$$i(t) = C\frac{dv(t)}{dt} \tag{4.7}$$

The current at time $t = t_k$ is

$$i_k = C\frac{dv}{dt}\bigg|_{t=t_k} \tag{4.8}$$

Here again the voltages and currents will be known only at discrete equally spaced intervals in time. Hence the slope of the voltage vs. time curve may at best be approximated only from a knowledge of some past values of v. The simplest such approximation is

$$\frac{dv}{dt}\bigg|_{t=t_k} = \frac{1}{\Delta t}(v_k - v_{k-1}) \tag{4.9}$$

Hence the capacitor current is approximated by

$$i_k = \frac{C}{\Delta t}(v_k - v_{k-1}) \tag{4.10}$$

The above equation describes the voltage-current relationship of the circuit shown in Fig. 4.3. Note that the conductance is unchanging as long as Δt remains unchanged. As with the inductor, special note should be taken of the relative directions of

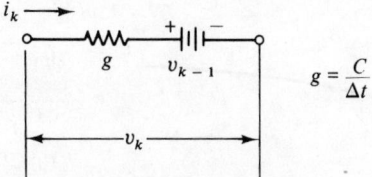

Fig. 4.3 Model of a Capacitor

voltages and currents in the model and of the fact that these currents and voltages are not branch quantities but that they refer to the capacitor element only.

This capacitor model is an essential building block of the transient analysis program assembled in Sec. 4.5.

4.1c Computational Examples

To illustrate the above two computational models and to provide a set of check problems for the program to be developed in the following sections, consider the circuit in Fig. 4.4. It is desired to calculate the node voltages for this network if at time $t = 0$ the voltage across the capacitor is EO = 10 volts as indicated.

In Fig. 4.4 the input data for the program to be discussed in Sec. 4.5 are also

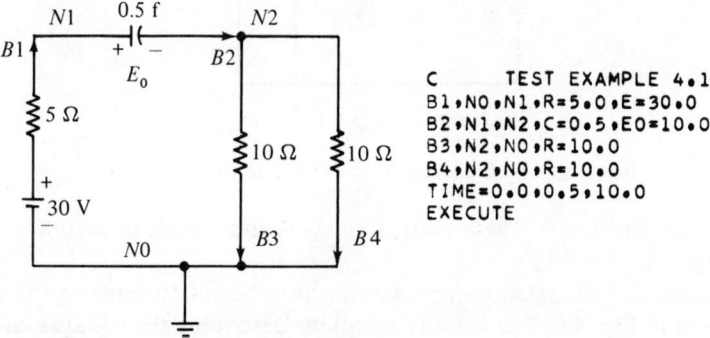

Fig. 4.4 Test Example 4.1

indicated. Note that the word TIME instead of FREQUENCY will be used and that a new type identifier EO is also shown. For inductor initial current, IO will be also added to the different type numbers. This will increase the total number of "element" types used in Subroutines STORE and INPUT to twelve. Note that the input section of the free-form routine of Sec. 3.12 must be modified only slightly for this type of transient program.

The node voltages may be verified to be

$$V_1 = 30 - 10e^{-t/RC}$$
$$V_2 = 10e^{-t/RC}$$

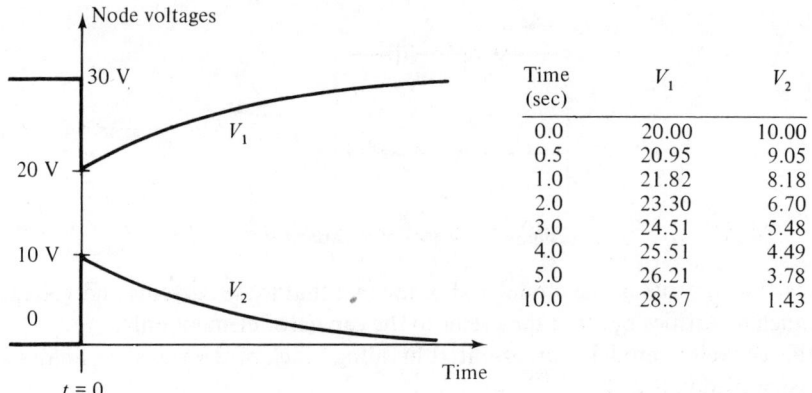

Fig. 4.5 Solution for Test Example 4.1

Time (sec)	V_1	V_2
0.0	20.00	10.00
0.5	20.95	9.05
1.0	21.82	8.18
2.0	23.30	6.70
3.0	24.51	5.48
4.0	25.51	4.49
5.0	26.21	3.78
10.0	28.57	1.43

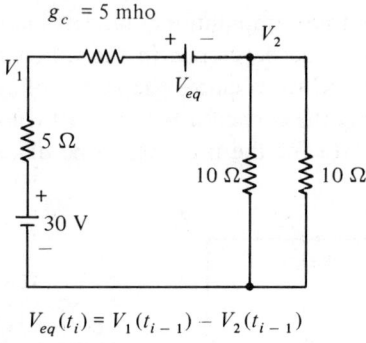

$$V_{eq}(t_i) = V_1(t_{i-1}) - V_2(t_{i-1})$$

Fig. 4.6 Equivalent Circuit for Test Example 4.1

with $RC = 5.0$ and $t > 0$. These plotted voltages and a table of pertinent values are given in Fig. 4.5.

Application of the relationships developed in Sec. 4.1b leads to the equivalent circuit shown in Fig. 4.6. The various equations describing the voltages and currents are

$$g_C = \frac{C}{\Delta t} = \frac{0.5}{0.1} = 5 \text{ mho}$$

$$V_{eq} = V_1^{(i-1)} - V_2^{(i-1)}$$

$$I_C = \frac{30 - V_{eq}}{10 + \frac{1}{g_C}}$$

$$V_1 = 30 - 5 \cdot I_C$$

$$V_2 = 5 \cdot I_C$$

$$t = t^{(i-1)} + \Delta t$$

The superscript i was omitted in all terms in the above equations (it belongs on all variables except on g_C). To obtain the voltage-time relations for V_1 and V_2, the above equations are iterated starting with the initial voltages on V_1 and V_2:

$$V_1(0) = 20 \text{ V}$$
$$V_2(0) = 10 \text{ V}$$

These iterations are best summarized as shown in Table 4.1.

Table 4.1 Calculation of Voltages for Ex. 4.1

Time	V_{eq}	I_C	V_1	V_2	V_b
0.0			20.00*	10.00*	10.00*
0.1	10.00	1.961	20.20	9.80	10.39
0.2	10.39	1.922	20.39	9.61	10.78
0.3	10.78	1.885	20.58	9.42	11.15
0.4	11.15	1.848	20.76	9.24	11.52
0.5	11.52	1.811	20.94	9.06	11.89
.					
1.0	13.26	1.641	21.80	8.20	13.59
.					
2.0	16.27	1.346	23.27	6.73	16.54
.					
5.0	22.42	0.743	26.28	3.72	22.57
.					
10.0	27.18	0.276	28.62	1.38	27.24

*Separately obtained from initial conditions.

Comparison of values appearing in Table 4.1 and those indicated in Fig. 4.5 shows a very close agreement even though the time increments taken were relatively large.

Another fact that must be pointed out is that the calculations must start out with a determination of the initial conditions existing in the network prior to application of the time-iterations. Thus a transient analysis program usually starts with a calculation of these initial conditions (often termed *equilibrium conditions*). Such conditions are determined from a knowledge of the independent generator values, the initial voltages on capacitors, and initial currents through inductors. Care must be taken that such initial conditions are consistent. For instance, the capacitor branch in the above example could have been replaced by two parallel 1-farad capacitors.

Consistency of initial conditions demands then that the initial voltages across the two separate capacitors be equal.

To obtain the initial conditions, each capacitor could be replaced by a voltage source of strength equal to the initial voltage on the capacitor. Similarly, initial currents in inductors will appear as current sources in place of the respective inductors. Such unaccompanied sources can be handled by methods outlined in Sec. 3.9 (provided such sources are consistent; see also Problem 4.3). It is usually more expedient, however, to add a small resistance in series with the voltage source that replaces the capacitor and a small conductance in parallel with the current source that represents the initial current in the inductor. Such values shall be called "SHORT" and "OPEN" in the program discussed in Sec. 4.5; these values will be set in the program to 0.001 ohm and 10 megohms respectively. These components will not materially change the actual initial voltage and current values; they will, however, ensure that the networks being analyzed do in fact possess equilibrium solutions. Should such equilibrium solutions contain unexpectedly large voltages or currents, usually a set of inconsistent initial conditions has been specified.

As a second example, the voltages and currents in Fig. 4.7 shall be calculated. The solution of this network is given by the equations

$$I_{b_2}(t) = 3 - 2e^{-t/T}$$

$$V_1(t) = 15 + 10e^{-t/T}$$

$$V_2(t) = 15 - 10e^{-t/T}$$

with $T = L/R_T = 1$ since $R_T = 10$ ohms is the total resistance that exists in the current loop in this network. These solutions are plotted in Fig. 4.8 and a table of values is also shown for selected times.

The inductor model is used to derive the equivalent circuit shown in Fig. 4.9. The voltages and currents are calculated from the following relationships:

$$V_{eq} = \frac{I_{b_2}^{(i-1)} \times 20.}{\Delta t} + V_{b_2}^{(i-1)}$$

$$I_{b_2} = \frac{30. + V_{eq}}{10. + \frac{20.}{\Delta t}}$$

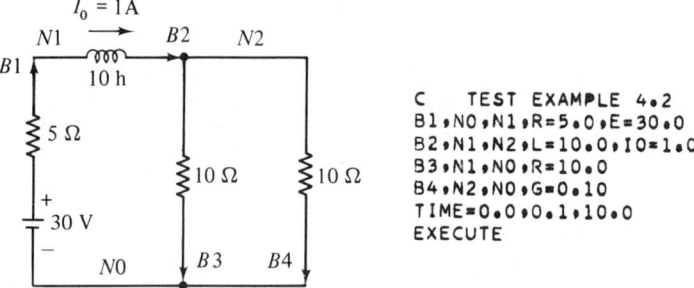

Fig. 4.7 Test Example 4.2

Sec. 4.1 Models for Energy Storage Elements 137

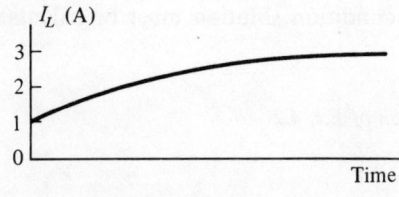

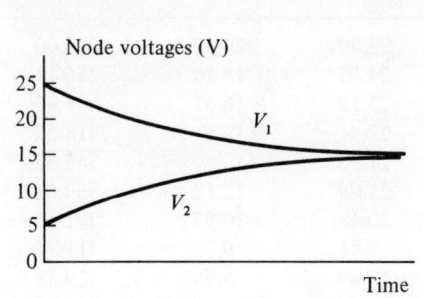

Time	I_L	V_1	V_2
0.0	1.000	25.00	5.00
0.25	1.443	22.78	7.22
0.50	1.788	21.06	8.94
1.00	2.264	18.68	11.32
2.50	2.836	15.82	14.18
5.00	2.987	15.07	14.93

Fig. 4.8 Solution of Example 4.2

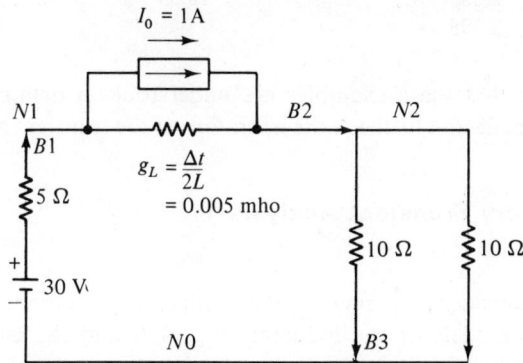

Fig. 4.9 Equivalent Circuit for Example 4.2

$$V_2 = I_{b_2} \times 5.0$$
$$V_1 = 30. - 5.0 \times I_{b_2}$$
$$V_{b_2} = V_1 - V_2$$
$$t^{(i)} = t^{(i-1)} + \Delta t$$

As before, the superscript i was omitted (it belongs on all variables above); iteration is started with the initial conditions

$$I_L(0) = 1.0$$
$$V_1(0) = 25.0$$
$$V_2(0) = 5.0$$

The iterations are summarized in Table 4.2; comparison of these numbers with the numbers shown in Fig. 4.8 shows a good approximation to the "true" solution. As discussed in the previous example, an initial condition solution must be calculated before the iterations proceed.

Table 4.2 Calculation of Ex. 4.2

Time	I_b	V_2	V_1	V_b	V_{eq}
0.00	1.000	5.00	25.00	20.00	220.00
0.10	1.190	5.95	24.05	18.10	256.19
0.20	1.363	6.81	23.19	16.37	288.93
0.30	1.519	7.59	22.41	14.81	318.56
0.40	1.660	8.30	21.70	13.40	345.36
0.50	1.787	8.93	21.06	12.12	369.61
0.60	1.903	9.51	20.48	10.97	391.56
0.70	2.007	10.04	19.96	9.93	411.41
0.80	2.102	10.51	19.49	8.98	429.38
0.90	2.188	10.94	19.06	8.12	445.63
1.00	2.265	11.32	18.68	7.35	460.33
1.10	2.335	11.67	18.33	6.65	473.63
1.20	2.398	—	—	—	—

It is important that these examples are understood in detail because they are used to develop the code and to check the operation of the transient analysis programs.

4.2 A Rudimentary Transient Analysis Program

We shall now consider the structure of a rudimentary transient analysis program. The computational models of the inductor (Fig. 4.2) and the capacitor (Fig. 4.3) replace each of the active components by a conductance and a variable source. Hence the network to be analyzed is a simple DC network with the values of I_g and V_g (the independent current and voltage sources) changing at each time step. For each increment in time the values of I_g and V_g are constants; thus the equations developed in Chapter 3 apply directly. The pertinent equations are

$$\mathbf{V_n} = [\mathbf{A Y_b A}^T]^{-1} \mathbf{A}(\mathbf{I_g} - \mathbf{Y_b V_g}) \tag{4.11}$$

$$\mathbf{V_b} = \mathbf{A}^T \mathbf{V_n} \tag{4.12}$$

$$\mathbf{V_e} = \mathbf{V_b} - \mathbf{V_g} \tag{4.13}$$

$$\mathbf{I_e} = \mathbf{Y_e V_e} \tag{4.14}$$

Note that Eq. (4.11) is merely a rewritten form of Eq. (3.65). In the above, $\mathbf{I_g}$ includes the inductor equivalent current and $\mathbf{V_g}$ includes the capacitor equivalent voltage. Since the equivalent DC problem has to be solved at every time increment, it is more economical to invert the node-admittance matrix, re-evaluate $\mathbf{I_g}$ and $\mathbf{V_g}$ each time, and carry out the product indicated in Eq. (4.11). To set the current sources

Sec. 4.2   A Rudimentary Transient Analysis Program

in the inductor equivalent circuits, the element currents must be calculated, but only for the various inductors. Similarly, the capacitor equivalent voltage sources are set from the element voltages.

In order to set the initial values of inductor currents and capacitor voltages, two new vectors may be added to the program. To set the initial current and initial voltage into these arrays, the same convention can be used as was used for setting $\mathbf{I}_g$ and $\mathbf{V}_g$ values. Two new type numbers are needed: For this chapter a value of IX = 11 is chosen for initial capacitor voltage and IX = 12 is taken for initial inductor current. In each of these two (LLIC and CCIC respectively) the last element currents and voltages must be entered before time is incremented. Note that in order to facilitate the calculation of $\mathbf{Y}_b\mathbf{V}_g$ and that of $\mathbf{I}_e$, the matrix $\mathbf{Y}_b$ should be stored separately. Also, the value of $\Delta t/2L$ should be stored for the inductors. Pointer arrays could be used to indicate which branches have inductors and which have capacitors, much like the pointer array IB was used in Sec. 3.6. Initial time, time increment, and final time to which this transient solution is to be carried can be read in together with the values of M (number of branches) and N (number of independent nodes) at the beginning of the program.

A flowchart of this rudimentary transient analysis program is given in Fig. 4.10. The reader is urged to code it using his own DC analysis program DCAN1 (see Problem 3.3) and program INVRT given in Appendix A (see also Sec. 2.8). As an alternate program DCAN2 may be modified. Note that the program is cumbersome, particularly with regard to storage requirements, and is not capable of an initial condition solution. The usefulness of the program is further limited to circuits not having conflicting initial conditions. Such occur when two or more capacitors of different initial voltage are in a loop in the network or when two or more inductors having different initial currents are connected to a common node having no other current path to it. Circuits of these types will be considered carefully in Chapter 5.

The various arrays occurring in Fig. 4.10 have the same meanings as before (see Sec. 3.3); additional arrays not occurring before are:

(1) LLIC: the array of equivalent current for the inductor model; initially, this array will contain the initial value of the inductor current; this array is also referred to as $\mathbf{I}_{L0}$ in the flowchart; notice that LLIC must be declared a floating-point array.

(2) CCIC: the array of capacitor equivalent voltage sources; initially, the initial voltages on the capacitor are stored here; the array is also referred to as $\mathbf{V}_{C0}$ in the flowchart.

(3) IL and IC: arrays that contain a marker (such as +1) in locations whose subscripts are equal to the branch number containing an inductor or a capacitor respectively; otherwise the entries are zero; these markers are used to recalculate the entries of $\mathbf{I}_{L0}$ and $\mathbf{V}_{C0}$ at each new time step.

(4) GL: an array that contains $\Delta t/2L$ for each inductor; the corresponding entry in IL will be a marker; the value of this conductance is used in resetting $\mathbf{I}_{L0}$ at each new time step.

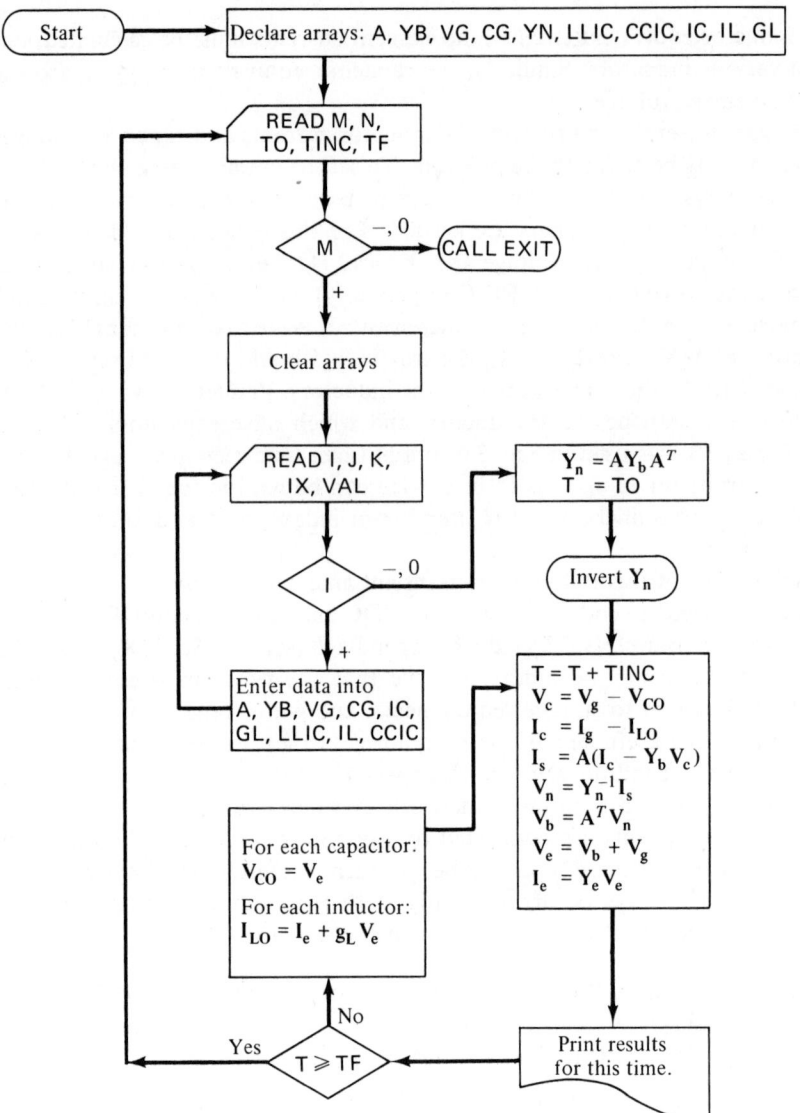

Fig. 4.10 Flowchart of Simple Transient Analysis Program

It is not absolutely necessary to calculate every entry in the vectors V_c, I_c, V_b, V_e, and I_e, but for ease of programming, it may be preferable to declare these five arrays in addition to the ones indicated in Fig. 4.10 and carry out the vector operations indicated on every member of the vectors. Further note that in Fig. 3.15, I_s was put in column $N + 1$ of Y_n; here a new vector may be called for. Depending on the coding details, some or all of the arrays discussed here may be needed; the reader is urged to develop his own simple transient analysis program.

Note that this program cannot handle mutual inductances.

The program outlined in Fig. 4.10 may be made more efficient in utilizing memory by using the DC analysis program developed in Sec. 3.7 as the starting point (see Problem 4.8).

4.3 General Equations

In this section we shall derive fully the equations for transient analysis. The simple models of energy storage elements introduced in Sec. 4.1 will be taken as the basis for the derivations. The general branch described in Fig. 3.10 is taken as the model for each branch.

4.3a Pairwise Mutual Inductance

In Sec. 4.1 mutual inductance was ignored. Let us assume that branches p and s, containing the inductors L_p and L_s are coupled by means of a (single) mutual inductance M. The voltage-current relationships are

$$v_p = L_p \frac{di_p}{dt} + M \frac{di_s}{dt}$$
$$v_s = L_s \frac{di_s}{dt} + M \frac{di_p}{dt} \tag{4.15}$$

The above two relationships may be solved to yield i_p and i_s:

$$i_p(t) = \frac{L_s}{L_p L_s - M^2} \int_0^t v_p(t')\,dt' - \frac{M}{L_p L_s - M^2} \int_0^t v_s(t')\,dt' + i_p(0)$$
$$i_s(t) = \frac{L_p}{L_p L_s - M^2} \int_0^t v_s(t')\,dt' - \frac{M}{L_p L_s - M^2} \int_0^t v_p(t')\,dt' + i_s(0) \tag{4.16}$$

Approximating the integrals in the above by trapezoids (as in Sec. 4.1) results in the relations

$$i_p(t_k) = \frac{\Delta t}{2L_1}[v_p(t_{k-1}) + v_p(t_k)] - \frac{\Delta t}{2M_1}[v_s(t_{k-1}) + v_s(t_k)] + i_p(t_{k-1})$$
$$i_s(t_k) = \frac{\Delta t}{2L_2}[v_s(t_{k-1}) + v_s(t_k)] - \frac{\Delta t}{2M_1}[v_p(t_{k-1}) + v_p(t_k)] + i_s(t_{k-1}) \tag{4.17}$$

where $L_1 = (L_p L_s - M^2)/L_s$, $L_2 = (L_p L_s - M^2)/L_p$, and $M_1 = (L_p L_s - M^2)/M$.

The terms in the above may be re-arranged to correspond to the form given in Eq. (4.5):

$$i_p(t_k) = \frac{\Delta t}{2L_1}v_p(t_k) - \frac{\Delta t}{2M_1}v_s(t_k) + \left[\frac{\Delta t}{2L_1}v_p(t_{k-1}) - \frac{\Delta t}{2M_1}v_s(t_{k-1}) + i_p(t_{k-1})\right]$$
$$i_s(t_k) = \frac{\Delta t}{2L_2}v_s(t_k) - \frac{\Delta t}{2M_1}v_p(t_k) + \left[\frac{\Delta t}{2L_2}v_s(t_{k-1}) - \frac{\Delta t}{2M_1}v_p(t_{k-1}) + i_s(t_{k-1})\right] \tag{4.18}$$

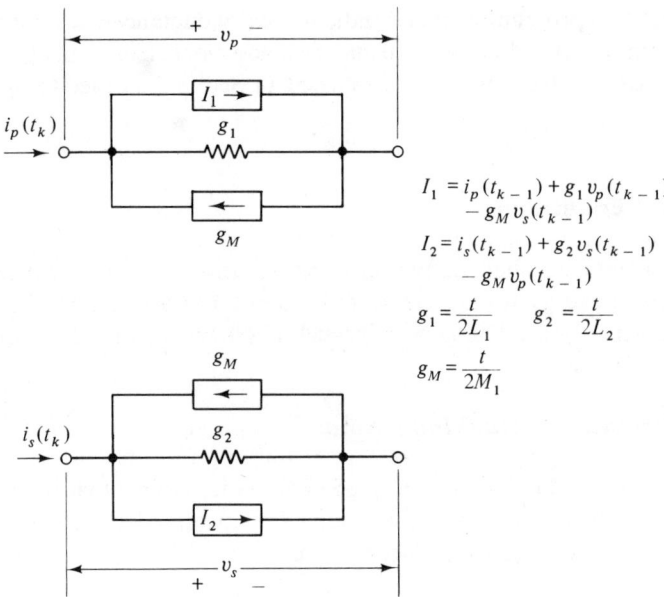

Fig. 4.11 Model of Pairwise Mutual Inductance

$$I_1 = i_p(t_{k-1}) + g_1 v_p(t_{k-1}) - g_M v_s(t_{k-1})$$

$$I_2 = i_s(t_{k-1}) + g_2 v_s(t_{k-1}) - g_M v_p(t_{k-1})$$

$$g_1 = \frac{t}{2L_1} \quad g_2 = \frac{t}{2L_2}$$

$$g_M = \frac{t}{2M_1}$$

From the above, a model for pairwise mutual inductances may be derived as was done in Fig. 4.6 for single inductances. The resultant model is shown in Fig. 4.11. Note that the model is valid only for two inductances (L_p and L_s) coupled by a mutual inductance (M).

The model contains a conductance for the primary and for the secondary inductor (g_1 and g_2 in Fig. 4.11); the values of these conductances depend not only on the value of the inductances that they represent but also on the coupling inductance. A second effect of the coupling is the presence of a dependent source g_M which injects directly the voltage changes from the secondary side to the primary one (and vice versa). Finally, the current sources I_1 and I_2 represent the past history of the voltages across the two coupled coils.

Although the model shown in Fig. 4.11 is inadequate to represent arbitrarily coupled coils, it does lend itself for an easy addition of some mutual inductance capability in a simple transient analysis program. Inclusion of this model in the program outlined in Fig. 4.10 is left as an exercise.

4.3b Equations for an Arbitrary Linear Network

In the following let us consider a linear network made up of M branches of the kind shown in Fig. 3.10. Let each branch contain a single passive element, such as a resistance, a capacitance, or an inductance. In addition, let each inductor be coupled

Sec. 4.3   General Equations   143

to any other inductor in the network. We shall now develop the node-voltage equations for such a network by using the capacitor and inductor models introduced in Sec. 4.1.

Since each branch contains only one passive component, the element currents $\mathbf{I}_e$ and element voltages $\mathbf{V}_e$ may be separated as follows:

$$\mathbf{I}_e = \begin{bmatrix} \mathbf{I}_L \\ \mathbf{I}_R \\ \mathbf{I}_C \end{bmatrix} \quad \text{and} \quad \mathbf{V}_e = \begin{bmatrix} \mathbf{V}_L \\ \mathbf{V}_R \\ \mathbf{V}_C \end{bmatrix} \tag{4.19}$$

where $\mathbf{I}_L$ and $\mathbf{V}_L$ refer to the currents and voltages of the inductor elements, $\mathbf{I}_R$ and $\mathbf{V}_R$ to those of resistors, and $\mathbf{I}_C$ and $\mathbf{V}_C$ to those of capacitors.

The voltages and currents of the various types of elements are related as follows.
For resistances

$$\mathbf{I}_R = [1/R]\mathbf{V}_R \tag{4.20}$$

where $[1/R]$ is a diagonal matrix of the conductances of the elements.

For capacitances

$$\mathbf{I}_C = [C]\frac{d\mathbf{V}_C}{dt} \tag{4.20a}$$

$$= [C]\frac{1}{\Delta t}(\mathbf{V}_{C_i} - \mathbf{V}_{C_{i-1}}) \tag{4.20b}$$

where $[C]$ is a diagonal matrix of capacitance element values, Δt is the time step, and $\mathbf{V}_{C_i}$ are the capacitor voltages at $t = t_i$.

For inductances the voltage-current relationships are

$$\mathbf{V}_L = [L]\frac{d}{dt}\mathbf{I}_L \tag{4.21}$$

where $[L]$ is a matrix of inductance values including mutual inductances. As in Sec. 4.1 the above is solved for $\mathbf{I}_L$:

$$\mathbf{I}_L = [L]^{-1}\int_0^t \mathbf{V}_L\, dt + \mathbf{I}_{L0} \tag{4.22}$$

where $\mathbf{I}_{L0} = \mathbf{I}_L$ at $t = 0$. (We are assuming that the analysis is started at time $t = 0$.) For $t = t_k$ the above is rewritten in the recursion form:

$$\begin{aligned}\mathbf{I}_{L_k} &= [L]^{-1}\frac{\Delta t}{2}(\mathbf{V}_{L_k} + \mathbf{V}_{L_{k-1}}) + \mathbf{I}_{L_{k-1}} \\ &= [L]^{-1}\frac{\Delta t}{2}\mathbf{V}_{L_k} + [L]^{-1}\frac{\Delta t}{2}\mathbf{V}_{L_{k-1}} + \mathbf{I}_{L_{k-1}}\end{aligned} \tag{4.23}$$

In the last form the first term in the sum on the right side of the equation represents a set of equivalent conductances (possibly mutual conductances) and the last two terms represent equivalent current sources for each of the branches containing an inductor.

Hence the element currents at time t_k are given by

$$\mathbf{I}_{e_k} = \begin{bmatrix} \frac{\Delta t}{2}[L]^{-1} & 0 & 0 \\ 0 & [1/R] & 0 \\ 0 & 0 & \frac{1}{\Delta t}[C] \end{bmatrix} \begin{bmatrix} \mathbf{V}_{L_k} \\ \mathbf{V}_{R_k} \\ \mathbf{V}_{C_k} \end{bmatrix}$$

$$+ \begin{bmatrix} \frac{\Delta t}{2}[L]^{-1} \mathbf{V}_{L_{k-1}} + \mathbf{I}_{L_{k-1}} \\ 0 \\ -\frac{1}{\Delta t}[C]\mathbf{V}_{C_{k-1}} \end{bmatrix}$$

$$= \mathbf{Y}_e \mathbf{V}_{e_k} + \mathbf{I}_{p_{k-1}} \tag{4.24}$$

The above equation defines the element-admittance matrix $\mathbf{Y}_e$ and the vector $\mathbf{I}_{p_{k-1}}$ which incorporates all "past history" of the inductors and capacitors in the network.

If one follows the development of the general node equations outlined in Sec. 3.2, one will see that the branch (b), independent sources (g), element (e), and dependent sources (d) voltages and currents are (see also Fig. 3.10):

$$\mathbf{I}_b + \mathbf{I}_g = \mathbf{I}_e + \mathbf{I}_d \tag{4.25}$$

$$\mathbf{V}_b + \mathbf{V}_g = \mathbf{V}_e + \mathbf{V}_d \tag{4.26}$$

As before, the currents in the dependent current sources are made up of linear combinations of voltages and currents in elements located in other branches:

$$\mathbf{I}_d = \mathbf{B}^*\mathbf{I}_e + \mathbf{G}\mathbf{V}_e \tag{4.27}$$

where $\mathbf{B}^*$ represents the set of current controlled current sources (betas), $\mathbf{G}$ represents the set of voltage controlled current sources (g_m), with $\mathbf{B}^*$ and $\mathbf{G}$ having zeros on the main diagonals.

At time $t = t_k$

$$\mathbf{I}_{d_k} = \mathbf{B}^*\mathbf{I}_{e_k} + \mathbf{G}\mathbf{V}_{e_k} \tag{4.28}$$

Note that in the above it is assumed that the element voltages and currents are *instantaneously* injected into the to-branches. If the dependent current sources contain any inherent lags, the above relationship must be modified.

From Eq. (4.24), the element voltage may be expressed in terms of the element current

$$\mathbf{V}_{e_k} = \mathbf{Y}_e^{-1}\mathbf{I}_{e_k} - \mathbf{Y}_e^{-1}\mathbf{I}_{p_{k-1}} \tag{4.29}$$

The inverse of the element-admittance matrix is the element-impedance matrix:

$$\mathbf{Z}_e = \mathbf{Y}_e^{-1} = \begin{bmatrix} \frac{2}{\Delta t}[L] & 0 & 0 \\ 0 & [R] & 0 \\ 0 & 0 & \Delta t[1/C] \end{bmatrix} \tag{4.30}$$

where $[L]$ is the matrix of inductances, including mutual inductances, $[R]$ is the diagonal matrix of resistances, and $[1/C]$ is the diagonal matrix of inverse capacitances, sometimes termed *elastances*.

Thus the dependent currents become

$$\mathbf{I}_{d_k} = [\mathbf{B}^* + \mathbf{GZ_e}]\mathbf{I}_{e_k} - \mathbf{GZ_e}\mathbf{I}_{p_{k-1}} \quad (4.31)$$

Note that a considerable simplification of the last relationship is obtained if the voltage controlled current sources (the set $\mathbf{G}$) are dependent only on resistors. In this case the last term of Eq. (4.31) is zero and the dependent current sources do not have to "remember" that past history of the controlling elements.

The branch currents at time $t = t_k$ are obtained from Eqs. (4.31) and (4.25):

$$\mathbf{I}_{b_k} = [\mathbf{U} + \mathbf{B}^* + \mathbf{GZ_e}]\mathbf{I}_{e_k} - \mathbf{GZ_e}\mathbf{I}_{p_{k-1}} - \mathbf{I}_{g_k} \quad (4.32)$$

with $\mathbf{U} =$ an M by M unit matrix.

Similarly, to the above development the dependent voltages are related to the controlling branch voltages and currents, assuming instantaneous action in the dependent sources:

$$\mathbf{V}_{d_k} = \mathbf{D}^*\mathbf{V}_{e_k} + \mathbf{R}_t\mathbf{I}_{e_k} \quad (4.33)$$

where $\mathbf{D}^*$ is the set of voltage controlled voltage sources (set of $[\mu]$) and $\mathbf{R}_t$ is the set of current controlled voltage sources, with $\mathbf{D}^*$ and $\mathbf{R}_t$ containing only zeros on their main diagonals.

Again with the help of Eq. (4.24) the dependent voltages are related to the controlling branch voltages:

$$\mathbf{V}_{d_k} = (\mathbf{D}^* + \mathbf{R}_t\mathbf{Y}_e)\mathbf{V}_{e_k} + \mathbf{R}_t\mathbf{I}_{p_{k-1}} \quad (4.34)$$

The element voltages are given from Eqs. (4.34) and (4.26):

$$\mathbf{V}_{e_k} = (\mathbf{U} + \mathbf{D}^* + \mathbf{R}_t\mathbf{Y}_e)^{-1}(\mathbf{V}_{b_k} + \mathbf{V}_{g_k} - \mathbf{R}_t\mathbf{I}_{p_{k-1}}) \quad (4.35)$$

By Kirchhoff's current law $\mathbf{AI}_b = 0$ at all values of time; hence from Eqs. (4.32) and (4.25)

$$\mathbf{A}(\mathbf{U} + \mathbf{B}^* + \mathbf{GZ_e})(\mathbf{Y}_e\mathbf{V}_{e_k} + \mathbf{I}_{p_{k-1}}) - \mathbf{AGZ_e}\mathbf{I}_{p_{k-1}} - \mathbf{AI}_{g_k} = 0 \quad (4.36)$$

The terms in the last expression are re-arranged

$$\mathbf{A}(\mathbf{U} + \mathbf{B}^* + \mathbf{GZ_e})\mathbf{Y}_e\mathbf{V}_{e_k} + \mathbf{A}(\mathbf{U} + \mathbf{B}^*)\mathbf{I}_{p_{k-1}} = \mathbf{AI}_{g_k} \quad (4.37)$$

The element voltages are related to the branch voltages from Eq. (4.35):

$$\mathbf{A}(\mathbf{U} + \mathbf{B}^* + \mathbf{GZ_e})\mathbf{Y}_e(\mathbf{U} + \mathbf{D}^* + \mathbf{R}_t\mathbf{Y}_e)^{-1}(\mathbf{V}_{b_k} + \mathbf{V}_{g_k} - \mathbf{R}_t\mathbf{I}_{p_{k-1}}) \\ + \mathbf{A}(\mathbf{U} + \mathbf{B}^*)\mathbf{I}_{p_{k-1}} = \mathbf{AI}_{g_k} \quad (4.38)$$

The branch-admittance matrix $\mathbf{Y}_b$ is now defined

$$\mathbf{Y}_b = (\mathbf{U} + \mathbf{B}^* + \mathbf{GZ_e})\mathbf{Y}_e(\mathbf{U} + \mathbf{D}^* + \mathbf{R}_t\mathbf{Y}_e)^{-1} \quad (4.39)$$

Note that this form is the same as was derived in Sec. 3.2. It appears that both $\mathbf{Y}_e$ and $\mathbf{Z}_e$, the inverse of $\mathbf{Y}_e$, are needed to calculate the branch-admittance matrix, but since

$$\mathbf{A}_2^{-1}\mathbf{A}_1^{-1} = (\mathbf{A}_1\mathbf{A}_2)^{-1} \quad (4.40)$$

for A_1 and A_2 being arbitrary invertible matrices, Y_b may be expressed in the form

$$Y_b = (U + B^* + GZ_e)[(U + D^*)Z_e + R_t]^{-1} \qquad (4.41)$$

This last form does not require the element-admittance matrix Y_e and is therefore preferable to Eq. (4.39).

The network node equations are now obtained from Eqs. (4.38), (3.7), and (3.9):

$$AY_b A^T V_{n_k} = A[I_{g_k} - Y_b V_{g_k} - (U + B^* - Y_b R_t) I_{p_{k-1}}] \qquad (4.42)$$

The above equation is now solved for V_{n_k}. The entire transient analysis consists of stepping k from 1 to K where $K\Delta t$ equals or exceeds t_f, the "final" time for which the analysis is to be performed. Note that $I_{p_{k-1}}$ incorporates the voltages and currents for each inductor calculated at the previous time step and the previously calculated voltage across the capacitors [see Eq. (4.24)]. Note that the vector I_s used in Chapter 3 must also include a contribution from I_p in the transient calculations. The overall analysis equations are Eqs. (4.42), (4.41) and (4.30), (4.19), (4.20) and (4.21). The derivations assumed that the branches were numbered such that the inductors are in a group of branches, followed by resistors and then by branches containing capacitors. The final equations, however, are such that, with the sole exception of Eq. (4.24), when $I_{p_{k-1}}$ is calculated using $[L]^{-1}$, it is immaterial how the branches are numbered. The calculation of $I_{p_{k-1}}$ may be done separately (such as in a subroutine) and the values of the equivalent currents are then distributed to the proper branches.

Note that in the above it is not assumed that V_g and I_g are constants; instead, it is assumed only that at a given time step V_g and I_g can be evaluated (again in a separate subroutine) and regarded as constants for that time interval.

Finally, we remark that the vector I_p, which is evaluated at $t = t_k$, can take various forms, depending on what kind of approximation is used for integration of Eq. (4.21) and for obtaining the approximate value of the derivative in Eq. (4.20a). For these other approximations all that is required for I_p is that proper approximations be substituted into the forms which are to be used in place of the one shown in Eq. (4.24).

The formulas above clearly indicate how the node voltages may be calculated at time t_k using the vector $I_{p_{k-1}}$. At time $t = 0$ the vector I_p does not exist. Therefore another scheme must be used for the first solution (the initial condition). For this solution one can solve the given network with the circuit being replaced by a suitable equivalent: The capacitor initial voltages become voltage sources, and the inductor initial currents become independent current sources. Ideally, no passive components should be used with these sources, but this may change the network topology. Thus, in practice, one uses a very small series resistance with the initial capacitor voltage in lieu of the capacitors and a very small conductance in parallel with the inductor initial current in lieu of the inductors. The resultant circuit equations are then solved for the node voltages, thus resulting in an initial condition solution. Then the capacitor voltages and inductor currents are solved for; these form the basis for calculating the first I_p so that the normal circuit equations may be formed for $t = \Delta t$. Subsequent solutions are obtained by applying the procedure outlined above.

Sec. 4.4   Time-Variant Sources   147

A number of extensions of the above procedures are indicated in the problems. As long as no special requirements exist for streamlining computer code or for conserving storage space, the equations above may be assembled relatively easily into a general linear transient analysis program. In coding such a program, provisions should be made for the inclusion of time-variant sources as discussed in the next section.

4.4 Time-Variant Sources

The inclusion of time-variant sources, as part of the input data for a transient analysis program, makes the input data a relatively long stream of numbers. It is very convenient to use some free-form input scheme, such as was discussed in Sec. 3.11. In the following we assume that all data to the transient analysis program developed here will be input through such a free-form routine.

We now discuss the data necessary to provide time-varying sources in a transient analysis program together with the mechanics of implementing the inputting of the necessary information for those sources. For each value of time when the solution is calculated these sources must be evaluated and their numerical values used in calculating the vector $\mathbf{I}_s$. Two decisions that must be made in relation to these questions are:

(1) What kinds of sources shall be considered?

(2) How should these sources be input to the program and stored?

Three types of sources will account for the bulk of time-dependent sources encountered:

(1) Sinusoidal sources.

(2) Periodic (non-sinusoidal) sources.

(3) Non-periodic sources.

Of each type both voltage and current sources will be required; hence a total of six different time-varying sources will be allowed in the transient analysis program developed here. The six sources may be specified by the following input data:

(1) Sinusoidal
$$E \# \text{SIN } A, T, t_0, B$$
$$I \# \text{SIN } A, T, t_0, B$$

(2) Periodic
$$E \# P\, t_1, t_2, v_1, \ldots\ldots$$
$$I \# P\, t_1, t_2, v_1, \ldots\ldots$$

(3) Non-periodic
$$E \# N\, t_1, v_1, t_2, v_2, \ldots\ldots$$
$$I \# N\, t_1, v_1, t_2, v_2, \ldots\ldots$$

To identify a time-varying source, one could design the code such that the letter E (for voltage source) or I (for current source) shall appear as the first non-blank character on an input data card. Following the identifier (E or I) the branch number (#) shall be on the card, either separated from the identifier by any non-numeric string of characters, or immediately next to the identifier. Thus the source type and the branch number in which the source is located are defined. The source is treated as an independent voltage or current source in the basic branch defined in Fig. 3.10; if the identifier is E, the voltage source is treated as being in *series* with any other v_g existing in the branch; if the identifier is I, the current source is treated as being in *parallel* with any other i_g existing in that branch. Polarities of such time-dependent voltage and current sources are the same as the polarities of independent sources in that branch.

Coding considerations for three kinds of sources will now be discussed.

1. Sinusoidal sources (SIN). Such a source is identified by the letter S following the branch number.

Assuming an identifier of E, the time-dependent voltage of a sinusoidal source may be defined as

$$e(t) = A \sin\left[\frac{2\pi(t - t_0)}{T}\right] + B \tag{4.43}$$

Hence four constants are required to completely characterize such a source:

(1) T = period.

(2) A = amplitude.

(3) t_0 = time-offset for origin of sine wave.

(4) B = D.C.-offset of periodic wave.

These four numbers are generally floating-point values; the last two may be omitted if their value is zero. These quantities are indicated in Fig. 4.12(a).

2. Periodic Sources (P). Such a source is identified by the letter P followed by the branch number. Values for a periodic source will be entered by a string of floating-point numbers. The first value, t_1, is the time period for which a string of values $(v_1, v_2, \ldots, v_n)$ will be supplied, as shown in Figure 4.12(b). The number of such different values is

$$n = \left\lceil \frac{t_1}{t_2} + 0.5 \right\rceil \tag{4.44}$$

with $\lceil x \rceil$ denoting the integer part of x.

Periodicity demands that $v_{n+1} = v_1$. Hence the values to be stored for a periodic source will be $t_1, t_2, v_1, \ldots, v_{n+1}$ for a total of $(n + 3)$ values. The first value v_1 is for $t = 0$.

Sec. 4.4   Time-Variant Sources   **149**

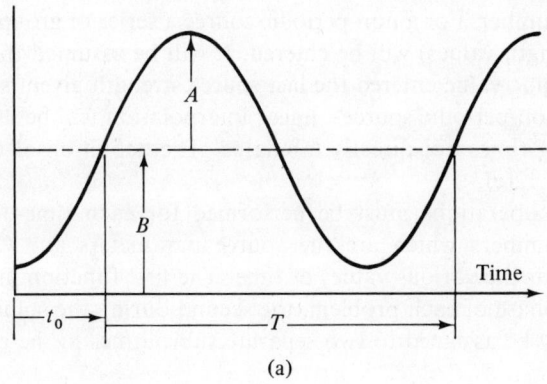

(a)

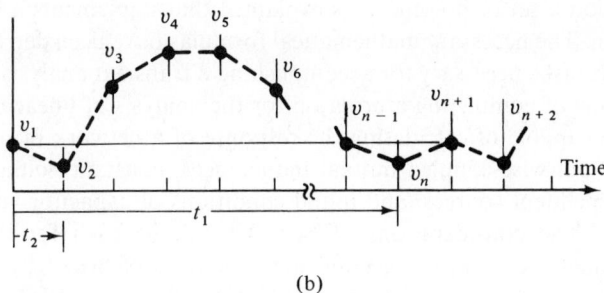

(b)

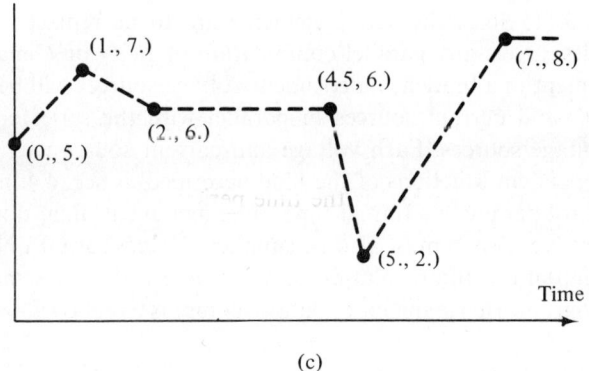

(c)

Fig. 4.12 Independent Sources for Transient Analysis Program: (a) Sinusoidal Source; (b) Periodic Source; (c) Nonperiodic Source

3. Non-periodic Sources (N). These sources shall be identified by the letter N after the branch number. For a non-periodic source a series of groups of two numbers (time, source strength values) will be entered. It will be assumed that for time larger than the highest time value entered the last source strength given is being held. Both for periodic and non-periodic sources, linear interpolation will be used to find source strength values for times not directly tabulated. The definition of the data values is indicated in Fig. 4.12(c).

Two distinct operations must be performed for each time-dependent source: (1) storing the numbers which are the source parameters and (2) evaluating the source strengths for the various values of time. The first function must be performed during the input phase of each problem, the second during the solution phase. These two functions may be assigned to two separate subroutines in the program.

4.5 A Basic Transient Analysis Program

In this section a set of flowcharts is explained that implements a basic transient analysis program. The necessary mathematical formulas have been derived in previous sections. Various tasks necessary for accomplishing a transient analysis are assembled here with the view of completing a program for the analysis of linear networks. This program will be capable of calculating the response of a network of passive components, including pairwise-coupled mutual inductances, constant voltage and current sources, time-dependent sources, and initial conditions of capacitor voltages and inductor currents. The considerations of Sec. 4.4 will be used for time-dependent sources, the inductor and capacitor equivalent elements of Sec. 4.1 will replace the inductors and the capacitors, and mutual inductors will be modeled as in Sec. 4.3a (pairwise-coupled inductors, replaced by equivalent inductors and dependent sources).

The input to the program shall consist of data supplied in freeform format as discussed in Sec. 3.11. Since the active elements are to be replaced by equivalent conductances and sources, any parallel combination of R, L, or C may be used for the dependent element in a branch; independent voltage sources will be in series with such an "element" and current sources in parallel with the series combination of "element" and voltage sources. Each voltage and current source may be a constant value and time-dependent functions of the kind described in Sec. 4.4 in series with it.

Solution control parameters (initial time, time increment, final time) and output control parameters (i.e., VN2) must also be supplied. Values for OPEN and SHORT must be used in initial condition solutions; their numerical values may be set optionally in this program. Hence the input data structure is as follows:

(A) Required solution control card: The values of initial time, time increment, and final time must appear in this respective order on a card after the identifier TIME =. The letter T must be the first character on the card; the equal sign must precede the value of initial time. The three numerical values may appear anywhere on the card

Sec. 4.5 A Basic Transient Analysis Program 151

in contiguous columns with any non-numeric separator. The values may be any integer or floating-point format.
Examples:

```
TIME = 0.,1.5/3.
T = -0.1E-2, 1.5E-4  3.7E-3
```

(B) Optional solution control cards: The values of OPEN and SHORT (by default 10MEG and 0.01) may be reassigned:

```
OPEN = 1.E+6
SHORT=0.1E-2
```

The letters O or S must be the first non-blank characters; the equal sign must precede the integer or floating-point value.

(C) Comment card: Any card whose first column contains a 'C'. Such a card will be listed as part of the input, but it will be ignored otherwise by the program.

(D) Output specification card: The first two non-blank characters must be 'OU' (for OUTPUT); the equal sign (' = ') must be followed by any combination of delimiters made up of two letters, followed by an integer of one or two digits. The first letter may be V, I, or P (for voltage, current, or power); the second letter may be N, B, E, or C (for node, branch, element, or component); the integer denotes the corresponding circuit node or branch number, for example,

```
OUT=VN12,BC3,PB2,VN4
```

specifies node voltage 12, branch current 3, power in branch 2, and node voltage 4 to be printed for every value of time. Up to nine such specifications may be given; the output will consist of a single line of print containing the value of time followed by the values of the variables specified for printing. If no output specification is given, node voltage 1 is printed by default.

(E) Component description cards: Three integers, each one or two digits separated by any non-numeric field, followed by a specification of an element.

(1) For passive components and fixed sources: The element specification may be any one of the combination

```
G,R,L,C,E,I,EO,IO
```

("oh" or "zero" are usable for the last two specifications) followed by ' = ', followed by an integer of floating-point value. Any combination is allowable.
Example:

```
3,2,4,R=1.5E3,C=4.5E-6,E0=15.3
```

describes branch 3, from node 2, to node 4 containing a resistor of 1.5 kΩ, a capacitor of 4.5 μf with an initial voltage on the capacitor of 15.3 V.

(2) Time-dependent sources: These are identified by a string of numbers as defined in Sec. 4.4; the first non-blank character must be an E (for a voltage source) or an I (for a current source).

(3) Controlled sources: Same format as for passive elements; the first integer is taken as a sequence number for the source. The identifier A = (for BETA =), GM = must precede the value of β or g_m. The controlling branch is the second integer, the controlled branch the third integer.

(4) Mutual inductances: Same format as 2, just preceding. The identifier M = must precede the value of the mutual inductance.

(5) Controlled voltage sources are not allowed.

(F) Execution is initiated by the letters EX as the first two non-blank characters on a card.

(G) The program may be terminated by EN (for END) as the first two non-blank characters on a card.

(H) The order of input cards of type A to E is immaterial. All columns between 1 and 72 may be used for input data. Columns 73 through 80 are ignored by the program.

Figure 4.13 presents a flowchart for the overall analysis task. The program uses a set of arrays, such as $\mathbf{Y_n}$ and $\mathbf{I_s}$, which must be declared in the program. In writing the actual code whenever a new array is used, it is put in a DIMENSION statement at the beginning of this routine and is also initialized with the appropriate DO-loops or DATA statements. It is also preferable to use a COMMON statement for all arrays so that arguments may be omitted in the various subroutine calls.

Part of the initialization is the setting of the numerical values of an "open circuit" (here set to 10 megohms) and a "short circuit" (here set to 0.01 ohms). These values are used within the program to calculate an initial condition solution for the network. The integer KSTART = 1 is used to flag such an initial condition solution; the subroutine SETUP, used for setting up the node-admittance matrix $\mathbf{Y_n}$, uses the value of OPEN to replace any inductance and the value of SHORT to replace any capacitance. The network is then analyzed as a simple DC network by calculating all node voltages, branch voltages, and currents due to the values of the initial capacitor voltages and inductor currents, including the values of all time-dependent sources at the initial time. The technique shown in the flowchart consists of calculating $\mathbf{I_s}$ and solving for $\mathbf{V_n} = \mathbf{Y_n^{-1} I_s}$.

The various subroutines and their functions shown in Fig. 4.13 are as follows:

(a) SUBROUTINE INPUT scans input data cards and stores all data necessary for the analysis. Circuit element descriptions, values, and interconnection data are input in the form described earlier in this section. Transient analysis also requires the initial time TO, the time increment TINC and the final time TF. Optionally, values for OPEN and SHORT may be input and a set of specifications for output printing (editing) may be given.

Another function of this subroutine is to recognize time-variant source descriptions and to prepare the appropriate arrays for these numbers (see also Sec. 4.4).

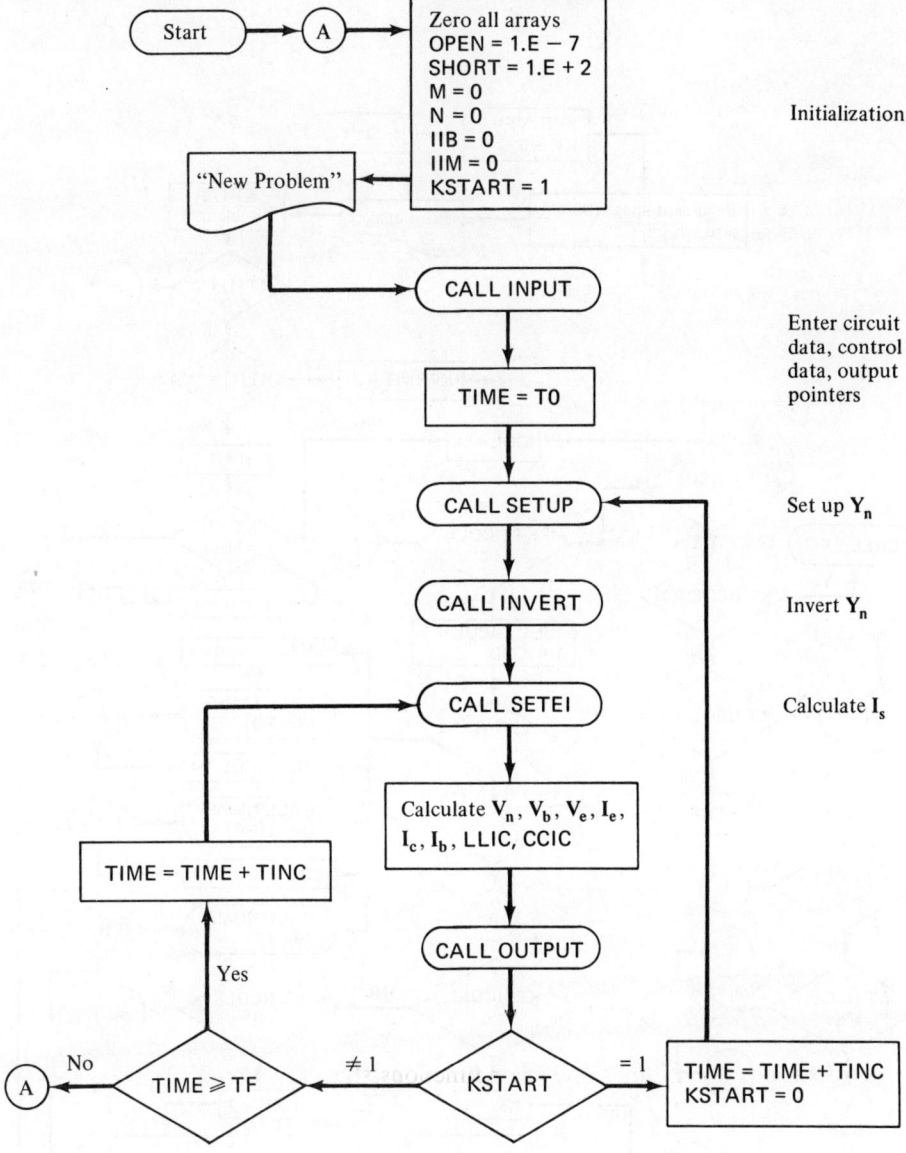

Fig. 4.13 Flowchart of Main Routine TRALP

The flowchart for this routine is shown in Fig. 4.14; the various subroutines used are as follows:

(i) SUBROUTINE NUMB obtains the value of a general floating-point number from a free-field input card. This subroutine is listed in Listing 3.7. Element values,

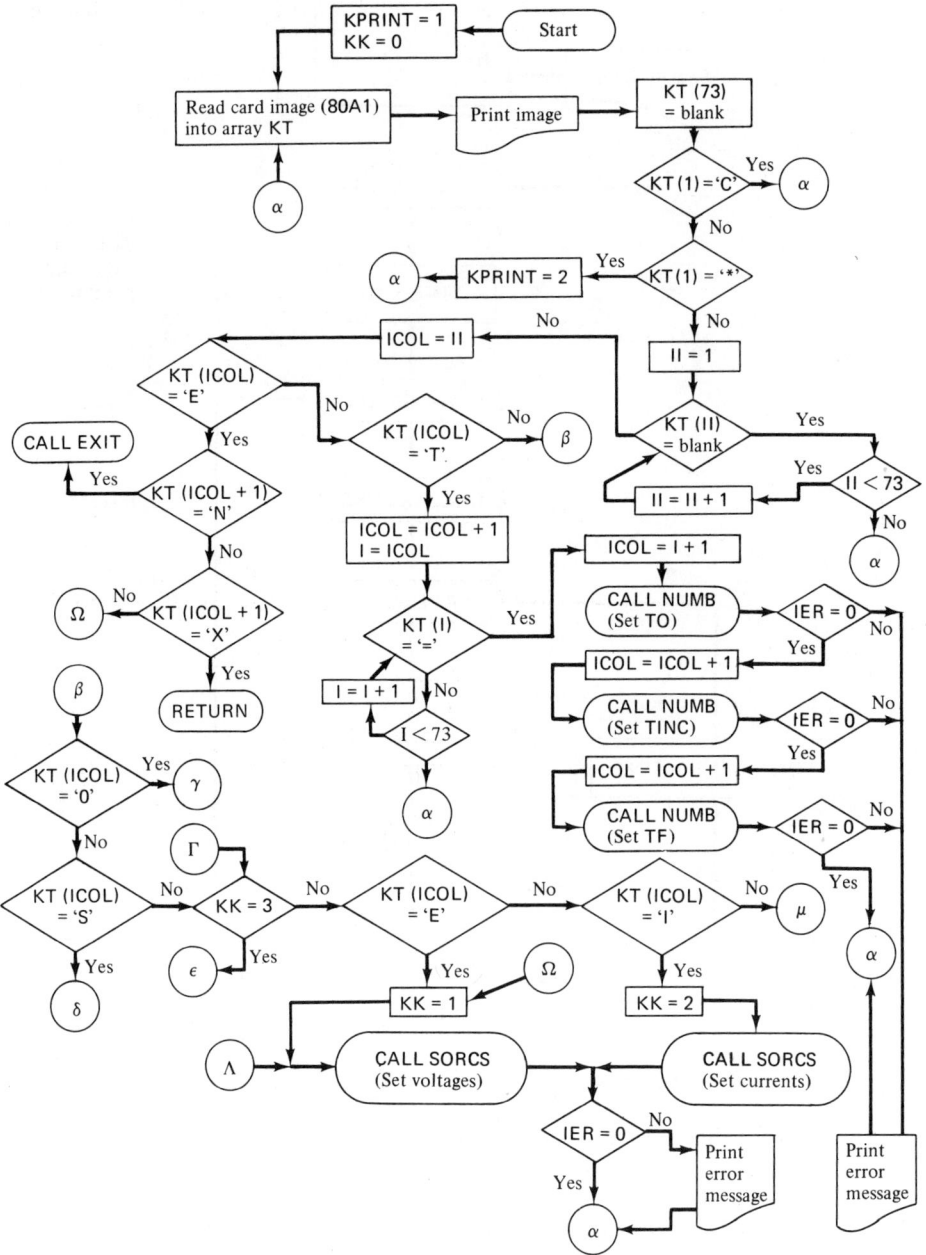

Fig. 4.14 Flowchart of Subroutine INPUT

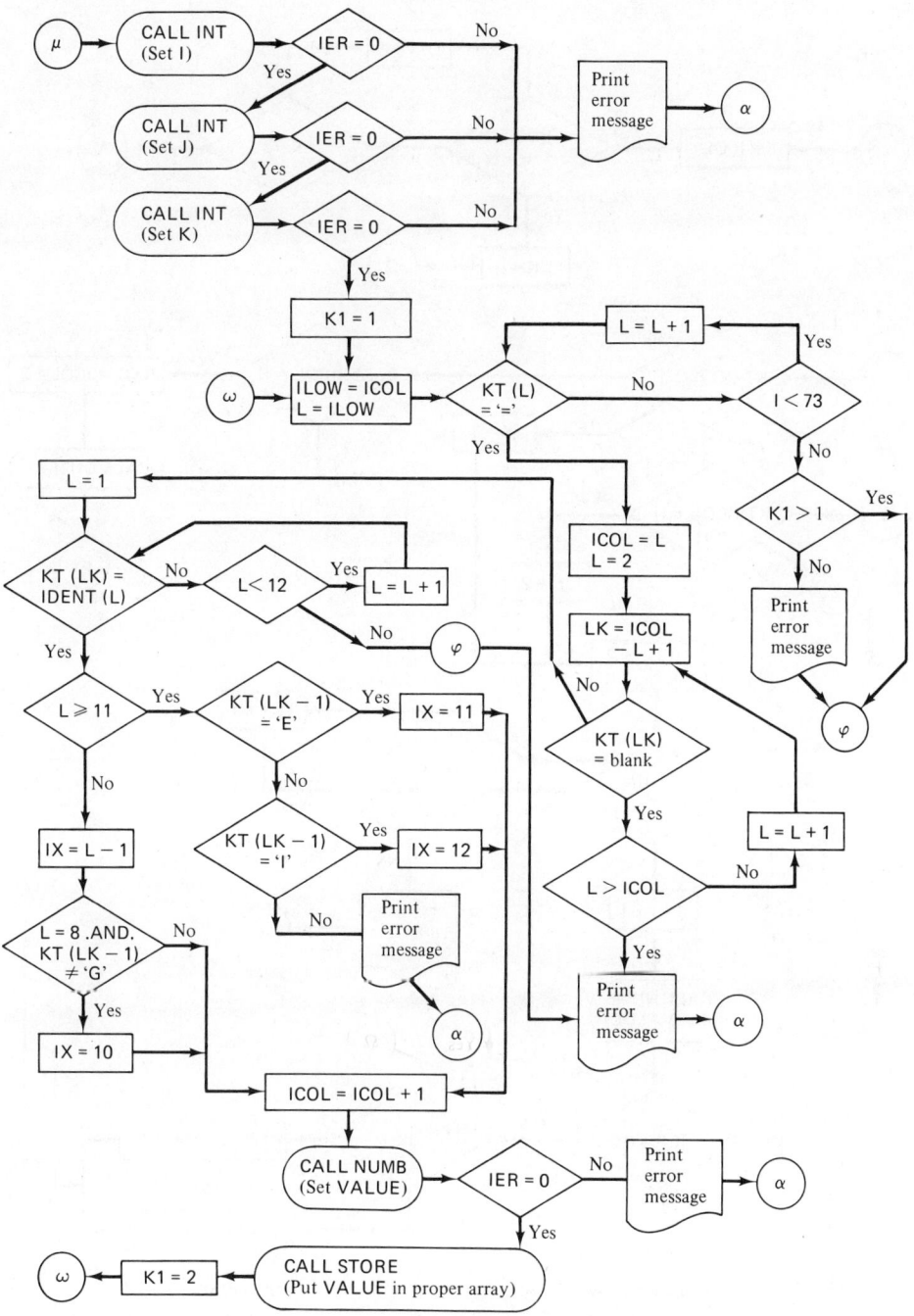

Fig. 4.14 Flowchart of Subroutine INPUT—*Cont*.

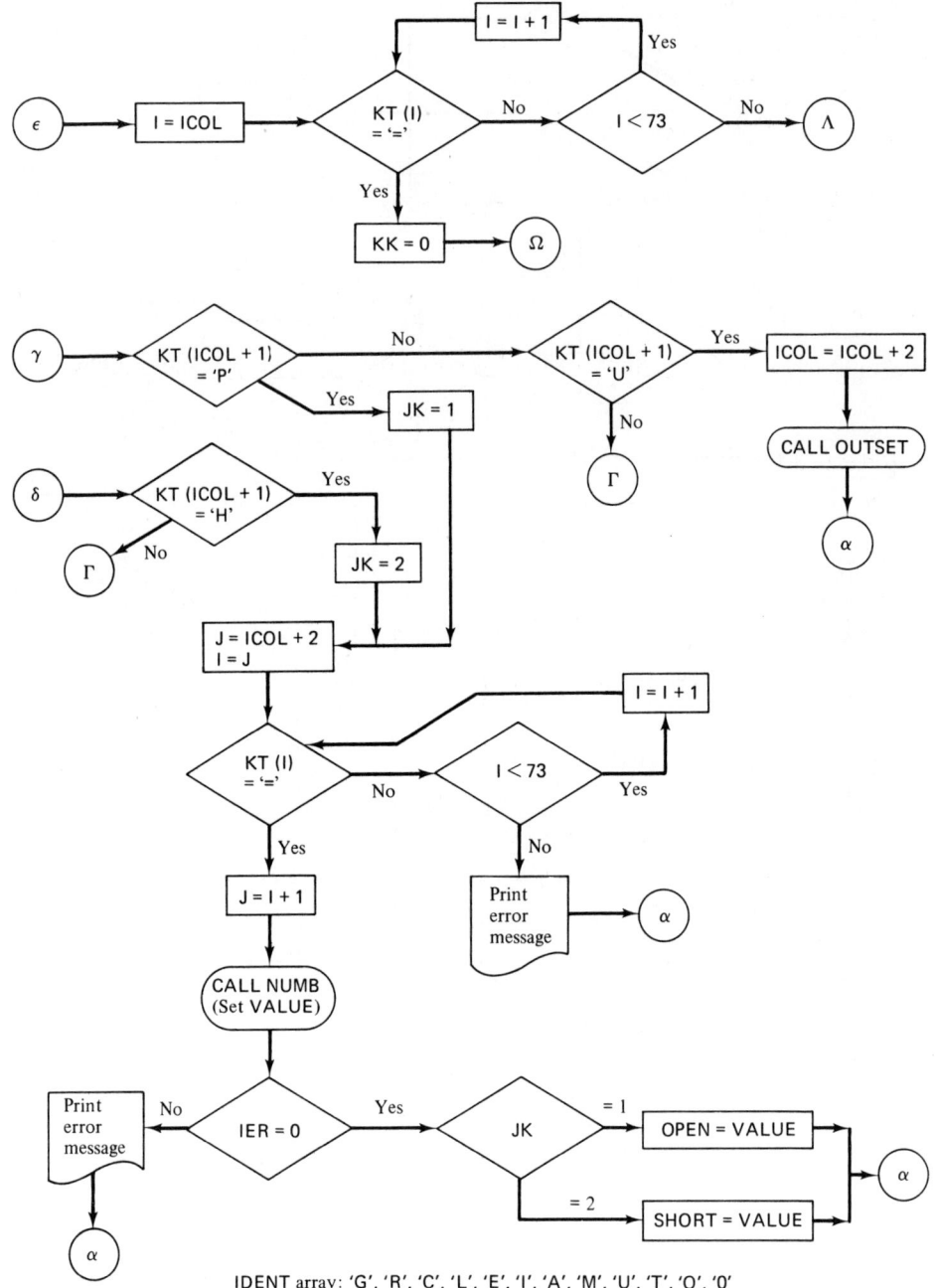

Fig. 4.14 Flowchart of Subroutine INPUT—*Cont*.

values for OPEN, SHORT, TO, TINC, and TF are identified with this routine. SUBROUTINE INT is used by this routine.

(ii) SUBROUTINE INT reads an integer value from a free-form data card. This routine is listed in Listing 3.7 and is used for identifying branch and node numbers.

(iii) SUBROUTINE STORE is used to store an identified value in the element arrays and array locations. The structure of the subroutine closely resembles the storing instructions used in Program ACAN 2 shown in Listing 3.3 (statements 240 through 451).

(iv) SUBROUTINE SORCS stores data on time-dependent sources into the arrays describing such sources. There are two separate arrays used in the storage of time-dependent sources data:

(1) An array NSORC containing four numbers for each source describing what kind of source (1 = votage source; 2 = current source); what branch number the source is located in; what type of source (1 = sinusoidal; 2 = periodic; 3 = non-periodic); and what is the last value entered for this source in an array of floating-point values SORC. For the Lth such source, the floating-point values needed to describe the source data (see Sec. 4.4) will be located between SORC(I) and SORC(J) where I = NSORC (4*(L−1)) + 1 and J = NSORC (4*L).

(2) The array SORC that contains the actual values describing the sources.

Note that the array NSORC is similar to the "guide array" IG used in Sec. 3.8 and that the array SORC is the exact counterpart of the vector Y of that section.

This subroutine must be able to recognize source data descriptions that may extend over many input data cards. The flowchart of this routine is shown in Fig. 4.15.

(v) SUBROUTINE OUTSET is a routine that recognizes a set of output flags. This routine is flowcharted in Fig. 4.16. The routine scans the input card image left to right until the first equal sign is located. After that the first alphabetic V, I, or P is identified (all others are ignored); the next character then must be N or B or E or C. The valid combinations of these characters are used to set the number KSS as shown in Table 4.3.

Table 4.3 *Print-Flag Value Used in First column of Array* KPR

PB = 1	IB = 4	VB = 7	VN = 10
PE = 2	IE = 5	VE = 8	
PC = 3	IC = 6	VC = 9	

The print flag is stored in column 1 of the array KPR; column 2 contains the corresponding node or branch number, found by subroutine INT. Up to nine such print flags may be used; hence KPR is DIMENSIONed as KPR(9, 2). These flags are used in subroutine OUTPUT to produce the required output quantities.

If no output is specified or can be specified, KPR(1, 1) is set to zero; subroutine OUTPUT will then produce VN(1) as the default output.

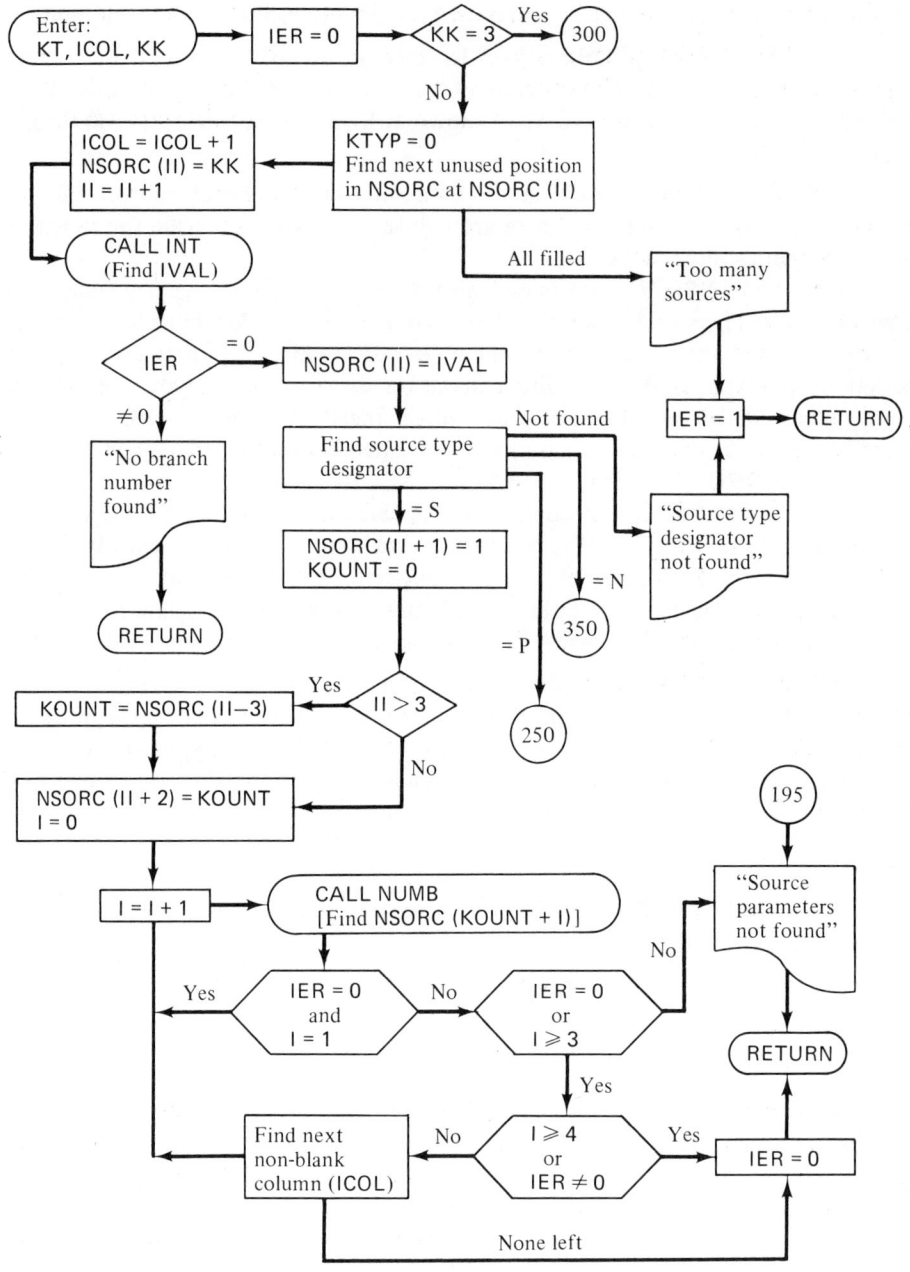

Fig. 4.15 Flowchart of Subroutine SORCS

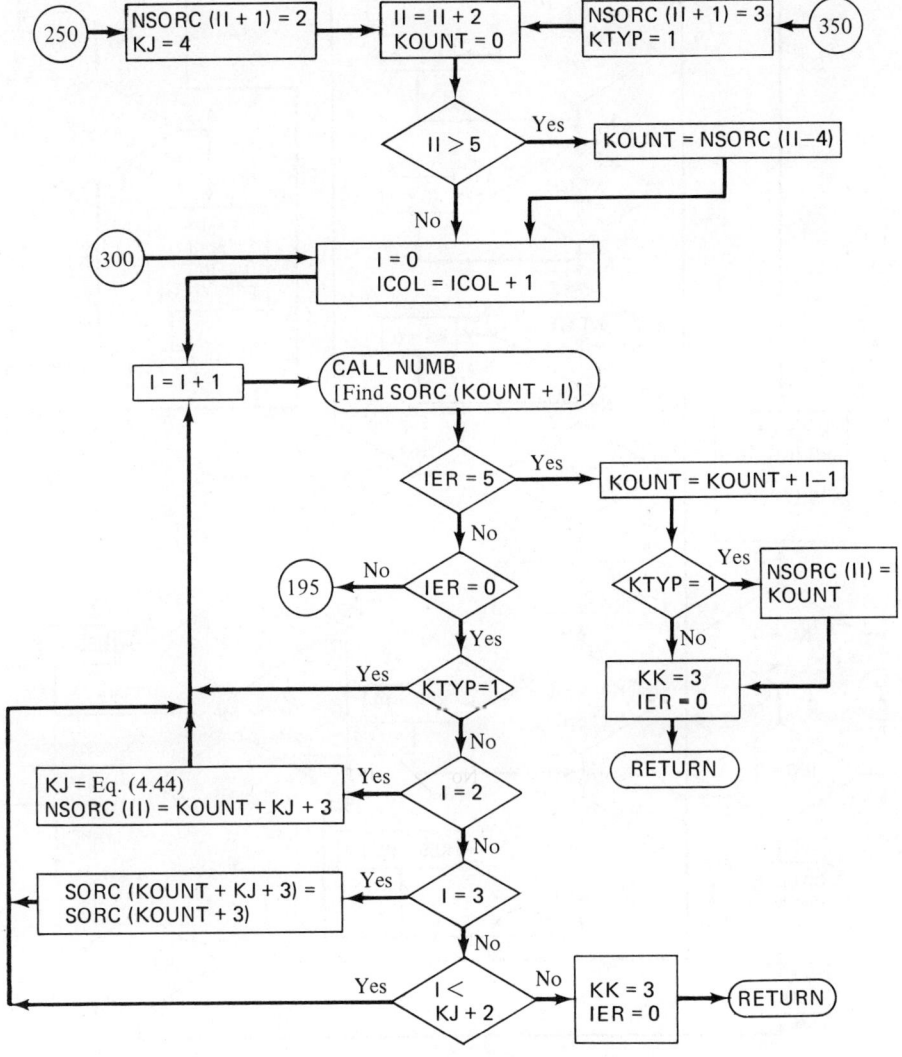

Fig. 4.15 Flowchart of Subroutine SORCS—*Cont*.

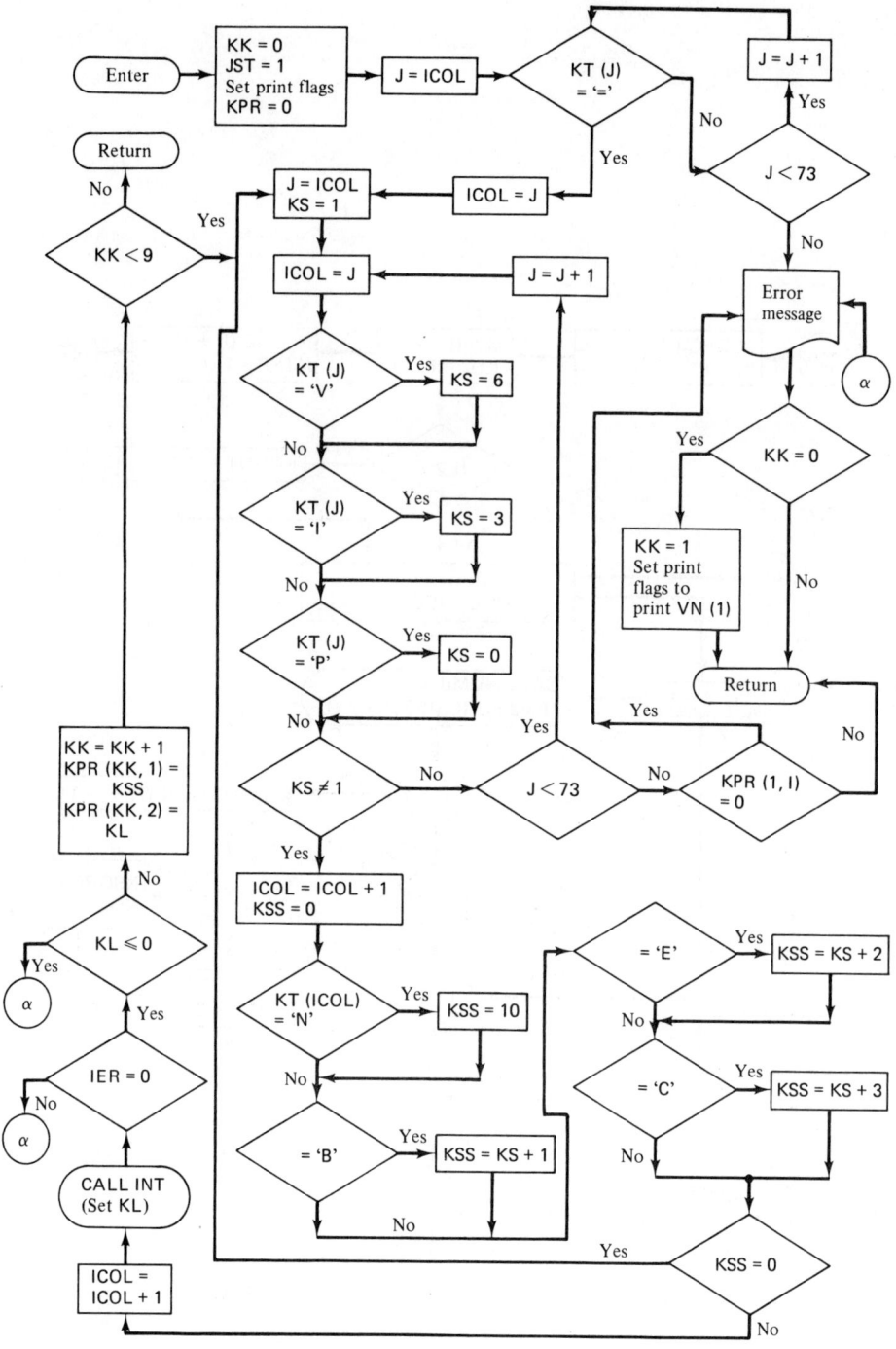

Fig. 4.16 Flowchart of Subroutine OUTSET

The organization of subroutine INPUT is such that additional input forms may be handled by appropriately modifying array IDENT and extending the chain of comparisons for the newly added forms at point ⑦ in the program, second page, Fig. 4.14. The flowcharted program does identify μ(MU) and r_t(RT) components (produces IX = 9 and IX = 8 respectively); Subroutine STORE ignores these elements, and no provision is made elsewhere in the program for these components.

(b) SUBROUTINE SETUP prepares the appropriate entries for the Y_n and I_s matrices. For the initial condition solution, a large conductance value (SHORT) must be used for the capacitors and a very small conductance (OPEN) must be used for the inductors. The initial condition solution is calculated for KSTART = 1.

The various Y_n and I_s entries are obtained exactly as they were in the AC analysis program detailed in Sec. 3.5 (see Listing 3.3); the convention used in that section for handling β and g_m sources is retained here (negative controlled branch number for g_m). The only difference is that prior to setting up any node equations the mutual inductors are replaced by their controlled source equivalents as discussed in Sec. 4.3a. This latter operation will change some of the LL entries, create new entries in IB and B (their meaning is the same as those used in Listing 3.3), and modify N and M.

This routine is partially flowcharted in Fig. 4.17.

(c) Subroutine INVERT inverts the YN array. The routine listed in Appendix A may be used. It is advisable, however, to double-precision the routine. Significant round-off errors may be encountered even with moderate circuits since the range of numerical values (SHORT to OPEN) is very large. The routine can be re-arranged so that its input and output are single-precision but all internal operations are performed on double-precision values. Provisions must be made to provide adequate storage for the YN matrix in double-precision during the execution of this program.

(d) Subroutine SETEI evaluates the time-dependent sources at the current value of TIME and stores the contributions of these sources in the vector T. The data structure for storing information about the sources is discussed in Sec. 4.4; the vector T in the program is the vector of current generators I_s (see Fig. 4.18).

(e) Calculation of V_n and other intermediate results is done as outlined in Fig. 4.10; the equivalent generators LLIC and CCIC are reset using the models shown in Figs. 4.2 and 4.3.

(f) Subroutine OUTPUT prepares one output print line. The data are presented in tabular form of up to ten columns of print. The first column is the value of TIME, the next columns are as specified on the OUTPUT card; the array KPR is used to hold the output flags. Because of its simplicity no flowchart of this routine is presented.

(g) A check is made to see if the final time has been reached; if not, time is incremented by TINC and the analysis repeated. This is shown at the bottom of Fig. 4.13.

Coding of this routine is left as an exercise. Including subroutines NUMB, INT, and INVERT, the listing contains approximately 1,000 FORTRAN statements.

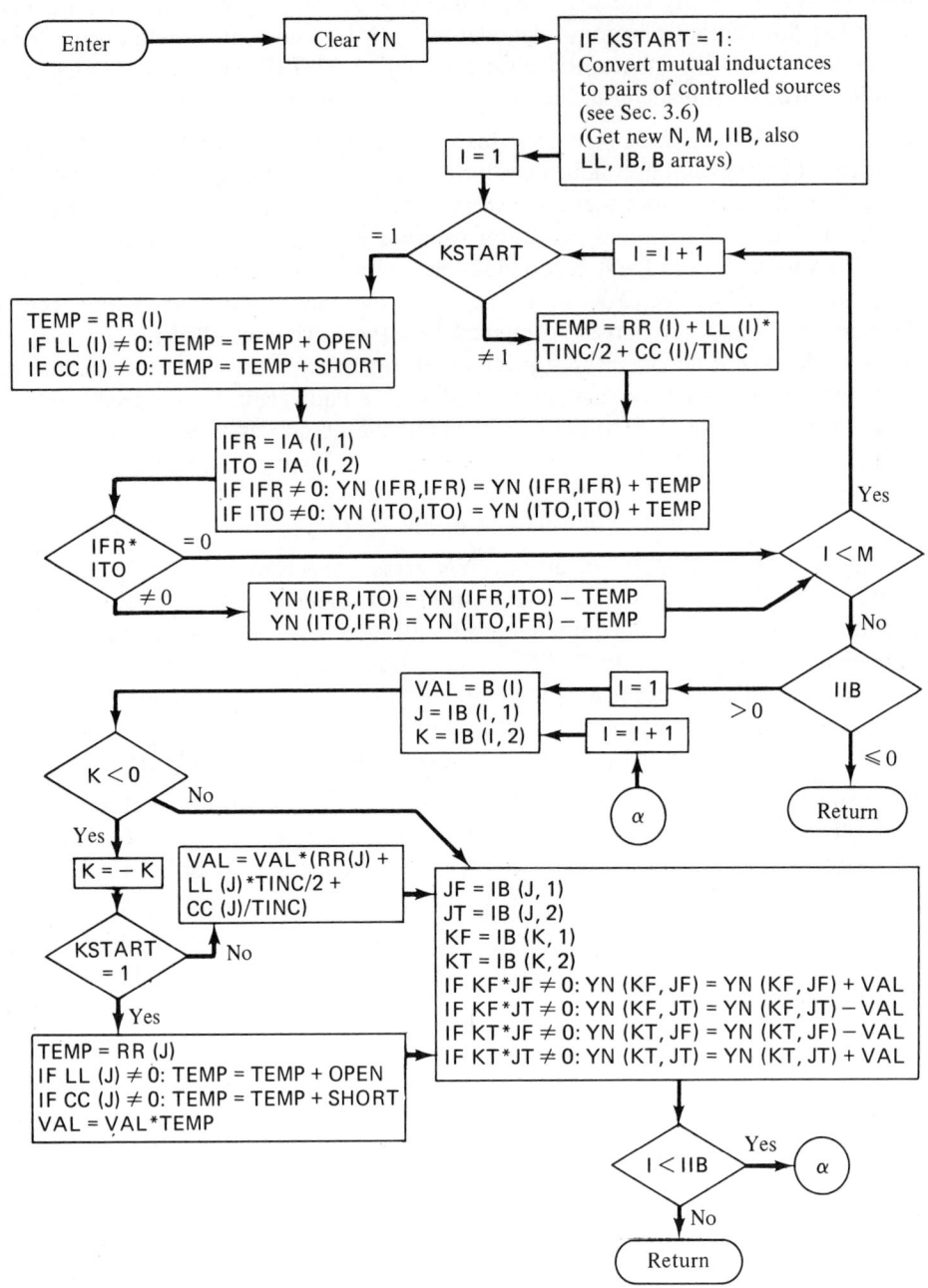

Fig. 4.17 Flowchart of Subroutine SETUP

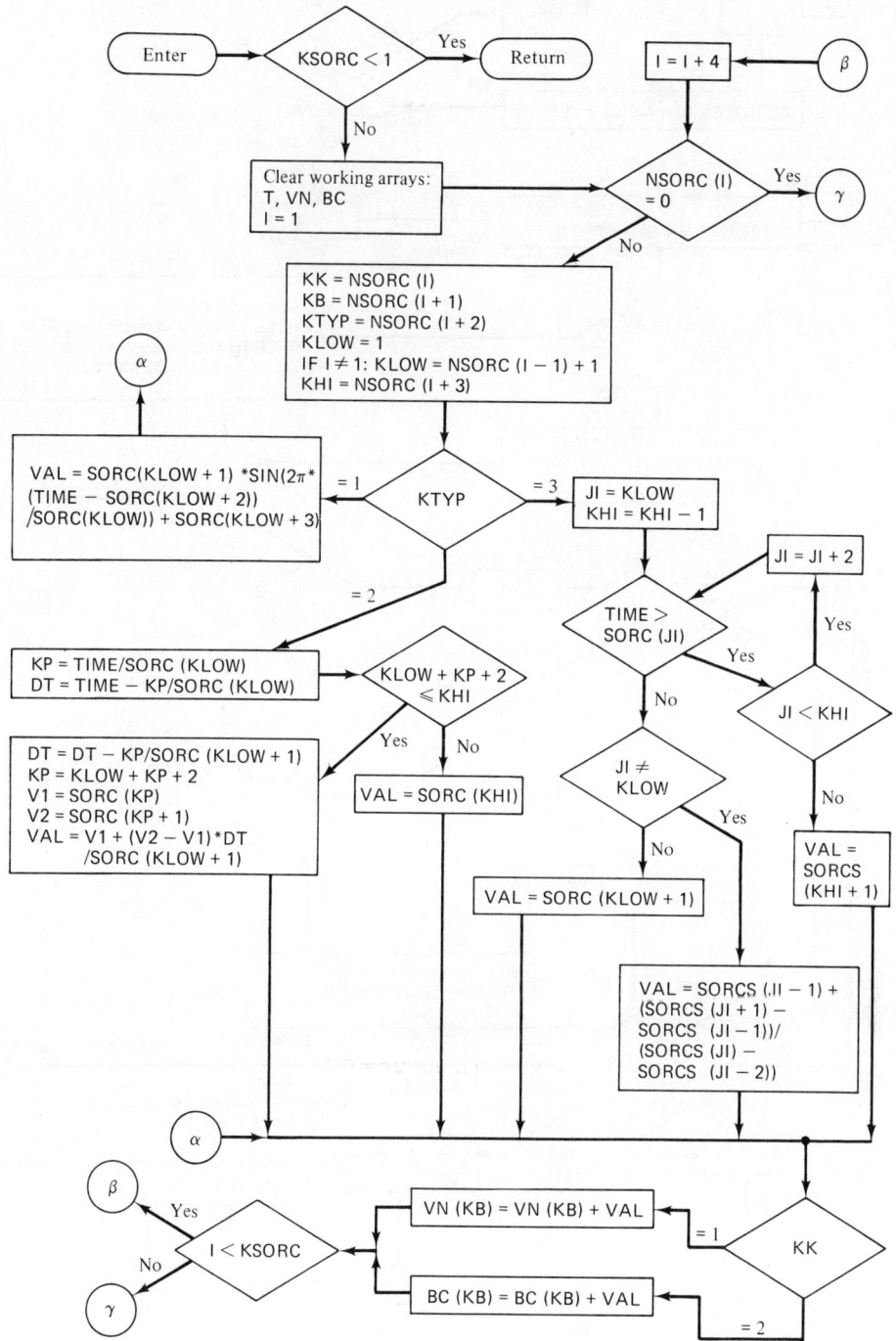

Fig. 4.18 Flowchart of Subroutine SETEI

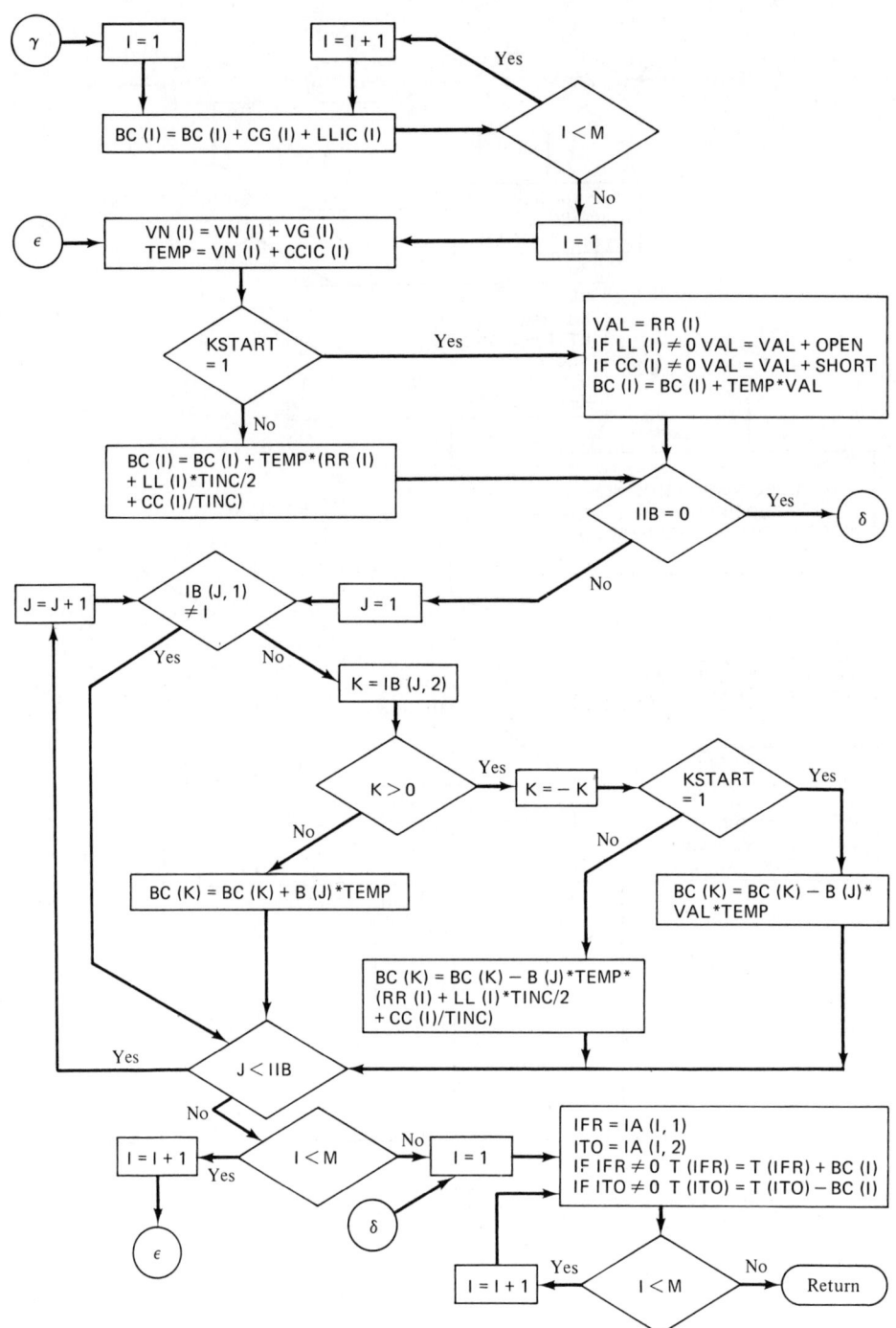

Fig. 4.18 Flowchart of Subroutine SETEI (cont.)

Sec. 4.6   Final-Value Solutions   **165**

EXAMPLES

(1) The input data usable by this routine for Ex. 4.1 is shown in Fig. 4.4. Note the use of **B** and **N** as delimiters; the program ignores them.
(2) The input data for Ex. 4.2 is shown in Fig. 4.7.
(3) The input data for Ex. 3.8 (Sec. 3.5) is as follows:

```
TIME=0.,1.E-5,1.E-3
B1,N0,N1,R=100.,E=1.E-3
B2,N1,N2,C=20.E-6,E0=0.
B3,N2,N0,L=0.05,I0=0.
B4,N0,N3,L=0.10
B5,N3,N0,C=1.2E-5
B6,N3,N0,R=200.
B7,0-4 R=5E3
B8 4-5 C=2.E-8
BRANCH 9 FROM 5 TO 0, RESISTOR=10E3
B10 (0-6) R=47.0E+3
11/6/7/C=1.E-9
12/7/0/R=470.E+3
1/3/4/M=0.05
1/6/7 BETA= 80
2/9/10 GM=6.E-3
OUTPUT=VN(7),VN(3),VB2
EX
```

Note that mutual inductances and controlled current sources are numbered separately. OPEN and SHORT could also have been specified.

4.6 Final-Value Solutions

The equations developed in Sec. 4.4 and the program outlined in Sec. 4.5 can be used, with minor modifications, to calculate the node voltages (and hence all voltages and currents in the network) after all transients have died out. Formally, the final value of V_n is

$$V_n(\infty) = \lim_{t \to \infty} (AY_b A^T)^{-1} A(I_g - Y_b V_g) \qquad (4.45)$$

The value of $V_n(\infty)$, the final value of V_n, will exist in a stable network (where all poles are in the left-half plane; see Sec. 5.6) provided that the generators will have a constant value as $t \to \infty$. Hence I_g and V_g may not contain any sinusoidal or periodic sources; only non-periodic sources can lead to a final-value solution.

The program can be easily modified for the calculation of such final values. This condition requires that currents through inductors and voltages across capacitors do not change. We ensure this by setting the inductor equivalent resistance to zero and the capacitor equivalent resistance to infinity. In the program the values of SHORT and OPEN would be substituted for these respectively.

In addition, a new keyword would have to be used. The word EQUILIBRIUM (or EQ as the first two non-blank characters on an input line) could be used for such a purpose; a check should be made for the absence of sinusoidal and periodic sources (array NSORC can not contain 1 or 2 in the proper locations, locations 1, 5, 9, . . .). The network is then analyzed as a DC network, akin to the initial condition solution characterized by KSTART = 1 in the flowchart of the transient analysis program shown in Fig. 4.13. For the purposes of this modification it may be advantageous to designate KSTART=2 as the final-value solution, exchange the values of OPEN and SHORT, and then do an *initial value* solution.

In actually coding this change the most complicated part is to modify Subroutine INPUT to accept the new keyword EQ (for EQUILIBRIUM). A few other changes are also necessary when the variable KSTART is examined at various points in the solution process. Modifications must be made to properly handle the new case of KSTART=2.

Normally, it is too complicated to examine in detail the question of existence of a set of final values for V_n. The usual practice is to make cursory checks on the I_g and V_g sources and calculate whatever might result from the substitution of SHORT for inductors and OPEN for capacitors. The burden of ensuring the validity of the resultant numbers is left for the user.

Coding details are left as an exercise.

4.7 Piecewise Linear Networks

In carrying out calculations on the time-domain analysis of circuits it is quite easy to alter the value of any number of components. From Fig. 4.13 one can easily see that the computational loop for KSTART = 1 can be used at any time to restart the analysis procedure if any parameter is to be altered. For example, if the time step were to be changed, Subroutine SETUP would have to be called anew, such that the new equivalent conductances for capacitors and inductors are obtained. At that same time one could just as easily change the circuit components too. One would merely have to hold off destroying the last set of computed values at point A in the flowchart. Thus when the calculations are completed, one returns to Subroutine INPUT, searches for a keyword (such as CONTINUE, or at least CO) as the first non-blank characters on an input card. Only when the first input card is not such a CONTINUE would one start with the initialization task in the subroutine.

Note that one does not need to do a new initial condition solution (KSTART = 1) unless new voltages are introduced on the capacitors and/or new currents are set in the inductors. One proceeds merely to use the last set of computed capacitor voltages and inductor currents as the solution at the previous time point; this automatically generates the initial condition solutions from which subsequent node voltages are calculated by the recurrence formula Eq. (4.42).

The mechanism for changing the network component values at the end of the time interval TO to TF is a trivial task. A new "initial conditions" solution is started when the changes in the circuit data involve capacitor voltages and/or inductor currents; otherwise the program steps remain unchanged except for one extra call to Subroutine SETUP. Details for such an implementation are left as an exercise (see Problem 4.22).

A more comprehensive change to the program involves changing circuit element values as a function of voltages or currents within the circuit. Note that the component values depend on quantities that are solutions of the network equations; this makes the network being analyzed a nonlinear one. For example, a diode can be considered as an ON-OFF device; the internal impedance of the device in the ON condition may be a few hundred ohms; in the OFF condition it may be several megohms. The direction of current flow determines which value one would use; however, in either condition the circuit to be analyzed is a linear circuit. If a network contains such diodes for each change in currents through the branches containing the diodes, a new linear circuit is analyzed until the current changes in another diode (or possibly in the same diode); the procedure is repeated until the analysis for the time interval is completed.

We will now consider some details in implementing a scheme that allows branch element values to be switched depending on the current direction of a branch in the network. Such a capability is akin to having a polarized relay of instantaneous response in the network. If the controlled element and the controlling (monitored) branch current refer to the same branch, a simple diode may be simulated; otherwise we shall have a general switching capability for the network.

One of the difficulties with this scheme is finding the instance when the new conductance value is to be substituted, i.e., finding the time point at which the current reverses. Obviously, when the network equations are solved at discrete time points only, the instant of zero current will most likely be between two solution points. Some procedure has to be used to subdivide the basic time interval **TINC** whenever the sign of the current changes between two successive solution points. Also, a limit must be set on just how small an error in the time one is willing to tolerate when searching for the switching time. Similar to the values **OPEN** and **SHORT**, this too can be made a parameter in the program.

Upon detection of the fact that a current direction has changed through a branch one can try to use linear interpolation to find the time at which the current was zero. Let the branch current in question at time t_k be i_k. Then the time when the current was zero is

$$t_z = t_k + \Delta t \frac{1}{1 - \frac{i_{k+1}}{i_k}} \tag{4.46}$$

Hence a new time step of

$$\Delta t' = \Delta t \frac{i_k}{i_k - i_{k+1}} \tag{4.47}$$

should be used in a new analysis for the time interval t_k and $t_k + \Delta t'$. The above formulas can be re-applied until the difference between two successive $\Delta t'$ values falls below a threshold value. Normally, such threshold may be $10^{-3} \Delta t$, and a few iterations will establish the switching time to this accuracy; the actual number of trials does depend on the behavior of the current in the interval (t_k, t_{k+1}).

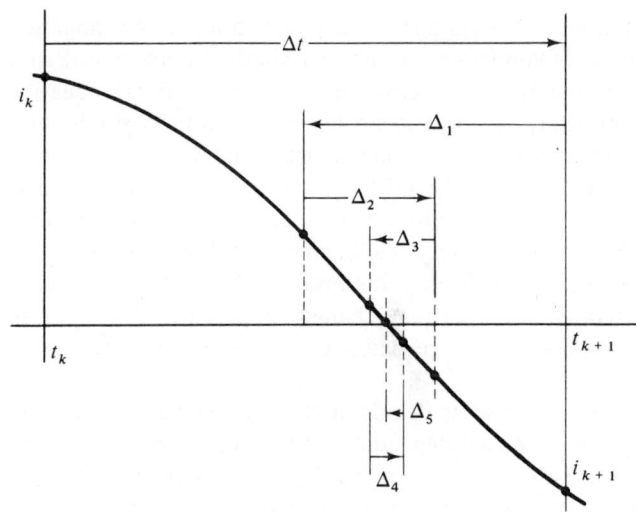

Fig. 4.19 Interval-Halving Procedure

A less sophisticated approach is to use interval halving. Successively Δt is divided in two each time the currents are different at the beginning and end of the time interval as illustrated in Fig. 4.19.

The interval halving procedure consists of halving the time step between the last two currents (which must differ in sign), recalculating the current, and discarding that solution point which has the same sign as the newly calculated current. The procedure is then re-applied to the reduced interval. In this manner a time interval reduction of 10^{-3} can be obtained in ten steps ($2^{10} = 1024$). Fig. 4.19 illustrates five such steps.

The major advantage of the interval-halving procedure is that it is little affected by round-off errors that can overwhelm the denominator in Eq. (4.47) and always converges in ten steps to an interval of $10^{-3} \Delta t$. The procedure is also easier to implement.

Note that when an interval is changed all that is really necessary to calculate the new node voltages is a change in some of the values in the $\mathbf{Y_b}$ and $\mathbf{I_p}$, Eqs. (4.24) and (4.42), as well as reevaluation of $\mathbf{I_g}$ and $\mathbf{V_g}$ at the new time point. This can be accomplished in the program flowcharted in Fig. 4.13 by re-entering Subroutine SETUP with new TINC, as is done with the first time-step solution (KSTART = 1 in Fig. 4.13). Some computer time may be saved by changing such elements only and not restarting the calculation of $\mathbf{Y_b}$ from a zeroed-out matrix. Such programming details are, however, not necessary to writing a capable code.

The format of setting up such element switching is relatively easy to define, but it leads to substantial changes in the code outlined before. Obviously, it will be necessary to store two values for each element that may be altered; programming simplicity will dictate that this should be accomplished by setting up doubly dimensioned arrays **RR**, **LL**, and **CC**. Conceptually, it is, of course, possible to switch all circuit components, although identification in the input may be a more complicated matter. With passive components one needs merely refer to the branch number; if we then restrict the element in any branch to a single passive component, the element is uniquely described with just a single number. Thus for the moment let us pursue just this type of switched element specification.

Suppose that we identify switching specifications with the keyletter S as the first non-blank character in an input. In Subroutine INPUT, Fig. 4.14, provision must be made to distinguish SHORT from S, probably by requiring that the specification for SHORT shall consist of at least the letters SH. We may then use a branch number (any one-or two-digit integer) as the next item for the number of the branch whose element current value is to be monitored. The choice of element current instead of branch current is consistent with the usage adopted for controlling branches in Chapter 3. Next, we may follow with a list of branch numbers (any sequence of one- or two-digit integers) which refer to the element values in those branches. If those elements will contain two values (possible if the INPUT subroutine is modified for this), then we have the mechanism for switching the element values as the current changes.

An additional convenience may be added. Whenever the current direction is positive, we can take the first value of the elements; otherwise the second value is used. We may add a flag to the switch statement which will indicate that for negative element current the first element value is to be used, i.e., invert the direction of the current sensing. The keywords ON and OFF could be used.

An example of such a switch statement would be

$$S3,5,6,7,8, \text{ ON} \tag{4.48}$$

indicating that the element current in branch 3 is to be monitored, and if it is positive, the first value entered for the elements in branches 5, 6, 7, and 8 are to be used. When the current direction reverses, the second set of values for the elements in branches 5, 6, 7, and 8 are to be used.

Obviously, a great deal of programming is involved in implementing these ideas. In Fig. 4.13 (flowchart of the main analysis program) a routine (or a block of statements) must be added after the calculation of V_n, V_b, etc. to check if any current directions have changed from the last solution time. If switching is to be done, the time increment must be modified and Subroutine SETUP is reentered. The solution is repeated, a new check is made, until it is ascertained that no further switchings are to be made; only then is an output printed.

Obviously, such check code must have access to the information that occurs on the Switch statements; hence those numbers must be stored in appropriate arrays upon input. Some of the necessary changes in Subroutine INPUT have been discussed earlier.

The necessary code for the entire capability is very tedious; however, various widely used computer programs use the implementation discussed here [1, 2, 6].

We shall not pursue the details of implementation further here; the interested reader is referred to the literature.

4.8 Sparse Matrix Analysis

In Sec. 4.4 we developed the various equations necessary for transient analysis. We note that several of the relationships involve matrices containing a great number of zeros. For example, Z_e in Eq. (4.30) is largely a diagonal matrix; the inverse, Y_e, will contain additional non-zeros off the main and diagonal only where the inductance matrix was originally located and only when more than two branches are to be coupled by mutual inductances. As in Chapter 3, the B and D matrices are very sparse and the node-admittance matrix Y_n is usually very sparse also. Hence both sides of Eq. (4.42), the major relation for transient analysis, will contain very sparse matrices. Furthermore, the structure of these matrices will not change when element values are altered, as in Sec. 4.7.

Solution methods for sparse matrix equations have been studied intensively. Very powerful methods exist for the solution of such equations; they may be applied to the equations derived here [7, 8]. However, a simple procedure such as the one that was derived in Sec. 3.8 may be used to provide some speed-up in the solution process. Particularly, the calculation of the right side of Eq. (4.42), a vector, benefits from sparse matrix procedures:

$$V_{n_k} = (AY_bA^T)^{-1}A[I_{g_k} - Y_bV_{g_k} - (U + B^* - Y_bR_t)I_{p_{k-1}}] \qquad (4.42)$$

Here the matrices B^* and R_t are almost always highly sparse; hence the contribution of $I_{p_{k-1}}$ could be calculated without recourse to a standard matrix multiplication.

Unfortunately, the inversion of Y_n usually leads to nearly full matrices (very few zero entries). Typically, the product $Y_n^{-1}A$ is not sparse; no great advantage arises from utilizing sparse matrix techniques for this part of the calculations. The matrix $Y_n^{-1}A$ should be calculated and used to step the solution forward in time in order to reduce the total computational cost. Sparse matrix procedures typically help in solving a set of simultaneous linear equations. One can, of course, re-solve the equations described by Eq. (4.42) at each time value and take advantage of sparse matrix techniques, but in most moderate-sized networks, bypassing sparse matrix techniques does not incur great additional computational costs. However, if the circuit is such that many switchings occur in the course of the solution, one may want to resort to solving the simultaneous equations at each instant and not incur the cost of producing the inverse of Y_n.

In most applications the use of sparse matrix procedures is indicated only in producing $I_{p_{k-1}}$; unless there is a very large number of changes in the circuit element values, the term $Y_n^{-1}A$ should be calculated and used without resort to sparse matrix techniques. For analyses requiring very frequent component switchings, a program that bypasses the inversion of the node-admittance matrix is indicated. The reader may want to outline and implement such a program (see Problems 4.26 and 4.27).

4.9 Iteration on Element Values

Often it is desirable to repeat an analysis with some element values changed. The program discussed thus far may be used for such re-analysis if an entirely new set of input cards is used to respecify the network. Considerable computer time may be saved if the input data are not destroyed, the changes are entered in the input file, and the problem solution is repeated.

We can implement this by using a special keyword, such as **MODIFY** (or at least **MO**) on an input card. This card would have to follow the **EXECUTE** card of the previous input file. In the flowchart, Fig. 4.13, a special section of code must be added after the completion of the program execution, at point A, which checks whether or not a **MODIFY** statement follows. This code will also have to check for **CONTINUE** statements as discussed in Sec. 4.7.

In implementing the **MODIFY** and the **CONTINUE** statements one is merely concerned with putting the transient analysis program into a loop which controls some of the solution parameters. For the **MODIFY** statement an entire re-analysis is specified. One is often interested in the effects of changing parameters over a range of values akin to the frequency changes discussed in Sec. 3.5. Hence one could allow three values for a component to be specified: the lowest value, the increment, and the highest value. For example, the specification of a 100-ohm resistor in branch 2 which is to be changed in 15 ohm steps up to 175 ohms could be

$$B2, \ 3-4, R=100. (15.)175.$$

if branch 2 is connected from node 3 to node 4. Note that a special delimiter "(" (i.e., open parenthesis) was used to indicate that another floating-point number follows; such special delimiters are an easy way to signal a modified version of Subroutine INPUT that another floating-point value is to be read.

The mechanics of implementation require that the above statement set a pointer which points to branch 2, $IX = 1$ and step the value of the resistance in $RR(2)$ from 100 ohms to 175 ohms in increments of 15 ohms (six steps). For each step the input section of the main routine is bypassed, and analysis is started with the initial condition solution at time **TO**.

A series of such modifications could be allowed. Since each is of the nature of a **DO**-loop, such series of modifications in effect set up nested **DO**-loops. Although the implementation is not too difficult, extreme care must be exercised in using such a program so that the analysis task specified does not become computationally prohibitive. (Three nested statements each with 10 steps require 1,000 executions of the analysis task; the amount of output might also become a problem.)

No attempt is made here to implement such a program. The reader is urged that should he attempt to write such a code, he should carefully consider the volume of output data and employ some summarizing or editing functions on the results in order to reduce the output to manageable size.

4.10 Summary

In this chapter we derived all pertinent equations for the time-domain solution of arbitrary linear networks. The method of solution was based on the solution of the node voltages in the network for equally spaced time increments. The procedure rests on replacing energy storage elements in the network by a constant equivalent conductance and a source whose magnitude depends on the voltages and currents through the equivalent conductance.

Since the energy storage elements voltage-current relationships involve first-order derivatives, the method presented in this chapter is a solution method for such equations. The particular integration method shown is a trapezoidal integration for the inductor currents and a secant approximation for the capacitor voltages.

A set of examples was calculated through carefully at the beginning of this chapter; a computer program was outlined for solving the transient response of circuits of the same element types as those used in the programs discussed in Chapter 3. Extensions of this program were discussed. A more comprehensive program may be constructed by a direct application of the formulas developed in Sec. 4.4; such a program would allow the inclusion of arbitrary mutual inductances and controlled voltage sources. An extension can also be made to allow Switch capabilities in the network.

The problems contain a set of extensions to the ideas presented in this chapter. The reader is urged to at least outline solutions to them.

REFERENCES

1. JENSEN, R. W., and M. D. LIEBERMAN, *The IBM Electronic Circuit Analysis Program.* Prentice-Hall, Inc., Englewood Cliffs, N.J., 1968.

2. LEVIN, H., *Introduction to Computer Analysis: ECAP for Technicians and Engineers.* Prentice-Hall, Inc., Englewood Cliffs, N.J., 1970.

3. L. D. MILLIMAN, W. A. MASSERA, and R. H. DICKHAUT, *CIRCUS—A Digital Computer Program for Transient Analysis of Electronic Circuits.* Report AD-346-1. Harry Diamond Laboratories, Washington, D.C., 1967.

4. BOWERS, J. C., and S. R. SEDORE, *SCEPTRE: A Computer Program for Circuit and System Analysis.* Prentice-Hall, Inc., Englewood Cliffs, N.J., 1971.

5. JOHNSON, E. D., et al., *Transient Radiation Analysis by Computer Program (TRAC).* Harry Diamond Laboratories, Washington, D.C., Contract No. DAA 639-68-C-0041, 1968.

6. IBM Corporation, 1620 Electronic Circuit Analysis Program (ECAP), User's Manual, H20-0170-1, IBM Corporation, Technical Publications Department, White Plains, N.Y., 1965.

7. WILLOUGHBY, R. A., ed., *Proceedings of the Symposium on Sparse Matrices and Their Applications.* IBM Corporation, Thomas J. Watson Research Center, Yorktown Heights, N.Y., RA-1 (#11707).

8. HACHTEL, G. D., R. K. BRAYTON, and F. G. GUSTAVSON, "The Sparse Tableau Approach to Network Analysis and Design", *IEEE Trans. on Circuit Theory*, Vol. CT-18, No. 1. January (1971) p. 101.

PROBLEMS

4.1 Write the node-voltage equations and solve for the two node voltages in Ex. 4.1.

4.2 Write and solve the node-voltage equations for the circuit in Fig. P4.1.

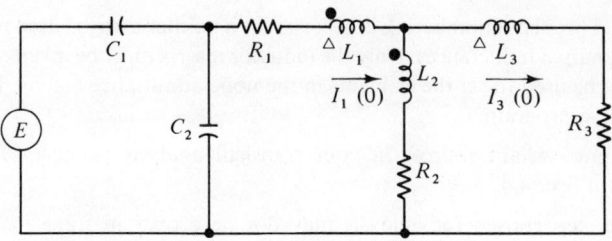

Fig. P4.1

$E = 10 \sin 2\pi \cdot 10^5 t$
$C_1 = 0.02 \ \mu f, \ E(0) = 0$
$C_2 = 0.05 \ \mu f, \ E(0) = 0$
$L_1 = 4 \ \text{mh}, \ I(0) = 2 \ \text{ma}; \ R_1 = 1000 \ \Omega \quad \bullet M_{12} = 1 \ \text{mh}$
$L_2 = 4 \ \text{mh}, \ I(0) = 0; \quad R_2 = 750 \ \Omega$
$L_3 = 6 \ \text{mh}, \ I(0) = 2 \ \text{ma}; \ R_3 = 2000 \ \Omega \quad \triangle M_{13} = 2 \ \text{mh}$

4.3 In Problem 4.2 the capacitors and the voltage source form a loop. The initial voltages on C_1 and C_2 add up to the source voltage. What difficulties arise when they do not? One can always force compatible initial conditions by insisting that no pure capacitors (nor pure inductors) be allowed in the analysis. If each capacitor is made to have a small resistor in series with it, show that no inconsistent initial capacitor voltages will arise. A similar arrangement with parallel inductor leakage resistances will eliminate initial condition inconsistencies with inductors.

4.4 Write and solve the node-voltage equations for the circuit shown in Fig. 4.8.

4.5 Carry out the coding indicated in Fig. 4.10; start with your program developed in Problem 3.3.

4.6 Use your program to verify the results shown in Tables 4.1 and 4.2.

4.7 In order to conserve memory the branch-admittance matrix need not be calculated, but the node-admittance matrix can be calculated directly as outlined in Sec. 3.7. Use your program developed in Problem 3.9 as the basis for such a transient analysis program.

4.8 The transient analysis programs in Problems 4.6 and 4.7 will not handle unaccompanied sources. Review the material presented in Sec. 3.9 and apply it to your above two programs.

4.9 Add pairwise mutual inductance capability to your rudimentary transient analysis program of Problem 4.5 or 4.7 by adding controlled sources as indicated in Fig. 4.11.

4.10 A complete transient analysis program can be based on Eqs. 4.42, 4.41, 4.30, 4.19, 4.20, and 4.21 and an initial condition solution. Assemble such a program starting with your program of Problem 3.3 (or the program shown in Listing 3.1). Assume that the user is required to number the inductor branches with the lowest branch numbers.

4.11 Write a pre-processor program that accepts network branch numbers in any order, re-arranges them so that the program in Problem 4.10 may be used, and re-assigns the proper original branch numbers for output at each time step.

4.12 Computer memory requirements may be eased if the program developed in Problem 4.7 is used as a basis for implementation in Problem 4.10. Code such a program.

4.13 No special branch renumbering is necessary if a pointer array is used to identify inductors and mutual inductances. Only the inductor matrix must be inverted and the pointer array can be used to set the elements in the node-admittance matrix. Design and complete such a program.

4.14 Include time-variant sources in your transient analysis package. The procedure is outlined in Sec. 4.4.

4.15 The time step chosen for analysis may not be proper in some instances. Devise a procedure that reduces (halves) the time step if the node voltages change by more than a suitable amount (say 10%) and increases the time step for essentially constant voltages. Normally, some care must be taken not to change the time step too frequently because it is computationally expensive.

4.16 The time-variant sources discussed in Sec. 4.4 do not allow instantaneous function changes, as with a square-wave. Design a process that would allow such changes.

4.17 Sec. 4.5 outlines a very capable transient analysis program. Coding details are given and flowcharts are presented. Code this program (approximately 1,000 statements, but this includes modifications to INVRT, Problems 3.31, 4.13, and 4.14).

4.18 Add a plotting capability (see Appendix B) to your program of Problem 4.17.

4.19 Modify your transient analysis program (Problem 4.17) to handle arbitrary mutual inductance couplings.

4.20 Voltage controlled sources may be incorporated in the transient analysis package by use of Eq. 4.41 and subsequent formation of Y_n from Y_b. Modify the flow of the program developed in Sec. 4.5 for the inclusion of these sources.

4.21 Include an EQUILIBRIUM (Final Value) solution in your transient analysis package. Check for absence of periodic and sinusoidal sources; print a message when no such solution can exist. (See Sec. 4.6 for details.)

4.22 Modify your transient analysis program (or the program flowcharted in Sec. 4.5) to include a capability of changing components at the end of the specified time period TF. Use the keyword CONTINUE (the letters CO should suffice) for this capability as indicated in Sec. 4.7. Check for introduction of voltages and currents that invalidate the last solution as the starting values for the new time interval.

4.23 The addition of a Switch capability (i.e., a current-direction sensing capability) makes the transient analysis package capable of analyzing nonlinear circuits as well. Outline the code changes necessary in the program flowcharted in Sec. 4.5 for inputting switch data, for monitoring the solution, and for establishing initial condition solutions. These changes are extensive; details are given in Sec. 4.7.

4.24 Code the Switch capability for your transient analysis program.

4.25 Diode and transistor characteristics can be approximated with linear segments (References 1, 2, 6). The Switching capability can be extended to model these devices by using a prestored model for transistor and diode data. Outline the code changes necessary to include this "stored-model" capability in your transient analysis program.

4.26 Implement the techniques discussed in Sec. 3.8 to the calculation of I_p in Eq. 4.42.

4.27 Since I_g and V_g usually contain very few non-zero components sparse matrix ideas can be used to speed up the calculation of the right side of Eq. 4.42, except for the inversion. Outline and implement such a scheme.

4.28 Outline and implement a MODIFY feature (see Sec. 4.9) for changing specific parameters of an input data set.

4.29 Outline and implement a stepping MODIFY feature as discussed in Sec. 4.9. Output volume will dictate the use of plotted output summaries.

STATE-SPACE ANALYSIS

5

Linear, lumped-element networks are made up of components whose voltage-current relationship at an instant of time can be described by numerical constant (voltage of a voltage source), a simple time-independent equation ($e = iR$; $i_s = \beta i_p$) or by a first-order linear differential equation

$$\left(y_L = L\frac{di_L}{dt}; \quad i_C = C\frac{dv_C}{dt}\right).$$

All these relationships can be regrouped into the form

$$\frac{d\mathbf{x}}{dt} = \mathbf{Ax} + \mathbf{Bu} \qquad \mathbf{x}(0) = \mathbf{x}_0 \tag{5.1}$$

where **A** and **B** are matrices with numerical coefficients, **x** is a collection of inductor currents and capacitor voltages, and **u** is a vector of input values (forcing functions).

The items to be computed for a given network may be the capacitor voltages and inductor currents, in which case calculation of **x** completes finding the desired information on the networks. Usually, however, some combination of internal variables and inputs is desired as the output from the network calculation. Let the desired outputs be elements of the **y** vector; then

$$\mathbf{y} = \mathbf{Cx} + \mathbf{Du} \tag{5.2}$$

with **C** and **D** being matrices containing numerical coefficients.

In the majority of practical circuits the above relations are adequate to describe the network behavior, but derivatives of the input must be considered occasionally. In these latter cases the above two equations must be modified, but the procedures discussed in this chapter will apply.

Obviously, the vectors **x** and **y** can contain all the information about the currents and voltages of the elements of the network. The differential equations in Eq. (5.1) describe time-dependent quantities and the linear equations in Eq. (5.2) describe combinations of those quantities with the inputs. The former gives the "state" of the currents in inductors and voltages across capacitors, the latter the resistor and dependent source voltage and/or current values. Formally, the state of a linear lumped parameter system is defined as that minimal set of variables which fully describe the behavior of that system for an interval of time from a knowledge of the values of the state variables at the beginning of that time interval and the values of the forcing variables (inputs) during that time interval. Note that in this definition it is implicit that linear combinations of **x** and **u** will yield all information about the network in the time interval in question. Also, in a linear network the state variables are not removed during the interval for which the analysis is made (as a switch might do) and hence the number of variables necessary to describe the network at various times is an invariant. Note also that the "minimal set" requires that the **A** matrix refer only to independent variables **x**.

This chapter details methods for the efficient formulation of the matrices appearing in Eqs. (5.1) and (5.2). The process used can easily be applied to simple linear networks made up of elements of the kind shown in Table 3.5. More detailed procedures for formulating the state equations for more general problems are given in References 1, 2, and 3 at the end of this chapter.

Methods for the solution of the state equations are then discussed. It should be observed that this amounts to no more than the solution of a set of first-order differential equations followed by matrix multiplications. General methods for solving this kind of equation abound in the literature [6, 7].

Finally, we discuss pole-zero analysis and consider numerical aspects of transient analysis.

5.1 General Relationships

The state equations for a given network can always be obtained by re-arrangement of the basic element voltage-current relationships and by applying Kirchhoff's laws to the network, but this procedure can be very tedious.

EXAMPLE 5.1

Find the state equations for the network shown in Fig. 5.1; E_{out} is the desired output.

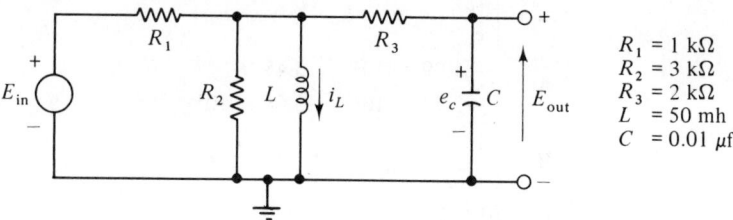

Fig. 5.1 Circuit for Example 5.1

Solution. The basic time-dependent element v-i relationships are

$$e_1 = e_L = L\frac{di_L}{dt}$$

$$i_c = C\frac{de_c}{dt}$$

The output equation is

$$E_{\text{out}} = e_c$$

Application of Kirchhoff's current law yields

$$\frac{e_1 - E_{\text{in}}}{R_1} + \frac{e_1}{R_2} + i_L + \frac{e_1 - e_c}{R_3} = 0$$

$$i_c = \frac{e_1 - e_c}{R_3}$$

The above equations may be re-arranged:

$$L\frac{di_L}{dt}\left[\frac{1}{R_1} + \frac{1}{R_2} + \frac{1}{R_3}\right] = -i_L + \frac{1}{R_3}e_c + \frac{1}{R_1}E_{\text{in}}$$

$$C\frac{de_c}{dt} = -\frac{1}{R_3}e_c + \frac{1}{R_3}\cdot\frac{1}{\frac{1}{R_1} + \frac{1}{R_2} + \frac{1}{R_3}}\left[-i_L + \frac{1}{R_3}e_c + \frac{1}{R_1}E_{\text{in}}\right]$$

Finally, the state equations may be written down

$$\begin{bmatrix}\frac{di_L}{dt} \\ \frac{de_c}{dt}\end{bmatrix} = \begin{bmatrix} \dfrac{-1}{L\left[\frac{1}{R_1} + \frac{1}{R_2} + \frac{1}{R_3}\right]} & \dfrac{1}{R_3 L\left[\frac{1}{R_1} + \frac{1}{R_2} + \frac{1}{R_3}\right]} \\ \dfrac{-1}{CR_3\left[\frac{1}{R_1} + \frac{1}{R_2} + \frac{1}{R_3}\right]} & \dfrac{-1}{C\left[R_3 + \dfrac{R_1 R_2}{R_1 + R_2}\right]} \end{bmatrix}\begin{bmatrix}i_L \\ e_c\end{bmatrix}$$

$$+ \begin{bmatrix} \dfrac{1}{R_1 L\left[\frac{1}{R_1} + \frac{1}{R_2} + \frac{1}{R_3}\right]} \\ \dfrac{R_3}{C(R_1 R_2 + R_1 R_3 + R_2 R_3)} \end{bmatrix} E_{\text{in}}$$

$$E_{\text{out}} = e_c$$

The matrices **A**, **B**, **C**, **D** and vectors **x**, **u**, and **y** can be deduced from the last two expressions.

Obviously, usually numerical values are used in all calculations. For the circuit element values given the various vectors and matrices appearing in Eqs. (5.1) and (5.2) become

$$\mathbf{x} = \begin{bmatrix}i_L \\ e_c\end{bmatrix} \quad \mathbf{u} = E_{\text{in}} \quad \mathbf{y} = E_{\text{out}}$$

$$\mathbf{A} = \begin{bmatrix}-1.09 \times 10^4 & 5.45 \\ -2.72 \times 10^7 & -3.64 \times 10^4\end{bmatrix}$$

$$\mathbf{B} = \begin{bmatrix}10.9 \\ 2.72 \times 10^4\end{bmatrix}$$

$$\mathbf{C} = [0 \quad 1]$$

$$\mathbf{D} = [0]$$

The reader is urged to verify the above values.

In calculating the coefficients of the state equation matrices, one can always proceed as above. In any but the most trivial circuits the algebraic labor involved is horrendous and the concomitant probability of numerical error very high. In the following section a method is presented for calculating the coefficients of the state equations directly.

5.2 Reduction to Repeated DC Analysis

For the general time-invariant network let the number of input variables be N, the number of state variables be M, and the number of output variables be J. Let the state equations, Eqs. (5.1) and (5.2), be coalesced into the single matrix equation

$$\mathbf{p} = \mathbf{Q}\mathbf{z} \tag{5.3}$$

with

$$\mathbf{p} = \begin{bmatrix} \dot{\mathbf{x}} \\ \hline \mathbf{y} \end{bmatrix}; \quad \mathbf{z} = \begin{bmatrix} \mathbf{x} \\ \hline \mathbf{u} \end{bmatrix}$$

and

$$\mathbf{Q} = \begin{bmatrix} \mathbf{A} & \mathbf{B} \\ \hline \mathbf{C} & \mathbf{D} \end{bmatrix}$$

with $\mathbf{p}$ being a vector of $(M + J)$ elements, $\mathbf{z}$ being a vector of $(M + N)$ elements, and $\mathbf{Q}$ being a matrix of $(M + J)$ rows and $(M + N)$ columns. Note that this matrix and vectors are merely a collection of the elements of the original state equations without any re-arrangements in the elements.

The internal variables in the network $\mathbf{x}$ are the capacitor voltages $\mathbf{V_C}$ and the inductor currents $\mathbf{I_L}$:

$$\mathbf{x} = \begin{bmatrix} \mathbf{V_C} \\ \hline \mathbf{I_L} \end{bmatrix} \tag{5.4}$$

The derivative of $\mathbf{x}$ can in turn be related to the capacitor currents and inductor voltages. For each of the K capacitors in the network

$$i_{C_k} = C_k \frac{dV_{C_k}}{dt} \quad (k = 1, \ldots, K) \tag{5.5}$$

and hence

$$\mathbf{I_C} = \mathbf{C}\frac{d}{dt}\mathbf{V_C} \tag{5.6}$$

with

$\mathbf{I_C}$ = a K element vector of the capacitor currents;
$\mathbf{V_C}$ = a K element vector of the corresponding capacitor voltages;
$\mathbf{C}$ = a diagonal matrix whose elements are

$$c_{ij} = \delta_{ij} C_j$$

with C_j being the value of the jth capacitance.

The derivative of the capacitor voltages is given by direct inversion of the above

$$\frac{d}{dt}\mathbf{V_C} = [\mathbf{C}]^{-1}\mathbf{I_C} = [\mathbf{S}]\mathbf{I_C} \tag{5.7}$$

The matrix $\mathbf{S}$, the inverse capacitance matrix, is known as the elastance matrix.

Let the circuit contain P inductances, which may be inductively coupled. The inductance voltages are given by summing the voltage developed across the inductance by the primary current and the voltages developed through the mutual inductances by the secondary currents:

$$v_{L_j} = L_j \frac{di_{L_j}}{dt} + \sum_{\substack{k=1 \\ k \neq j}}^{P} M_{jk} \frac{d}{dt} i_{L_k} \tag{5.8}$$

For the set of all inductor voltages and currents the above becomes the matrix relation

$$\mathbf{V_L} = \mathbf{L} \frac{d}{dt} \mathbf{I_L} \tag{5.9}$$

with
 $\mathbf{V_L}$ = a P-element vector of the inductor voltages;
 $\mathbf{I_L}$ = a P-element vector of the corresponding inductor currents;
 $\mathbf{L}$ = a matrix of P rows and P columns, containing on the main diagonal the self-inductances and in the off-diagonal terms the mutual inductances:

$$l_{ij} = M_{ij} = l_{ji}$$

where M_{ij} is the mutual inductance between branches i and j.

The matrix $\mathbf{L}$ is symmetric and can contain either positive or negative terms off the main diagonal.

Inversion of Eq. (5.9) gives a relationship for the derivative of the inductance currents:

$$\frac{d}{dt} \mathbf{I_L} = [\mathbf{L}]^{-1} \mathbf{V_L} = [\mathbf{H}] \mathbf{V_L} \tag{5.10}$$

The inverse inductance matrix $\mathbf{H}$ is usually not given any special name.

The derivative of the state variables becomes

$$\dot{\mathbf{x}} = \begin{bmatrix} \mathbf{S} & 0 \\ 0 & \mathbf{H} \end{bmatrix} \begin{bmatrix} \mathbf{I_C} \\ \mathbf{V_L} \end{bmatrix} = \mathbf{R} \begin{bmatrix} \mathbf{I_C} \\ \mathbf{V_L} \end{bmatrix} \tag{5.11}$$

with $\mathbf{R}$, a matrix of $(P + K)$ rows and $(P + K)$ columns, being a collection of the elements of the matrices $\mathbf{S}$ and $\mathbf{H}$. We note that in the absence of mutual inductance terms, $\mathbf{R}$ is a diagonal matrix containing inverse capacitance and inverse inductance values on the main diagonal. When mutual inductances are present, the inductance matrix must be inverted and this inverse must be used in parts of the $\mathbf{R}$ matrix. Also, the ease of computation of the elements of $\mathbf{R}$ (and of $\mathbf{Q}$) is eased considerably if the capacitor voltages are grouped together and the inductor currents collected in another group.

We now observe that the elements of the matrix $\mathbf{Q}$ may be obtained by the relation

$$q_{ij} = p_i \quad \text{if} \quad z_k = \delta_{kj} \quad (k = 1, \ldots, M + J) \tag{5.12}$$

The above states that the jth column of $\mathbf{Q}$ is the vector $\mathbf{P}$ for a solution of the network with the kth element of $\mathbf{z}$ replaced by a source of unit strength and all other elements of $\mathbf{z}$ set to zero. This is accomplished in a network by:

Sec. 5.2 Reduction to Repeated DC Analysis 181

(1) Capacitors being replaced by shorts.

(2) Inductors being opened.

(3) Input voltages being shorted.

(4) Input currents being deleted.

(5) Except for one element of **z**, which is replaced by a source of unit strength ($1V$ source for input voltages and for capacitors; $1A$ for input currents and for inductors).

(6) Analyses being carried out such that each state variable and each input is eventually replaced by a source once.

The elements of p_i are the capacitor currents divided by the respective capacitance values and the set of inductor currents pre-multiplied by the inverse inductance matrix **H**. Obviously, numerical, instead of algebraic, solutions are indicated.

The following examples illustrate this procedure.

EXAMPLE 5.2

Use the method just discussed to obtain the elements of the state equations for Ex. 5.1, Fig. 5.1.

Solution. The first step in any state-space analysis is to establish the elements of the vectors **x**, **u**, and **y**. Following the various relations above we collect the capacitor voltages and inductor currents into a vector. These will be the elements of **x** provided the analysis does not fail to yield unique numerical values for the **A** matrix thus defined. The case in which difficulties of this nature occur will be discussed in Ex. 5.4.

For this problem we take as in Ex. 5.1:

$$\mathbf{x} = \begin{bmatrix} i_L \\ v_c \end{bmatrix} \qquad \mathbf{u} = E_{\text{in}} \qquad y = E_{\text{out}}$$

To obtain the first column of the **Q** matrix we apply Eq. (5.12) with $j = 1$. Thus the elements of **z** become

$$\mathbf{z}_1 = \begin{bmatrix} \mathbf{x} \\ \cdots \\ \mathbf{u} \end{bmatrix}_{j=1} = \begin{bmatrix} i_L \\ v_c \\ E_{\text{in}} \end{bmatrix}_{j=1} = \begin{bmatrix} 1 \\ 0 \\ 0 \end{bmatrix}$$

Hence the elements of the vector $\mathbf{p}_1$ are obtained from the auxiliary circuit shown in Fig. 5.2.

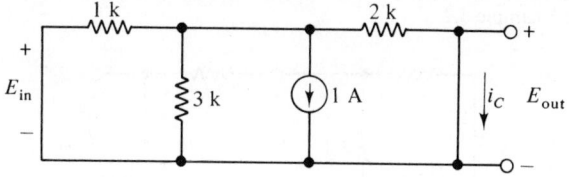

Fig. 5.2 Auxiliary Circuit for Calculating First Column of **Q** for Example 5.2

The elements of the p_j vector are

$$\mathbf{p}_j = \begin{bmatrix} \dfrac{di_L}{dt} \\ \dfrac{dv_c}{dt} \\ E_{\text{out}} \end{bmatrix}_j = \begin{bmatrix} \dfrac{1}{L} v_L \\ \dfrac{1}{C} i_c \\ E_{\text{out}} \end{bmatrix}_j$$

where j corresponds to the element number of the source of unit strength in $\mathbf{z}$. In the circuit shown in Fig. 5.2 $j = 1$; hence we are calculating $\mathbf{p}_1$ from this circuit.

The voltage developed across the current source is

$$v_L = -1 \cdot (2^k // 1^k // 3^k) = -545 \text{ V}$$

The current i_c is given by

$$i_c = \frac{e_L}{2^k} = -0.272 \text{ A}$$

The output voltage is

$$E_{\text{out}} = 0$$

Hence the elements of the first column of $\mathbf{Q}$ are

$$\mathbf{p}_1 = \begin{bmatrix} \dfrac{-545}{.05} \\ \dfrac{-0.272}{10^{-8}} \\ 0 \end{bmatrix} = \begin{bmatrix} -1.09 \times 10^4 \\ -2.72 \times 10^7 \\ 0 \end{bmatrix}$$

In a similar way for $j = 2$ and $j = 3$ we are led to the analysis of the auxiliary circuits shown in Figs. 5.3 and 5.4.

The calculations of elements of $\mathbf{p}_2$ and $\mathbf{p}_3$ proceed as above. The results are

$$\mathbf{p}_2 = \begin{bmatrix} 5.45 \\ -3.64 \times 10^4 \\ 1.0 \end{bmatrix}; \quad \mathbf{p}_3 = \begin{bmatrix} 10.9 \\ 2.73 \times 10^4 \\ 0 \end{bmatrix}$$

We note from the above example that calculation of the coefficients of the state equation reduces to the analysis of simple DC networks. These networks contain shorts in place of the

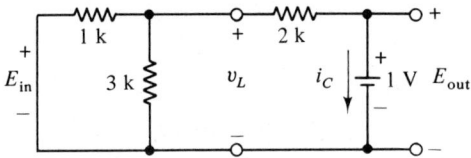

Fig. 5.3 Auxiliary Circuit for Calculating Second Column of $\mathbf{Q}$ for Example 5.2

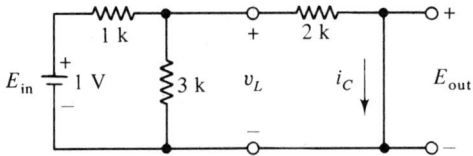

Fig. 5.4 Auxiliary Circuit for Calculating Third Column of $\mathbf{Q}$ for Example 5.2

Sec. 5.2   Reduction to Repeated DC Analysis   183

original circuit's capacitors and opens in place of the original inductors. Such (often trivial) circuits may be analyzed readily (almost) by inspection.

Automation of this analysis may be accomplished by using a low resistance value for the capacitors (say 0.01 ohms) and a very large value (such as 10 Meg) for the inductors and using the DC analysis procedures discussed in detail in Chapter 3. The analysis program of Sec. 3.3 may be used with little change; one must merely move a unit source into the branches containing capacitors and inductors (one at a time) and select the appropriate elements from the $\mathbf{I}_e$ and $\mathbf{V}_e$ vectors from which the entries in $\mathbf{Q}$ may be obtained directly.

The small resistor values substituted for the capacitors and the small conductances used for the inductors can lead to severe numerical inaccuracies. For this reason the procedure outlined in Sec. 3.9 is preferable because it calculates voltages and currents in circuits containing actual zero conductances and zero resistances.

The next example calculates the state equations for a circuit containing mutual inductances.

EXAMPLE 5.3

Calculate the state equations for the network shown in Fig. 5.5.

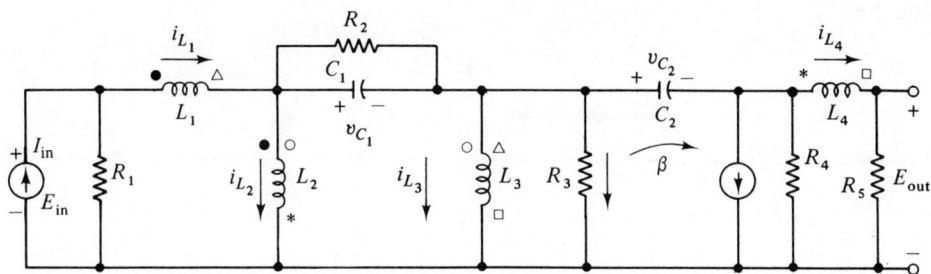

Fig. 5.5   Circuit for Example 5.3

Input: I_{in}
Outputs: E_{out}, E_{in}
$C_1 = 0.05\ \mu f$ $C_2 = 0.001\ \mu f$
$R_1 = 12\ k\Omega$ $R_2 = 30\ k\Omega$ $R_3 = 200\ \Omega$
$R_4 = 25\ k\Omega$ $R_5 = 10\ k\Omega$
$L_1 = 8\ mh$ $L_2 = 15\ mh$ $L_3 = 20\ mh$ $L_4 = 80\ mh$
● $= M_{12} = 2\ mh$ △ $= M_{13} = 4\ mh$
○ $= M_{23} = 10\ mh$ * $= M_{24} = 12\ mh$ □ $= M_{34} = 15\ mh$
$\beta = 80$

Solution. The network contains a number of coupled coils; hence Eq. (5.11) must be applied. The vectors $\mathbf{x}$, $\mathbf{u}$, and $\mathbf{y}$ are:

$$\mathbf{x} = \begin{bmatrix} v_{C_1} \\ v_{C_2} \\ i_{L_1} \\ i_{L_2} \\ i_{L_3} \\ i_{L_4} \end{bmatrix}; \qquad \mathbf{u} = I_{in}; \qquad \mathbf{y} = \begin{bmatrix} E_{out} \\ E_{in} \end{bmatrix}$$

The matrices **S** and **L** are:

$$\mathbf{S} = \begin{bmatrix} 1/C_1 & 0 \\ 0 & 1/C_2 \end{bmatrix} \qquad \mathbf{L} = \begin{bmatrix} L_1 & M_{12} & -M_{13} & 0 \\ M_{12} & L_2 & M_{23} & -M_{24} \\ -M_{13} & M_{23} & L_3 & M_{34} \\ 0 & -M_{24} & M_{34} & L_4 \end{bmatrix}$$

Numerical values are used to calculate the inverse inductance matrix **H** and the elastance matrix **S** (slide-rule accuracy is given):

$$\mathbf{S} = \begin{bmatrix} 2.00 \times 10^7 & 0 \\ 0 & 1.00 \times 10^9 \end{bmatrix} \qquad \mathbf{H} = \begin{bmatrix} 547 & -592 & 558 & -192 \\ -592 & 863 & -752 & 270 \\ 558 & -752 & 721 & -248 \\ -192 & 270 & -248 & 99.5 \end{bmatrix}$$

The vectors **p** and **z** are formed from **x**, **u**, and **y**; the **Q** matrix has eight rows and seven columns. To calculate the first column of the **Q** matrix, capacitor C_1 is replaced by a one-volt source, C_2 is shorted, all inductors are replaced by open circuits and the input current source is opened (replaced by a source of zero value). The resultant auxiliary circuit is shown in Fig. 5.6.

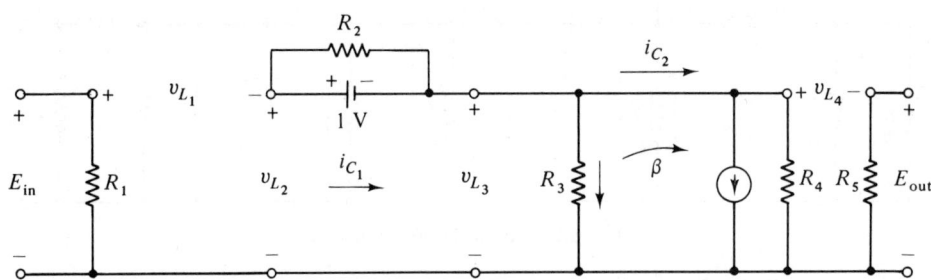

Fig. 5.6 Auxiliary Circuit for Calculating First Column of **Q** for Example 5.3

Note the dependent sources must be retained in the auxiliary circuit. From the auxiliary circuit the inductor voltages, capacitor currents, and output voltage are given by inspection:

$$i_{C_1} = -3.33 \times 10^{-5} \qquad i_{C_2} = 0.00$$
$$v_{L_1} = -1.00 \qquad v_{L_2} = 1.00$$
$$v_{L_3} = 0.00 \qquad v_{L_4} = 0.00$$
$$E_{\text{out}} = 0.00 \qquad E_{\text{in}} = 0.00$$

The above are the entries of the vector $\begin{bmatrix} \mathbf{I_C} \\ \hline \mathbf{V_L} \end{bmatrix}$ occurring in Eq. (5.11). Pre-multiplication by the **R** matrix made up of **S** and **H** given earlier yields the entries in the first column of the **Q** matrix (the vector $\mathbf{p}_1$ in the previous example):

$$\mathbf{p}_1 = \begin{bmatrix} \mathbf{R} \begin{bmatrix} \mathbf{I_C} \\ --- \\ \mathbf{V_L} \end{bmatrix} \\ -------- \\ E_{\text{out}} \\ \\ E_{\text{in}} \end{bmatrix}_{j=1} = \begin{bmatrix} -667 \\ 0 \\ -1139 \\ 1455 \\ -1310 \\ 462 \\ 0 \\ 0 \end{bmatrix}$$

Setting up the state equations for this circuit is completed by analyzing six similar auxiliary circuits. By use of the **S** and **H** matrices the **A**, **B**, **C**, **D** matrices are obtained directly and are listed below.

$$\mathbf{A} = \begin{bmatrix} -667 & 0 & 2.00 \times 10^7 & -2.00 \times 10^7 & 0 & 0 \\ 0 & -495 & 9.88 \times 10^8 & -9.88 \times 10^8 & -9.88 \times 10^8 & 1.23 \times 10^7 \\ -1139 & 192.1 & -6.57 \times 10^6 & 1910 & 1910 & 1.92 \times 10^6 \\ 1455 & -269.9 & 7.11 \times 10^6 & -2320 & -2320 & -2.70 \times 10^6 \\ -1310 & 248.0 & -6.71 \times 10^6 & 2070 & 2070 & 2.48 \times 10^6 \\ 462 & -99.5 & 2.31 \times 10^6 & -775 & -775 & -9.95 \times 10^5 \end{bmatrix}$$

$$\mathbf{B} = \begin{bmatrix} 0 \\ 0 \\ 6.56 \times 10^6 \\ -7.11 \times 10^6 \\ 6.69 \times 10^6 \\ -2.30 \times 10^6 \end{bmatrix}$$

$$\mathbf{C} = \begin{bmatrix} 0 & 0 & 0 & 0 & 0 & 1.00 \times 10^4 \\ 0 & 0 & -1.20 \times 10^4 & 0 & 0 & 0 \end{bmatrix}$$

$$\mathbf{D} = \begin{bmatrix} 0 \\ 1.20 \times 10^4 \end{bmatrix}$$

These coefficients may, of course, be calculated directly from the basic *v-i* relations for the elements and from the equations which result from applying Kirchhoff's laws to the network; any attempt to verify the above results in this way will show that the numerical work becomes rather excessive.

We shall now discuss a modification to the above procedure for cases in which it fails to yield finite numerical values for the elements of the **Q** matrix.

5.3 Detection of Surplus Variables

The reduction of the state-space analysis to a series of DC analyses is a quick and easy way to obtain the numerical coefficients once the **x**, **u**, and **y** vectors have been identified. Implicit in the definition of the state variables of the network is the fact that these variables must be independent of another. For example, if one of the variables could be expressed as any combination of the remaining other variables, then at least one variable can be eliminated from the set contained in **x**, and a smaller set of differential equations will suffice for the analysis of the network. Such a situation arises when elements are connected such that a loop of capacitors and voltage sources are formed or when a node has only current sources and inductors connected to it. In these cases at least one capacitor (or at least one inductor, as the case might be) is a surplus element for the purposes of the analysis and is removed from the vector **x**.

When the manual procedure of drawing the various auxiliary circuits is used, the occurrence of a surplus variable is signaled by a loop of shorts connecting the terminals of a 1-V source which was substituted for a capacitor or an input voltage source or by the appearance of a 1-A source which is not connected on one end to anything. Thus surplus variables can be spotted by inspection. When a program such as the one given in Listing 3.1 is used for finding the state equation coefficients, the auxiliary networks will contain small resistors instead of shorts and small conductances instead of opens. Note that all auxiliary networks derived for a circuit will be of the same topology in this case (topology changes occur for auxiliary networks when actual shorts and opens are used); the analysis for various forcing sources merely changes the equivalent current vector $\mathbf{I_s}$. Hence inversion of the $\mathbf{Y_n}$ matrix followed by post-multiplication with the several $\mathbf{I_s}$ vectors is the fastest way to solve the set of auxiliary circuit problems. The occurrence of a surplus variable in this process is signaled by the appearance of a very high current through one of the shorts which represents a capacitor or by a very high voltage which appears across the high resistance which represents an inductor. Correspondingly high currents or voltages will occur in the source causing the circuit response.

We note that such large values of current or voltage are the result of forcing current through an "open circuit" or applying voltage across a "short circuit." The resultant large values must be larger than what would result merely through the normal resistors in the network. For instance, in Ex. 5.3 the value of 10,000 volts is calculated for the element c_{16}; this number results from putting the 1-A current of inductor L_4 through resistor R_5. Although this number is large, it does not signify a surplus variable. Had we used a version of the DC program in Sec. 3.3 with an open circuit denoted by 10 Meg, a surplus variable would have been signaled by voltage values in the order of 10^7 V. In any one situation the "open circuit" value should be chosen several orders of magnitude larger than the highest circuit impedance value. Similarly, "short circuit" value must be chosen several orders of magnitude lower than any other resistance in the network.

Once a surplus variable has been detected it must be removed from the vector **x**. The initial conditions on the elements of **x** must be such that no contradiction exists among them: the initial conditions must be compatible. For example, if two capacitors were connected in parallel, the two initial voltages must be the same. If the circuit initial conditions were specified in violation of these conditions, an incompatible set of initial conditions exists and the analysis cannot be performed. A respecification of the circuit initial conditions and/or reformulation of the problem is then in order.

Once the minimal set of state variables is found, or once the circuit is reformulated to contain the minimal set of state variables, the solution of the state equations for this minimal network is carried out. The elements of the solution vector for the minimal network are then distributed appropriately among the elements of the original network.

The following example illustrates some of the points raised above.

EXAMPLE 5.4

Calculate the coefficients of the state equations of the network shown in Fig. 5.7 with all capacitor voltages and currents equaling zero at time $t = 0$.

Solution. The initial state vector is made up of the four capacitor voltages and the three inductor currents. Note that the inductors are coupled with the polarities indicated on the diagram. The auxiliary network for finding the **Q** matrix column corresponding to v_{C_1} is shown in Fig. 5.8.

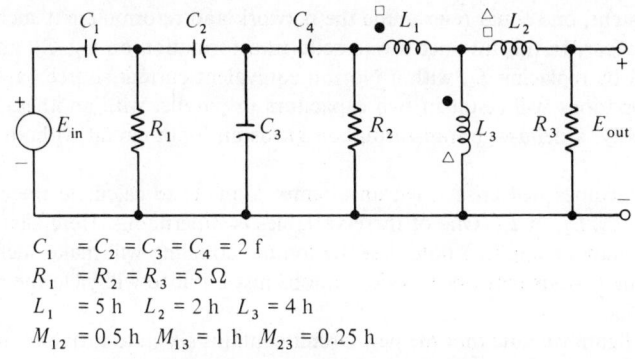

$C_1 = C_2 = C_3 = C_4 = 2 \text{ f}$
$R_1 = R_2 = R_3 = 5 \text{ }\Omega$
$L_1 = 5 \text{ h} \quad L_2 = 2 \text{ h} \quad L_3 = 4 \text{ h}$
$M_{12} = 0.5 \text{ h} \quad M_{13} = 1 \text{ h} \quad M_{23} = 0.25 \text{ h}$

Fig. 5.7 Circuit for Example 5.4

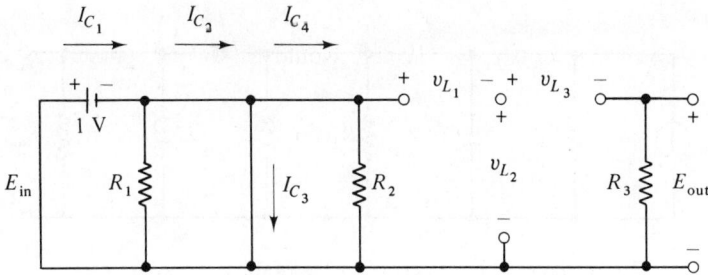

Fig. 5.8 Auxiliary Circuit for Calculating First Column of **Q** Matrix for Example 5.4

188 State-Space Analysis Ch. 5

We note immediately that E_{in}, C_1, C_2, and C_3 form a loop of shorts; hence one of the elements C_1, C_2, or C_3 is a surplus variable. It matters little which of the variables is considered the surplus one; if an automated procedure is employed, a default algorithm may be used to eliminate the element with the largest branch number.

Once surplus variable is detected a check should be made for consistency of the initial conditions. In this case we check if at $t = 0$ the voltages

$$E_{in} + v_{C_1} + v_{C_2} + v_{C_3} = 0$$

With zero initial voltages on the capacitors, the above requires that the input voltage be zero at $t = 0$.

If we take v_{C_3} as the surplus variable, the auxiliary equation

$$v_{C_3} = -v_{C_1} - v_{C_2} - E_{in}$$

will yield the voltage across C_3. Note that this equation is of the form of an output equation, a row in Eq. (5.2).

Thus the new state equations will contain v_{C_1}, v_{C_2}, v_{C_4}, i_{L_1}, i_{L_2}, and i_{L_3}. For the purposes of setting up the Q matrix, C_3 could be removed from the network, but such removal must be done carefully since the current through C_3 must be properly accounted for in the analysis.

Once the surplus variable is identified and it has been established that the initial conditions are consistent, one could re-examine the network and reformulate it such that no surplus variables occur. In this instance a wye-delta transformation on C_2, C_3, and C_4 may be made, followed by replacing E_{in} with a Norton equivalent current source paralleled by C_1. These transformations will result in two capacitors in parallel with another. They may be combined. Finally, a delta-wye transformation yields an input circuit without surplus variables.

Similar difficulties will arise when an attempt is made to calculate the columns of **Q** associated with L_1, L_2, or L_3. One of these variables is superfluous. Here again a wye-delta transformation may be applied; note that the mutual coupling will make such transformation tedious. The various network transformations just outlined will yield the revised circuit shown in Fig. 5.9.

From this figure we note that the new circuit contains six state variables, none of which appear surplus. We also note that the new input contains a derivative of the original input

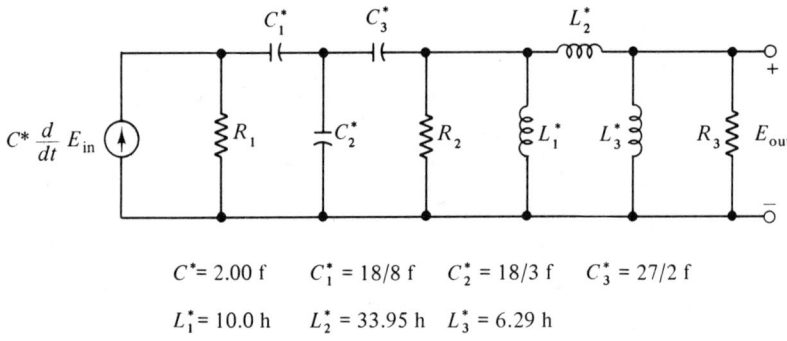

$C^* = 2.00$ f $C_1^* = 18/8$ f $C_2^* = 18/3$ f $C_3^* = 27/2$ f

$L_1^* = 10.0$ h $L_2^* = 33.95$ h $L_3^* = 6.29$ h

Fig. 5.9 Revised Circuit for Example 5.4

voltage. Thus the original state equations given in Eqs. (5.1) and (5.2) should be generalized:

$$\dot{\mathbf{x}} = \mathbf{A}\mathbf{x} + \sum_{k=0}^{\infty} \mathbf{B}_k \frac{d^k}{dt^k}\mathbf{u}; \qquad \mathbf{x}(0) = \mathbf{x}_0$$

$$\mathbf{y} = \mathbf{C}\mathbf{x} + \sum_{k=0}^{\infty} \mathbf{D}_k \frac{d^k}{dt^k}\mathbf{u} \qquad (5.13)$$

with $\mathbf{B}_k$ and $\mathbf{D}_k$ being zero matrices for $k \geq K$ and K being the number of surplus variables that may be factored out from the network so that the remaining elements contain no further surplus variables. Note that the order of $\mathbf{A}$ (number of state variables) plus K cannot exceed the number of capacitors plus inductors:

$$\text{Order }(\mathbf{A}) + K \leq \#C + \#L$$

For the example at hand the various matrices are

$$\mathbf{A} = \begin{bmatrix} -0.0333 & -0.0333 & 0. & 0. & 0. & 0. \\ -0.0333 & -0.0667 & 0.0333 & -0.167 & -0.167 & 0. \\ 0. & 0.0148 & -0.0148 & 0.074 & 0.074 & 0. \\ 0. & 0.1000 & -0.1000 & 0. & 0. & 0. \\ 0. & 0.0295 & -0.0295 & 0. & -0.147 & 0.147 \\ 0. & 0. & 0. & 0. & 0.796 & -0.796 \end{bmatrix}$$

$$\mathbf{B}_0 = \begin{bmatrix} 0. \\ 0. \\ 0. \\ 0. \\ 0. \\ 0. \end{bmatrix}; \quad \mathbf{B}_1 = \begin{bmatrix} 0.333 \\ 0.333 \\ 0. \\ 0. \\ 0. \\ 0. \end{bmatrix}; \quad \mathbf{x} = \begin{bmatrix} v_{c_1}^* \\ v_{c_2}^* \\ v_{c_3}^* \\ i_{L_1}^* \\ i_{L_2}^* \\ i_{L_3}^* \end{bmatrix} \quad \mathbf{D}_0 = [0.] \quad \mathbf{D}_1 = [0.]$$

$$\mathbf{y} = E_{\text{out}} \quad \mathbf{u} = E_{\text{in}}$$

$$\mathbf{C} = [0. \quad 0. \quad 0. \quad 0. \quad 5.00 \quad -5.00]$$

The reader will note that row (and column) 3 is a linear combination of rows (columns) 1 and 2; similarly row (column) 6 is a linear combination of rows (columns) 4 and 5. Thus two more variables can be eliminated resulting in a 4×4 $\mathbf{A}$ matrix and two state variables associated with capacitors and two with inductors. The reader is asked to derive this minimal set of equations, the true state equations.

The difficulties associated with the detection and elimination of surplus variables cannot occur if the elements in the network contain realistic parasitic resistances in the various capacitors and inductors. Also, such circuits are incapable of generating a true differential of an input; hence $\mathbf{B}_k = \mathbf{0}$ in Eq. (5.13) for $k \geq 1$ for all networks having parasitic resistors. Although the analysis of circuits with mathematical capacitors and inductors is a very desirable capability, the majority of networks can be analyzed adequately by the methods developed in the previous section. Surplus variables can usually be eliminated by a reformulation of the network once these variables are identified.

For the remainder of this chapter we shall discuss methods and procedures related to state equations of the form stated originally in Eqs. (5.1) and (5.2). Procedures have been worked out for the general case and are discussed in [3]; the computer program required is complicated and executes relatively slowly. The program described in this book does not cover the general case fully, but it does lead to procedures that are easily implemented and that indicate when surplus variables occur or when the procedure breaks down. In these cases a careful re-analysis of the problem is indicated; usually a reformulation of the problem will alleviate the difficulty found in the analysis.

5.4 Time-Domain Solution

In this section a simple procedure is presented for the calculation of the time function represented by the state equation

$$\dot{\mathbf{x}} = \mathbf{A}\mathbf{x} + \mathbf{B}\mathbf{u} \qquad \mathbf{x}(0) = \mathbf{x}_0 \qquad (5.1)$$

with $\mathbf{A}$ being an M by M matrix, $\mathbf{B}$ being an M by N matrix, and $\mathbf{u}$, the input vector, being a function of time. The associated output equation, Eq. (5.2), is ignored in this section.

As in the case where $\mathbf{x}$ is a simple one-dimensional variable, the solution may be constructed from two parts: the transient solution and the forced response:

$$\mathbf{x}(t) = \mathbf{x}_t + \mathbf{x}_f \qquad (5.14)$$

The transient part, $\mathbf{x}_t$, is the solution of the vector differential equation

$$\dot{\mathbf{x}} = \mathbf{A}\mathbf{x} \qquad \mathbf{x}(0) = \mathbf{x}_0 \qquad (5.15)$$

This solution exists for $t > 0$ and may be written as [4]

$$\mathbf{x}_t = \mathbf{h}(t)\mathbf{x}_0 \qquad (5.16)$$

where $\mathbf{h}(t)$ is a unique M by M matrix which satisfies the matrix differential equation

$$\frac{d\mathbf{h}(t)}{dt} = \mathbf{A}\mathbf{h}(t) \qquad \mathbf{h}(0) = \mathbf{U} \qquad (5.17)$$

and $\mathbf{U}$ being an M by M unit matrix. The function $\mathbf{h}(t)$, is known as the impulse response of the network.

We shall now find a series expansion for the function $\mathbf{h}(t)$. Assume that

$$\mathbf{h}(t) = \sum_{j=0}^{\infty} \mathbf{c}_j t^j \qquad (5.18)$$

with the $\mathbf{c}_j$ being M by M matrices. The derivative of the above is

$$\frac{d\mathbf{h}}{dt} = \sum_{j=1}^{\infty} j\mathbf{c}_j t^{j-1} \qquad (5.19)$$

Combining Eqs. (5.17), (5.18), and (5.19) gives the following relationships by equating like powers of t:

$$(j+1)\mathbf{c}_{j+1} = \mathbf{A}\mathbf{c}_j \qquad j = 0, \ldots$$
$$\mathbf{c}_0 = \mathbf{U} \qquad (5.20)$$

Hence the power series expansion for $\mathbf{h}(t)$ becomes

$$\mathbf{h}(t) = \mathbf{U} + \mathbf{A}t + \frac{1}{2}\mathbf{A}^2 t^2 + \frac{1}{3!}\mathbf{A}^3 t^3 + \cdots \qquad (5.21)$$

Note that the above series is the Maclaurin series for the exponential function, and hence

$$\mathbf{h}(t) = e^{\mathbf{A}t} \tag{5.22}$$

By post-multiplying Eq. (5.22) by $\mathbf{x}_0$ we immediately get Eqs. (5.16) and (5.15). Hence the vector $\mathbf{x}_t$ given above is the solution of the differential equation with $\mathbf{h}(t) = \exp(\mathbf{A}t)$.

Uniqueness and uniform convergence for all finite $\mathbf{A}$ and t are proven elsewhere [4]. The exponential matrix is the exact analogy of the scalar exponential function; it obeys all the familiar formalities of an exponential:

$$e^{\mathbf{A}(t+p)} = e^{\mathbf{A}t} \cdot e^{\mathbf{A}p} = e^{\mathbf{A}p} \cdot e^{\mathbf{A}t} \tag{5.23}$$

$$[e^{\mathbf{A}t}]^{-1} = e^{-\mathbf{A}t} \tag{5.24}$$

$$\frac{d}{dt} e^{\mathbf{A}t} = \mathbf{A} \cdot e^{\mathbf{A}t} = e^{\mathbf{A}t} \cdot \mathbf{A} \tag{5.25}$$

$$\int_0^t e^{\mathbf{A}x} \, dx = e^{\mathbf{A}t} \cdot \mathbf{A}^{-1} \Big|_0^t$$

$$= (e^{\mathbf{A}t} - \mathbf{U})\mathbf{A}^{-1} \tag{5.26}$$

As in the case for the scalar exponential function, the matrix exponential never vanishes; hence its inverse exists (for all finite $\mathbf{A}$ and t). The properties listed above may be proven readily and are proven in [4].

We proceed to use the matrix exponential in the solution of the basic state equation, Eq. (5.1), which may be rewritten as

$$\frac{d\mathbf{x}}{dt} - \mathbf{A}\mathbf{x} = \mathbf{B}\mathbf{u} \tag{5.27}$$

Pre-multiplying the above by the matrix exponential $\exp(-\mathbf{A}t)$, we get

$$e^{-\mathbf{A}t}\left(\frac{d\mathbf{x}}{dt} - \mathbf{A}\mathbf{x}\right) = e^{-\mathbf{A}t}\mathbf{B}\mathbf{u}(t) \tag{5.28}$$

The left side of the above expression is $d/dt[\exp(-\mathbf{A}t) \cdot \mathbf{x}]$. Hence upon integration

$$e^{-\mathbf{A}t}\mathbf{x} = \mathbf{x}_0 + \int_0^t e^{-\mathbf{A}p}\mathbf{B}\mathbf{u}(p) \, dp \tag{5.29}$$

with $\mathbf{x}_0$ being the initial condition vector, the values of $\mathbf{x}$ at time $t = 0$.

The solution becomes

$$\mathbf{x}(t) = e^{\mathbf{A}t}\mathbf{x}_0 + e^{\mathbf{A}t}\int_0^t e^{-\mathbf{A}p}\mathbf{B}\mathbf{u}(p) \, dp \tag{5.30}$$

The above is the complete formal solution to the state equations. For a non-trivial $\mathbf{A}$ and $\mathbf{B}$ (any network with more than a very few components will produce such matrices) the above can be evaluated only numerically; the usual case is to calculate the values of $\mathbf{x}(t)$ at equally spaced intervals of time. Let these time intervals occur T apart:

$$t = 0, T, 2T, \ldots, nT$$

In any given situation the choice of T is critical. For economy one wishes to choose T large enough so that few calculations need be made (and be paid for), but accuracy demands small steps. We shall discuss the choice of T in some detail later; for now let us assume that a satisfactory value for T can be used.

The value of the function $\mathbf{x}$ at time $t = nT$ is

$$\mathbf{x}(nT) = \mathbf{x}_{nT} = e^{\mathbf{A}nT}\mathbf{x}_0 + e^{\mathbf{A}nT}\int_0^{nT} e^{-\mathbf{A}p}\mathbf{B}\mathbf{u}(p)\,dp \tag{5.31}$$

and the value of $\mathbf{x}$ at the end of the next time interval is

$$\begin{aligned}\mathbf{x}_{(n+1)T} &= e^{\mathbf{A}(n+1)T}\mathbf{x}_0 + e^{\mathbf{A}(n+1)T}\int_0^{(n+1)T} e^{-\mathbf{A}p}\mathbf{B}\mathbf{u}(p)\,dp\\ &= e^{\mathbf{A}T}\mathbf{x}_{nT} + e^{\mathbf{A}(n+1)T}\int_{nT}^{(n+1)T} e^{-\mathbf{A}p}\mathbf{B}\mathbf{u}(p)\,dp\end{aligned} \tag{5.32}$$

The integral occurring in the last expression cannot generally be evaluated in a closed form without great difficulty. However, if $\mathbf{u}(t)$ can be approximated in the interval $nT \leq t \leq (n+1)T$ by a constant:

$$\mathbf{u}^*_{(1)} = \mathbf{u}(nT) \tag{5.33}$$

or

$$\mathbf{u}^*_{(2)} = \tfrac{1}{2}\{\mathbf{u}(nT) + \mathbf{u}[(n+1)T]\} \tag{5.34}$$

then the integral becomes

$$\int = \left[\int_{nT}^{(n+1)T} e^{-\mathbf{A}p}\,dp\right]\mathbf{B}\mathbf{u}^* \tag{5.35}$$

where $\mathbf{u}^*$ is either $\mathbf{u}^*_{(1)}$ or $\mathbf{u}^*_{(2)}$ as given above. The integral can be evaluated easily:

$$\int = \left[e^{-\mathbf{A}nT} - e^{-\mathbf{A}(n+1)T}\right]\mathbf{A}^{-1}\mathbf{B}\mathbf{u}^* \tag{5.36}$$

Thus the value of the function $\mathbf{x}$ at $t = (n+1)T$ becomes

$$\mathbf{x}_{(n+1)T} = e^{\mathbf{A}T}\mathbf{x}_{nT} + (e^{\mathbf{A}T} - \mathbf{U})\mathbf{A}^{-1}\mathbf{B}\mathbf{u}^* \tag{5.37}$$

The above is a recursion formula for calculating the values of the elements of $\mathbf{x}$ at the end of an interval of duration T from a knowledge of the values of the elements of that vector at the beginning of the interval and a knowledge of the forcing function $\mathbf{u}$ during that interval. Obviously, if $\mathbf{u}(t)$ changes markedly during the interval being considered, the approximation given above is not very accurate; hence the accuracy of calculating $\mathbf{x}(t)$ depends, at least partially, on how accurately $\mathbf{u}^*$ approximates the actual forcing function.

We note that the matrix exponential is a square matrix (M rows and M columns), the matrix $\mathbf{B}$ is an M row and N column matrix, and that Eq. (5.37) may be rewritten as

$$\mathbf{x}_{(n+1)T} = \mathbf{F}\mathbf{x}_{nT} + \mathbf{G}\mathbf{u}^* \tag{5.38}$$

with $\mathbf{G}$ being an M row, N columns matrix. Since both $\mathbf{F}$ and $\mathbf{G}$ may be calculated solely from a knowledge of the state equations (the $\mathbf{A}$ and the $\mathbf{B}$ matrices) and the time step T, both $\mathbf{F}$ and $\mathbf{G}$ may be calculated prior to application of the recursion relation given in Eq. (5.38). Once these matrices are calculated, the successive values of **x** are obtained from simple matrix multiplications.

There remains the problem of calculating the elements of the exponential matrix **F**. Obviously, direct application of Eq. (5.20) is impossible; one must find a finite approximation for this function.

Let the Maclaurin expansion be written in two parts:

$$e^{\mathbf{A}T} = \sum_{j=0}^{K} \frac{1}{j!} \mathbf{A}^j T^j + \sum_{i=K+1}^{\infty} \frac{1}{i!} \mathbf{A}^i T^i$$

$$= \mathbf{H} + \mathbf{R} \tag{5.39}$$

Since the above series expansion is uniformly convergent for all finite **A** and T, the series will be terminated at the Kth term such that the elements of the remainder matrix **R** become term-by-term negligible with respect to the **H** matrix. For an accuracy of δ we shall have

$$\delta |h_{ij}| \geq |r_{ij}| \tag{5.40}$$

Thus a way must be found to estimate the magnitude of the elements of the matrix **R**. One way to do this is to use the *matrix norm*.

Suppose that we wanted to establish an estimate of the magnitude of the square of an arbitrary matrix **Q** of L rows and L columns:

$$q_{ij}^2 = \sum_{k=1}^{L} q_{ik} q_{kj} \tag{5.41}$$

We note that a limit to the magnitude of the elements of the square of the matrix **Q** may be obtained directly:

$$|q_{ij}^2| \leq \sum_{k=1}^{L} |q_{ik}| \cdot |q_{kj}| \tag{5.42}$$

If the largest element of the **Q** matrix is q^*:

$$q^* = \max_{i,j} |q_{ij}| \tag{5.43}$$

then the inequality involving $|q_{ij}^2|$ may be rewritten as

$$|q_{ij}^2| \leq \left(\sum_{k=1}^{L} |q_{ik}| \right) \cdot q^* \tag{5.44}$$

The largest of the elements of the matrix $\mathbf{Q}^2$ is then given by

$$\max_{i,j} |q_{ij}^2| \leq \left[\max_{i} \left(\sum_{k=1}^{L} |q_{ik}| \right) \right] \cdot q^* \tag{5.45}$$

The bracketed expression above must be larger than q^*:

$$q^* \leq \max_{i} \left(\sum_{k=1}^{L} |q_{ik}| \right) \tag{5.46}$$

Thus the magnitude of the individual elements of the square of the matrix can be estimated from the inequality

$$\max_{i,j} |q_{ij}^2| \leq \|\mathbf{Q}\|^2 \tag{5.47}$$

with

$$\|\mathbf{Q}\| = \max_{i} \sum_{k=1}^{L} |q_{ik}| \tag{5.48}$$

known as the norm of the matrix $\mathbf{Q}$.

The matrix norm is a number which is useful in bounding the values of the elements of a matrix when limiting values are necessary. There are several possible matrix norms; for example one such norm would be the sum of the absolute values of all elements. For a detailed discussion of the matrix norm the reader is referred to [4]. The norm used in this section is easily calculated and leads to a reasonable estimate of the remainder terms $\mathbf{R}$.

By induction, the magnitude of the elements of the nth power of the matrix $\mathbf{Q}$ may be estimated from

$$\max_{i,j} |q_{ij}^n| \leq \|\mathbf{Q}\|^n \tag{5.49}$$

The above allows calculation of an upper limit to the magnitude of the elements of the remainder matrix $\mathbf{R}$ occurring in Eq. (5.39). We observe that

$$\max |r_{ij}| \leq \sum_{k=K+1}^{\infty} \frac{\|\mathbf{A}\|^k T^k}{k!} \tag{5.50}$$

Note that the above infinite series may be rewritten as

$$\frac{\|\mathbf{A}\|^{K+1} T^{K+1}}{(K+1)!} \left(1 + \frac{\|\mathbf{A}\|T}{K+2} + \frac{\|\mathbf{A}\|^2 T^2}{(K+2)(K+3)} + \cdots \right)$$

$$\leq \frac{\|\mathbf{A}\|^{K+1} T^{K+1}}{(K+1)!} \left(1 + \frac{\|\mathbf{A}\|T}{K+2} + \left[\frac{\|\mathbf{A}\|T}{K+2}\right]^2 + \left[\frac{\|\mathbf{A}\|T}{K+2}\right]^3 + \cdots \right) \tag{5.51}$$

The inequality above is obtained readily by noting that each term is replaced by a larger term in the second series, which is of the form of a geometric series. Thus the last series may be summed in a closed form. Therefore the magnitude of the elements of the remainder matrix $\mathbf{R}$ are limited by

$$\max_{i,j} |r_{ij}| \leq \frac{\|\mathbf{A}\|^{K+1} T^{K+1}}{(K+1)!} \frac{1}{1 - \frac{\|\mathbf{A}\|T}{K+2}} \tag{5.52}$$

Hence the above formula may be used to estimate the elements of the remainder after summing K terms.

In summary then, the solution of the state equations and the associated output equation, Eq. (5.2), is given by the following steps:

(1) Invert the $\mathbf{A}$ matrix.

(2) Multiply $\mathbf{A}^{-1}$ by $\mathbf{B}$ and store the result in $\mathbf{B}$ (this conserves memory space; the $\mathbf{B}$ matrix is never needed after this step).

(3) Calculate the matrix norm $\|\mathbf{A}\|$.

(4) Form the exponential matrix $\mathbf{F}$:
    (a) Calculate $\mathbf{F} = \mathbf{U} + \mathbf{A}T$.
    (b) Set $K = 2$.

(c) Calculate the limiting value of the magnitude of the elements of the remainder from Eq. (5.52).

(d) For each element of **F** check if the convergence condition is met.

(e) If all of the elements of **F** meet the convergence conditions, go to Step 5.

(f) Increase K by one.

(g) If K exceeds a preset maximum, go to Step 12.

(h) Calculate the next term in the power expansion $(1/K!)[\mathbf{A}T]^K$ and add it to F.

(i) Go to Step 4(c).

(5) Set $t = 0$ (or to the value of time where the solution is to be started).

(6) Print the initial values of **x** and of **y**.

(7) Set $t = t + T$.

(8) Calculate $\mathbf{q} = \mathbf{B}\mathbf{u}^*(t)$.

(9) Calculate $\mathbf{x}_{new} = \mathbf{F}\mathbf{x}_{old} + \mathbf{G}\mathbf{u}^*$ and $\mathbf{y} = \mathbf{C}\mathbf{x} + \mathbf{D}\mathbf{u}^*$.

(10) Print the new **x** and the new **y**.

(11) If $t \leq t_{\text{final}}$, go to Step 7.

(12) Stop.

Note that the vectors $\mathbf{x}_{new}$ and $\mathbf{x}_{old}$ are physically in the same memory locations.

A listing of a program which implements the above steps is given in Listing 5.1.

```
          DIMENSION A(15,15),B(15,5),C(10,15),D(10,5),U(5),Y(10)
          DIMENSION X(15),H(15,15),T(15),AI(15,15),XX(15)
          NOYS=10
          NOUS=5
          NOXS=15
C             NO. OF INPUTS = 5
C             NO. OF OUTPUTS = 10
C             NO. OF INTERNAL STATE VARIABLES = 15
   10     READ 20,NOX,NOU,NOY,MAX,ACC,AMN,DT,TMAX
   20     FORMAT (3I2,I3,5E10.0)
          IF (NOX.EQ.0) CALL EXIT
          PRINT 15,NOX,NOU,NOY,MAX,ACC,AMN,DT,TMAX
   15     FORMAT ('1'/5X,'NUMBER OF STATE VARIABLES =',I5/
         15X,'NUMBER OF INPUT VARIABLES =',I5/
         25X,'NUMBER OF OUTPUT VARIABLES=',I5/
         35X,'MAXIMUM NUMBER OF TERMS ALLOWED FOR E**(A*TIME-INCREMENT) =',
         4I3/5X,'DESIRED ACCURACY IN SUMMATION =',1PE12.4/
         55X,'SMALLEST VALUE CHECKED FOR =',1PE15.4/
         65X,'TIME INCREMENTS =',1PE12.4/5X,'MAXIMUM TIME =',1PE15.4)
          READ 30,(X (I),I=1,NOX)
   30     FORMAT (8E10.0)
          PRINT 32,(I,X (I),I=1,NOX)
   32     FORMAT (//5X,'INITIAL CONDITIONS VECTOR'/(5(I4,' - ',1PE12.4,5X)))
          PRINT 34
   34     FORMAT (// 5X,'A - MATRIX IS'/)
          DO 40 I=1,NOX
          READ 30,(A(I,J),J=1,NOX)
   35     FORMAT (' ROW ',I3/(5(1X,I3,' - ',1PE12.4,5X)))
```

Listing 5.1 Program for the Evaluation of the State Equations

```
   40   PRINT 35,I,(J,A(I,J),J=1,NOX)
        PRINT 42
   42   FORMAT(//5X,'B - MATRIX IS'/)
        DO 50 I=1,NOX
        READ 30,(B(I,J),J=1,NOU)
   50   PRINT 35,I,(J,B(I,J),J=1,NOU)
        PRINT 52
   52   FORMAT(//5X,'C - MATRIX IS'/)
        DO 60 I=1,NOY
        READ 30,(C(I,J),J=1,NOX)
   60   PRINT 35,I,(J,C(I,J),J=1,NOX)
        PRINT 62
   62   FORMAT (//5X,'D - MATRIX IS'/)
        DO 70 I=1,NOY
        READ 30,(D(I,J),J=1,NOU)
   70   PRINT 35,I,(J,D(I,J),J=1,NOU)
C       INPUT COMPLETED, START CALCULATIONS.
        KNT=10
        DO 80 I=1,NOX
        DO 80 J=1,NOX
   80   AI(I,J)=A(I,J)
        CALL INVRT(AI,NOXS,NOX,DET,T)
        IF (DET.NE.0.) GO TO 100
        PRINT 90
   90   FORMAT (//10X,'** INVERSION OF A-MATRIX FAILED. **')
        GO TO 10
  100   DO 130 J=1,NOU
        DO 120 I=1,NOX
        T(I)=0.
        DO 120 K=1,NOX
  120   T(I)=T(I)+AI(I,K)*B(K,J)
        DO 130 I=1,NOX
  130   B(I,J)=T(I)
        TIME=0.
C       A-INVERSE * B IS IN B , SET UP E**(A*TIME-INCREMENT)
        CALL EAT (A,NOX,NOXS,DT,H,T,ACC,AMN,IER,MAX,AI)
        IF (IER.NE.0) GO TO 10
        PRINT 142
  142   FORMAT (///10X,'THE STATE TRANSITION MATRIX IS'/)
        DO 145 I=1,NOX
  145   PRINT 35,I,(J,H(I,J),J=1,NOX)
        PRINT 147
  147   FORMAT('1')
  150   CALL EVALU(U,NOU,TIME)
        DO 152 I=1,NOY
        Y(I)=0.
        DO 151 J=1,NOX
  151   Y(I)=Y(I)+C(I,J)*X(J)
        DO 152 J=1,NOU
  152   Y(I)=Y(I)+D(I,J)*U(J)
        KNT=KNT+1
        IF (KNT.LT.10) GO TO 156
        KNT=0
        PRINT 154,TIME,(I,X(I),I=1,NOX)
  154   FORMAT(/' TIME=',1PE15.5/ '    STATES',6(I5,' - ',1PE11.4)/
       1(8X,6(I5,' -',1PE11.4)))
        PRINT 155,(I,Y(I),I=1,NOY)
  155   FORMAT('   OUTPUT',6(I5,' - ',1PE11.4)/(8X,6(I5,' - ',1PE11.4)))
  156   CONTINUE
        IF (TIME.GE.TMAX) GO TO 10
        TIME=TIME+DT
        DO 160 I=1,NOX
        XX(I)=0.
        DO 160 J=1,NOU
  160   XX(I)=XX(I)+B(I,J)*U(J)
```

Listing 5.1—*Cont.*

```
      DO 170 I=1,NOX
      T(I)=-XX(I)
      DO 170 J=1,NOX
170   T(I)=T(I) + H(I,J)*(X(J)+XX(J))
      DO 180 I=1,NOX
180   X(I)=T(I)
      GO TO 150
      END
      SUBROUTINE RAISE (A,N,NS,DT,KK,H,T,AP)
      DIMENSION A(NS,NS),H(NS,NS),T(N),AP(NS,NS)
      B=DT/FLOAT(KK)
      DO 20 J=1,N
      DO 10 I=1,N
      T(I)=0.
      DO 10 K=1,N
10    T(I)=T(I)+AP(J,K)*A(K,I)*B
      DO 20 K=1,N
20    AP(J,K)=T(K)
C         PRODUCED NEXT TERM IN POWER SERIES.
      DO 30 I=1,N
      DO 30 J=1,N
30    H(I,J)=H(I,J)+AP(I,J)
      RETURN
      END
      SUBROUTINE NORM(A,N,NA,IROW,ANORM)
      DIMENSION A(NA,NA)
      IROW=0
      ANORM=0.
      DO 50 I=1,N
      AMX=0.
      DO 20 J=1,N
20    AMX=AMX+ABS(A(I,J))
      IF (AMX.LT.ANORM) GO TO 50
      AMX=ANORM
      IROW=I
50    CONTINUE
      RETURN
      END
      SUBROUTINE CHECK (H,N,NS,ACC,RIJN,AMN,IER)
      DIMENSION H(NS,NS)
      SIZE=RIJN/ACC
      DO 10 I=1,N
      DO 10 J=1,N
      IF((ABS(H(I,J)).LE.AMN).OR.(ABS(H(I,J)).GE.SIZE)) GO TO 10
      IER= 1
      GO TO 20
10    CONTINUE
      IER=0
C         CONVERGENCE MET IF IER.EQ.0 UPON RETURN.
20    RETURN
      END
      SUBROUTINE EAT(A,N,NS,DT,H,T,ACC,AMN,IER,MAX,AP)
      DIMENSION A(NS,NS),H(NS,NS),T(N),AP(NS,NS)
      CALL NORM(A,N,NS,IROW,ANORM)
      DO 20 I=1,N
      DO 10 J=1,N
      H(I,J)=A(I,J)*DT
10    AP(I,J)=H(I,J)
20    H(I,I)=H(I,I)+1.0
C         H = UNIT MATRIX + A*DT
      RIJO=(ANORM*DT)**2/2.0
```

Listing 5.1—*Cont.*

```
          DO 30 I=2,MAX
          CALL RAISE (A,N,NS,DT,I,H,T,AP)
C             NEXT TERM IN SERIES ADDED TO H
          CAY=I+1
          RIJO=RIJO*ANORM*DT/CAY
          RIJN=RIJO/(1.-ANORM*DT/(CAY+1.))
          CALL CHECK (H,N,NS,ACC,RIJN,AMN,IER)
          IF(IER.EQ.0) GO TO 50
   30     CONTINUE
          PRINT 40,MAX,ACC,AMN
   40     FORMAT(//5X,  'E**(A*DT) DID NOT CONVERGE IN ',I3,' ITERATIONS',/
         110X,'DESIRED ACCURACY =',1PE12.4,'  LEAST VALUE CHECKED =',1PE12.4
         2/10X,'CALCULATIONS STOPPED.')
   50     RETURN
          END
          SUBROUTINE EVALU (U,N,T)
C             EXAMPLE 5.5
          DIMENSION U(N)
   10     U(1)=1.0
   99     RETURN
          END
```

Listing 5.1—*Cont.*

The various major variables in the program are:

      A = the **A** matrix; allocated NOXS by NOXS locations.
      B = the **B** matrix; allocated NOXS by NOUS locations.
      C = the **C** matrix; allocated NOYS by NOXS locations.
      D = the **D** matrix; allocated NOYS by NOUS locations.
  NOXS = number of rows allocated for A (as well as B).
  NOYS = number of rows allocated for C (as well as D).
  NOUS = number of outputs allowed for.
     X = the initial condition vector of NOX values.
   NOX = number of state variables (NOX $\leq$ NOXS).
   NOU = number of input functions (NOU $\leq$ NOUS).
   NOY = number of output functions (NOY $\leq$ NOYS).
    DT = time step.
  TMAX = largest value of time for which calculations are to be made.
   ACC = accuracy to which the elements of **F** are to be approximated (δ in the above description).
   MAX = maximum number of terms allowed for power series expansion in **F** [used in Step 4(g) above].
   AMN = smallest size of **F** still checked for convergence (usually a small value, for example 1. E $-$ 30).

The actual values of **u***(*t*) must be supplied from a separate subroutine EVALU. A dummy routine is given in Listing 3.1.

The structure of the input data required is evident from the READ statements (statements 10 through 70); an echo checking printout is provided. Note that each row of each matrix starts on a new card.

EXAMPLE 5.5

In Ex. 5.1 the state equations for a simple network were derived. These were

$$\frac{d}{dt}\begin{bmatrix} i_L \\ v_c \end{bmatrix} = \begin{bmatrix} -1.09 \times 10^4 & 5.45 \\ -2.72 \times 10^7 & -3.64 \times 10^4 \end{bmatrix} \begin{bmatrix} i_L \\ v_c \end{bmatrix} + \begin{bmatrix} 10.9 \\ 2.72 \times 10^4 \end{bmatrix} E_{\text{in}}$$

$$E_{\text{out}} = \begin{bmatrix} 0 & 1.00 \end{bmatrix} \begin{bmatrix} i_L \\ v_c \end{bmatrix} + [0]E_{\text{in}}$$

For zero initial conditions and a step input voltage of magnitude 1 volt, calculate the response.

Solution. The program given in Listing 5.1 requires a subroutine for the calculation of $u(t)$; the one given in the listing is the one for this example.

The calculations were carried out with a step size of 0.0001 sec; part of the output is reproduced in Listing 5.2.

```
   NUMBER OF STATE VARIABLES  =      2
   NUMBER OF INPUT VARIABLES  =      1
   NUMBER OF OUTPUT VARIABLES=       1
   MAXIMUM NUMBER OF TERMS ALLOWED FOR E**(A*TIME-INCREMENT) = 55
   DESIRED ACCURACY IN SUMMATION =   1.0000E-05
   SMALLEST VALUE CHECKED FOR =      1.0000E-07
   TIME INCREMENTS =  1.0000E-04
   MAXIMUM TIME =     5.0000E-03

   INITIAL CONDITIONS VECTOR
1 -    0.0000E-01         2 -    0.0000E-01

     A - MATRIX IS

ROW   1
1 -   -1.0900E 04         2 -    5.4500E 00

ROW   2
1 -   -2.7200E 07         2 -   -3.6400E 04

     B - MATRIX IS

ROW   1
1 -    1.0900E 01

ROW   2
1 -    2.7200E 04

     C - MATRIX IS

ROW   1
1 -    0.0000E-01         2 -    1.0000E 00

     D - MATRIX IS

ROW   1
1 -    0.0000E-01

         THE STATE TRANSITION MATRIX IS

ROW   1
1 -   -2.3715E-01         2 -   -7.4392E-04
```

Listing 5.2 Results of Example 5.5

```
ROW  2
  1 -   3.7128E 03          2 -   3.2436E 00

TIME=    4.19963E-05
STATES    1 -  4.2584E-04    2 -  4.2484E-01
OUTPUT    1 -  4.2484E-01

TIME=    4.29962E-05
STATES    1 -  4.3436E-04    2 -  4.2487E-01
OUTPUT    1 -  4.2487E-01

TIME=    4.39960E-05
STATES    1 -  4.4280E-04    2 -  4.2468E-01
OUTPUT    1 -  4.2468E-01
```

Listing 5.2—*Cont.*

The accuracy may be readily estimated in this instance. The solution is found analytically to be

$$i_L(t) = 0.001 + 1.14e^{-27650t} - 2.12e^{-19650t}$$
$$v_c(t) = 3.40(e^{-19650t} - e^{-27650t})$$

As can be verified from the results in Listing 5.2, the agreement between the computed results and the analytic results is quite acceptable.

The procedure discussed in this chapter, or some other differential equation integration routine, may be used to solve the state equations. All that remains in order to automate this process is to derive a procedure for obtaining these equations in the first place. The next section describes such a method.

5.5 Program for the Formulation of the State Equations

We now outline a program that will generate the state equation coefficients of a linear network made up of components of the type used in the programs of Chapters 3 and 4. The components are thus the ones originally listed in Table 3.5 with the exception of μ and r_t; also mutual inductances will be limited to pairwise coupled branches. In the problems we suggest ways to extend the programs to more general networks. The material in this section was taken from [8].

In the main, the program follows the processes used in Sec. 3.4 for AC analysis. The input of the data must conform to two basic requirements:

(1) It must be simple and logical to use.

(2) It must provide a description of the network in such a way that program manipulation of this data does not become excessive.

In order to achieve a balance between program coding and user coding the fixed data format of Sec. 3.4 has been utilized and extended slightly to accommodate the new program objectives. Three types of input cards are utilized.

(1) General Network Size Descriptors: Contain the number of independent nodes N, the number of branches M.

(2) Branch Descriptors: Contain the number of the branch being described, the two node numbers between which it is connected, the type of element contained in the branch, the value of the element. The standard branch is defined in Fig. 3.10.

(3) Output Descriptors: Contain the variables that the user wants in the y vector. Fig. 5.10 shows the simplified input data flow.

Optionally, a print-level indicator flag can be set (on the network size descriptor card) which can be used to print intermediate results so that program execution may be checked.

When all branch descriptor cards are read, the program sets up the YB matrix and includes all controlled sources. The node numbers are stored in the IA matrix (dimensioned M by 2) and the controlled branch numbers are stored in the IB matrix (dimensioned IBB by 2, where IBB is the number of controlled branches).

Capacitors are replaced by small resistances (10^{-10} ohms) and inductors are replaced by large resistances (10^{+15} ohms). These substitutions perform the open circuiting and short circuiting as necessary without changing the network topology. Capacitor and inductor locations are stored in the ISTATE and JSTATE vectors respectively.

The program shown in Listing 5.3 does not allow arbitrary mutual inductances. Pairwise mutual inductances can be replaced by their equivalents as shown in Sec. 4.3a (Fig. 4.11). The program does incorporate an automatic conversion of mutual inductances in this form.

As each branch descriptor card is read and interpreted, the branch description is printed for checking. The source vectors are also generated and printed for checking.

The method for setting up the state equations was developed in Sec. 5.2. It consists of replacing each energy storage element in the network and each input by a unit source and calculating the resulting voltages and currents in all energy storage elements at each output. Hence the same network must be solved with a series of forcing functions. The DC analysis program of Sec. 3.3 solves the network node equations for a single forcing function (a single set of I_g and V_g values). We may modify it so that a series of I_s vectors is generated, one for each reactive component and one for each input; these are then assembled in the matrix equations

$$\begin{aligned}\mathbf{Y_n V_n^{(1)}} &= \mathbf{I_s^{(1)}} \\ \mathbf{Y_n V_n^{(2)}} &= \mathbf{I_s^{(2)}} \\ &\vdots \\ \mathbf{Y_n V_n^{(k)}} &= \mathbf{I_s^{(k)}}\end{aligned} \quad (5.53)$$

There will be $k = N + M$ equations (N = number of input variables, M = number of state variables; see Sec. 5.2). These equations can each be solved for the set of node voltages $\mathbf{V_n^{(k)}}$ for the kth equivalent source.

A reduction in the computer time required results if the coefficients of the above equations are assembled in the augmented matrix

$$\mathbf{Y_{na}} = [\mathbf{Y_n} \,|\, \mathbf{I_{s_1}} \,|\, \mathbf{I_{s_2}} \,|\, \cdots \,|\, \mathbf{I_{s_k}}] \quad (5.54)$$

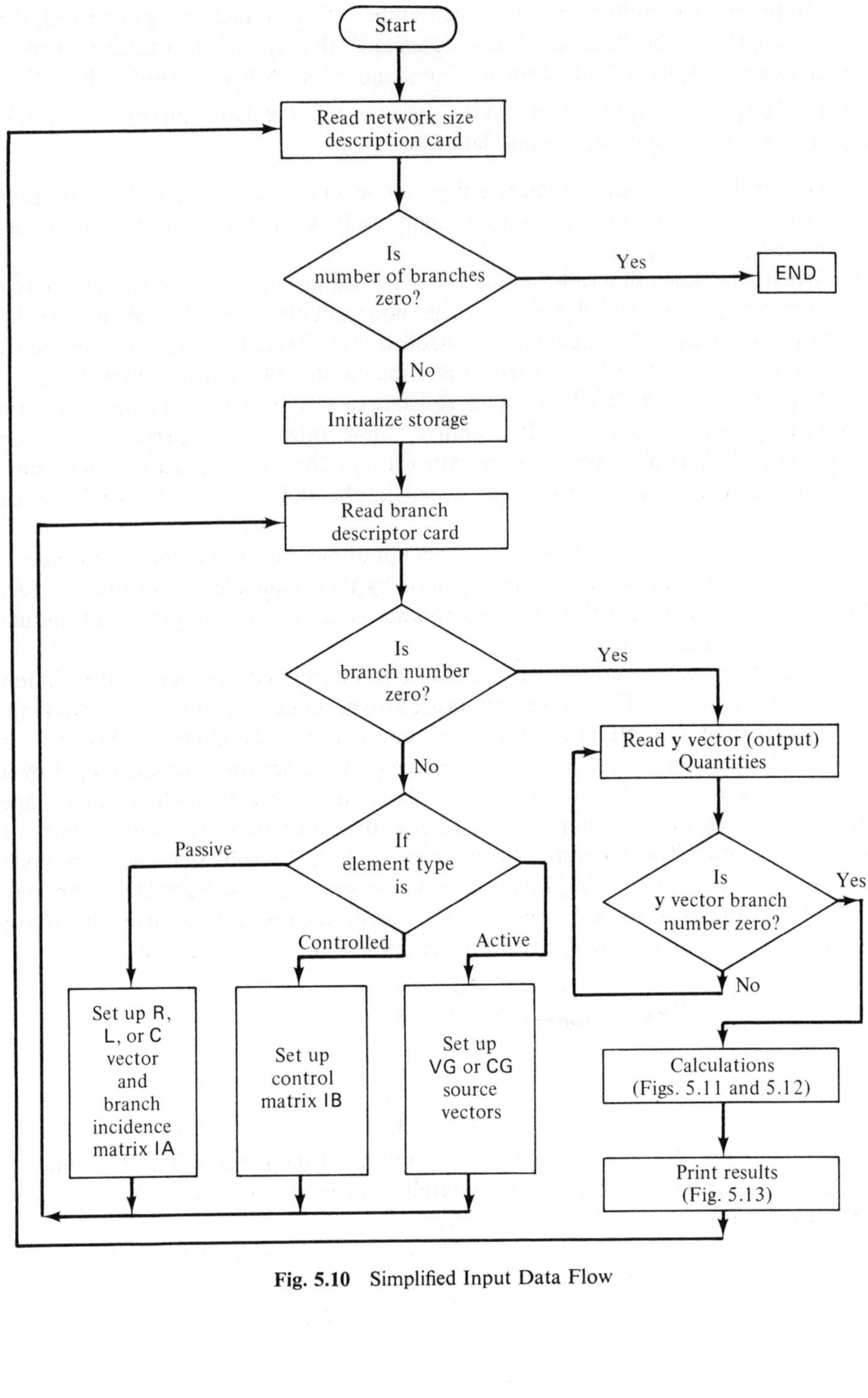

Fig. 5.10 Simplified Input Data Flow

```
C            A STATE VARIABLE ANALYSIS PROGRAM, ARBITRARY MUTUAL INDUCTANCE
C              PRODUCES THE A,B,C AND D MATRICES
      IMPLICIT REAL*8 (A-H,O-Z)
      REAL*8 LL
      DIMENSION IA(15,2),R(15),IB(15,2),LL(15),CC(15),VG(15),CG(15),
     1ISTATE(15),JSTATE(15),INPV(15),JNPV(15),BC(15,15),YB(15,15),
     2Q(15,15),BCP(15,15),VBP(15,15),T(15),IIT(15),B(15),YBB(15),
     3XMUT(15),ML(15,2)
      EQUIVALENCE (IIT(1),T(1))
      NNK=15
C         INPUT M,N,TRACE
C         M = NUMBER OF BRANCHES (UP TO 15)
C         N = NUMBER OF INDEPENDENT BRANCHES (UP TO 15)
  100 READ 110,M,N,ITRACE
  110 FORMAT (3I3)
      IF (M.LE.0) GO TO 1350
      PRINT 120,M,N
  120 FORMAT ('1   NEW PROBLEM.'/'   NUMBER OF BRANCHES=',I3,5X,'NUMBER O
     1F INDEPENDENT NODES=',I3//' INPUT CARDS ARE'// ' BRANCH/   FROM  T
     20  TYPE  VALUE'/' NUMBER'/)
      N2=N+1
      KPART=1
C         REQ = VALUE OF EQUIVALENT RESISTOR USED FOR CAPACITORS.
      REQ=1.0E+10
C         RELQ = VALUE OF RESISTOR USED FOR INDUCTORS.
      RELQ=1.0E-15
C         ZERO OUT ALL VARIABLES.
      DO 130 I=1,15
      CG(I)=0.D0
      VG(I)=0.D0
      R(I)=0.D0
      B(I)=0.D0
      DO 130  J=1,15
      Q(I,J)=0.D0
  130 YB(I,J)=0.D0
      ISC=0
      ISL=0
      INV=0
      INC=0
      IIB=0
      MUTUAL=0
C         INPUT I,J,K,IX,VAL
C         I = BRANCH NUMBER OR SEQUENCE NUMBER (IX=6 OR 7)
C             FOR IX=10  IGNORED IF MUTUAL INDUCTANCE SPECIFIED.
C         I = 0 STARTS CALCULATIONS.
C         J = 'FROM' NODE NUMBER, OR CONTROLLING BRANCH NO. IF IX= 6 OR 7
C         K = 'TO' NODE NUMBER, OR CONTROLLING BRANCH NO. IF IX = 6 OR 7
C             FOR IX = 10 (MUTUAL INDUCTANCE) J = PRIMARY BRANCH NUMBER,
C             K = SECONDARY BRANCH NUMBER.
C         IX= ELEMENT TYPE
C             0 CONDUCTANCE (MHOS)
C             1 RESISTANCE (OHMS)
C             2 CAPACITANCE (MICROFARADS)
C             3 INDUCTANCE (HENRIES)
C             4 INDEPENDENT VOLTAGE SOURCE (VOLTS)
C             5 INDEPENDENT CURRENT SOURCE (AMPS)
C             6 CURRENT DEPENDENT CURRENT SOURCE (BETA) (DIMENSIONLESS)
C             7 VOLTAGE DEPENDENT CURRENT SOURCE (MU) (MHOS)
C             8,9 NOT USED
C             10 MUTUAL INDUCTANCE BETWEEN BRANCHES K AND J (HENRIES)
  140 READ 150,I,J,K,IX,VAL
  150 FORMAT (4I3,G10.0)
      IF (I.EQ.0) GO TO 310
      PRINT 180, I,J,K,IX,VAL
```

Listing 5.3 State Analysis Program

```
  180 FORMAT (I5,I8,I5,I5,1PE14.5)
      IF (IX.NE.0) GO TO 210
      R(I)=VAL
      GO TO 300
  210 GO TO (220,230,240,250,260,290,280,270,270,270),IX
C         RESISTORS
  220 R(I)=1./VAL
      GO TO 300
C         CAPACITORS
  230 R(I)=REQ
      VAL=VAL*1.D-6
      CC(I)=1./VAL
      ISC=ISC+1
      ISTATE(ISC)=I
      GO TO 300
C         INDUCTORS
  240 R(I)=RELQ
      LL(I)=VAL
C         STORE INDUCTANCE VALUE.
      ISL=ISL+1
      JSTATE(ISL)=I
      GO TO 300
C         VOLTAGE SOURCES
  250 VG(I)=VAL
      INV=INV+1
      INPV(INV)=I
      GO TO 140
C         CURRENT SOURCES
  260 CG(I)=VAL
      INC=INC+1
      JNPV(INC)=I
      GO TO 140
C         MUTUAL INDUCTANCES
  270 MUTUAL=MUTUAL+1
      XMUT(MUTUAL)=VAL
C         NO COEFFICIENT OF COUPLING ALLOWED.
      ML(MUTUAL,1)=J
      ML(MUTUAL,2)=K
      GO TO 140
C         TRANSCONDUCTANCE GM
  280 K=-K
C         CONTROLLED CURRENT BETA
  290 IIB=IIB+1
      B(IIB)=VAL
      IB(IIB,1)=-K
      IB(IIB,2)=J
      GO TO 140
C     SET UP IA MATRIX
  300 IA(I,1)=J
      IA(I,2)=K
      GO TO 140
C         THE ENTIRE INPUT SECTION TO HERE CAN BE REPLACED BY
C         A FREE FORMAT INPUT ROUTINE.
  310 IF (((ISL.EQ.0).AND.(ISC.EQ.0)) GO TO 1330
C         SET UP YB MATRIX
C             RESISTORS
      DO 340 I=1,M
      J=IA(I,1)
      K=IA(I,2)
      IF((J*K).EQ.0) GO TO 320
      YB(J,K)=YB(J,K)-R(I)
      YB(K,J)=YB(K,J)-R(I)
  320 IF (K.EQ.0) GO TO 330
      YB(K,K)=YB(K,K)+R(I)
```

Listing 5.3—*Cont.*

```
      330 IF (J.EQ.0) GO TO 340
          YB(J,J)=YB(J,J)+R(I)
      340 CONTINUE
          IF (IIB.EQ.0) GO TO 370
C             CONTROLLED CURRENT SOURCES.
          DO 360 I=1,IIB
          VAL=B(I)
          J=IB(I,2)
          K=IB(I,1)
          IF (K.GT.0) GO TO 350
          K=-K
          VAL=VAL*R(J)
C             ALL SOURCES GM SOURCES NOW.
      350 JF=IA(J,1)
          JT=IA(J,2)
          KT=IA(K,2)
          KF=IA(K,1)
C             GENERATE YN ENTRIES.
          IF((KF*JF).NE.0) YB(KF,JF)=YB(KF,JF)+VAL
          IF ((KT*JT).NE.0) YB(KT,JT)=YB(KT,JT)+VAL
          IF((JF*KT).NE.0) YB(KT,JF)=YB(KT,JF)-VAL
          IF((JT*KF).NE.0) YB(KF,JT)=YB(KF,JT)-VAL
      360 CONTINUE
C             GENERATE FORMATS FOR OUTPUT
      370 CONTINUE
C             PRINT CS - VECTOR
          PRINT 390,(CG(I),I=1,M)
      390 FORMAT (//'  THE VECTOR OF INDEPENDENT CURRENTS IS'/(1X,1P8E14.5))
          PRINT 400,(VG(I),I=1,M)
      400 FORMAT (//'  THE VECTOR OF INDEPENDENT VOLTAGES IS'/(1X,1P8E14.5))
C             READY TO SET UP CONSTANTS MATRIX.
C             ISL = NUMBER OF INDUCTORS
C             ISC = NUMBER OF CAPACITORS
C             INC = NUMBER OF INDEPENDENT CURRENT SOURCES
C             INV = NUMBER OF INDEPENDENT VOLTAGE SOURCES
C             IINC = TOTAL CURRENT FACTORS
C             IINV = TOTAL VOLTAGE FACTORS
C             IJK = NUMBER OF COLUMNS IN CONSTANTS MATRIX
          IINC=ISL+INC
          IINV=ISC+INV
          IJK=IINV+IINC
          MIJK=IINV+ISL
          DO 620 MM=1,IJK
          IFLAGV=0
          IF (MM.GT.IINV) GO TO 570
          IF (MM.GT.ISC) GO TO 470
C             CAPACITORS
      460 ICOL=ISTATE(MM)
          JM=MM
          GO TO 480
C             VOLTAGES
      470 ICOL=INPV(MM-ISC)
C             SET IFLAGV FOR USE LATER
          IFLAGV=1
          JM=MM+ISL
      480 DO 490 LK=1,M
      490 YBB(LK)=0.
          YBB(ICOL)=R(ICOL)
C             YBB IS A GIVEN COLUMN OF YB, ASSEMBLED SEPARATELY


          IF (IIB.EQ.0) GO TO 510
          DO 505 IL=1,IIB
```

Listing 5.3—*Cont.*

```
          JJ=IB(IL,2)
          IF (JJ.NE.ICOL) GO TO 505
          KK=IB(IL,1)
          VAL=B(IL)
          IF (KK.GT.0) GO TO 500
          KK=-KK
          VAL=VAL*R(JJ)
      500 YBB(KK)=YBB(KK)+VAL
      505 CONTINUE
      510 DO 520 ILK=1,N
      520 BC(ILK,JM)=0.
          DO 560 I=1,M
C             REVERSE SIGN OF VOLTAGE SOURCES
          IF (IFLAGV.EQ.1)YBB(I)=-YBB(I)
          IBF=IA(I,1)
          IBT=IA(I,2)
          IF (IBF.GT.0) BC(IBF,JM)=BC(IBF,JM)+YBB(I)
          IF (IBT.GT.0) BC(IBT,JM)=BC(IBT,JM)-YBB(I)
      560 CONTINUE
          GO TO 620
      570 IF (MM.GT.MIJK) GO TO 610
C             INDUCTORS
          ICOL=JSTATE(MM-IINV)
          JM=MM-INV
          XXU=1.
      590 DO 600 I=1,N
      600 BC(I,JM)=0.
          IBF=IA(ICOL,1)
          IBT=IA(ICOL,2)
          IF (IBF.GT.0) BC(IBF,JM)=-XXU
          IF (IBT.GT.0) BC(IBT,JM)=XXU
          GO TO 620
C             CURRENT SOURCES
      610 ICOL=JNPV(MM-MIJK)
          JM=MM
          XXU=-1.
          GO TO 590
      620 CONTINUE
C             CONSTANTS MATRIX COMPLETE, EACH COLUMN CORRESPONDS
C             TO EITHER A STATE VARIABLE OR TO AN INDEPENDENT SOURCE.
C             PRINT YB- MATRIX IF TRACE LEVEL IS .GT. 3
          IF (ITRACE.LT.3) GO TO 670
          PRINT 630
      630 FORMAT (//'   THE YN-MATRIX IS'/)
          DO 640 I=1,N
      640 PRINT 645,(YB(I,J),J=1,N)
      645 FORMAT (1X,1P8E14.5/(2X,1P8E14.5))
          PRINT 650
      650 FORMAT (//'   THE CURRENTS MATRIX IS'/)
          DO 660 I=1,N
      660 PRINT 645,(BC(I,J),J=1,IJK)
C             GENERATE AUGMENTED YB-MATRIX.
      670 NIJK=IJK+N
          DO 680 K=N2,NIJK
          DO 680 I=1,N
      680 YB(I,K)=BC(I,K-N)
          IF(ITRACE.LT.2) GO TO 710
          PRINT 690
      690 FORMAT (//'   THE AUGMENTED NODE-ADMITTANCE MATRIX IS'/)
          DO 700 I=1,N
      700 PRINT 645,(YB(I,J),J=1,NIJK)
C             CALL SOLUTION ROUTINE FOR SIMULTANEOUS EQUATIONS WITH MANY
      710 IF (ITRACE.LT.1) GO TO 740
```

Listing 5.3—*Cont.*

```
C           PRINT NODE VOLTAGES
C              AN ANALYSIS PACKAGE TO DETERMINE EXCESS STATE VARIABLES
C                COULD BE ADDED AT THIS POINT.
      PRINT 720
  720 FORMAT (//'   THE NODE VOLTAGE VECTORS ARE'/)
      DO 730 I=1,N
  730 PRINT 645,(YB(I,J),J=N2,NIJK)
C          CALCULATE BRANCH VOLTAGES.
  740 DO 800 K=N2,NIJK
      DO 750 I=1,M
  750 VBP(I,K-N)=0.
      DO 790 I=1,M
      IBF=IA(I,1)
      IBT=IA(I,2)
      IF (IBF.GT.0) VBP(I,K-N)=VBP(I,K-N)+YB(IBF,K)
      IF (IBT.GT.0) VBP(I,K-N)=VBP(I,K-N)-YB(IBT,K)
  790 CONTINUE
  800 CONTINUE
      IF (ITRACE.LT.4) GO TO 830
      PRINT 810
  810 FORMAT (//'   THE BRANCH VOLTAGES ARE'/)
      DO 820 I=1,M
  820 PRINT 645,(VBP(I,K),K=1,IJK)
  830 ISS=ISC+ISL
C          GENERATE BRANCH CURRENTS.
C          ISS = TOTAL NUMBER OF STATE VARIABLES
      ISV=ISS+INV
      DO 890 K=1,IJK
      IF (K.LE.ISC) GO TO 840
      IF ((K.GT.ISS).AND.(K.LE.ISV)) GO TO 850
      GO TO 860
  840 I=ISTATE(K)
      VBP(I,K)=VBP(I,K)-1.D0
      GO TO 860
  850 I=INPV(K-ISS)
      VBP(I,K)=VBP(I,K)+1.D0
  860 DO 870 I=1,M
  870 BCP(I,K)=R(I)*VBP(I,K)
      IF (IIB.LE.0) GO TO 890
      DO 880 I=1,IIB
      VAL=B(I)
      IR=IB(I,1)
      J=IB(I,2)
      IF (IR.GT.0) GO TO 880
      IR=-IR
      VAL=VAL*R(J)
  880 BCP(IR,K)=BCP(IR,K)+VAL*VBP(J,K)
  890 CONTINUE
      IF (ISC.LE.0) GO TO 905
      DO 900 K=1,ISC
      II=ISTATE(K)
      JF=IA(II,1)
      DO 900 L=1,ISC
      BCP(II,L)=0.
      DO 895 I=1,M
      IF (I.EQ.II) GO TO 895
      KF=IA(I,1)
      KT=IA(I,2)
      IF (KF.EQ.JF) BCP(II,L)=BCP(II,L)-BCP(I,L)
      IF (KT.EQ.JF) BCP(II,L)=BCP(II,L)+BCP(I,L)
  895 CONTINUE
  900 CONTINUE
  905 IF (ITRACE.LT.4) GO TO 930
```

Listing 5.3—*Cont.*

```
          PRINT 910
 910      FORMAT (//'   THE BRANCH CURRENT MATRIX IS'/)
          DO 920 I=1,M
 920      PRINT 645,(BCP(I,K),K=1,IJK)
C         GENERATE A AND B MATRICES.
 930      IF (ISC.LE.0) GO TO 950
          DO 940 I=1,ISC
          IJ=ISTATE(I)
          DO 940 J=1,IJK
 940      A(I,J)=BCP(IJ,J)*CC(IJ)
 950      IF (ISL.LE.0) GO TO 1010
          IF (MUTUAL.GT.0) GO TO 970
C         INVERSION NOT NEEDED.
          DO 960 I=1,ISL
          JJ=JSTATE(I)
          IJ=I+ISC
          DO 960 J=1,IJK
 960      A(IJ,J)=VBP(JJ,J)/LL(JJ)
C         ALL DONE.
          GO TO 1010
 970      DO 990 I=1,ISL
C         CLEAR SPACE FOR INVERSION.
          DO 980 J=1,ISL
 980      YB(I,J)=0.
          IJ=JSTATE(I)
 990      YB(I,I)=LL(IJ)
C         PUT IN SELF-INDUCTANCE.
          DO 993 I=1,MUTUAL
          JJ=ML(I,1)
          KK=ML(I,2)
          DO 992 II=1,ISL
          IF (JJ.EQ.JSTATE(II)) J=II
          IF (KK.EQ.JSTATE(II)) K=II
 992      CONTINUE
          YB(J,K)=XMUT(I)
 993      YB(K,J)=XMUT(I)
C         INDUCTANCE MATRIX IS COMPLETE.
          IF (ITRACE.LE.6) GO TO 996
          PRINT 994
 994      FORMAT (//'   THE INDUCTANCE MATRIX IS'/)
          DO 995 I=1,ISL
 995      PRINT 645,(YB(I,J),J=1,ISL)
 996      CALL DINVRT (YB,NNK,ISL,DET,YBB)
C         INDUCTANCE MATRIX INVERTED.
          IF (ITRACE.LE.6) GO TO 999
          PRINT 997
 997      FORMAT (//'   THE INVERSE INDUCTANCE MATRIX IS'/)
          DO 998 I=1,ISL
 998      PRINT 645,(YB(I,J),J=1,ISL)
 999      DO 1000 I=1,ISL
          II=I+ISC
          DO 1000 J=1,IJK
          A(II,J)=0.
          DO 1000 K=1,ISL
          KK=JSTATE(K)
1000      A(II,J)=A(II,J)+YB(I,K)*VBP(KK,J)
C         A AND B MATRICES COMPLETE
C         TOLERANCE ANALYSIS, FREQUENCY AND TIME DOMAIN ANALYSIS
C         AND SENSITIVITY ANALYSIS CAN BE ADDED HERE.


C         READY FOR OUTPUT QUANTITIES  (Y-VECTOR VALUES)
```

Listing 5.3—*Cont.*

```
C           READ IN  BRANCH NUMBER, BRANCH VARIABLE TYPE
C                1 = VOLTAGE, 2 = CURRENT
C                DIRECTION OF VARIABLE CORRESPONDS TO INPUT BRANCH SENSE
 1010 KEY=0
      PRINT 1020
 1020 FORMAT (//'    THE OUTPUT QUANTITIES ARE'/)
 1030 READ  1040,I,ITYPE
 1040 FORMAT (2I3)
      IF (I.LE.0) GO TO 1120
      GO TO (1060,1090),ITYPE
 1060 PRINT 1070,I
 1070 FORMAT (' VOLTAGE ACROSS BRANCH',I4)
      KEY=KEY+1
      IROW=ISS+KEY
      DO 1080 J=1,IJK
 1080 Q(IROW,J)=VBP(I,J)
      GO TO 1030
 1090 PRINT 1100,I
 1100 FORMAT (' CURRENT THROUGH BRANCH',I3)
      KEY=KEY+1
      IROW=ISS+KEY
      DO 1110 J=1,IJK
 1110 Q(IROW,J)=BCP(I,J)
      GO TO 1030
C         READY TO PRINT RESULTS
 1120 PRINT 1130
 1130 FORMAT (//' THE STATE VARIABLES ARE (X-VECTOR)'/)
      DO 1180 I=1,ISS
      IF (I.GT.ISC) GO TO 1160
 1140 J=ISTATE(I)
      PRINT 1150,J
 1150 FORMAT (' CAPACITOR VOLTAGE - BRANCH ',I3)
      GO TO 1180
 1160 J=JSTATE(I-ISC)
      PRINT 1170,J
 1170 FORMAT (' INDUCTOR CURRENT  - BRANCH ',I3)
 1180 CONTINUE
      PRINT 1190
 1190 FORMAT (//' THE INPUT VARIABLES ARE (U-VECTOR)'/)
      INPUT=INV+INC
      DO 1240 I=1,INPUT
      IF (I.GT.INV) GO TO 1220
      J=INPV(I)
      PRINT 1210,J
 1210 FORMAT (' VOLTAGE SOURCE - BRANCH',I3)
      GO TO 1240
 1220 J=JNPV(I-INV)
      PRINT 1230,J
 1230 FORMAT (' CURRENT SOURCE - BRANCH',I3)
 1240 CONTINUE
      IBCOL=ISS+1
      PRINT 1250
 1250 FORMAT (//'   THE A-MATRIX IS'/)
      DO 1260 I=1,ISS
 1260 PRINT 645,(Q(I,J),J=1,ISS)
      PRINT 1270
 1270 FORMAT (//' THE B-MATRIX IS'/)
      DO 1280 I=1,ISS
 1280 PRINT 645,(Q(I,J),J=IBCOL,IJK)
      PRINT 1290
 1290 FORMAT (//' THE C-MATRIX IS'/)
      IRONE=ISS+1
      DO 1300 I=IRONE,IROW
 1300 PRINT 645,(Q(I,J),J=1,ISS)
      PRINT 1310
```

Listing 5.3—Cont.

```
 1310 FORMAT (//' THE D-MATRIX IS'/)
      DO 1320 I=IRONE,IROW
 1320 PRINT 645,(Q(I,J),J=IBCOL,IJK)
C        ALL RESULTS ARE PRINTED
      GO TO 100
 1330 PRINT 1340,M,N,ISL,ISC,MUTUAL
 1340 FORMAT (//' *** THE NETWORK CONTAINS NO STATE VARIABLES ***'/
     1' M=',I4,' N=',I4,' ISL=',I4,' ISC=',I4,' MUTUAL=',I4)
      GO TO 100
 1350 CALL EXIT
      END


      SUBROUTINE SOLVEK (A,NA,N,D,IX,KKK)
C        SOLUTION OF SIMULTANEOUS LINEAR EQUATIONS WITH KKK FORCING
C        VECTORS.  GAUSS ELIMINATION WITH PIVOT SEARCHING AND ROW/COLUMN
C        EXCHANGES.
C        A    = AUGMENTED ARRAY OF COEFFICIENTS
C        NA   = STORAGE DIMENSION OF A
C        N    = NUMBER OF EQUATIONS
C        D    = VALUE OF THE CALCULATE DETERMINANT
C        IX   = TEMPORARY VECTOR OF AT LEAST N ELEMENTS
C        KKK  = NUMBER OF FORCING VECTORS AUGMENTED IN A
C               (COEFFICIENTS ARE N ROWS BY N+KKK COLUMNS)
C        THE SOLUTION VECTORS ARE PRODUCED IN COLUMNS N+1 THROUGH N+KKK
C        THE INPUT ARRAY IS DESTROYED
      IMPLICIT REAL*8(A-H,O-Z)
      DIMENSION A(NA,NA),IX(N)
      NK=N+KKK
      N1=N+1
      NN=N-1
      D=1.D0
      DO 100 I=1,N
  100 IX(I)=I
      DO 210 L=1,NN
      LL=L+1
      AMAX=0.D0
      IM=L
      JM=L
      DO 120 I=L,N
      DO 120 J=L,N
      IF (AMAX.GE.DABS(A(I,J))) GO TO 120
      AMAX=DABS(A(I,J))
      IM=I
      JM=J
  120 CONTINUE
C        PIVOT AT IM,JM START EXCHANGE IF NECESSARY
      IF (IM.LE.L) GO TO 150
      DO 140 J=L,NK
      T=A(IM,J)
      A(IM,J)=A(L,J)
  140 A(L,J)=T
      D=-D
  150 IF (JM.LE.L) GO TO 180
      DO 170 I=1,N
      T=A(I,JM)
      A(I,JM)=A(I,L)
  170 A(I,L)=T
      D=-D
      IT=IX(JM)
      IX(JM)=IX(L)
```

Listing 5.3—*Cont.*

```
C             FORCING VECTORS.
        CALL SOLVEK (YB,NNK,N,DET,IIT,IJK)
  190   A(L,K)=A(L,K)/AMAX
        DO 200 J=LL,N
        DO 200 K=LL,NK
  200   A(J,K)=A(J,K)-A(J,L)*A(L,K)
  210   CONTINUE
        DO 220 K=N1,NK
  220   A(N,K)=A(N,K)/A(N,N)
        D=D*A(N,N)
C             FORWARD COURSE COMPLETE, START RETURN COURSE.
        DO 230 KK=N1,NK
        DO 230 I=1,NN
        II=N-I
        III=II+1
        DO 230 K=III,N
  230   A(II,KK)=A(II,KK)-A(II,K)*A(K,KK)
C             UNSCRAMBLE VARIABLES.
        DO 280 I=1,N
        DO 270 J=1,N
        IF (I.NE.IX(J)) GO TO 270
        IF (I.EQ.J) GO TO 280
        DO 260 KK=N1,NK
        T=A(J,KK)
        A(J,KK)=A(I,KK)
  260   A(I,KK)=T
        IT=IX(J)
        IX(J)=IX(I)
        IX(I)=IT
        GO TO 280
  270   CONTINUE
  280   CONTINUE
        RETURN
        END
```

Listing 5.3—*Cont.*

and then the Gauss elimination scheme (with pivoting) is applied keeping row equality over all elements in each row. The corresponding node voltages $V_n^{(j)}$ will replace the current vector $I_s^{(j)}$. For a network of NN independent nodes, the above results in NN equations in k sets of NN unknowns.

A simplified flowchart for the above process is shown in Fig. 5.11. The various subroutines are as follows.

Subroutine SOLVEK is a routine to solve Eq. (5.54). Its basic structure is Gaussian elimination with pivot searching to form the lower triangular unit matrix and a backward substitution scheme to produce the VN matrix. The main advantages of SOLVEK are that it requires only one reduction of the YN matrix to produce the K node-voltage vectors and it can be called at a later point in the program to solve a normal set of simultaneous equations if program extensions so require.

The node voltages are thus solved for each of the forcing functions. The K node-voltage vectors are combined in the VN matrix and one matrix multiplication by A^T will yield VB, a branch-voltage matrix where each column corresponds to setting successive $z_k = 1$. In this program this matrix multiplication is replaced by a search through the packed incidence matrix IA whose function is exactly the function of the IA matrix of Sec. 3.7 [see Eq. (3.103)].

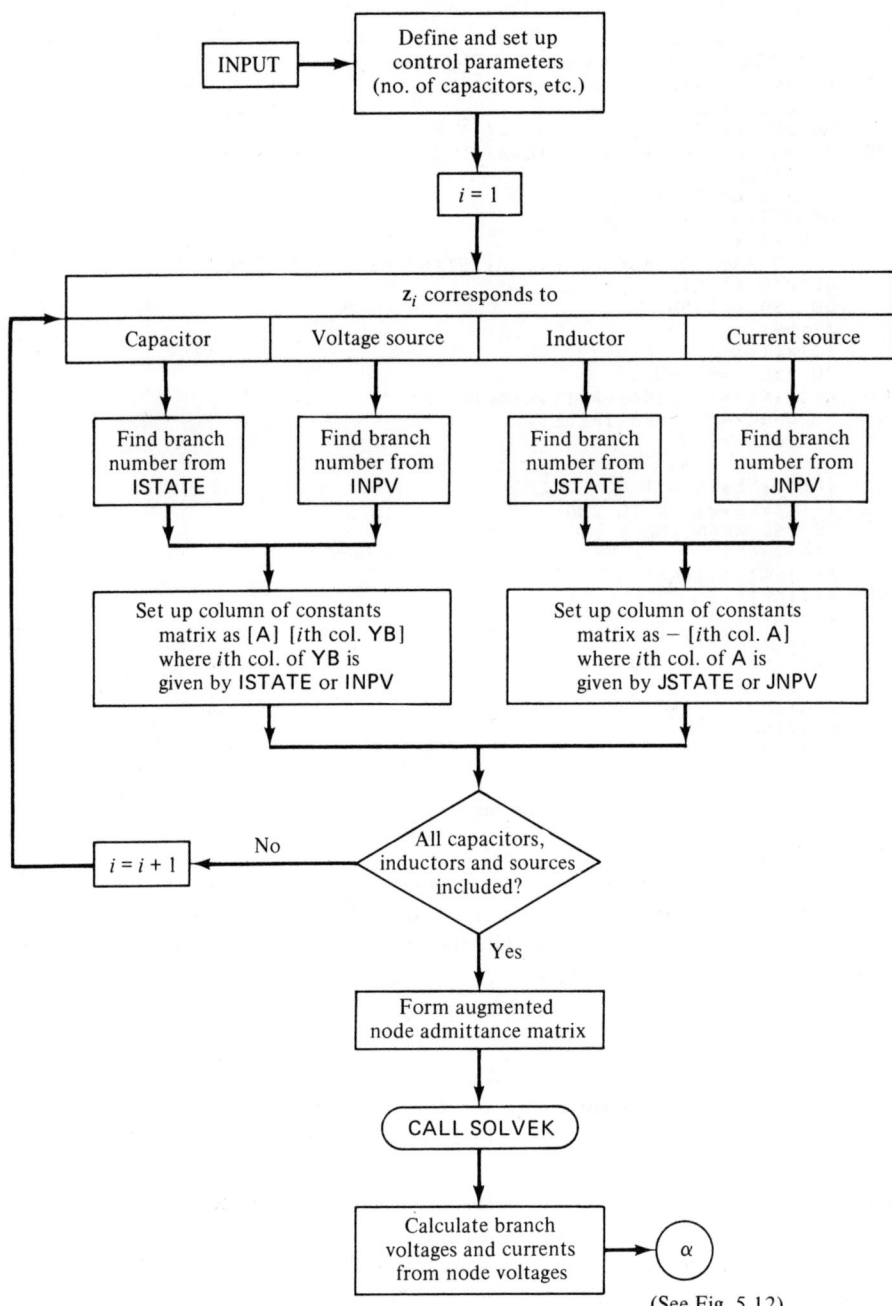

Fig. 5.11 Generation of Augmented Node Equations

The branch current is found by deriving the element voltages from the branch voltages and multiplying by the branch admittance (Ohm's law). The element voltage is found by subtracting any source voltage from the corresponding branch voltage. Any one column of VB can have only one source voltage so that one subtraction per column is necessary to produce the element voltage. The location of the $z_k = 1$ corresponding to voltage sources (that is, capacitors and input voltages sources) is stored in the input routine as the ISTATE and INPV vectors. The admittance of each branch is stored, on input, in the R vector. Then the branch currents BC corresponding to any $z_k = 1$ are found by

$$BC = VB * R$$

where BC is the matrix of branch currents; each column k corresponds to the branch current for $z_k = 1$ [see Eq. (5.3)].

At this point the four sets of quantities necessary to generate the **Q** matrix, namely, the C, L, VB, and BC matrices, have been calculated. Actual generation of the **Q** matrix entries is then only a data manipulation problem. Note that matrices C and L contain the reciprocal values of the capacitors and inductors respectively.

The rows of the **Q** matrix corresponding to the A and B matrices can be found from the VB and BC matrices as follows:

(1) If $z_k = 1$ corresponds to setting $v_{c_k} = 1$, then Kth row of $\mathbf{Q} = C_k \times K$th row of BC.

(2) If $z_k = 1$ corresponds to setting $V_{G\kappa} = 1$, then Kth row of $\mathbf{Q} = K$th row of VB.

(3) If $z_k = 1$ corresponds to setting $i_{L\kappa} = 1$, then Kth row of $\mathbf{Q} = L_k \times K$th row of VB.

(4) If $z_k = 1$ corresponds to setting $I_{G\kappa} = 1$, then Kth row of $\mathbf{Q} = K$th row of BC.

The generation of these entries of **Q** is shown in the flowchart of Fig. 5.12.

In order to facilitate the use of the equations generated by the program, the results are printed in such a form that they can be easily used. Since the state variable equations are most frequently stated as in Eqs. (5.1) and (5.2), the output should conform to this form. The three network vectors (**x**, **u**, **y**) are listed in the same order as they relate to the matrices and the branch number and either the voltage or current variable is specified. Immediately following are the A, B, C, and D matrices, in that order, and in matrix form.

The output section is provided in modular form and can easily be modified to provide either punched cards for input to other analysis programs or can be input directly into attached analysis subroutines. The data flow for the output section is shown in Fig. 5.13.

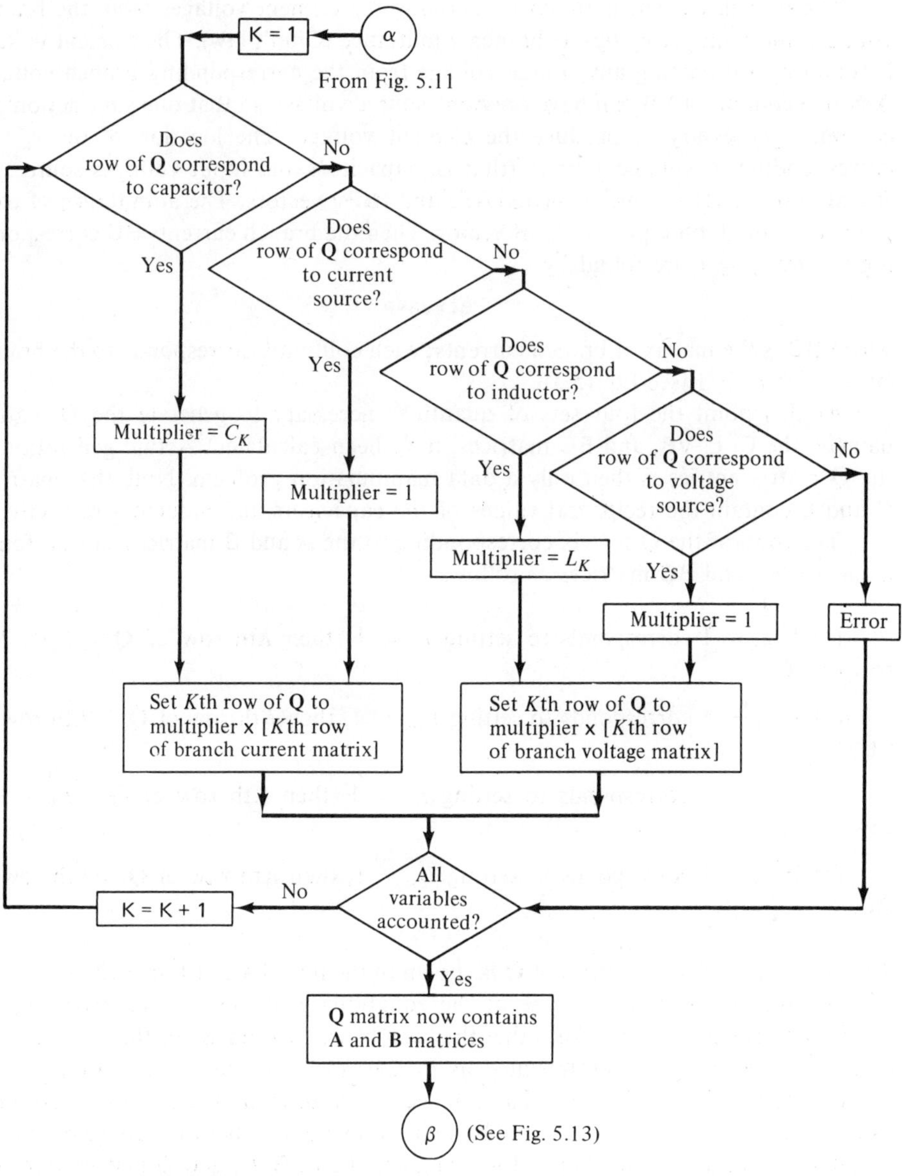

Fig. 5.12 Generation of **A** and **B** Matrices

In order to facilitate program development, modifications and debugging a print-level indicator, ITRACE, is incorporated in the program. By setting this variable to different values the user can obtain additional information on the execution of the program. Table 5.1 shows the various levels and the printed outputs they enable.

Sec. 5.5 Program for the Formulation of the State Equations   **215**

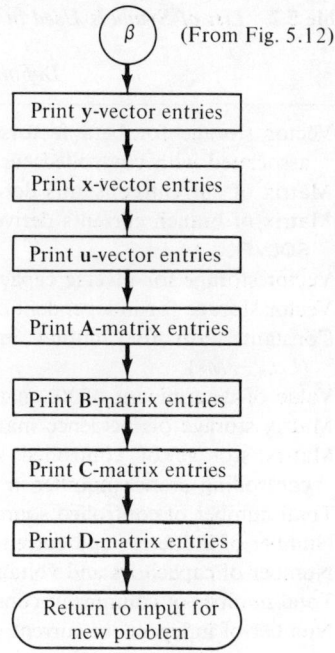

Fig. 5.13 Simplified Output Program Flowchart

Table 5.1 *Printing Levels*

Value of ITRACE	*Printed Results*
0	No additional output (default value)
1	Node voltages
2	Augmented node-admittance matrix and (1)
3	Node-admittance matrix, current matrix, and (2)
4	Branch voltages, branch currents, and (3)

The default value for ITRACE is zero and its modification is by an input value to the source program on the network size description card.

An additional aid is the marking of extension points within the source program shown in Listing 5.3. These extension points are marked by comment cards and are the points at which subroutine CALL statements or additional code may be included to increase program capabilities. The points and suggested extension topics discussed in the problems are by no means unique or complete.

In order to facilitate modifications and understanding of this program, a list of of the **FORTRAN** variables used is given in Table 5.2. The program is limited by the following factors:

Table 5.2 *List of Symbols Used in Program*

Symbol	Definition
B	Vector storage for beta factors or transconductance factors associated with controlled current sources
BC	Matrix of current constants derived and sent to SOLVEK
BCP	Matrix of branch currents derived from results returned from SOLVEK
CC	Vector storage for inverse capacitance values
CG	Vector storage for independent current sources
DELM	Constant term for mutual inductance denominator term $(L_1 L_2 - M^2)$
DET	Value of determinant of YB matrix returned by SOLVEK
IA	Matrix storage of incidence matrix in packed form
IB	Matrix storage of controlled source branch numbers and controlling branch number in packed form
IIB	Total number of controlled sources
IINC	Number of inductors and current sources
IINV	Number of capacitors and voltage sources
IJK	Total number of columns in constants matrix
INC	Number of independent current sources
INPV	Vector of voltage source branch numbers
INV	Number of independent voltage sources
ISC	Number of capacitors
ISL	Number of inductors
ISS	Total number of state variables
ISTATE	Vector of capacitor branch numbers
ITRACE	Print-level control parameter
ITYPE	Voltage or current control parameter used in input cards
IX	Element type designator used in input
JNPV	Vector of current source branch numbers
JSTATE	Vector of inductor branch numbers
LL	Vector of inverse inductances
M	Total number of branches in the network
ML	Paired list of inductor numbers that are mutually coupled
MUTUAL	Total number of mutual elements in the network
N	Number of nodes in the network
NIJK	Total number of columns in augmented branch-admittance matrix
Q	Common storage area of A, B, C, and D matrices
R	Vector of branch admittances
VAL	Element value used on input cards
VBP	Vector of branch voltages
VG	Vector of voltage sources
YB	Storage array for various calculations
YBB	Temporary storage for one vector of YB
XMUT	Vector of mutual inductance values

(1) The network to be analyzed must be a non-degenerate RLC network. Thus only equations of the kind shown in Eqs. (5.1) and (5.2) can be calculated; networks which lead to derivatives of the input [see Eq. (5.13)] will lead to erroneous results. However, the majority of networks of interest are of the non-degenerate type.

(2) No controlled voltage sources can be handled. Controlled voltage sources (μ and r_t terms) should be converted to controlled current sources as shown in Fig. 3.14.

(3) Capacitor loops and inductor trees will lead to meaningless results and will show up as very large numbers in the Q matrix.

(4) Numerical difficulties result because of the assumption that a short circuit may be represented by a resistance of 10^{-10} ohms and an open circuit by a resistance of 10^{+18} ohms. Although these assumptions do not limit the types of circuits that can be analyzed, they can cause a reduction of accuracy to approximately the sixth decimal place in some small networks and possible additional degradation of accuracy in larger networks. Because of the large range of numbers used, double-precision arithmetic is indicated.

Possible extensions and modifications to the program may include time and frequency response calculation capabilities which may be added with no changes in the existing program. With minor changes, certain cases of degenerate networks can be handled although extensive changes would be necessary to make the program truly general in nature. An additional analysis subroutine should be added to enable the program to detect and modify capacitor loop and inductor tree combinations. The input task is simplified greatly by replacing the existing input section with a free-format input routine.

EXAMPLE 5.6

It is desired to find the state equations for small signal analysis of the circuit shown in Fig. 5.14. The output quantities are the input current (I_{IN}) and the output voltage (E_{OUT}).

Solution. If it is assumed that the circuit is linear, then E_{OUT} and I_{IN} are not affected by V_{BB}. The equivalent circuit for analysis is shown in Fig. 5.15 and the branch and node numbers used in the program are indicated. Note that the transistor is replaced by a very simple linear equivalent circuit, normally valid for small signals.

The circuit must be analyzed with each capacitor and the voltage source replaced one at a time by a unit voltage source and the remaining ones short circuited. The **Q** matrix can be established by columns. The entries of the first column are found from

$$\begin{bmatrix} q_{11} \\ q_{21} \\ q_{31} \\ q_{41} \\ q_{51} \end{bmatrix} = \begin{bmatrix} \frac{1}{C_{B2}} i_{B2} \\ \frac{1}{C_{B7}} i_{B7} \\ \frac{1}{C_{B9}} i_{B9} \\ i_{B2} \\ v_{B10} \end{bmatrix} \quad \text{where} \quad \begin{cases} v_{B2} = 1 \\ v_{B7} = v_{B9} = V_{IN} = 0 \end{cases}$$

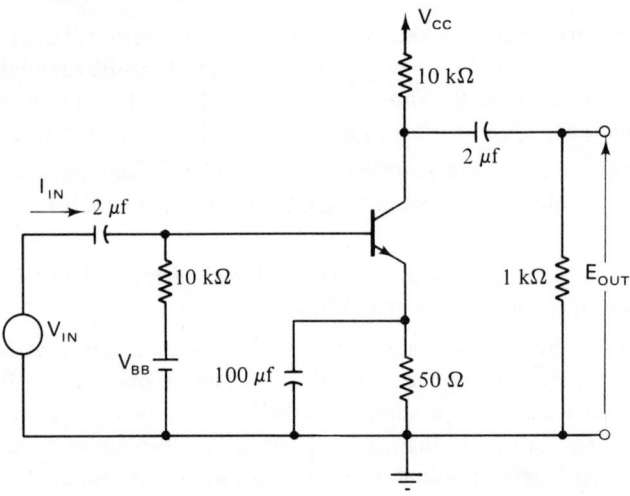

Fig. 5.14 Transistor Amplifier with Capacitance Coupling

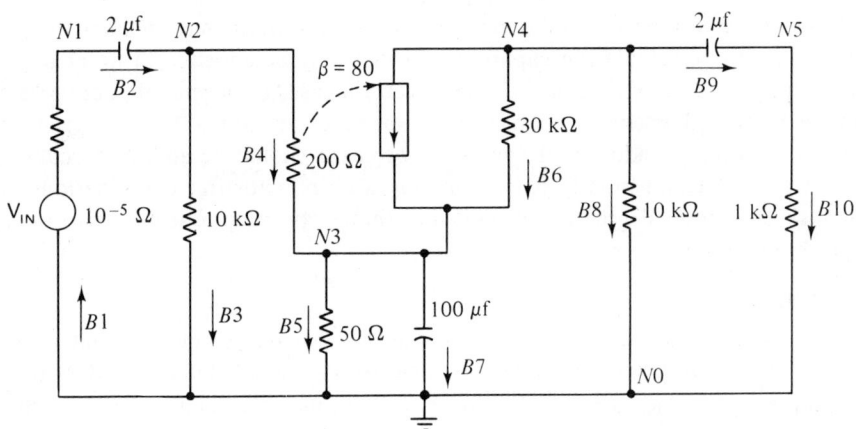

Fig. 5.15 Equivalent Circuit for the Transistor Amplifier

The entries may be calculated directly from an auxiliary circuit. The values for the first column of **Q** are

$$q_{11} = \frac{1}{C_{B2}} i_{c_{B2}} = \frac{1}{2 \times 10^{-4}} \times \left(\frac{-1}{10^k} + \frac{-1}{200}\right) = -2.55 \times 10^3$$

$$q_{21} = \frac{1}{C_{B7}} i_{c_{B7}} = \frac{1}{10^{-4}} \times \left(\frac{-1}{200} + \left(\frac{-1}{200} \times 80\right) \frac{30^k//10^k//1^k}{10^k//1^k}\right) = -3.932 \times 10^3$$

$$q_{31} = \frac{1}{C_{B9}} i_{c_{B9}} = \frac{1}{2 \times 10^{-6}} (-) \left(\frac{-1}{200}\right)(80) \cdot \frac{30^k//10^k//1^k}{1^k} = 1.765 \times 10^6$$

$$q_{41} = i_{B2} = i_{c_{B2}} = \frac{-1}{10^k} + \frac{-1}{200} = -5.100 \times 10^{-3}$$

$$q_{51} = v_{B10} = i_{B9} \times 1^k = 3.529 \times 10^2$$

Sec. 5.5 Program for the Formulation of the State Equations

The entries in the second column are calculated as in the first, except that the applied voltages are $v_{B2} = 0$, $v_{B7} = v_{B9} = 0$, and $V_{IN} = 0$. The applied voltages for the third column have $v_{B9} = 1$ and for the fourth, $V_{IN} = 1$. The problem is then reduced to a series of four DC circuit analyses. The final state equations are:

$$\frac{d}{dt}\begin{bmatrix} v_{c_{B2}} \\ v_{c_{B7}} \\ v_{c_{B9}} \end{bmatrix} = \begin{bmatrix} -2.55 \times 10^3 & -2.50 \times 10^3 & 0 \\ -3.93 \times 10^3 & -4.13 \times 10^3 & 0.294 \\ 1.76 \times 10^5 & 1.76 \times 10^5 & -5.88 \times 10^1 \end{bmatrix} \begin{bmatrix} v_{c_{B2}} \\ v_{c_{B7}} \\ v_{c_{B9}} \end{bmatrix} + \begin{bmatrix} 2.55 \times 10^3 \\ 3.93 \times 10^3 \\ -1.76 \times 10^5 \end{bmatrix} V_{IN}$$

$$\begin{bmatrix} I_{IN} \\ E_{OUT} \end{bmatrix} = \begin{bmatrix} -5.10 \times 10^{-3} & -5.00 \times 10^{-3} & 0 \\ 3.53 \times 10^2 & 3.53 \times 10^2 & -0.118 \end{bmatrix} \begin{bmatrix} v_{c_{B2}} \\ v_{c_{B7}} \\ v_{c_{B9}} \end{bmatrix} + \begin{bmatrix} 5.10 \times 10^{-3} \\ -3.53 \times 10^2 \end{bmatrix} V_{IN}$$

These results may be compared with the computer generated output shown in Listing 5.4, the results of execution of the program shown in Listing 5.3.

Note the use of a small resistor in series with V_{IN}; also V_{IN} was arbitrarily set to one since it will appear as a multiplicative variable in the results.

```
          NEW PROBLEM.
          NUMBER OF BRANCHES= 10
          NUMBER OF INDEPENDENT NODES=  5

          INPUT CARDS ARE

          BRANCH    FROM   TO   TYPE   VALUE
          NUMBER
             1        0     1     1    1.00000D-05
             1        0     1     4    1.00000D 00
             2        1     2     2    2.00000D 00
             3        2     0     1    1.00000D 04
             4        2     3     1    2.00000D 02
             5        3     0     1    5.00000D 01
             6        4     3     1    3.00000D 04
             7        3     0     2    1.00000D 02
             8        4     0     1    1.00000D 04
             9        4     5     2    2.00000D 00
            10        5     0     1    1.00000D 03
             1        4     6     6    8.00000D 01

          THE OUTPUT QUANTITIES ARE

          CURRENT THROUGH BRANCH   2
          VOLTAGE ACROSS BRANCH   10

          THE STATE VARIABLES ARE (X-VECTOR)

          CAPACITOR VOLTAGE - BRANCH    2
          CAPACITOR VOLTAGE - BRANCH    7
          CAPACITOR VOLTAGE - BRANCH    9

          THE INPUT VARIABLES ARE (U-VECTOR)

          VOLTAGE SOURCE - BRANCH    1

          THE A-MATRIX IS

          -2.54914D 03   -2.50000D 03   -7.35831D-12
          -3.93231D 03   -4.13263D 03    2.94332D-01
           1.76499D 05    1.76499D 05   -5.84255D 01
```

Listing 5.4 Results of the Analysis of the Circuit of Fig. 5.15

5.6 Poles and Zeros

The state equations determine a set of input-output relations for a network. The equations themselves are a set of linear first-order differential equations. We may solve such equations by means of the Laplace transformation.

Let the transforms of the vectors $\mathbf{x}(t)$, $\mathbf{u}(t)$, and $\mathbf{y}(t)$ be $\mathbf{X}(s)$, $\mathbf{U}(s)$, and $\mathbf{Y}(s)$ respectively. The dimensions of these transform vectors are the same as the dimensions of the original set of variables. Since the Laplace transform of the derivative is

$$\mathcal{L}\left(\frac{d\mathbf{x}}{dt}\right) = s\mathbf{X}(s) - \mathbf{x}_0 \tag{5.55}$$

the transformed state equations become

$$\begin{aligned} s\mathbf{X}(s) - \mathbf{x}_0 &= \mathbf{A}\mathbf{X}(s) + \mathbf{B}\mathbf{U}(s) \\ \mathbf{Y}(s) &= \mathbf{C}\mathbf{X}(s) + \mathbf{D}\mathbf{U}(s) \end{aligned} \tag{5.56}$$

Formally, the above may be solved for the input-output relations:

$$\mathbf{Y}(s) = \mathbf{C}[s\mathbf{I} - \mathbf{A}]^{-1}(\mathbf{B}\mathbf{U}(s) + \mathbf{x}_0) + \mathbf{D}\mathbf{U}(s) \tag{5.57}$$

where $\mathbf{I}$ denotes the unit matrix (to avoid confusion with $\mathbf{U}$, which is used here as the input variable set). The formal solution, $\mathbf{y}(t)$, is the inverse Laplace transformation of the above set.

The matrix inverse appearing above contains not only numbers but also a set of Laplace transform variables. We shall now study this matrix, the transform of the state transition matrix, Eq. (5.22), in some detail.

The inverse, if it exists, of an arbitrary square matrix $\mathbf{G}$ may be written as in [13]

$$\mathbf{G}^{-1} = \frac{\operatorname{adj} \mathbf{G}}{\det \mathbf{G}} \tag{5.58}$$

where det $\mathbf{G}$ is the determinant of $\mathbf{G}$ and adj $\mathbf{G}$ is the adjoint of $\mathbf{G}$.

Basic properties of *determinants* are presumed to be familiar to the reader. For our purposes it suffices to observe that the determinant is a sum of products of the elements of the matrix such that each product term contains one element from each row and each column of the matrix without any repetitions. Thus for an N by N matrix $\mathbf{A}$ the determinant of $(s\mathbf{I} - \mathbf{A})$ is a polynomial in s of degree N with the highest polynomial coefficient equal to one:

$$\det(s\mathbf{I} - \mathbf{A}) = q(s) = \sum_{i=0}^{N} c_i s^i \tag{5.59}$$

with $c_N = 1$.

One can obtain the polynomial directly by the following rules:

(1) Multiplying any row (or column) of a matrix multiplies the value of the determinant by that number.

(2) Adding to any row (or column) of a matrix any linear combination of other rows leaves the value of the determinant unchanged.

(3) Exchanging any two rows (or two columns) reverses the algebraic sign of the matrix.

(4) The value of the determinant of a matrix having all zeros below (or above) the main diagonal is the product of the main-diagonal elements.

The above rules can be used to obtain the determinant of any square matrix. The procedure is exactly the one used for Gauss elimination (Sec. 2.4); the rules apply to numeric matrices as well as to matrices with algebraic elements as in $(s\mathbf{I} - \mathbf{A})$.

EXAMPLE 5.7

$$\det \begin{bmatrix} s-3 & 2 & 4 & 5 \\ 2 & s-1 & 0 & 2 \\ 1 & 0 & s-5 & 3 \\ 3 & 2 & 0 & s-4 \end{bmatrix}$$

$$= (s-3) \det \begin{bmatrix} 1 & \frac{2}{s-3} & \frac{4}{s-3} & \frac{5}{s-3} \\ 0 & (s-1) - \frac{4}{s-3} & \frac{-8}{s-3} & 2 - \frac{10}{s-3} \\ 0 & \frac{-2}{s-3} & (s-5) - \frac{4}{s-3} & 3 - \frac{5}{s-3} \\ 0 & 2 - \frac{6}{s-3} & \frac{12}{s-3} & (s-4) - \frac{15}{s-3} \end{bmatrix}$$

$$= s^4 - 13s^3 + 32s^2 + 139s - 399$$

Of course, the intermediate reduction steps are very tedious to carry out; an expansion by minors is probably easier with algebraic terms [13].

The polynomial $q(s)$ can be factored by means of any root finding program, such as Muller's method [12, 14]. Root finding of high-degree polynomials is a complicated subject in its own right; a great many numerical procedures are in use with Muller's method and Lehmer's method [15] yielding good results for circuit-analysis problems.

If the roots of the polynomial $q(s) = 0$ are s_i, then

$$q(s) = \prod_{i=1}^{N} (s - s_i) \qquad (5.60)$$

is an alternate form of the determinant of $\mathbf{G}$. The numbers s_i are the eigenvalues of the matrix $\mathbf{G}$. We observe that

$$q(s_i) = q(s)\Big|_{s=s_i} = 0 \qquad (5.61)$$

In turn, this requires that

$$\det(s_i \mathbf{I} - \mathbf{A}) = 0 \qquad (5.62)$$

which results in an indeterminate value for $\mathbf{Y}(s)$. Such values of s_i are, of course, the poles of the network function. We note that the poles are common for each of the individual outputs $\mathbf{Y}_i$ and for the state variables $\mathbf{X}_i$ with possible cancellation of individual poles from the adjoint. In general, the values s_i are complex numbers.

The *adjoint* appearing in Eq. (5.58) is both simpler and more complicated than the determinant. We observe that the determinant of a matrix is a one-dimensional quantity, either a number or an algebraic expression. Since the inverse demands that

$$\mathbf{G} \cdot \frac{\text{adj } \mathbf{G}}{\det \mathbf{G}} = \mathbf{G} \cdot \mathbf{G}^{-1} = \mathbf{I} = \mathbf{G}^{-1} \cdot \mathbf{G} = \frac{\text{adj } \mathbf{G}}{\det \mathbf{G}} \cdot \mathbf{G} \qquad (5.63)$$

it is obvious that the adjoint must be a matrix of the same dimensions as the original matrix. If we denote the elements of the adjoint as

$$\text{adj } \mathbf{G} = {}^N[G_{ij}] \qquad (5.64)$$

then the G_{ij} are given as determinants made up of elements of the original matrix $\mathbf{G}$ with row j and column i deleted [13] and with alternating sign:

$$G_{ij} = (-1)^{i+j} \det [\mathbf{G}_{ji}] \qquad (5.65)$$

where $\mathbf{G}_{ji}$ is the original matrix with row j and column i removed.

Since the removal of a row and a column from the original matrix $(s\mathbf{I} - \mathbf{A})$ will remove at least one term containing s (on the main diagonal, when $i = j$), each of the adjoint's elements become polynomials with at least one lower power than the determinant. If we denote

$$\text{adj } (s\mathbf{I} - \mathbf{A}) = {}^N[p(s)_{ij}] \qquad (5.66)$$

with

$$p(s)_{ij} = \det [(s\mathbf{I} - \mathbf{A})_{ji}] \cdot (-1)^{i+j} \qquad (5.67)$$

then each of the polynomials is of degree $(N - 1)$ at most. The set of these polynomials may be rewritten in the form

$$\text{adj } (s\mathbf{I} - \mathbf{A}) = \sum_{i=0}^{N-1} \mathbf{P}_i s^i \qquad (5.68)$$

where each of the coefficients $\mathbf{P}_i$ is an N by N matrix of real number elements.

The determination of the matrices $\mathbf{P}_i$ and the coefficients c_i of the polynomial $q(s)$ are related problems. An elegant procedure exists for the determination of both sets, the Souriau-Frame algorithm [16]. The procedure requires approximately N^4 steps and is strongly affected by numerical round-off and therefore should normally be carried out in multiple-precision arithmetic. The procedure is based on calculating the trace, that is the sum of the main-diagonal elements, of various related matrices obtained from the following recursion relations:

$$\begin{aligned} & & \mathbf{P}_{N-1} &= \mathbf{I} \\ c_{N-1} &= -\text{tr }(\mathbf{A}) & & \\ & & \mathbf{P}_{N-2} &= \mathbf{P}_{N-1}\mathbf{A} + c_{N-1}\mathbf{I} \\ c_{N-2} &= -\tfrac{1}{2} \text{tr }(\mathbf{P}_{N-2}\mathbf{A}) & & \\ & & \mathbf{P}_{N-3} &= \mathbf{P}_{N-2}\mathbf{A} + c_{N-2}\mathbf{I} \\ c_1 &= -\tfrac{1}{N-1} \text{tr }(\mathbf{P}_1\mathbf{A}) & & \\ & & \mathbf{P}_0 &= \mathbf{P}_1\mathbf{A} + c_1\mathbf{I} \\ c_0 &= -\tfrac{1}{N} \text{tr }(\mathbf{P}_0\mathbf{A}) & & \\ & & 0 &= \mathbf{P}_0\mathbf{A} + c_0\mathbf{I} \end{aligned} \qquad (5.69)$$

Note that a check is obtained in the last step of this process.

For N state variables, M inputs, and K outputs, the input-output "transfer function" becomes a set of K by M matrices; the initial conditions are set to zero for obtaining transfer functions:

$$\mathbf{Y}(s) = \left[\frac{1}{q(s)} \sum_{i=0}^{N-1} \mathbf{CP}_i s^i \mathbf{B} + \mathbf{D}\right] \mathbf{U}(s) = \mathbf{T}(s)\mathbf{U}(s) \tag{5.70}$$

where the brackets denote the set of transfer functions. There will be one such transfer function, each a ratio of polynomials in s, for each of the input-output pairs:

$$t_{jk}(s) = \frac{1}{q(s)} \sum_{n=0}^{N} p_{n_{jk}} s^n = \frac{p_{jk}(s)}{q(s)} \tag{5.71}$$

where

$p_{n_{jk}}$ = the j, k element of $\mathbf{CP}_n \mathbf{B} + \mathbf{D}$;
$p_{jk}(s)$ = the polynomial in s constructed from the $p_{n_{jk}}$;
$t_{jk}(s)$ is the transfer function between the kth input and the jth output.

As any polynomial $p_{jk}(s)$ may be factored

$$p_{jk}(s) = K_{jk} \sum_{i=1}^{N^*} (s - s_i^{(j,k)}) \tag{5.72}$$

where

$s_i^{(j,k)}$ = the zeros of the transfer function t_{jk};
K_{jk} = the coefficient of the highest power of s in $p_{jk}(s)$;
N^* = at most N, its value for any t_{jk} is to be determined from the Souriau-Frame procedure.

We now illustrate the above steps by using Ex. 5.1 (see Sec. 5.1 and Ex. 5.2).

EXAMPLE 5.8

Find the poles and the zeros of the transfer function of the circuit shown in Fig. 5.1.

Solution. From Ex. 5.1:

$N = 2; \quad M = 1; \quad K = 1$

$$\mathbf{P}_1 = \begin{bmatrix} 1 & 0 \\ 0 & 1 \end{bmatrix}$$

$c_1 = -\text{tr}(\mathbf{A})$

$\quad = -\text{tr}\begin{bmatrix} -1.09 \times 10^4 & 5.45 \\ -2.72 \times 10^7 & -3.64 \times 10^4 \end{bmatrix}$

$\quad = 4.73 \times 10^4$

$$\mathbf{P}_0 = \begin{bmatrix} 1 & 0 \\ 0 & 1 \end{bmatrix} \begin{bmatrix} -1.09 \times 10^4 & 5.45 \\ -2.72 \times 10^7 & -3.64 \times 10^4 \end{bmatrix} + 4.73 \times 10^4 \begin{bmatrix} 1 & 0 \\ 0 & 1 \end{bmatrix}$$

$$= \begin{bmatrix} 3.64 \times 10^4 & 5.45 \\ -2.72 \times 10^7 & 1.09 \times 10^4 \end{bmatrix}$$

$c_0 = -\frac{1}{2} \text{tr}(\mathbf{P}_0 \mathbf{A})$

$\quad = -\frac{1}{2} \text{tr}\begin{bmatrix} -5.45 \times 10^8 & 0 \\ 0 & -5.45 \times 10^8 \end{bmatrix}$

$\quad = 5.45 \times 10^8$

It is easily seen that the check matrix $\mathbf{P}_0\mathbf{A} + c_0\mathbf{I}$ is in fact a zero matrix.
From the above

$$(s\mathbf{I} - \mathbf{A})^{-1} = \frac{1}{s^2 + 4.73 \times 10^4 s + 5.45 \times 10^8} \left[\begin{bmatrix} 1 & 0 \\ 0 & 1 \end{bmatrix} s + \begin{bmatrix} 3.64 \times 10^4 & 5.45 \\ -2.72 \times 10^7 & 1.09 \times 10^4 \end{bmatrix} \right]$$

The poles are immediately obvious from

$$s^2 + 4.73 \times 10^4 s + 5.45 \times 10^8 = 0$$
$$s_1 = -1.986 \times 10^4$$
$$s_2 = -2.743 \times 10^4$$

For the transfer function there results

$$\mathbf{Y}(s) = \frac{1}{s^2 + 4.73 \times 10^4 s + 5.45 \times 10^8} \begin{bmatrix} 0 & 1 \end{bmatrix} \left[\begin{bmatrix} 1 & 0 \\ 0 & 1 \end{bmatrix} s + \begin{bmatrix} 3.64 \times 10^4 & 5.45 \\ -2.72 \times 10^7 & 1.09 \times 10^4 \end{bmatrix} \right]$$
$$\begin{bmatrix} 10.9 \\ 2.72 \times 10^4 \end{bmatrix} + [0] \mathbf{U}(s)$$

where the matrices $\mathbf{B}$, $\mathbf{C}$, and $\mathbf{D}$ from Ex. 5.1 were used. The above calculations are carried out and the results are

$$\mathbf{Y}(s) = \frac{2.72 \times 10^4 s}{(s + 1.986 \times 10^4)(s + 2.743 \times 10^4)}$$

Thus there is a zero at the origin.

Because of the simple input-output relationship the calculations yield a very simple ratio of polynomials. Application to the circuit in Ex. 5.3 (Fig. 5.5) shows the power of this method far more effectively.

EXAMPLE 5.9

Find the poles and zeros of the circuit used in Ex. 5.5.

Solution.
$$N = 6; \quad M = 1; \quad K = 2$$

Thus the resultant $\mathbf{T}(s)$ transfer matrix will have two rows and one column; the inversion of the $(s\mathbf{I} - \mathbf{A})$ matrix will require the Souriau-Frame algorithm for a six by six matrix. The determinant will be a polynomial in s of degree 6 and the adjoint elements will have a degree of up to 5. Since the $\mathbf{D}$ matrix is not zero, a set of ratios of polynomials for $\mathbf{T}(s)$ can be expressed as

$$\mathbf{Y}(s) = \frac{1}{\det(s\mathbf{I} - \mathbf{A})} [\mathbf{C} \operatorname{adj}(s\mathbf{I} - \mathbf{A})\mathbf{B} + \det(s\mathbf{I} - \mathbf{A})\mathbf{D}]\mathbf{U}(s)$$

and thus the numerator polynomials of $t_{ij}(s)$ may be of order up to 6.

The required steps are best carried out by a computer program (see Problem 5.24); salient results of the algorithm are shown in Listing 5.5. Note that the numbers printed are of limited precision; the program uses quadruple-precision arithmetic on the IBM 370/165 system. A straightforward application of the Souriau-Frame algorithm will result in severe round-off errors. The data shown were obtained by using a normalization constant in the calculations. The original inversion is replaced by

$$(s\mathbf{I} - \mathbf{A})^{-1} = k^{-1}\left(\frac{s}{k}\mathbf{I} - \frac{1}{k}\mathbf{A}\right)^{-1} = k^{-1}(z\mathbf{I} - \mathbf{A}_1)^{-1}$$

```
NUMBER OF STATE VARIABLES =  6
NUMBER OF INPUTS =  1
NUMBER OF OUTPUTS=  2

THE A-MATRIX IS

-6.67000D+02   0.            2.00000D+07  -2.00000D+07   0.            0.
 0.           -4.95000D+02   9.88000D+08  -9.88000D+08  -9.88000D+08   1.23000D+07
-1.13900D+03   1.92100D+02  -6.57000D+03   1.91000D+03   1.91000D+03   1.92000D+06
 1.45500D+03  -2.69900D+02   7.11000D+06  -2.32000D+03  -2.32000D+03  -2.70000D+06
-1.31000D+03   2.48000D+02  -6.71000D+06   2.07000D+03   2.07000D+03   2.48000D+06
 4.62000D+02  -9.95000D+01   2.31000D+06  -7.75000D+02  -7.75000D+02  -9.95000D+05

THE B-MATRIX IS

 0.
 0.
 6.56000D+06
 7.11000D+06
 6.69000D+06
-2.30000D+06

THE C-MATRIX IS

 0.            0.            0.           0.            0.            1.00000D+04
 0.            0.           -1.20000D+04  0.            0.            0.

THE D-MATRIX IS

 0.
 1.20000D+04

DETERMINANT COEFFICIENTS IN DECREASING POWERS OF S

 1.00000D+00   7.56641D+06   1.95337D+12  -2.29657D+16   5.25796D+21   3.22345D+26
 8.05029D+28

RE-NORMALIZE WITH    6.57106D+04

DETERMINANT COEFFICIENTS IN DECREASING POWERS OF S

 1.00000D+00   1.15148D+02   4.52391D+02  -8.09421D+01   2.82018D+02   2.63114D+02
 1.00000D+00

COEFFICIENTS OF   S**1

 1.05718D+01  -2.75008D-01   4.89753D+04  -3.64655D+05  -4.31604D+05   7.94915D+03
-6.21834D+03   2.31712D+02   6.47883D+05  -1.26830D+03  -2.09597D+06  -3.60495D+05
-1.53112D-03   4.80780D-05   8.08801D+00   3.31510D+01   2.96095D+01  -5.08773D-01
 8.62586D-01   5.82783D-05   1.14841D+01   4.03006D+01   2.84200D+01  -1.63104D+01
```

Listing 5.5

and the eigenvalues of $\mathbf{A}_1$, the z_i, are calculated. Using a suitable number k, such as the square root of the trace of $\mathbf{A}_1$, restricts the resultant numbers to a smaller range than the original $\mathbf{A}$ matrix would give. We note that the resultant polynomials will be in powers of z and must be rescaled to give actual poles and zeros. The calculations are thus performed for $s = kz$:

$$\mathbf{Y}(z) = \frac{1}{\det(z\mathbf{I} - \mathbf{A}_1)}[\mathbf{C} \text{ adj}(z\mathbf{I} - \mathbf{A}_1)\mathbf{B} + \det(z\mathbf{I} - \mathbf{A}_1)\mathbf{D}]\mathbf{U}(z)$$

with the poles located at

$$\det(z_i\mathbf{I} - \mathbf{A}_1) = 0 \qquad s_i = kz_i$$

and the zeros scaled the same way.

```
-8.53244D-01  1.70674D-02 -6.62342D+01  9.94956D+01  2.34511D+02  1.86617D+02
 6.23181D-01 -2.31989D-02 -1.70340D+01  2.38629D+01  5.46995D+01  3.88526D+01
```

COEFFICIENTS OF   S**0

```
 4.76467D-20 -3.13125D-04 -1.53079D+03  1.52539D+03  4.74307D+03  4.72877D+03
 1.55123D-20  2.54520D-01  9.36741D+05 -1.58946D+06 -3.48210D+06 -2.55833D+06
-3.90402D-23  6.17093D-08  2.54524D-01 -2.53460D-01 -7.19478D-01 -6.14348D-01
 3.28553D-03  7.21520D-08  3.05576D-01 -3.04332D-01 -8.77659D-01 -7.72052D-01
-3.28553D-03  6.60532D-03 -1.68803D+00  2.82745D+00  6.24433D+00  4.63479D+00
 8.58931D-24 -2.55056D-05 -9.37929D+01  1.59063D+02  3.48738D+02  2.56665D+02
```

THE CHECK MATRIX IS

```
-3.55350D-18  3.67706D-21 -1.11414D-16 -1.37999D-17  7.02134D-19  3.67356D-17
-1.53636D-18 -3.26351D-18 -8.10520D-15 -6.73716D-16 -6.68995D-16  2.69133D-15
 3.50427D-21 -6.13383D-22  1.63951D-17 -9.69960D-20 -1.08878D-19 -6.12940D-18
 2.83942D-21 -5.07525D-22  1.53153D-17 -3.58110D-18 -5.49808D-20  5.07373D-18
-2.73917D-21  4.92150D-22 -1.50557D-17  4.81295D-20 -3.47820D-18  4.91928D-18
-6.68939D-22  1.01396D-22 -4.06100D-18  1.75924D-20  2.02067D-20 -2.51877D-18
```
THE ADJOINT COEFFICIENTS ARE
COEFFICIENTS OF   S**5

```
1.00000D+00  0.           0.           0.           0.           0.
0.           1.00000D+00  0.           0.           0.           0.
0.           0.           1.00000D+00  0.           0.           0.
0.           0.           0.           1.00000D+00  0.           0.
0.           0.           0.           0.           1.00000D+00  0.
0.           0.           0.           0.           0.           1.00000D+00
```

COEFFICIENTS OF   S**4

```
 1.15137D+02  0.           3.04365D+02 -3.04365D+02  0.           0.
 0.           1.15140D+02  1.50356D+04 -1.50356D+04 -1.50356D+04  1.87185D+02
-1.73336D-02  2.92343D-03  1.51636D+01  2.90669D-02  2.90669D-02  2.92190D+01
 2.21426D-02 -4.10741D-03  1.08202D+02  1.15112D+02 -3.53063D-02 -4.10893D+01
-1.99359D-02  3.77413D-03 -1.02114D+02  3.15018D+02  1.15179D+02  3.77413D+00
 7.03083D-03 -1.51422D-03  3.51542D+01 -1.17941D-02 -1.17941D-02  1.00005D+02
```

COEFFICIENTS OF   S**3

```
 4.39207D+02  2.13994D+00  2.83206D+04 -3.50242D+04  1.95930D+01  2.13994D+04
-2.92484D+02  5.00207D+02  1.42936D+05 -1.73071D+06 -1.73071D+06  5.08384D+05
-5.71662D-02  5.41531D-05  2.29782D+00 -3.85838D+01 -4.38595D+01  1.07777D+02
 3.84942D-01 -9.43772D-02  1.41042D+02  5.06973D+02  6.13218D+01 -9.48248D+02
 2.59937D-01  7.88700D-02 -1.70795D+02 -5.04646D+01  3.95859D+02  7.91244D+02
 9.36760D-02 -4.86439D-02 -1.95437D+01  2.04696D+01  2.26096D+01 -3.49798D+01
```

COEFFICIENTS OF   S**2

```
-2.19963D+02  2.87199D+01 -4.19413D+04 -1.65693D+05 -3.20137D+04  2.88724D+05
-2.71932D+03  1.40159D+02  4.77103D+05 -7.42716D+06 -7.51618D+06  2.36648D+06
-2.71932D+03  1.40159D+02  4.77103D+05 -7.42716D+06 -7.51618D+06  2.36648D+06
-1.16357D-02 -1.96698D-03  1.44738D+01  1.66198D+01 -7.79548D-01 -1.91034D+01
 8.87544D-01 -2.02124D-03 -1.45037D+02  1.22023D+03  1.41834D+03 -5.48495D+01
-4.82982D-01  3.28354D-03  1.15778D+02 -1.10679D+03 -1.26685D+03  5.69071D+01
 1.01315D-01 -3.71610D-03 -3.24343D+01  7.02702D+02  7.31214D+02 -1.30881D+02
```
THE NUMERATOR OF Y(S) IS
COEFFICIENT OF S**6

```
0.
1.20000D+04
```

COEFFICIENT OF S**5   COEFFICIENT OF S**2

```
-2.30000D+10                      -1.61346D+11
```

Listing 5.5—*Cont.*

```
    -7.87186D+10                    -1.86042D+11
 COEFFICIENT OF S**4              COEFFICIENT OF S**1

     6.03828D+09                    -4.82887D+10
    -3.87085D+11                    -1.99334D+11
 COEFFICIENT OF S**3              COEFFICIENT OF S**0

    -4.46579D+11                    -3.49538D+10
     7.79350D+10                    -8.57657D+08

 NUMERATOR POLYNOMIAL FOR INPUT   1 AND OUTPUT   1 OF DEGREE   5

 -2.30000D+10   6.03828D+09  -4.46579D+11  -1.61346D+11  -4.82887D+10  -3.49538D+10

 NUMERATOR POLYNOMIAL FOR INPUT   1 AND OUTPUT   2 OF DEGREE   6

  1.20000D+04  -7.87186D+10  -3.87085D+11   7.79350D+10  -1.86042D+11  -1.99334D+11
 -8.57657D+08
```
Listing 5.5—*Cont.*

We note from the results that the circuit is unstable (has poles in the right-half plane). In this respect this circuit is atypical of the circuits that one is usually asked to analyze.

Although the process described above is easy to program, it has several distinct disadvantages:

(1) The recurrence relations for the Souriau-Frame algorithm involve raising the **A** matrix to successively higher powers. This is subject to round-off errors.

(2) To determine any roots of $\det(s\mathbf{I} - \mathbf{A}) = 0$, the entire set of c_i must be obtained. (The set $\mathbf{P}_i$ is obtained simultaneously.)

(3) The matrices $\mathbf{P}_i$ must be stored during the operation of the algorithm. This implies a storage requirement of N^3 locations, namely, N matrices of size N by N.

(4) The process requires approximately N^4 multiplications and additions.

We note that this process establishes the inverse of $(s\mathbf{I} - \mathbf{A})$ as a ratio of polynomials: The denominator is the sum form of the determinant, Eq. (5.59), and the numerator is a sum of matrices $\mathbf{P}_i$, Eq. (5.68):

$$\phi(s) = [s\mathbf{I} - \mathbf{A}]^{-1} = \frac{1}{\sum_{i=0}^{N} c_i s^i} \cdot \sum_{i=0}^{N-1} \mathbf{P}_i s^i \tag{5.73}$$

An alternate form is to express $\phi(s)$ as a partial fraction expansion:

$$\phi(s) = \sum_{i=1}^{N} \frac{\mathbf{M}_i}{s - s_i} \tag{5.74}$$

where the s_i are the roots of $q(s)$, Eq. (5.60), and the **M** are matrices of dimensions N by N. In order to do such an expansion, the values s_i must be known; the determination of these values in turn means that an eigenvalue analysis of **A** had been made.

On the surface it appears that there will be N matrices $\mathbf{M}_i$ each of dimension N by N. We shall see that it will be possible to store these matrices in far simpler form and thus alleviate the storage disadvantage inherent in Eq. (5.68).

Although it is not intended to go through a thorough discussion of eigenvalue determination, a brief description of eigenvalue calculations will now be given. In order to perform state equation analysis of large networks, particularly those with widely separated eigenvalues (values s_i), the reader is urged to study the materials in [7] [12].

Consider the matrix equation

$$\lambda \mathbf{z} = \mathbf{H}\mathbf{z} \tag{5.75}$$

where $\mathbf{H}$ is an N by N matrix of real numbers and λ is an undetermined parameter (a number) and $\mathbf{z}$ is an arbitrary vector. Formally, the solution of the above is suggested by

$$\mathbf{z} = [\lambda \mathbf{I} - \mathbf{H}]^{-1} \cdot [0] = \frac{\text{adj}\,(\lambda \mathbf{I} - \mathbf{H})}{\det\,(\lambda \mathbf{I} - \mathbf{H})} \cdot [0] \tag{5.76}$$

Thus for $\mathbf{z} \neq \mathbf{0}$ the determinant must vanish. For an N by N matrix there will be N values of $\lambda, (\lambda_1, \lambda_2, \ldots, \lambda_N)$, for which the value of the determinant goes to zero. These values are called the eigenvalues of the matrix $\mathbf{H}$.

Let us assume that these values are distinct. For repeated roots the subsequent analysis must be modified although it remains substantially unchanged. If the roots are distinct, they may be ordered such that

$$|\lambda_1| \geq |\lambda_2| \geq |\lambda_3| \ldots, \geq |\lambda_N|$$

with

$$\lambda_i \neq \lambda_{i+1}$$

We remark that the λ_i are complex numbers in general and that complex roots appear in conjugate pairs for a real matrix $\mathbf{H}$.

Let us assume that the square matrix $\mathbf{H}$ can be factored in the form

$$\mathbf{H} = \mathbf{P}\mathbf{\Gamma}\mathbf{P}^{-1} \tag{5.77}$$

where $\mathbf{P}$ and $\mathbf{\Gamma}$ are N by N matrices with complex elements. We may rewrite Eq. (5.75) as follows since λ is a number

$$\mathbf{P}[\lambda \mathbf{I}]\mathbf{P}^{-1}\mathbf{z} = \mathbf{P}\mathbf{\Gamma}\mathbf{P}^{-1}\mathbf{z} \tag{5.78}$$

which in turn results in

$$\lambda \mathbf{I} - \mathbf{\Gamma} = 0 \tag{5.79}$$

Since the λ can take on the values λ_i given above, a solution of the above is

$$\gamma_{ii} = \lambda_i \tag{5.80}$$

implying that $\mathbf{\Gamma}$ is a diagonal matrix with the elements being the eigenvalues λ_i of the matrix $\mathbf{H}$. We shall denote $\mathbf{\Gamma}$ by $\mathbf{\Lambda}$:

$$\mathbf{H} = \mathbf{P} \begin{bmatrix} \lambda_1 & & & & 0 \\ & \lambda_2 & & & \\ & & \cdot & & \\ & & & \cdot & \\ 0 & & & & \lambda_N \end{bmatrix} \mathbf{P}^{-1} \tag{5.81}$$

This form, the Jordan canonical form of the matrix, is valid as long as the λ_i are distinct; for repeated values of λ_i, the matrix Λ will contain some ones on the upper diagonal. Note that pairs of complex values of λ_i constitute two distinct roots. Repeated roots are not normally found in RC and RLC networks; they will be omitted from most of the discussion that follows. The reader is referred to the literature for a more exhaustive analysis [17].

We remark that the matrix $\mathbf{P}$ is an N by N matrix with complex coefficients if Λ contains complex elements. As a consequence of Eq. (5.81) we may write $\mathbf{H}$ in the form

$$\mathbf{H} = \sum_{i=1}^{N} \lambda_i \mathbf{P}_i \mathbf{P}_i^{-1} \qquad (5.82)$$

where

$\mathbf{P}_i^{-1} = $ a matrix of zeros except for row i which contains the ith row of $\mathbf{P}^{-1}$;

$\mathbf{P}_i = $ a matrix of zeros except for column i which contains the ith column of $\mathbf{P}$.

We now present the power method for the determination of λ_1 and the corresponding $\mathbf{P}_1$ and $\mathbf{P}_1^{-1}$. The advantage of this method is its simplicity, but more powerful methods exist and the reader is urged to use his computer center's resources in affecting eigenvalue analyses, particularly if large matrices are involved. The scope of this book simply does not permit any more than an exposition of the underlying ideas and the discussion of relatively simple numerical procedures.

Consider an arbitrary non-zero column vector $\mathbf{r}_0$ having N elements, which we shall choose as real numbers for ease of computations. We post-multiply $\mathbf{H}$ by the sequence vectors as follows:

$$\begin{aligned}
\mathbf{H}\mathbf{r}_0 &= \mathbf{P}\Lambda\mathbf{P}^{-1}\mathbf{r}_0 = \mathbf{r}_1 \\
\mathbf{H}\mathbf{r}_1 &= \mathbf{P}\Lambda^2\mathbf{P}^{-1}\mathbf{r}_0 = \mathbf{r}_2 \\
&\vdots \\
\mathbf{H}\mathbf{r}_{k-1} &= \mathbf{P}\Lambda^k\mathbf{P}^{-1}\mathbf{r}_0 = \mathbf{r}_k
\end{aligned} \qquad (5.83)$$

We now note that because of the diagonal structure of Λ

$$\Lambda^k = \begin{bmatrix} \lambda_1^k & & & 0 \\ & \lambda_2^k & & \\ & & \ddots & \\ 0 & & & \lambda_N^k \end{bmatrix} \qquad (5.84)$$

In the following let the elements of $\mathbf{P}$ be

$$\mathbf{P} = [p_{ij}] \qquad (i, j = 1, \ldots, N) \qquad (5.85)$$

and let the elements of P^{-1} be

$$\mathbf{P}^{-1} = [q_{ij}] \qquad (i, j = 1, \ldots, N) \qquad (5.86)$$

The sequence of $\mathbf{r}_0 \ldots \mathbf{r}_k$ may be used to find the λ_1 values, but several cases must be distinguished.

A. Eigenvalue of Distinct Magnitude. This occurs for

$$|\lambda_1| > |\lambda_2| > |\lambda_3| > \cdots > |\lambda_N|$$

In this case $|\lambda_1|^k \gg |\lambda_i|^k$, $i = 2, \ldots, N$ and

$$\lim_{k \to \infty} \mathbf{\Lambda}^k = \lambda_1^k \begin{bmatrix} 1 & & & 0 \\ & 0 & & \\ & & \ddots & \\ 0 & & & 0 \end{bmatrix} \quad (5.87)$$

The limiting value of $\mathbf{H}^k$ becomes

$$\lim_{k \to \infty} \mathbf{H}^k = \lambda_1^k \begin{bmatrix} p_{11} & \cdots & p_{1N} \\ \vdots & & \vdots \\ p_{N1} & \cdots & p_{NN} \end{bmatrix} \begin{bmatrix} 1 & & 0 \\ & 0 & \\ & & \ddots \\ 0 & & 0 \end{bmatrix} \begin{bmatrix} q_{11} & \cdots & q_{1N} \\ \vdots & & \vdots \\ q_{N1} & \cdots & q_{NN} \end{bmatrix} = \lambda_1^k \mathbf{G} \quad (5.88)$$

with the elements of $\mathbf{G}$ given by

$$g_{ij} = p_{i1} q_{1j} \quad (5.89)$$

Let the elements of the arbitrary vector $\mathbf{r}_0$ be

$$\mathbf{r}_0 = [r_{01} \quad r_{02} \ldots r_{0N}]^T \quad (5.90)$$

Then the limiting value for $\mathbf{H}^k$ becomes

$$\lim_{k \to \infty} \mathbf{H}^k \mathbf{r}_0 = \lambda_1^k \mathbf{v} \quad (5.91)$$

with the elements of $\mathbf{v}$ being

$$v_i = \sum_{j=1}^{N} p_{i1} q_{1j} r_{0j} = p_{i1} \sum_{j=1}^{N} q_{1j} r_{0j} \quad (5.92)$$

and thus independent of k. Therefore the ratio of any two corresponding elements in $\mathbf{r}_{k-1}$ and $\mathbf{r}_k$ becomes

$$\lim_{k \to \infty} \frac{\mathbf{r}_{k,i}}{\mathbf{r}_{k-1,i}} = \lambda_1 \quad (5.93)$$

We note from Eq. (5.92) that the second form shows a summation independent of i; therefore

$$v_i = \begin{bmatrix} p_{11} \\ p_{21} \\ \vdots \\ p_{N1} \end{bmatrix} \sum_{j=1}^{N} q_{1j} r_{0j} = \alpha \mathbf{P}_1 \quad (5.94)$$

where α is a proportionality factor. Hence the sequence of iterations gives λ_1 and a column proportional to $\mathbf{P}_1$.

B. Distinct Pair of Complex Eigenvalues. This occurs for
$$|\lambda_1| = |\lambda_2| > |\lambda_3| \geq \cdots \geq |\lambda_N|$$
In this case

$$\lim_{k \to \infty} \mathbf{H}^k = \begin{bmatrix} p_{11}\lambda_1^k & p_{12}\lambda_2^k & 0 & \cdots \\ p_{21}\lambda_1^k & p_{22}\lambda_2^k & 0 & \cdots \\ \vdots & \vdots & \vdots & \\ \vdots & \vdots & \vdots & \\ p_{N1}\lambda_1^k & p_{N2}\lambda_2^k & 0 & \cdots \end{bmatrix} \begin{bmatrix} q_{11} & \cdots & q_{1N} \\ \vdots & & \vdots \\ \vdots & & \vdots \\ \vdots & & \vdots \\ q_{N1} & \cdots & q_{NN} \end{bmatrix} \quad (5.95)$$

and the corresponding vector $\mathbf{r}_k$ has the elements

$$r_{k,i} = \lambda_1^k p_{i1} \sum_{j=1}^{N} q_{1j} r_{0j} + \lambda_2^k p_{i2} \sum_{j=1}^{N} q_{2j} r_{0j} \quad (5.96)$$

Again, the elements of the vector contain the eigenvalues and the eigenvectors; the relationship is more involved, but it may be solved directly. We first determine a relationship between corresponding elements of successive iterations $\mathbf{r}_k$:

$$\mathbf{r}_{k,i} + \mu_1 \mathbf{r}_{k+1,i} + \mu_2 \mathbf{r}_{k+2,i} = 0 \quad (5.97)$$

where the coefficients μ_1 and μ_2 must be determined. If we express the eigenvalues as

$$\lambda_1 = a + jb; \qquad \lambda_2 = a - jb \quad (5.98)$$

and use Eq. (5.96) for $\mathbf{r}_{k,i}$, there results from the real and the imaginary parts of Eq. (5.97)

$$\mu_1 = \frac{-2a}{a^2 + b^2}; \qquad \mu_2 = \frac{1}{a^2 + b^2} \quad (5.99)$$

Hence the recurrence relation between successive iterations $\mathbf{r}_k$ becomes

$$(a^2 + b^2)\mathbf{r}_{k,i} - 2a\mathbf{r}_{k+1,i} + \mathbf{r}_{k+2,i} = 0 \quad (5.100)$$

which in turn can be evaluated for the real and the imaginary parts of the eigenvalues using two elements of the $\mathbf{r}$ vectors:

$$a = \frac{1}{2} \frac{r_{k+2,j} r_{k,i} - r_{k+2,i} r_{k,j}}{r_{k,i} r_{k+1,j} - r_{k+1,i} r_{k,j}} \quad (5.101)$$

$$b^2 = \frac{r_{k+2,j} r_{k+1,i} - r_{k+2,i} r_{k+1,j}}{r_{k,i} r_{k+1,j} - r_{k+1,i} r_{k,j}} - a^2 \quad (5.102)$$

The eigenvectors can be determined within a constant, as before

$$C_1 \mathbf{P}_1 = \frac{1}{\lambda_1 - \lambda_2} (\lambda_2 \mathbf{r}_k - \mathbf{r}_{k+1}) \quad (5.103)$$

$$C_2 \mathbf{P}_2 = \frac{1}{\lambda_2 - \lambda_1} (\lambda_1 \mathbf{r}_k - \mathbf{r}_{k+1}) \quad (5.104)$$

The rows of $\mathbf{P}^{-1}$, the $\mathbf{P}_i^{-1}$ occurring in Eq. (5.82), may be determined quite analogously to the above process by starting with an arbitrary row vector $\mathbf{t}_0$ and premultiplying $\mathbf{H}$ by a sequence of row vectors similar to Eq. (5.83):

$$t_0 H = t_0 P \Lambda P^{-1} = t_1$$
$$t_1 H = t_0 P \Lambda^2 P^{-1} = t_2$$
$$\cdot$$
$$\cdot$$
$$\cdot$$
$$t_{k-1} H = t_0 P \Lambda^k P^{-1} = t_k$$

(5.105)

It can again be shown that for an isolated eigenvalue λ_1 the sequence of row vectors converges to λ_1, as in Eq. (5.93), and that of row vectors t_k approaches a multiple of the row vector P_1^{-1}, similar to Eq. (5.94). Since the product of $P \cdot P^{-1}$ must result in a unit matrix, the two vectors obtained from the sequence of t vectors and r vectors Eq. (5.83) must satisfy

$$\alpha P_1^{-1} \cdot \beta P_1 = 1 \qquad (5.106)$$

from which a factor for $\alpha \beta$ may be determined. It is, however, customary to define the lengths of each of the vectors as one, thus resulting in normalized unit eigenvectors.

To determine other eigenvalues and eigenvectors of a given matrix, once an eigenvalue-eigenvector combination is found, it is subtracted from the original matrix by means of Eq. (5.82) and the next highest eigenvalue-eigenvector is calculated. Because of the accumulated round-off the process is not accurate for other than the first few eigenvalues.

We shall now demonstrate the above with numerical examples.

EXAMPLE 5.10

Write the following matrix in Jordan normal form.

$$H_1 = \begin{bmatrix} 2 & 4 & 6 \\ 3 & 9 & 15 \\ 4 & 16 & 36 \end{bmatrix}$$

Solution. We form the sequence of r vectors:

$$\begin{bmatrix} 2 & 4 & 6 \\ 3 & 9 & 15 \\ 4 & 16 & 36 \end{bmatrix} \cdot \begin{bmatrix} 1 \\ 0 \\ 0 \end{bmatrix} = \begin{bmatrix} 2 \\ 3 \\ 4 \end{bmatrix}$$

$$H_1 \begin{bmatrix} 2 \\ 3 \\ 4 \end{bmatrix} = \begin{bmatrix} 28 \\ 93 \\ 200 \end{bmatrix}, \text{ etc.}$$

The sequence becomes

r_0	r_1	r_2	r_3	r_4	r_5	r_6
1	2	40	1652	72220	3168284	139022284
0	3	93	3957	173289	7602981	333616161
0	4	200	8848	388448	17045632	747963584

Sec. 5.6 Poles and Zeros

The ratios of corresponding terms to the previous ones become

r_0	r_1	r_2	r_3	r_4	r_5	r_6
—	—	20	41.30	43.71	43.870	43.8794
—	—	31	42.55	43.79	43.875	43.8796
—	—	50	44.24	43.90	43.881	43.8801

From the above sequence we conclude that $\lambda_1 = 43.880$. We also note that since ratios of numbers are of interest, the sequence could be normalized by any constant at every step; the only requirement is that the vectors be compared with the last normalization constant properly considered. If we always normalize on the same elements, say the first one, the eventual normalization constant will be the eigenvalue, if it is isolated. We repeat the above normalizing of the first element to one each time.

	r_0	r_1	r_2	r_3	r_4	r_5	r_6
Elements	1	1	1	1	1	1	1
	0	1.50	2.325	2.395	2.399	2.400	2.400
	0	2.00	5.000	5.356	5.379	5.380	5.380
Factor	1	2.00	20.0	41.30	43.72	43.870	43.880

The last table was calculated to four places of precision and shows, to that accuracy, a vector proportional to $\mathbf{P}_1$. In order to make it a unit-length vector, we divide each element by the square root of the sum of the squares of the elements, namely 5.975:

$$\mathbf{P}_1 = [0.1674, 0.4016, 0.9004]^T$$

The row $\mathbf{P}_1^{-1}$ is calculated similarly:

		Elements		Factor
$t_0 =$	1	0	0	—
$t_1 =$	1	2	3	—
$t_2 =$	1	3.50	7.20	20
$t_3 =$	1	3.65	7.69	41.3
$t_4 =$	1	3.658	7.723	43.71
$t_5 =$	1	3.659	7.726	43.866
$t_6 =$	1	3.659	7.726	43.881

Again, the vector length is normalized to one:

$$\mathbf{P}_1^{-1} = [0.1162 \quad 0.4251 \quad 0.8976]$$

As a check, the product $\mathbf{P}_1^{-1}\mathbf{P}_1$ is calculated. The result is 0.9984, a close check to the expected value of one.

Next, we form the factor of $\mathbf{H}_1$ corresponding to this highest eigenvalue and subtract it from the original matrix:

$$\begin{bmatrix} 2 & 4 & 6 \\ 3 & 9 & 15 \\ 4 & 16 & 36 \end{bmatrix} - 43.88 \begin{bmatrix} 0.1674 \\ 0.4016 \\ 0.9004 \end{bmatrix} [0.1162 \quad 0.4251 \quad 0.8976]$$

$$= \begin{bmatrix} 1.1468 & 0.8783 & -0.5919 \\ 0.9525 & 1.5088 & -0.8189 \\ -0.5905 & -0.7952 & 0.5341 \end{bmatrix}$$

This matrix is now iterated as above. The results are as follow (note the arbitrary vector):

	$\mathbf{r}_0$	$\mathbf{r}_1$	$\mathbf{r}_2$	$\mathbf{r}_3$	$\mathbf{r}_7$	$\mathbf{r}_8$
Elements	1	1.0000	1.0000	1.0000	1.0000	1.0000
	-1	0.3052	1.0518	1.2505	1.2895	1.2895
	-1	0.3828	-0.6321	-0.7217	-0.7401	-0.7401
Factor	—	0.8604	1.6415	2.4482	2.7173	2.7175

The corresponding unit eigenvector is

$$\mathbf{P}_2 = [0.5581 \quad 0.7197 \quad -0.4130]^T$$

Similar iterations can be made for $\mathbf{P}_2^{-1}$, λ_3, $\mathbf{P}_3$, and $\mathbf{P}_3^{-1}$. The results are

$$\mathbf{H}_1 = \begin{bmatrix} 0.1674 & 0.5581 & 0.8047 \\ 0.4016 & 0.7197 & -0.5701 \\ 0.9004 & -0.4130 & 0.1658 \end{bmatrix} \begin{bmatrix} 43.880 & 0 & 0 \\ 0 & 2.7170 & 0 \\ 0 & 0 & 0.4025 \end{bmatrix}$$

$$\begin{bmatrix} 0.1162 & 0.4251 & 0.8977 \\ -0.5808 & -0.6978 & 0.4192 \\ 0.8139 & -0.5717 & 0.1037 \end{bmatrix}$$

EXAMPLE 5.11

Find the eigenvalues and eigenvectors of

$$\mathbf{H}_2 = \begin{bmatrix} 26 & -54 & 4 \\ 13 & -28 & 3 \\ 26 & -56 & 5 \end{bmatrix}$$

Solution. Several of the iterations are shown below:

	$\mathbf{r}_0$	$\cdots$	$\mathbf{r}_8$	$\mathbf{r}_9$	$\mathbf{r}_{10}$	$\mathbf{r}_{11}$
Elements	1		1.000000	1.000000	1.000000	1.000000
	0		0.519198	0.745279	0.475049	0.626199
	0		1.040140	1.000821	1.047818	1.021531
Multiplier	—	—	—	2.123874	-10.241790	4.538626

Sec. 5.6   Poles and Zeros   **235**

During the iterations the **r** vectors change signs rather wildly. This is an indication that complex roots are present. We apply Eqs. (5.101) and (5.102) to the first two rows of iterations 9, 10, and 11:

$$a = 0.999999$$
$$b = 5.000009$$

Identical results to five places are obtained for iterations 8, 9, and 10. Hence the two dominant complex roots are (to five places of accuracy)

$$\lambda_{1,2} = 1 \pm j5$$

The corresponding eigenvectors are obtained from Eqs. (5.102) and (5.104); they must be normalized such that their scalar product $\mathbf{P}_1^{-1}\mathbf{P}_1 = 1$. It is not difficult, but tedious, to show that $\lambda_3 = 1$ and that the matrix may be written in the form

$$\mathbf{H}_2 = \begin{bmatrix} 0.81649 & 0.64888 - j0.00996 & 0.64888 + j0.00996 \\ 0.40825 & 0.34932 - j0.06469 & 0.34932 + j0.06469 \\ 0.40825 & 0.67278 - j0. & 0.67278 + j0. \end{bmatrix} \cdot$$

$$\begin{bmatrix} 1 & 0 & 0 \\ 0 & 1+j5 & 0 \\ 0 & 0 & 1-j5 \end{bmatrix} \cdot$$

$$\begin{bmatrix} 2.54759 & -0.39198 & -2.25362 \\ -0.77293 - j3.86453 & 0.11896 + j8.32363 & 1.42690 - j0.59455 \\ -0.77293 + j3.86453 & 0.11896 - j8.32363 & 1.42690 + j0.59455 \end{bmatrix}$$

The first and third matrices are defined only within a multiplier such that their product is the unit matrix; here the eigenvector columns have been normalized to one.

Although the method presented above will work satisfactorily for moderate-sized matrices, in many instances round-off errors will often preclude its successful application. The problem of eigenvalue-eigenvector analysis (eigenanalysis) has been studied extensively and many good, but involved, procedures are in wide use. Here it is intended only to present some usable methods, to acquaint the reader with some of the problems, and to furnish a guide to further study [18, 19].

If only the eigenvalues λ_i are calculated by some process, the eigenvectors $\mathbf{u}_i$ can be obtained quickly from

$$[\lambda_i \mathbf{I} - \mathbf{A}]\mathbf{u}_i = [0] \qquad (5.107)$$

Here again we note that $\mathbf{u}_i$ can be determined only within a constant; therefore one of the components of $\mathbf{u}_i$ is arbitrarily set to 1, one of the equations in Eq. (5.107) is deleted, and the rest is solved for the $\mathbf{u}_i$ components. The vector length is then normalized to 1.

The inverse eigenvector (row) $\mathbf{v}_i$ is determined similarly from

$$\mathbf{v}_i[\lambda_i \mathbf{I} - \mathbf{A}] = [0] \qquad (5.108)$$

and the components of $\mathbf{v}_i$ are normalized such that

$$\mathbf{v}_i \mathbf{u}_i = 1 \qquad (5.109)$$

EXAMPLE 5.12

For the matrix used in Example 5.11 one of the roots is 1. Find the corresponding eigenvector.

Solution.

$$\begin{bmatrix} 25 & -54 & 4 \\ 13 & -29 & 3 \\ 26 & -56 & 4 \end{bmatrix} \begin{bmatrix} u_{11} \\ u_{21} \\ u_{31} \end{bmatrix} = [0]$$

Set $u_{31} = 1$ and delete equation 3:

$$u_{21} = 1$$
$$u_{11} = 2$$

The length is z:

$$z^2 = \sum_{i=1}^{3} u_{i1}^2 = 6$$

and the individual elements are divided by z:

$$\mathbf{u} = [0.81649 \quad 0.40825 \quad 0.40825]^T$$

The inverse eigenvector (row) is obtained from

$$\begin{bmatrix} 25 & 13 & 26 \\ -54 & -29 & -56 \\ 4 & 3 & 4 \end{bmatrix} \begin{bmatrix} v_1 \\ v_2 \\ v_3 \end{bmatrix} = [0]$$

Proceeding as above

$$\mathbf{v} = [-1.130435 \quad 0.173913 \quad 1.0]$$

The crossproduct is

$$\mathbf{v} \cdot \mathbf{u} = -0.44422$$

Hence the elements of $\mathbf{v}$ are scaled by -0.44422:

$$\mathbf{v} = [2.54759 \quad -0.39198 \quad -2.25362]$$

If the eigenvalues and eigenvectors of a matrix have been determined, these values can be stored in two N by N matrices and one vector of N elements. The expansion of $[s\mathbf{I} - \mathbf{A}]^{-1}$ into partial fractions is shown in Eq. (5.74):

$$\boldsymbol{\phi}(s) = \sum_{i=1}^{N} \frac{\mathbf{M}_i}{s - s_i} \tag{5.74}$$

where the s_i are the eigenvalues of $\mathbf{A}$.

The coefficients $\mathbf{M}_i$ are obtained from

$$\mathbf{M}_i = \lim_{s \to s_i} (s - s_i) \boldsymbol{\phi}(s) \tag{5.110}$$

Hence $\mathbf{M}_i$ becomes, by Eqs. (5.81)

$$\mathbf{M}_i = [s\mathbf{P}\mathbf{P}^{-1} - \mathbf{P}\boldsymbol{\Lambda}\mathbf{P}^{-1}]^{-1}(s - s_i)\Big|_{s=s_i}$$

$$= \left[\mathbf{P} \begin{bmatrix} s - s_1 & & & 0 \\ & s - s_2 & & \\ & & \ddots & \\ 0 & & & s - s_N \end{bmatrix} \mathbf{P}^{-1} \right]^{-1} (s - s_i) \Bigg|_{s=s_i} \tag{5.111}$$

$$= \mathbf{P}_i \mathbf{P}_i^{-1}$$

Hence the elements of $\mathbf{M}_i$ are

$$m_{jk} = p_{ji} r_{ik} \tag{5.112}$$

where r_{ik} is the i, k entry in $\mathbf{P}^{-1}$.

One advantage of this form is that the constituent components of all $\mathbf{M}_i$ can be kept in two N by N matrices, as opposed to the N different N by N matrices that result from the Souriau-Frame algorithm.

EXAMPLE 5.13

Using the data from Ex. 5.1 write the sum form of the solution to the state equations.

Solution. An eigenvalue analysis of the $\mathbf{A}$ matrix of Ex. 5.1 must be made. Note that the Souriau-Frame procedure is wasteful for finding the eigenvalues. An interation process was used here.

$$\mathbf{A} = \mathbf{P}\mathbf{\Lambda}\mathbf{P}^{-1} = \begin{bmatrix} -6.07886 \times 10^{-4} & -3.29614 \times 10^{-4} \\ 0.999999 & 1.000000 \end{bmatrix} \begin{bmatrix} -1.986 \times 10^4 & 0 \\ 0 & -2.743 \times 10^4 \end{bmatrix}$$

$$\cdot \begin{bmatrix} -3593.61 & -1.184504 \\ 3593.62 & 2.184509 \end{bmatrix}$$

$$[s\mathbf{I} - \mathbf{A}]^{-1} = \frac{1}{s + 1.986 \times 10^4} \begin{bmatrix} 2.18451 & 7.20043 \times 10^{-4} \\ -3593.61 & -1.18450 \end{bmatrix}$$

$$+ \frac{1}{s + 2.743 \times 10^4} \begin{bmatrix} -1.18451 & -7.20045 \times 10^{-4} \\ 3593.62 & 2.18451 \end{bmatrix}$$

The input-output relations become

$$\mathbf{Y}(s) = \left[\mathbf{C} \sum_{i=1}^{2} \frac{\mathbf{M}_i}{s - s_i} \mathbf{B} + \mathbf{D} \right] \mathbf{U}(s)$$

$$= \frac{-71388.7}{s + 1.986 \times 10^4} + \frac{98589.1}{s + 2.743 \times 10^4}$$

In the above, various results are given to many decimal places, not all of which are significant. The final result may be compared to the result of Ex. 5.8; because of round-off errors the numerator coefficient for s^0 is not identically equal to zero, but it becomes very close to it. If a zero for $\mathbf{Y}(s)$ is computed in this example, it is almost at zero, compared to the values of the poles. Thus a reasonable check is obtained.

The last examples show that if an eigenvalue is known, the corresponding eigenvector may be calculated and the response caused by that root can be computed independently. This is sometimes useful when it is desired to calculate specific frequency components of a response. Since the eigenvalue analysis (by means of existing programs available at the various computing centers) is normally the most accurate method available for pole-zero analysis, it is preferred to the Souriau-Frame procedure. The eigenvalue process also requires significantly less storage.

We have explored some problems in the calculation of the poles and the zeros of network functions. These problems are intimately tied to eigenvalue problems. The matrices involved are real, unsymmetric matrices, but if there are no controlled sources, symmetric matrices occur. A thorough study of the eigenvalue problem is

5.7 Frequency Response

The calculation of the AC steady-state response based on the state equations is quite simple. We note that for steady-state solutions the initial conditions are immaterial and that for a linear network all voltages and currents will be of the same frequency with only magnitude and phase angle changes between them.

For each of the NOU inputs a set of NOY y outputs will result; thus the u to y relation will, at each frequency, be a set of NOY by NOU matrices of complex numbers. The frequency response is normally a set of input-output relations. We set each of the inputs to one magnitude and zero phase angle, one at a time, and calculate the resultant outputs. Since there are NOY outputs, for each of the inputs there will result a vector of NOY values. For the entire set of inputs, the output is the collection of such vectors.

Formally, we have for an arbitrary input

$$\mathbf{u} = \begin{bmatrix} 1 & & 0 \\ & \cdot & \\ & \cdot & \\ 0 & & 1 \end{bmatrix} \begin{bmatrix} u_1 \\ \cdot \\ \cdot \\ u_{NOU} \end{bmatrix} e^{j\omega t} = \mathbf{IU}e^{j\omega t} \qquad (5.113)$$

where for convenience $\mathbf{U} = [u_1, \ldots, u_{NOU}]^T$ and the u_i represent the magnitudes of the individual input variables. ($\mathbf{I}$ represents the unit matrix.)

Let the state variables be

$$\mathbf{x} = \mathbf{X}e^{j\omega t} \qquad (5.114)$$

where $\mathbf{X}$ is a matrix of NOX by NOU complex quantities. The derivative becomes

$$\dot{\mathbf{x}} = j\omega \mathbf{X} e^{j\omega t} \qquad (5.115)$$

Correspondingly, the output becomes

$$\mathbf{y} = \mathbf{Y}e^{j\omega t} \qquad (5.116)$$

with $\mathbf{Y}$ a matrix of NOY by NOU complex elements.

Hence the state equations become

$$j\omega \mathbf{X} = \mathbf{AX} + \mathbf{BIU}$$
$$\mathbf{Y} = \mathbf{CX} + \mathbf{DIU} \qquad (5.117)$$

The solution of the above is

$$\mathbf{Y} = \mathbf{C}(j\omega \mathbf{I} - \mathbf{A})^{-1}\mathbf{BIU} \qquad (5.118)$$

The matrix on the right side, excluding $\mathbf{U}$, is the frequency response of the system, i.e., it defines the ratio of output response for each of the inputs. For a specific input excitation set $\mathbf{U}$ becomes a set of complex amplitudes and the output then can be calculated by the matrix multiplication indicated above. Of course, the $\mathbf{I}$ matrix as a multiplier could be omitted in the above derivation, but its inclusion points up the physical response concepts much more clearly. The calculations indicated in the last expression must be performed at each frequency (ω-value) separately.

We note that although the matrices **A**, **B**, **C**, and **D** are real, the inversion of ($j\omega\mathbf{I} - \mathbf{A}$) requires complex calculations. Since each inversion requires about $\frac{4}{3}$ NOX^3 number of operations (inversion based on Gauss-Jordan elimination, see Sec. 2.6), the number of operations becomes at each frequency of interest:

$$\begin{aligned}
&\text{Inversion:} && \tfrac{4}{3}\text{NOX}^3 \\
&\text{Multiplication by } \mathbf{B}: && \text{NOX}^2 \cdot \text{NOU} \\
&\text{Multiplication by } \mathbf{C}: && \text{NOX} \cdot \text{NOU} \cdot \text{NOY} \\
&\text{Total} = \text{NOX}(\text{NOU}(\text{NOX} + \text{NOY}) + \tfrac{4}{3}\text{NOX}^2)
\end{aligned}$$
(5.119)

In addition, the resultant NOY by NOU matrices must be stored (or else be printed as they are calculated.) If the response is sought for a given set of inputs, the number of operations can be reduced since in this case **BU** can be combined into a single vector. In this case the number of operations becomes:

$$\begin{aligned}
&\text{Inversion:} && \tfrac{4}{3}\text{NOX}^3 \\
&\text{Multiplication of } \mathbf{BU}: && \text{NOX} \cdot \text{NOU} \\
&\text{Multiplication by } \mathbf{BU}: && \text{NOX} \cdot \text{NOX} \\
&\text{Multiplication by } \mathbf{C}: && \text{NOY} \cdot \text{NOX} \\
&\text{Total} = \text{NOX}(\text{NOU} + \text{NOX} + \text{NOY} + \tfrac{4}{3}\text{NOX}^2)
\end{aligned}$$
(5.120)

Direct evaluation of the inverse in Eq. (5.118) at each frequency can become an expensive computational task. Alternately, the state transition matrix $[s\mathbf{I} - \mathbf{A}]^{-1}$ could be inverted by means of the methods in Sec. 5.6 and evaluated for $s = j\omega$. This latter method involves an eigenvalue analysis and is usually quite expensive also.

A third alternative is to resort to the frequency analysis method presented in Sec. 3.4. In this latter case, matrices at least as large as the **A** matrix have to be constructed at each frequency of interest and these matrices have to be manipulated. Although the calculation of the **A** matrix is time consuming, it has to be done only once and will result in the smallest matrix that has to be inverted compared to any direct node analyses.

Hence, unless very many frequency points are required, the inversion of the ($j\omega\mathbf{I} - \mathbf{A}$) matrix at each frequency will result in an acceptable computational task which is reasonable in running time and avoids the eigenvalue analysis difficulties. Questions of stability are yet to be answered if the poles are not determined.

The reader is urged to code a frequency analysis program based on Eq. (5.118) and requiring an inversion at each frequency (see Problem 5.28).

A more involved procedure requires the determination of the eigenvalues of the **A** matrix. Then the inverse of $[s\mathbf{I} - \mathbf{A}]$ is produced by either the Souriau-Frame algorithm or as a factored form. The inverse is then evaluated at each frequency of interest. The factored form is particularly useful in this context because of its lower storage

5.8 Stiff Differential Equations

Before this chapter is concluded we must discuss the very serious difficulties that occur in the time-domain solution of the state equations. The solution method shown in Sec. 5.4 is not an efficient one, but it has the advantage of being very straightforward. The basic problem in the solution of these equations is that of integrating a set of simultaneous linear differential equations. In a sense we could skip this section and merely make reference to the copious literature on this subject [7, 9, 10] and to the computer programs that exist in virtually all computing centers.

The basic difficulty in calculating an accurate solution to the state equation is that the choice of the time increment (DT in Listing 5.1) is dictated largely by the need for making the remainder terms, Eq. (5.52), acceptably small with a limited number of terms [K in Eq. (5.52)]. If K must be large, many powers of the **A** matrix must be calculated with serious accuracy problems arising. Conceptually, the accurate calculation of fast transients must be done with a time step smaller than the time constants of that fast response; however, once these transients have died out, the same small time steps will contribute significantly to round-off errors as the transients due to the lower time constants are being calculated. Roughly speaking, the initial parts of a unit-impulse response and the long-term parts of the same response should be calculated by using only the applicable time steps; hence the time step should be varied. In this manner, round-off errors can be avoided and computational effort can be greatly reduced [11].

Of course, the nature of the forcing function **Bu** will have a great deal to do with the appropriate time step that is to be used. Here again DT must be small enough to properly follow the salient features of the input function. A compromise between having to choose DT based both on the behavior of the state variables (**x**) and the nature of the forcing function (**u**) is to calculate the impulse response of the system and then convolute this response with the forcing function by using numerical methods only. The calculation of the impulse response is in turn best done by taking the derivative of the step response. We note that this requires the calculation of an entire step response for each state variable separately. The procedure is as follows.

Let **h** be the solution of

$$\dot{\mathbf{h}} = \mathbf{A}\mathbf{h} \qquad (5.121)$$

with

$$\mathbf{h}(0^-) = \mathbf{0}$$

$$\mathbf{h}(0^+) = \begin{bmatrix} 1 & & 0 \\ & \ddots & \\ 0 & & 1 \end{bmatrix} = \mathbf{U} \qquad (5.122)$$

Thus **h** is the initial condition response of the linear circuit with each initial variable value equal to 1. Note that **h** is an M by M matrix. It could be calculated relatively easily by the method to be discussed in this section and by any method of integration conveniently available (subject to numerical accuracy considerations). The matrix function **h** is available from the computational process implied above. For $\#T$ time values, **h** becomes a set of $\#T$ M by M matrices, stored normally on a mass-memory device (disk).

The solution of the state equations, Eq. (5.1), is given by convolution:

$$\mathbf{x}(t) = \mathbf{h}(t)\mathbf{x}_0 + \int_0^t \mathbf{h}(z)\mathbf{B}\mathbf{u}(t-z)\,dz \qquad (5.123)$$

The integral is normally evaluated by using a numerical convolution process. Also, accuracy considerations indicate that **h** should be stored and used in calculating the initial condition response. Since fast convolution processes require equal time steps, the time intervals for the stored values of $\mathbf{h}(t)$ must be equal for this process to become usable.

Some savings in storage are possible by storing **hB** instead of **h**. Note that either way a large amount of storage is needed, but the procedure requires only integration of a step response. Standard convolution routines may be used.

Another means of computing the solution is to perform a complete eigenvalue analysis of the matrix **A** and then vary the time step DT in Listing 5.1 as the various transients die out. This latter procedure is computationally very costly.

Alternately, a variable step-size integration procedure can be used which is designed for solving such systems of equations. The time-constant differences in most normal electronic circuits will be very great: The smallest time constants (often in nanoseconds) are associated with transistor parasitics and the larger ones (sometimes in milliseconds) are associated with bypass and coupling capacitors. Time constants differing by three orders of magnitude commonly occur. The only practical methods of integrating such differential equations are the implicit methods of integration [10] and they do require a great deal of numerical calculations.

Differential equations having a wide spread of time constants (or a wide spread of eigenvalues) are termed stiff differential equations. A large amount of work was done in devising procedures for the solution of such equations; there is no one best procedure, but a number of such procedures are discussed in [7].

Instead of looking into various integration methods, we shall discuss a far more fundamental process which is very easily related to the circuit behavior. We merely note that as a state variable approaches its final value (in a step-response calculation) its derivative approaches zero. Thus that variable no longer changes independently and therefore can be removed from the active set of state variables. The remaining set is smaller and it therefore involves less numerical work. Increased accuracy and faster solution times are achieved. We shall apply this process to the procedure developed in Sec. 5.4, a procedure which is not very accurate or elegant but one that relates well to the circuit behavior.

We shall now describe the method of removing variables from the state equations

$$\dot{\mathbf{x}} = \mathbf{A}\mathbf{x} + \mathbf{B}\mathbf{u} \qquad (5.1)$$

As the solution progresses, the derivatives are calculated. Supposing that the ith derivative is observed to be approaching zero:

$$\begin{bmatrix} \dot{x}_1 \\ \cdot \\ \cdot \\ \cdot \\ \dot{x}_i \\ \cdot \\ \cdot \\ \cdot \\ \dot{x}_M \end{bmatrix} = \begin{bmatrix} \dot{x}_1 \\ \cdot \\ \cdot \\ \cdot \\ 0 \\ \cdot \\ \cdot \\ \cdot \\ \dot{x}_M \end{bmatrix} = \begin{bmatrix} a_{11} & \cdots & a_{1i} & \cdots & a_{1M} \\ \cdot & & \cdot & & \cdot \\ \cdot & & \cdot & & \cdot \\ \cdot & & \cdot & & \cdot \\ a_{i1} & \cdots & a_{ii} & \cdots & a_{iM} \\ \cdot & & \cdot & & \cdot \\ \cdot & & \cdot & & \cdot \\ \cdot & & \cdot & & \cdot \\ a_{M1} & \cdots & a_{Mi} & \cdots & a_{MM} \end{bmatrix} \begin{bmatrix} x_1 \\ \cdot \\ \cdot \\ \cdot \\ x_i \\ \cdot \\ \cdot \\ \cdot \\ x_M \end{bmatrix} + \mathbf{B}\mathbf{u} \qquad (5.124)$$

Obviously, the ith variable may now be eliminated from the **A** matrix. One way to do this is to re-order the variables and eliminate the unchanging variable from the rest. The new variables are

$$\begin{bmatrix} \dot{x}_1 \\ \cdot \\ \cdot \\ \cdot \\ \dot{x}_{i-1} \\ \dot{x}_{i+1} \\ \dot{x}_M \\ \hline \dot{x}_i = 0 \end{bmatrix} = \begin{bmatrix} a'_{jk} & & & 0 \\ & & & \\ \hline a'_{i1} & \cdots & a'_{iM-1} & 1 \end{bmatrix} \begin{bmatrix} x_1 \\ \cdot \\ \cdot \\ \cdot \\ x_M \\ \hline x_i \end{bmatrix} + \begin{bmatrix} b'_{jk} \\ \hline b'_{ik} \end{bmatrix} \mathbf{u} \qquad (5.125)$$

where

$$\begin{aligned} a'_{jk} &= a_{jk} - \frac{a_{ik}}{a_{ii}} a_{jk} \qquad j \neq i \\ a'_{ik} &= \frac{a_{ik}}{a_{ii}} \\ b'_{jk} &= b_{jk} - \frac{b_{ik}}{a_{ii}} a_{ji} \qquad j \neq i \\ b'_{ik} &= \frac{b_{ik}}{a_{ii}} \end{aligned} \qquad (5.126)$$

Note that these are the formulas for a columnwise Gauss elimination procedure.

The analysis is then continued with the remaining variables ($M - 1$ in number). The variables are monitored and the next vanishing derivative is eliminated from the remaining $M - 1$ variables. Eventually all derivatives will vanish if the step response of a stable network is calculated; at that point the steady-state solution is obtained.

Computationally, the condition $\dot{x}_i = 0$ can not be reached because the round-off errors will prevent an exact zero. Hence the derivative is normally set to zero after it has become less than some predetermined value ϵ_i for a number of step sizes.

The conditions for deciding when a variable has in fact reached a zero derivative are of great importance for the accuracy of the subsequent calculations. The value for ϵ_i can be set *a priori* as a small parameter, but a somewhat better procedure is to set ϵ_i as a small fraction of the largest value of $\dot{x}_i$ calculated. This latter procedure assumes that the variable $\dot{x}_i$ will build up initially to some largest value and then will decay with or without oscillations toward zero. Such behavior is quite typical of stable damped networks.

On the other hand once a derivative is eliminated as $\dot{x}_i$ in Eq. (5.125), a subsequent change in $\mathbf{u}(t)$ might re-excite that variable, i.e., the derivative must continuously be monitored. If a derivative becomes non-zero, it may be re-introduced in the original equations. Two procedures may be done for this: the "undoing" of the Gauss elimination and starting again from the original equations eliminating only those variables that are still to be regarded as having reached a zero-derivative condition.

The Gauss elimination can be "undone" if the pertinent constants used for the elimination of the rows are kept within the reduced arrays. This is possible by using the locations where the reduction yields zeros or ones as storage locations for these constants. Although this process uses only as much storage as is required by the original $\mathbf{A}$ matrix (with possibly another pointer vector), it has the disadvantage of introducing round-off errors if a variable vanishes many times during the solution. Such repeated vanishing occurs when a periodic input is used that contains periodic magnitudes or slope changes, such as a triangular wave shape.

Somewhat better for round-off errors and easier to program is going back to the original equations each time a variable is detected as no longer having a zero derivative. A pointer array is kept on the variables that are to be eliminated from the $\mathbf{A}$ and the $\mathbf{B}$ matrices and when a variable is deleted from this pointer array a fresh reduction of the original equations is done. This requires, of course, that the original $\mathbf{A}$ and $\mathbf{B}$ matrices be stored (possibly on a mass-storage device) as well as the reduced matrices. The reduced matrices will be of the same dimensions as the original matrices.

We note that no matter how the state variables are calculated, the computation of the output ($\mathbf{y}$-vector values) proceeds exactly the same way. Basically, no reduction is possible in this phase of the computations, except for a possible detection of constant steady-state conditions, both on $\mathbf{x}$ and on $\mathbf{u}$.

Determination of a vanishing variable is only part of the problem of speeding up the calculations and improving the numerical results. As a variable is eliminated, the time constant associated with it is effectively removed from the circuit. One would like to then increase the time step so that it is compatible with the largest eigenvalue (smallest time constant) remaining in the set of independent variables. Unfortunately, this number is not known without having to perform some eigenvalue analysis to at least find the largest eigenvalue, if only approximately. If an eigenvalue analysis was performed, or if the characteristic polynomial is known (see Sec. 5.6) so that its roots can be obtained, the selection of the time step for integration becomes relatively easy.

If α is the real part of the most negative eigenvalue remaining, m is the accuracy to which the contribution of this root is to be calculated, and T is the minimum time step, then

$$T\alpha \leq \ln m \qquad (5.127)$$

will determine the largest time step usable. For a 0.01 % error $T\alpha \leq 9.2$; for a 0.005 % error $T\alpha \leq 9.9$. In addition, there will be round-off errors.

In the absence of an eigenvalue analysis one might try some "reasonable" time step which is established by the analyst from a knowledge of the circuit under investigation. Time is then stepped a few times and the resultant x-vector values are noted at each point. The interval is then halved and the x-vector values are again computed over the same trial range as before. Corresponding x values are then compared; a large discrepancy will normally indicate that the original time interval was too long. The process is repeated until successive halving does not change the computed x values appreciably. Note that round-off errors have to be accounted for in this process.

If interval halving does not change the corresponding solution values, the time interval was too short and the interval is doubled until a noticeable difference occurs. Backing up from this point results in a usable interval.

Obviously, a computationally easy method for bounding the largest eigenvalue is needed. For stable networks all poles will be in the left-half plane; two theorems from matrix algebra can be used to put loose limits on the leftmost eigenvalue [12].

1. Schur's theorem. The eigenvalues λ_i of a matrix $\mathbf{A}$ satisfy the inequality

$$\sum_{i=1}^{N} |\lambda_i|^2 \leq \operatorname{tr}(\mathbf{A}\mathbf{A}^T) \qquad (5.128)$$

where tr denotes the trace and $\mathbf{A}^T$ is the transpose of $\mathbf{A}$.

Let $|\operatorname{Re} \lambda_{\max}|$ be the real part of the leftmost pole. Then

$$|\operatorname{Re} \lambda_{\max}|^2 \leq \sum_{i=1}^{N} |\lambda_i|^2 \qquad (5.129)$$

and we immediately have

$$|\operatorname{Re} \lambda_{\max}| \leq \sqrt{\operatorname{tr}(\mathbf{A}\mathbf{A}^T)} \qquad (5.130)$$

The trace occurring in the above expression is simply the sum of the squares of the element of $\mathbf{A}$

$$\operatorname{tr}(\mathbf{A}\mathbf{A}^T) = \sum_{i=1}^{N} \sum_{j=1}^{N} a_{ij}^2 \qquad (5.131)$$

as can be verified directly. Eq. (5.130) gives an easily computed upper bound on $\lambda_{\max}$. Although the bound is not very "tight," it can be used to start the interval doubling process described above.

2. A lower bound on $\lambda_{\max}$ is obtained from

$$\operatorname{tr} \mathbf{A} = \sum_{i=1}^{N} \lambda_i \qquad (5.132)$$

Here we observe that for all eigenvalues in the left-half plane (stable networks)

$$\sum \lambda_i \leq N\lambda_{\max} \qquad (5.133)$$

Hence a lower bound on $\lambda_{\max}$ becomes

$$|\lambda_{\max}| \geq \frac{|\operatorname{tr} \mathbf{A}|}{N} \qquad (5.134)$$

Thus Eqs. (5.130) and (5.134) can be used to bound the leftmost eigenvalue.

Similarly, from Eq. (5.132), for networks having all eigenvalues in the left-half plane

$$\left|\sum_{i=1}^{N} \lambda_i\right| \geq |\operatorname{Re} \lambda_{\max}| \qquad (5.135)$$

and thus another bound on the leftmost eigenvalue is

$$|\operatorname{Re} \lambda_{\max}| \leq \operatorname{tr} \mathbf{A} \qquad (5.136)$$

For a bound on the leftmost eigenvalue, the expression

$$|\lambda_{\max}| = \max (\operatorname{tr} \mathbf{AA}^T, \operatorname{tr} \mathbf{A})$$

$$= \max \left(\sqrt{\sum_{i=1}^{N} \sum_{j=1}^{N} a_{ij}^2}, \sum_{i=1}^{N} a_{ii} \right) \qquad (5.137)$$

can be used for computing a trial integration time step.

The procedure outlined in Sec. 5.4 for the solution of the state equations is a relatively inefficient process, but it can be made more efficient by the methods described in this section. For various computationally efficient, but intricate, computer algorithms for the solution of stiff differential equations, the reader is referred to [7]. We now describe modifications to Listing 5.1 which will greatly increase its computational efficiency.

(1) The time increment DT is calculated from Eqs. (5.137) and (5.127) using the accuracy ACC at statement 130:

```
131     TR=0.
        TRR=0.
        DO 132 I=1,NOX
        TRR=TRR+A(I,I)
        NX(I)=0
        DO 132 J=1,NOX
132     TR=TR+A(I,J)**2
        TR=AMAX1(TR,TRR)
        DT=ALOG(ACC)/TR
```

The array NX will be used to count how often the corresponding derivative vanishes (see Step 3).

```
        PRINT 133, DT
133     FORMAT (///' THE TIME STEP IS =',1PE15.5)
```

(2) The derivatives of **x** are monitored prior to time being incremented at statement 155.

```
      DO 159 I=1,NOXO
      DX(I)=0.
      DO 156 J=1,NOXO
  156 DX(I)=DX(I)+A(I,J)*X(J)
      DO 157 J=1,NOU
  157 DX(I)=DX(I)+B(I,J)*U(J)
```

The variable NOXO is the number of original variables.

(3) The derivatives are checked for approaching zero. A check on NOX∗ACC is a reasonably small value:

```
      IF (ABS(DX(I)).GT.(NOX*ACC)) GO TO 158
      NX(I)=NX(I)+1
      GO TO 159
  158 NX(I)=0
  159 CONTINUE
```

(4) A check is made on whether or not the derivative has become sufficiently small enough number of times. Normally, NOX number of successive "zero" values should be sufficient to eliminate that variable:

```
      DO 159 I=1,NOXO
      IF (NX(I).LT.NOX) GO TO 1594
      CALL ELIM (A,B,NOXS,NOX,NOU,I,NA)
      GO TO 131
```

This code is inserted just before the time is stepped. Note that only one variable at a time is eliminated. Messages indicating that a variable is to be eliminated should also be printed.

Array NA is used to keep track of the original state variable rows. The subroutine ELIM is used to eliminate the ith variable from both the **A** and the **B** matrices by using Eq. (5.125). This subroutine will decrease NOX by one; checking if the previously eliminated variables are remaining at zero values must be done subsequently. This check is best done at Steps 2 and 3. In Step 2 all derivatives are calculated and in Step 4 a check is made that if a previously eliminated derivative is no longer zero it shall be added to the set:

```
 1594 IF (((I.GT.NOX).AND.(NX(I).NE.0)) GO TO 2000
 1598 CONTINUE
      . . . . . . .
```

The variable originally numbered NA(I) must be added to the active variables. The analysis should be restarted with only the proper variables eliminated.

(5) The code indicated in Steps 2 through 4 does not account for the fact that the original variables have been moved to other rows. The pointer array NA must be used to find the proper subscripts for the various array elements.

Coding the above steps is quite straightforward but very tedious. The resultant program listing is about twice the length of Listing 5.1. The coding details are left as an exercise.

5.9 Summary

This chapter introduced a simple procedure for the calculation of the coefficients of the simultaneous differential equations for a linear network and associated output equations. If this method is used, "normal" circuits may be analyzed with relative ease; extension to networks having surplus variables is possible. Procedures for the solution of these equations were then explored.

More comprehensive procedures exist [3]; the methods presented in this chapter should form the foundation for further work. The numerical procedures presented here are straightforward; again more comprehensive processes exist, but most of them ignore the fact that the equations studied here are linear differential equations. Consequently, the more general methods tend to be far more complicated and far more intricate.

The references indicate additional material in this subject. The problems contain a set of detailed exercises in expanding the material in this chapter.

REFERENCES

1a. BASHKOW, T. R., "The A-Matrix, A New Network Description," *IEEE Trans. on Circuit Theory*, Vol. CT-4 (September 1957).

1b. KUH, E. S., and R. A. ROHRER, "The State-Variable Approach to Network Analysis," *Proc. IEEE*, Vol. 53, No. 7 (July 1965).

2a. DESOER, C. A., and E. S. KUH, *Basic Circuit Theory*. McGraw-Hill Book Co., New York, 1969.

2b. ZADEH, L. A., and C. A. DESOER, *Linear System Theory*. McGraw-Hill Book Co., New York, 1963 (Chapter 5).

3. POTTLE, C., "Comprehensive Active Network Analysis by Digital Computer—A State-Space Approach," Proc. Third Allerton Conference, Urbana, Ill., October 1965.

4. BELLMAN, R., *Introduction to Matrix Analysis*. McGraw-Hill Book Co., New York, 1960.

5. LIOU, M. L., "A Novel Method for Evaluating Transient Responses," Proc. IEEE, Vol. 54, No. 1 (January 1966).

6. RALSTON, A., *A First Course in Numerical Analysis*. McGraw-Hill Book Co., New York, 1965.

7. GEAR, C. W., *Numerical Initial Value Problems in Ordinary Differential Equations*. Prentice-Hall, Inc., Englewood Cliffs, N.J., 1971.

8. DREW, R. S., "A Simplified Approach to the Analysis of Electrical Networks by Means of the State Variable Method." Master's thesis, North Carolina State University, Raleigh, N.C., 1970.

9. GEAR, C. W., "Simultaneous Numerical Solution of Differential-Algebraic Equations," *IEEE Trans. on Circuit Theory*, Vol. CT-18, No. 1 (January 1971), 89.

10. SANDBERG, I. W., and H. SCHICHMAN, "Numerical Integration on Systems of Stiff Nonlinear Differential Equations," *Bell System Technical Journal*, Vol. 47, (April 1968), 511–27.

11. RICHARD, S., et al., "On-line Reduction Method for Computer Analysis of High-order Electrical Circuits with Widely Separated Time Constants," *IEEE Trans. on Circuit Theory*, Vol. CT-19, No. 6 (November 1972), 664.

12. HAMMING, R. W., *Numerical Methods for Scientists and Engineers*, 2nd ed. McGraw-Hill Book Co., New York, 1973.

13. PIPES, L. A., *Matrix Methods for Engineering*. Prentice-Hall, Inc., Englewood Cliffs, N.J., 1963.

14. MULLER, D. E., "A Method for Solving Algebraic Equations Using an Automatic Computer," *MTAC*, Vol. 10, (1956), 208.

15. LEHMER, D. H., "A Machine Method for Solving Polynomial Equations," *Journal of the Association for Computing Machinery*, Vol. 2 (1961), 151.

16. FRAME, J. S., "Matrix Functions and Applications, Part IV: Matrix Functions and Constituent Matrices," *IEEE Spectrum*, Vol. 1, No. 6 (June 1964), 123–31.

17. FADDEEV, D. K., and V. N. FADDEEVA, *Computational Methods in Linear Algebra*. Freeman Press, San Francisco, Calif., 1963.

18. WILKINSON, J. H., *Rounding Errors in Algebraic Processes*. Prentice-Hall, Inc., Englewood Cliffs, N.J., 1963.

19. WILKINSON, J. H., and C. REINCH, *Handbook of Automatic Computation, Vol. 2, Linear Algebra*. Springer Verlag, Berlin, 1971.

PROBLEMS

5.1 Calculate the state equations for Ex. 5.1 by forming the node-voltage equations and re-arranging the terms into the proper form shown in Eq. (5.1).

5.2 Calculate the state equations for Ex. 5.3 starting with the appropriate node equations and equations involving the inductor voltages.

5.3 Use the DC analysis programs developed in Chapter 3 to derive the state equations for Exs. 5.1 and 5.2 using numerical values for the components. Use the method described in Sec. 5.2.

5.4 Derive the state equations for the circuit shown in Fig. P5.1. A DC analysis program, such as the one shown in Listing 3.1, transformation of dependent voltage sources to dependent current sources, and the use of the inverse inductance matrix represent one avenue of solution.

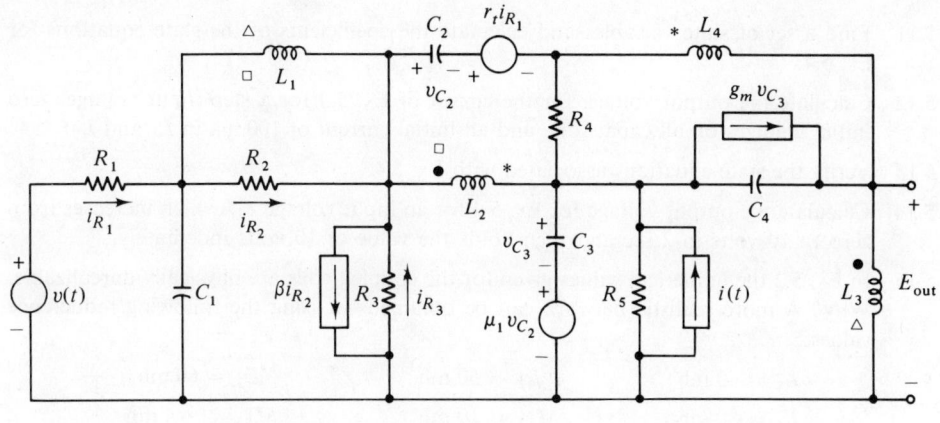

$L_1 = 100$ mh
$L_2 = 150$ mh
$L_3 = 200$ mh
$L_4 = 250$ mh
$R_1 = 30\Omega$
$R_2 = 80\Omega$
$R_3 = 200\,\Omega$

$L_{12} = 60$ mh □
$L_{13} = 120$ mh △
$L_{23} = 100$ mh ●
$L_{24} = 150$ mh *
$R_4 = 50\Omega$
$R_5 = 300\,\Omega$

$C_1 = 20\mu f$
$C_2 = 10\,\mu f$
$C_3 = 15\,\mu f$
$C_4 = 5\,\mu f$

$\beta = 40$
$g_m = 0.60$
$r_t = 10$
$\mu = 20$

Inputs: $i(t), v(t)$

Outputs: E_{out}, i_{R_1} (input impedance when $i(t) = 0$)

Fig. P5.1

5.5 Calculate the state equations for Ex. 3.8.

5.6 Using the program developed in Sec. 5.4, calculate the output voltage for a step input voltage in Ex. 5.1. Assume that the initial voltage on the capacitor is zero.

5.7 Calculate the response for a ramp input voltage $v(t) = Kt$ in Ex. 5.2 with zero initial conditions in the network.

5.8 The impulse response of a network is the output for an input occurring at time zero, of zero duration, infinite amplitude and area under the input wave equal to one and all initial conditions in the network equal to zero. The impulse response can also be obtained by differentiating the step response. Using the results of Problem 5.6, calculate the impulse response of the circuit shown in Ex. 5.1.

5.9 Following the method outlined in Problem 5.8, find the impulse responses of the network shown in Fig. P5.1.

5.10 Show that

$$\frac{d}{dt}e^{\mathbf{A}t} = \mathbf{A}e^{\mathbf{A}t} = e^{\mathbf{A}t}\mathbf{A}$$

5.11 Find a set of state variables and calculate the coefficients of the state equations for Ex. 5.4.

5.12 Calculate the output voltage for the circuit of Ex. 5.3 for a step input voltage, zero initial voltages on all capacitors, and an initial current of 100 ma in L_1 and L_3.

5.13 Verify the state equations associated with Ex. 5.4.

5.14 Calculate the output voltage for Ex. 5.4 for an input voltage E_{in} which increases from zero to 10 volts in 2 sec and then holds the value of 10 volts indefinitely.

5.15 In Ex. 5.3 the numerical values given for the coupled coils are physically unrealizable. Why? A more realistic network can be obtained by using the following inductance values:

$L_1 = 40$ mh $\qquad L_2 = 50$ mh $\qquad L_3 = 60$ mh

$L_4 = 80$ mh $\qquad M_{12} = 10$ mh $\qquad M_{13} = -4$ mh

$M_{23} = 15$ mh $\qquad M_{24} = -12$ mh $\qquad M_{34} = 10$ mh

Derive the **A**, **B**, **C**, and **D** matrices corresponding to the ones in Ex. 5.3.

5.16 Derive the analytic expression for the inductor current and the capacitor voltage for Ex. 5.5. Compare the computed results with the analytic results. Vary the step size in the integration and observe the accuracy obtainable. What is an "optimum" step size, i.e., one that produces acceptably accurate results using a reasonable amount of computer time. Your conclusions will depend on the word length of your computer. Why?

5.17 The state analysis program shown in Listing 5.3 can handle pairwise inductances only (see Sec. 3.4). Modify the program for arbitrarily coupled mutual inductances. You will need an inversion program in double precision; modify subroutine INVRT from Appendix A.

5.18 (a) Using your program from Problem 5.17, verify the results of Ex. 5.3. (b) Verify your results of Problem 5.4.

5.19 The calculation of currents through the 1-volt sources which represent the capacitor state variables is accomplished by dividing the voltage across the REQ resistor (set at 10^{-10} Ω) by the value of REQ. Since this voltage is near zero, the round-off errors will greatly influence this current. A far more accurate method is to sum all currents flowing into one of the terminals of the 1-volt source, excluding the current of the source, and then calculating the source current as the negative of this sum. Implement this method for the program obtained in Problem 5.17 or the one shown in Listing 5.3.

5.20 The calculation of the voltages across the inductor-replacing 1-amp sources is not subject to the kind of numerical errors explained in Problem 5.19. Why?

5.21 When the program shown in Listing 5.3 is used on a circuit containing surplus variables (as in Ex. 5.4), very large voltages and currents are encountered during the solution process. Write a subroutine which checks on such surplus variables and incorporate it into the program. Make your program eliminate the surplus variable with the highest branch number.

5.22 Check your modified program by using Ex. 5.4 as a test case.

5.23 Voltage controlled sources require, in the node-voltage analysis, the inversion of the voltage dependency matrix [U + D] (see Sec. 3.2). When arbitrary mutual inductances are allowed, an inversion program must be used anyway; thus adding the voltage controlled sources does not require much additional code. Using state variables as controlling branches can be tricky; restriction of controlling and controlled branches to resistors avoids these difficulties and still allows the analysis of most practical circuits. Modify your state analysis program for such a capability.

5.24 Code the Souriau-Frame algorithm described in Sec. 5.6 and add it to the state analysis program, Listing 5.3. Make your combined program and calculate poles and zeros of all transfer quantities.

5.25 The Souriau-Frame algorithm is very sensitive to round-off errors. Some relief can be obtained by calculating all poles and zeros with scaled frequency (scaled value of s). A reasonable, easy normalization constant is a fraction of the trace of the matrix **A**. Implement this change in your solution to Problem 5.24.

5.26 Devise a program to calculate the eigenvector of a matrix if the eigenvalues are known.

5.27 Write a program to produce the sum form of the system state transition matrix $\phi(s)$ [Eqs. (5.74) and (5.111)] from a knowledge of the eigenvalues and eigenvectors of the matrix **A**.

5.28 Implement a frequency analysis program based on Eq. (5.118). Note that for M inputs and K outputs at each frequency, a K by M matrix is produced.

5.29 The program produced in Problem 5.27 can be used as the basis of a frequency analysis program that avoids inversion at each frequency. Write such a program.

5.30 Verify the results of Ex. 3.8 by means of the program developed in either Problem 5.28 or 5.29.

5.31 A degree of automation in the integration can be achieved in Listing 5.1 by including the code shown in Step 1 of Sec. 5.8. Add this to your state analysis program and compare the results for Ex. 5.5.

5.32 Complete the automatic variable-elimination and variable-re-insertion scheme outlined in Sec. 5.9; include it in your version of the program shown in Listing 5.1.

5.33 Combine the state variable analysis program (Listing 5.2) and the solution of the state equations program (Listing 5.1) into a comprehensive time-response analysis program.

5.34 Use your program from Problem 5.25 and combine it with your state analysis program (Problems 5.23 and 5.21) into a flexible analysis package. Both frequency and time-response analysis generate much data; it is convenient to include a plotting package (see Appendix B) with your analysis programs.

TOLERANCE ANALYSIS

6

The preceding chapters dealt with the problem of calculating the circuit performance, usually the node voltages, from a knowledge of the inputs and the component values in a given circuit. This task assumes implicitly that the component values are known numbers. When the measured values of the circuit performance function (the node voltages, most commonly) are compared against the calculated (predicted) values, some discrepancy is expected since all component values in the actual network will vary from the nominal value by some amount. Normally, the circuit analysis precedes design of the circuit; usually, a detailed analysis is made because the performance has to be predicted before any laboratory work is done. Part of the analysis must be concerned with specifying the allowable variation of the circuit components, the component tolerance, for the design specification limits. The tolerance limits are specified by the state of the art (transistor betas may have no less than $\pm 30\%$ variation), by circuit performance requirements (output voltage must be within $\pm 5\%$), and by economic considerations (the cheapest components).

Evaluating circuit performance variations could conceptually be done by re-analyzing the network for each variation of the components. This method is, of course, very expensive in computer time; also for very close tolerances, such that may be encountered in analog-to-digital (and digital-to-analog) converters, the round-off errors in the computation may be masking the true contribution of the component variations. At the expense of yet more computer costs, one could double-precision all calculations in an effort to get away from these latter errors.

The computational task in finding the variations caused by small changes in the parameters is reminiscent of the numerical problems discussed in Sec. 2.7, namely, changes are sought of at least an order of magnitude smaller than the center values. Hence, as in Sec. 2.7, derivative approximations could be used to find these changes. We will therefore find the equations for the derivatives of the node voltages with respect to all possible parameters in the network. These derivatives can then be used to find the node-voltage changes caused by known variations in the parameters.

The calculation of the changes is further complicated by the fact that it is not usually clear what the parameter variations are. Rather, only the expected limits of the parameter variations may be known and the actual parameter value used in a given circuit is taken at random from a population whose statistical characteristics may be known only vaguely. For example, the exact distribution of 470-ohm $\pm$ 10% resistors is usually not known; it would require a very large number of measurements to establish such data for any one batch of these resistors that one might want to use in a production run. Furthermore, any other batch of the same nominal resistors may not have the same distribution.

An attempt to aid the design of reliable circuits in the face of such uncertainties in the component values is the technique of worst-case analysis. One merely specifies the permissible component variation, selects the worst possible set of components, and thus establishes the upper and the lower limits of the circuit performance function. Since such a combination of components has a near-zero probability of occurrence, the performance limits thus obtained are the most pessimistic estimates possible.

More realistic predictions of the circuit performance can be obtained by considering the statistical nature of the component variations. If these variations are known, or can be obtained relatively easily, quite accurate predictions of circuit performance variations can be made with relative ease. The theory underlying the calculations of such statistical variations is rooted deeply in probability theory; here no attempt is made to derive these bases. A cursory explanation of the necessary theory is presented, appropriate circuit relations are derived, and programming details are shown. The parameters are assumed to be independently variable through most of the developments; extension to correlated parameters is shown in Sec. 6.5.

The ultimate analysis of circuit performance variations is to use actual component distribution data and to simulate the behavior of a large number of replicas of the circuit, each with components chosen randomly from among the component distributions. Although this analysis is very costly, it is the only one that can account for arbitrary variations of component values. A large amount of input data is required for this type of analysis and a large amount of output data may be generated. The method faithfully (painstakingly is a better word) duplicates the actual process of construction of a network: the output results are no more accurate, however, than the component data entering the analysis. Sec. 6.6 discusses the details of this Monte Carlo analysis.

Examples are used throughout to illustrate the developments. A comprehensive example is used to summarize the material in this chapter.

The methods presented here form the basis of component tolerance selection in a design. Basically, one must invert the solution process shown here: the starting point is usually the permissible circuit performance variation and one must establish from this the allowable variations on the circuit components. The fundamental steps in this latter process are presented at the end of this chapter.

6.1 Derivatives and Sensitivity

The node voltages in a network are functions of the values of the driving sources (voltages and currents) and of the circuit components. We seek formulas for small changes in the node voltages as functions of small changes in all the circuit components and sources. In this chapter we shall call the circuit components and sources collectively by the name *parameters*.

The set of all node voltages, the node-voltage vector, $\mathbf{V_n}$, may be expanded in a multi-dimensional Taylor series:

$$\mathbf{V_n}(\mathbf{p_0} + \delta\mathbf{p}) = \mathbf{V_{n0}} + \Delta\mathbf{V_n}$$
$$= \mathbf{V_n}(\mathbf{p_0}) + \sum_{i=1}^{P} \frac{\partial \mathbf{V_n}}{\partial p_i} \delta p_i + \sum_{i=1}^{P} \sum_{j=1}^{P} \frac{\partial^2 \mathbf{V_n}}{\partial p_i \partial p_j} \delta p_i \delta p_j + \cdots \quad (6.1)$$

where

$\mathbf{p}$ = vector of all circuit parameters;
P = the number of circuit parameters, i.e., resistors, capacitors, inductors, controlled sources, independent sources;
$\mathbf{V_n}(\mathbf{p_0})$ = vector of node voltages when the circuit parameters take on the values $\mathbf{p_0}$;
δp_i = change (variation) in the ith circuit parameter;
$\mathbf{p_0}$ = vector of circuit components at the point about which the changes in $\mathbf{V_n}$ are to be calculated;
$\mathbf{V_{n0}} = \mathbf{V_n}(\mathbf{p_0})$;
$\Delta\mathbf{V_n}$ = vector of variations of the node voltages for variation $\delta\mathbf{p}$;
$\delta\mathbf{p}$ = vector of variations of all δp_i.

For small changes in the parameters the change in $\mathbf{V_n}$ can be approximated by the first derivatives, provided the neglected terms in the summations indicated in Eq. (6.1) are negligible:

$$\Delta\mathbf{V_n} = \sum_{i=1}^{P} \frac{\partial \mathbf{V_n}}{\partial p_i} \delta p_i \quad (6.2)$$

It is difficult to know whether or not Eq. (6.2) is accurate without actually computing the higher derivatives. However, for a wide class of networks the node voltages are smooth functions of the circuit parameters; for all linear networks all voltages (node and branch) and all currents in the network are linear functions of the source strengths. In these cases the derivatives yield good approximations to actual small changes in the node voltages.

Note that certain other types of circuits cannot be characterized accurately with the first derivatives alone: resonant, high-Q circuits and notch-filters (twin-tee filters) have wildly varying responses in the neighborhood of resonance. Variations in such networks can still be calculated quite accurately away from the resonance points by the methods developed here. However, whenever large response variations are suspected, the actual network response should be calculated and the differences should be computed directly.

In order to apply Eq. (6.2) to the calculation of node-voltage changes, the partial derivatives indicated must be evaluated at an "operating point" and the actual parameter changes about this point must be known. Although this section will develop the derivative formulas for the general node equations, it is quite possible to write out the node equations as algebraic expressions and form the derivatives directly. The following example illustrates the process.

EXAMPLE 6.1

Calculate the variations of the node voltages for a 1% increase in resistances R_1, R_2, and R_3 and a 2% decrease in the source voltage E as well as in R_4.

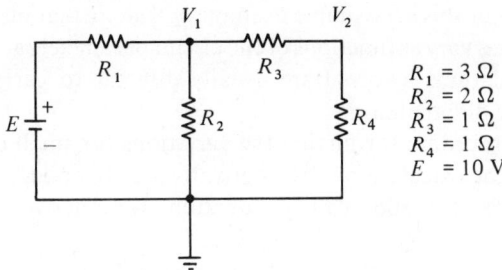

Fig. 6.1 Circuit for Example 6.1

Solution. The node voltage vector is

$$\begin{bmatrix} \dfrac{R_2(R_3 + R_4)}{R_1(R_2 + R_3 + R_4) + R_2(R_3 + R_4)} E \\ \dfrac{R_2(R_3 + R_4)}{R_1(R_2 + R_3 + R_4) + R_2(R_3 + R_4)} \cdot \dfrac{R_4}{R_3 + R_4} E \end{bmatrix} = \begin{bmatrix} 2.5 \\ 1.25 \end{bmatrix}$$

The derivatives may be calculated directly from the above expressions. These are

$$\frac{\partial \mathbf{V}_n}{\partial E} = \begin{bmatrix} \frac{1}{4} \\ \frac{1}{8} \end{bmatrix}$$

$$\frac{\partial \mathbf{V}_n}{\partial R_1} = \begin{bmatrix} \dfrac{-R_2(R_3 + R_4)(R_2 + R_3 + R_4)}{[R_1(R_2 + R_3 + R_4) + R_2(R_3 + R_4)]^2} E \\ \dfrac{-R_2(R_3 + R_4)(R_2 + R_3 + R_4)}{[R_1(R_2 + R_3 + R_4) + R_2(R_3 + R_4)]^2} \cdot \dfrac{R_4}{R_3 + R_4} E \end{bmatrix} = \begin{bmatrix} -\dfrac{5}{8} \\ -\dfrac{5}{16} \end{bmatrix}$$

Similarly,

$$\frac{\partial \mathbf{V}_n}{\partial R_2} = \begin{bmatrix} \frac{15}{32} \\ \frac{15}{64} \end{bmatrix}; \quad \frac{\partial \mathbf{V}_n}{\partial R_3} = \begin{bmatrix} \frac{15}{32} \\ -\frac{25}{64} \end{bmatrix}; \quad \frac{\partial \mathbf{V}_n}{\partial R_4} = \begin{bmatrix} \frac{15}{32} \\ \frac{55}{64} \end{bmatrix}$$

Hence the change in node voltages is

$$\Delta V_n = \begin{bmatrix} \frac{1}{4} \\ \frac{1}{8} \end{bmatrix}(-0.02) + \begin{bmatrix} -\frac{5}{8} \\ -\frac{5}{16} \end{bmatrix}(0.03) +$$

$$- \begin{bmatrix} \frac{15}{32} \\ \frac{15}{64} \end{bmatrix}(0.02) + \begin{bmatrix} \frac{15}{32} \\ -\frac{25}{64} \end{bmatrix}(0.01) + \begin{bmatrix} \frac{15}{32} \\ \frac{55}{64} \end{bmatrix}(-0.02)$$

$$= \begin{bmatrix} -\frac{410}{6400} \\ -\frac{325}{6400} \end{bmatrix} = \begin{bmatrix} -0.0641 \\ -0.0508 \end{bmatrix}$$

A direct calculation with the changed voltage and resistance values gives the changes in the node voltages as

$$\overset{*}{\Delta V}_n = \begin{bmatrix} 2.4362 \\ 1.1997 \end{bmatrix} - \begin{bmatrix} 2.5000 \\ 1.2500 \end{bmatrix} = \begin{bmatrix} -0.0638 \\ -0.0503 \end{bmatrix}$$

This example illustrates the fact that usually it is far simpler to recalculate the node equations when the changes in element values are known.

A careful study of this very simple example will show that intuitive feelings about how the node voltages vary as functions of the circuit parameters are difficult to acquire. Although the calculations involved are usually difficult to carry out, they require only the simplest of differentiations.

A systematic procedure for finding the variations for small changes in the parameters begins with the calculation of the derivatives of the node voltages with respect to all the parameters. The node-voltage equations were derived in Chapter 3. They are repeated here for reference.

$$\mathbf{A Y_b A}^T \mathbf{V_n} = \mathbf{A}(\mathbf{I_g} - \mathbf{Y_b V_g})$$

$$\mathbf{Y_b} = (\mathbf{U} + \mathbf{B}^* + \mathbf{G Z_e}) \mathbf{Y_e} (\mathbf{U} + \mathbf{D}^* + \mathbf{R_t Y_e})^{-1}$$

$$= (\mathbf{U} + \mathbf{B}^* + \mathbf{G Z_e})((\mathbf{U} + \mathbf{D}^*)\mathbf{Z_e} + \mathbf{R_t})^{-1}$$

where

$\mathbf{I_g}$ = independent current generators;
$\mathbf{V_g}$ = independent voltage generators;
$\mathbf{Z_e}$ = element matrix; contains all self-conductances, resistances, capacitors, inductors, and mutual inductances;
$\mathbf{B}^*$ = matrix of current controlled current sources;
$\mathbf{G}$ = matrix of voltage controlled current sources;
$\mathbf{D}^*$ = matrix of voltage controlled voltage sources;
$\mathbf{R_t}$ = matrix of current controlled voltage sources.

Normally, the vectors $\mathbf{I_g}$, $\mathbf{V_g}$ and matrices $\mathbf{B}^*$, $\mathbf{G}$, $\mathbf{D}^*$, and $\mathbf{R_t}$ are sparse.

We now derive the relationships for variations of $\mathbf{V_n}$ with respect to various circuit elements. We note that variations with respect to generators are easy to obtain; variations with respect to circuit elements, however, involve differentiations of $\mathbf{Y_b}$ which occur on both sides of the node-voltage expression.

I. *Independent Sources*

(a) Current Generator in Branch j

$$\frac{\partial V_n}{\partial I_{g_j}} = (AY_bA^T)^{-1} A \frac{\partial I_g}{\partial I_{g_j}} \qquad (6.3)$$

$$= (AY_bA^T)^{-1} A[\delta_j]$$

where δ_j is a vector containing a 1 on row j; otherwise the vector contains zero.

(b) Voltage Generator in Branch j

$$\frac{\partial V_n}{\partial V_{g_j}} = -(AY_bA^T)^{-1} AY_b[\delta_j] \qquad (6.4)$$

where δ_j is defined in Eq. (6.3).

II. *Passive Components and Controlled Sources*

Each of these circuit elements involve Y_b. For the moment, let one of these elements be denoted by p_i. The derivative of V_n with respect to this parameter is obtained by direct differentiation:

$$AY_bA^T \frac{\partial V_n}{\partial p_i} + A \frac{\partial Y_b}{\partial p_i} A^T V_n = -A \frac{\partial Y_b}{\partial p_i} V_g \qquad (6.5)$$

Hence the derivative is

$$\frac{\partial V_n}{\partial p_i} = -(AY_bA^T)^{-1} A \frac{\partial Y_b}{\partial p_i} (V_g + A^T V_n) \qquad (6.6)$$

Since the branch voltage is

$$V_b = A^T V_n \qquad (6.7)$$

the derivative may be written as

$$\frac{\partial V_n}{\partial p_i} = -(AY_bA^T)^{-1} A \frac{\partial Y_b}{\partial p_i} (V_g + V_b) \qquad (6.8)$$

The derivative of Y_b with respect to the various possible parameters must now be calculated.

(a) Resistor in Branch j

Such a resistor appears as the element in location j,j of the Z_e matrix; the value is R_j. The derivative of Y_b is

$$\frac{\partial Y_b}{\partial R_j} = \left(G \frac{\partial Z_e}{\partial R_j}\right)[(U+D^*)Z_e + R_t]^{-1}$$

$$-(U + B^* + GZ_e)[(U+D^*)Z_e + R_t]^{-1}(U+D^*) \frac{\partial Z_e}{\partial p_j} \cdot [(U+D)^*Z_e + R_t]^{-1}$$

$$= [G - Y_b(U+D^*)] \frac{\partial Z_e}{\partial R_j} [(U+D^*)Z_e + R_t]^{-1} \qquad (6.9)$$

The derivative of $\mathbf{Z}_e$ is simply 1 in location j, j and zero elsewhere:

$$\frac{\partial \mathbf{Z}_e}{\partial R_j} = [\delta_{jj}] \tag{6.10}$$

where $[\delta_{jj}]$ is an M by M matrix of zeros except in location j, j, where a 1 is located. Computations involving such a matrix are very simple to carry out.

The node-voltage variation becomes

$$\frac{\partial \mathbf{V}_n}{\partial R_j} = -(\mathbf{A}\mathbf{Y}_b\mathbf{A}^T)^{-1}\mathbf{A}[\mathbf{G} - \mathbf{Y}_b(\mathbf{U} + \mathbf{D}^*)][\delta_{jj}][(\mathbf{U} + \mathbf{D}^*)\mathbf{Z}_e + \mathbf{R}_t]^{-1}(\mathbf{V}_g + \mathbf{V}_b) \tag{6.11}$$

(b) Capacitor in Branch j

The development leading to Eq. (6.9) holds for this case also with R_j replaced by $1/(\sqrt{-1}\omega C_j)$; the derivative corresponding to Eq. (6.10) is

$$\frac{\partial \mathbf{Z}_e}{\partial C_j} = -\frac{1}{\sqrt{-1}\omega C_j^2}[\delta_{jj}] \tag{6.12}$$

where $\sqrt{-1}$ is the principal value of the square root of -1. The variation of $\mathbf{V}_n$ becomes

$$\frac{\partial \mathbf{V}_n}{\partial R_j} = \left(\frac{1}{\sqrt{-1}\omega C_j^2}\right)(\mathbf{A}\mathbf{Y}_b\mathbf{A}^T)^{-1}\mathbf{A}[\mathbf{G} - \mathbf{Y}_b(\mathbf{U} + \mathbf{D}^*)][\delta_{jj}][(\mathbf{U} + \mathbf{D}^*)\mathbf{Z}_e + \mathbf{R}_t]^{-1}(\mathbf{V}_g + \mathbf{V}_b) \tag{6.13}$$

(c) Inductor in Branch j

The variation of $\mathbf{Z}_e$ with respect to L_j is

$$\frac{\partial \mathbf{Z}_e}{\partial L_j} = \sqrt{-1}\omega[\delta_{jj}] \tag{6.14}$$

Hence the node-voltage variation becomes

$$\frac{\partial \mathbf{V}_n}{\partial L_j} = -\sqrt{-1}\omega(\mathbf{A}\mathbf{Y}_b\mathbf{A}^T)^{-1}\mathbf{A}[\mathbf{G} - \mathbf{Y}_b(\mathbf{U} + \mathbf{D}^*)][\delta_{jj}][(\mathbf{U} + \mathbf{D}^*)\mathbf{Z}_e + \mathbf{R}_t]^{-1}(\mathbf{V}_g + \mathbf{V}_b) \tag{6.15}$$

(d) Mutual Inductance Coupling Branches j and k

Here the variation of $\mathbf{Z}_e$ contains two terms:

$$\frac{\partial \mathbf{Z}_e}{\partial M_{jk}} = \sqrt{-1}\omega[\delta_{jk} + \delta_{kj}] \tag{6.16}$$

Thus the node-voltage variation becomes

$$\frac{\partial \mathbf{V}_n}{\partial M_{jk}} = -\sqrt{-1}\omega(\mathbf{A}\mathbf{Y}_b\mathbf{A}^T)^{-1}\mathbf{A}[\mathbf{G} - \mathbf{Y}_b(\mathbf{U} + \mathbf{D}^*)][\delta_{jk} + \delta_{kj}][(\mathbf{U} + \mathbf{D}^*)\mathbf{Z}_e \\ + \mathbf{R}_t]^{-1}(\mathbf{V}_g + \mathbf{V}_b) \tag{6.17}$$

(e) Current Controlled Current Source β_{jk}

Such an element appears in location j, k in the B^* matrix. The variation of $\mathbf{Y}_b$ becomes

$$\frac{\partial \mathbf{Y}_b}{\partial \beta_{jk}} = [\delta_{jk}][(\mathbf{U} + \mathbf{D}^*)\mathbf{Z}_e + \mathbf{R}_t]^{-1} \tag{6.18}$$

Sec. 6.1 Derivatives and Sensitivity

and the node-voltage variation is

$$\frac{\partial V_n}{\partial \beta_{jk}} = -(AY_bA^T)^{-1}A[\delta_{jk}][(U+D^*)Z_e + R_t]^{-1}(V_g + V_b) \quad (6.19)$$

(f) Voltage Controlled Current Source g_{jk}

Such an element appears in location j, k of the **G** matrix. The variation of Y_b becomes

$$\frac{\partial Y_b}{\partial g_{jk}} = [\delta_{jk}]Z_e[(U+D^*)Z_e + R_t]^{-1} \quad (6.20)$$

The node-voltage variation becomes

$$\frac{\partial V_n}{\partial g_{jk}} = -(AY_bA^T)^{-1}A[\delta_{jk}]Z_e[(U+D^*)Z_e + R_t]^{-1}(V_b + V_g) \quad (6.21)$$

(g) Current Controlled Voltage Source r_{jk}

Such an element appears in location j, k of the R_t matrix. The variation of Y_b becomes

$$\frac{\partial Y_b}{\partial r_{jk}} = -[U+B^* + GZ_e][(U+D^*)Z_e + R_t]^{-1}[\delta_{jk}][(U+D^*)Z_e + R_t]^{-1}(V_b + V_g)$$
$$= -Y_b[\delta_{jk}][(U+D^*)Z_e + R_t]^{-1}(V_b + V_g) \quad (6.22)$$

The node-voltage variation becomes

$$\frac{\partial V_n}{\partial r_{jk}} = (AY_bA^T)^{-1}AY_b[\delta_{jk}][(U+D^*)Z_e + R_t]^{-1}(V_g + V_b) \quad (6.23)$$

(h) Voltage Controlled Voltage Source μ_{jk}

Such an element appears in location j, k of the D^* matrix. The variation of Y_b becomes

$$\frac{\partial Y_b}{\partial \mu_{jk}} = -Y_b[\delta_{jk}]Z_e[(U+D^*)Z_e + R_t]^{-1} \quad (6.24)$$

The node-voltage variation becomes

$$\frac{\partial V_n}{\partial \mu_{jk}} = (AY_bA^T)^{-1}AY_b[\delta_{jk}]Z_e[(U+D^*)Z_e + R_t]^{-1}(V_g + V_b) \quad (6.25)$$

A comparison of the various node-voltage derivatives shows certain recurring matrix products. The expressions may be simplified by using the following:

$$P = (AY_bA^T)^{-1}A = Y_n^{-1}A \quad (6.26)$$

$$Q = [(U+D^*)Z_e + R_t]^{-1} \quad (6.27)$$

$$R = G - Y_b(U+D^*) \quad (6.28)$$

$$S = P \cdot R \quad (6.29)$$

$$V^* = Q(V_b + V_g) \quad (6.30)$$

Then the derivatives may be written concisely as

$$\frac{\partial \mathbf{V_n}}{\partial I_{g_j}} = \mathbf{P}[\delta_j] = j\text{th column of } \mathbf{P} \tag{6.31}$$

$$\frac{\partial \mathbf{V_n}}{\partial V_{g_j}} = -\mathbf{PY_b}[\delta_j] = j\text{th column of } (-\mathbf{PY_b}) \tag{6.32}$$

$$\frac{\partial \mathbf{V_n}}{\partial R_j} = -\mathbf{S}[\delta_{jj}]\mathbf{V}^* \tag{6.33}$$

$$\frac{\partial \mathbf{V_n}}{\partial C_j} = \frac{1}{\sqrt{-1}\omega C_j^2}\mathbf{S}[\delta_{jj}]\mathbf{V}^* \tag{6.34}$$

$$\frac{\partial \mathbf{V_n}}{\partial L_j} = -\sqrt{-1}\omega \mathbf{S}[\delta_{jj}]\mathbf{V}^* \tag{6.35}$$

$$\frac{\partial \mathbf{V_n}}{\partial M_{jk}} = -\sqrt{-1}\omega \mathbf{S}[\delta_{jk} + \delta_{kj}]\mathbf{V}^* \tag{6.36}$$

$$\frac{\partial \mathbf{V_n}}{\partial g_{jk}} = -\mathbf{P}[\delta_{jk}]\mathbf{Z_e}\mathbf{V}^* \tag{6.37}$$

$$\frac{\partial \mathbf{V_n}}{\partial \beta_{jk}} = -\mathbf{P}[\delta_{jk}]\mathbf{V}^* \tag{6.38}$$

$$\frac{\partial \mathbf{V_n}}{\partial r_{jk}} = \mathbf{PY_b}[\delta_{jk}]\mathbf{V}^* \tag{6.39}$$

$$\frac{\partial \mathbf{V_n}}{\partial \mu_{jk}} = \mathbf{PY_b}[\delta_{jk}]\mathbf{Z_e}\mathbf{V}^* \tag{6.40}$$

The delta matrices are defined in Eqs. (6.3) and (6.10).

Further significant simplifications can be obtained by noting the structure of the various δ-matrices. Some of the simplified relationships are

$$\frac{\partial \mathbf{V_n}}{\partial R_j} = -(j\text{th column of } \mathbf{S} * j\text{th element of } \mathbf{V}^*) \tag{6.41}$$

$$\frac{\partial \mathbf{V_n}}{\partial C_j} = (j\text{th column of } \mathbf{S} * j\text{th element of } \mathbf{V}^*)/\sqrt{-1}\omega C_j^2 \tag{6.42}$$

$$\frac{\partial \mathbf{V_n}}{\partial L_j} = -(j\text{th column of } \mathbf{S} * j\text{th element of } \mathbf{V}^*) * \sqrt{-1}\omega \tag{6.43}$$

Others may be obtained by careful examination of Eqs. (6.26) through (6.40). We shall return to these programming considerations later.

EXAMPLE 6.2

The formulas just derived are applied to the circuit and the variations given in Ex. 6.1.

Solution. We note that $\mathbf{I_g}$, $\mathbf{B}^*$, $\mathbf{G}$, $\mathbf{D}^*$, and $\mathbf{R_t}$ are zero in this case. The various circuit matrices of interest are

$$\mathbf{A} = \begin{bmatrix} -1 & 1 & 1 & 0 \\ 0 & 0 & -1 & 1 \end{bmatrix}$$

$$\mathbf{Z}_e = \begin{bmatrix} 3. & 0 & 0 & 0 \\ 0 & 2. & 0 & 0 \\ 0 & 0 & 1. & 0 \\ 0 & 0 & 0 & 1. \end{bmatrix} \qquad \mathbf{V}_g = \begin{bmatrix} 10. \\ 0 \\ 0 \\ 0 \end{bmatrix}$$

$$\mathbf{Y}_b = \begin{bmatrix} \frac{1}{3} & 0 & 0 & 0 \\ 0 & \frac{1}{2} & 0 & 0 \\ 0 & 0 & 1 & 0 \\ 0 & 0 & 0 & 1 \end{bmatrix}$$

$$\mathbf{A}\mathbf{Y}_b\mathbf{A}^T = \begin{bmatrix} \frac{11}{6} & -1 \\ -1 & 2 \end{bmatrix}$$

$$(\mathbf{A}\mathbf{Y}_b\mathbf{A}^T)^{-1} = \begin{bmatrix} \frac{3}{4} & \frac{3}{8} \\ \frac{3}{8} & \frac{11}{16} \end{bmatrix}$$

$$\mathbf{P} = \overset{4}{}2\begin{bmatrix} -\frac{3}{4} & \frac{3}{4} & \frac{3}{8} & \frac{3}{8} \\ -\frac{3}{8} & \frac{3}{8} & -\frac{5}{16} & \frac{11}{16} \end{bmatrix}$$

$$\mathbf{R} = -\mathbf{Y}_b$$

$$\mathbf{S} = \overset{4}{}2\begin{bmatrix} \frac{1}{4} & -\frac{3}{8} & -\frac{3}{8} & -\frac{3}{8} \\ \frac{1}{8} & \frac{3}{16} & \frac{5}{16} & \frac{11}{16} \end{bmatrix}$$

$$\mathbf{Q} = \mathbf{Y}_b$$

$$\mathbf{V}^* = \mathbf{Q}\left(\begin{bmatrix} -2.50 \\ 2.50 \\ 1.25 \\ 1.25 \end{bmatrix} + \begin{bmatrix} 10. \\ 0 \\ 0 \\ 0 \end{bmatrix}\right) = \mathbf{Q}\begin{bmatrix} 7.50 \\ 2.50 \\ 1.25 \\ 1.25 \end{bmatrix} = \begin{bmatrix} 2.50 \\ 1.25 \\ 1.25 \\ 1.25 \end{bmatrix}$$

$$\frac{\partial \mathbf{V}_n}{\partial R_1} = \begin{bmatrix} -\frac{5}{8} \\ -\frac{5}{16} \end{bmatrix}; \quad \frac{\partial \mathbf{V}_n}{\partial R_2} = \begin{bmatrix} \frac{15}{32} \\ \frac{15}{64} \end{bmatrix}; \quad \frac{\partial \mathbf{V}_n}{\partial R_3} = \begin{bmatrix} \frac{15}{32} \\ -\frac{25}{64} \end{bmatrix}; \quad \frac{\partial \mathbf{V}_n}{\partial R_4} = \begin{bmatrix} \frac{15}{32} \\ \frac{55}{64} \end{bmatrix}; \quad \frac{\partial \mathbf{V}_n}{\partial V_{g_1}} = \begin{bmatrix} \frac{1}{4} \\ \frac{1}{8} \end{bmatrix}$$

The formulas developed so far may be applied directly to the calculation of node-voltage variations in direct-current circuits, as was done in the above example. For AC analysis it is convenient to extend the above to the calculation of variations of magnitude and phase angle of the node voltages. Let the kth node voltage be V_k; then the result of an AC analysis will be a pair of real numbers A_k and B_k for this voltage:

$$V_k = A_k + \sqrt{-1}B_k \tag{6.44}$$

The magnitude of V_k becomes

$$|V_k| = \sqrt{A_k^2 + B_k^2} \tag{6.45}$$

The variation of $|V_k|$ may be obtained by differentiating the above

$$\Delta|V_k| = \frac{1}{|V_k|}\sum_{i=1}^{P}\left(A_k\frac{\partial A_k}{\partial P_i} + B_k\frac{\partial B_k}{\partial P_i}\right)\delta P_i \qquad (6.46)$$

We note that

$$\frac{\partial A_k}{\partial P_i} = \operatorname{Re}\frac{\partial V_k}{\partial P_i}$$

and that

$$\frac{\partial B_k}{\partial P_i} = \operatorname{Im}\frac{\partial V_k}{\partial P_i}$$

The phase angle equation is

$$\underline{/V_k} = \arctan\frac{B_k}{A_k} \qquad (6.47)$$

Hence the variation of the phase angle may be approximated by the first-order variations:

$$\Delta\underline{/V_k} = \frac{1}{|V_k|^2}\sum_{i=1}^{P}\left(A_k\frac{\partial B_k}{\partial p_i} - B_k\frac{\partial A_k}{\partial P_i}\right)\delta P_i \qquad (6.48)$$

Note that both variations require the availability of **A** and **B** and their derivatives for the various parameters P_i. (Here **A** is *not* the incidence matrix.)

The derivatives are obtained directly from the derivative of $\mathbf{V_n}$:

$$\frac{\partial \mathbf{V_n}}{\partial P_i} = \frac{\partial \mathbf{A}}{\partial P_i} + \sqrt{-1}\,\frac{\partial \mathbf{B}}{\partial P_i} \qquad (6.49)$$

Thus the kth element of the complex vector $\partial \mathbf{V_n}/\partial P_i$ will be the two required derivatives. The derivative of $\mathbf{V_n}$ is obtained by direct application of the formulas summarized in Eqs. (6.26) through (6.40).

The relative complexity of the above equations for the calculation of the changes in the magnitudes of the node voltages and the associated phase angle changes tempts one to compute the magnitudes and phases of the voltages before the change and subtract them from the voltage magnitudes and phases after the parameters are changed:

$$\Delta|\mathbf{V_n}| = |\mathbf{V_n} + \Delta\mathbf{V_n}| - |\mathbf{V_n}| \qquad (6.50)$$

$$\Delta\underline{/\mathbf{V_n}} = \underline{/\mathbf{V_n} + \Delta\mathbf{V_n}} - \underline{/\mathbf{V_n}} \qquad (6.51)$$

Great care must be used with the above; the changes sought are small in relation to the numbers that enter the subtractions. Such operations are highly susceptible to round-off accumulation (see Sec. 2.7). Normally, the above formulas should be avoided.

The advantage of Eqs. (6.46) and (6.48) is that they are purely algebraic. Another set of formulas for $\Delta|V_k|$ and $\Delta\underline{/V_k}$ may be obtained by regarding the changes as vector quantities. Then

$$\Delta|V_k| = |\Delta V_k|\cdot\cos(\alpha_2 - \alpha_1) \qquad (6.52)$$

Sec. 6.1   Derivatives and Sensitivity   263

where α_2 = phase angle of δV_k;
α_1 = phase angle of V_k.

Similarly, the phase angle change is:

$$\Delta \underline{/V_k} = \arctan \frac{|\Delta V_k| \sin(\alpha_2 - \alpha_1)}{|V_k| + \Delta |V_k|} \tag{6.53}$$

Normally, very small angles will be encountered; the small angle approximations for arc tan can be used to advantage. Hence

$$\Delta \underline{/V_k} \cong \frac{|\Delta V_k|}{|V_k|} \sin(\alpha_2 - \alpha_1) \tag{6.54}$$

may yield acceptable accuracy in most instances. Note that $\Delta \underline{/V_k}$ is calculated in radians; in order to get the change in degrees, the multiplier 180/3.14159 must be used.

When Eq. (6.53) appears in a FORTRAN-based computer program, the routine ATAN2 should be used, with the numerator being the first argument of ATAN2 and the denominator the second argument. Otherwise the computed angle may lie in the incorrect quadrant.

Another source of difficulty with the basic derivative relations is that V_{g_i} and I_{g_i} are complex. This shall be examined after the following example.

EXAMPLE 6.3

For the circuit shown in Fig. 6.2 calculate the change in magnitude and phase angle of the node voltages if each passive component (R_1, R_2, C, L) increases by 1%.

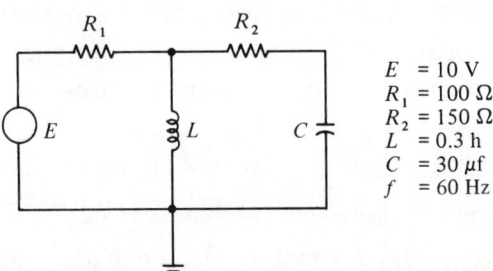

$E = 10$ V
$R_1 = 100\ \Omega$
$R_2 = 150\ \Omega$
$L = 0.3$ h
$C = 30\ \mu f$
$f = 60$ Hz

Fig. 6.2 Circuit for Example 6.3

Solution. As in Ex. 6.2, the various network matrices must be calculated. These calculations are summarized below; they were carried out on a slide rule. In this example $j = \sqrt{-1}$.

$$\mathbf{A} = \begin{bmatrix} -1 & 1 & 1 & 0 \\ 0 & -1 & 0 & 1 \end{bmatrix} \quad X_L = 2\pi \times 60 \times 0.3 = 113.1\ \Omega$$

$$X_C = \frac{1}{377 \times 30 \times 10^{-6}} = 88.4\ \Omega$$

$$\mathbf{Z}_e = \begin{bmatrix} 100 & 0 & 0 & 0 \\ 0 & 150 & 0 & 0 \\ 0 & 0 & j113.1 & 0 \\ 0 & 0 & 0 & -j88.4 \end{bmatrix}; \quad V_g = \begin{bmatrix} 10. \\ 0 \\ 0 \\ 0 \end{bmatrix}$$

$$\mathbf{Y_b} = \begin{bmatrix} 0.01000 & 0 & 0 & 0 \\ 0 & 0.00667 & 0 & 0 \\ 0 & 0 & -j0.00884 & 0 \\ 0 & 0 & 0 & j0.01131 \end{bmatrix}$$

$$\mathbf{Y_n} = \mathbf{A Y_b A}^T = \begin{bmatrix} 0.01667 - j0.00884 & -0.00667 \\ -0.00667 & 0.00667 + j0.01131 \end{bmatrix}$$

$$\mathbf{Y_n^{-1}} = (\mathbf{A Y_b A}^T)^{-1} = \begin{bmatrix} 57.82 + j22.90 & 24.94 - j19.39 \\ 24.94 - j19.39 & 36.64 - j81.52 \end{bmatrix}$$

$$\mathbf{V_n} = \begin{bmatrix} 5.782 + j2.290 \\ 2.494 - j1.939 \end{bmatrix} = \begin{bmatrix} 6.219 \; \underline{/21.6°} \\ 3.158 \; \underline{/-37.8°} \end{bmatrix}$$

$$\mathbf{Q} = \mathbf{Y_b}$$

$$\mathbf{V^*} = \mathbf{Y_b} \left(\begin{bmatrix} -5.782 - j2.290 \\ 3.288 + j4.229 \\ 5.782 + j2.290 \\ 2.494 - j1.939 \end{bmatrix} + \begin{bmatrix} 10.0 \\ 0.0 \\ 0.0 \\ 0.0 \end{bmatrix} \right) = \begin{bmatrix} 0.04218 - j0.02290 \\ 0.02193 + j0.02821 \\ 0.02024 - j0.05111 \\ 0.02193 + j0.02821 \end{bmatrix}$$

$$\mathbf{P} = \overset{4}{\underset{2}{\begin{bmatrix} -57.82 - j22.90 & 32.88 + j42.29 & 57.82 + j22.90 & 24.94 - j19.39 \\ -24.94 + j19.39 & -11.70 + j62.14 & 24.94 - j19.39 & 36.64 - j81.52 \end{bmatrix}}}$$

$$\mathbf{R} = -\mathbf{Y_b}$$

$$\mathbf{S} = \overset{4}{\underset{2}{\begin{bmatrix} 0.5782 + j0.2290 & -0.2193 - j0.2821 & -0.2024 + j0.5111 & -0.2193 - j0.2821 \\ 0.2494 - j0.1939 & 0.0780 - j0.4143 & 0.1714 + j0.2205 & -0.9220 - j0.4144 \end{bmatrix}}}$$

$$\frac{\partial \mathbf{V_n}}{\partial R_1} = \begin{bmatrix} -0.02964 + j0.00358 \\ -0.00608 + j0.01389 \end{bmatrix}; \quad \frac{\partial \mathbf{V_n}}{\partial R_2} = \begin{bmatrix} -0.00315 + j0.01236 \\ -0.01339 + j0.00688 \end{bmatrix}$$

$$\frac{\partial \mathbf{V_n}}{\partial C} = \begin{bmatrix} -36425 - j9277 \\ -103381 + j25132 \end{bmatrix}; \quad \frac{\partial \mathbf{V_n}}{\partial L} = \begin{bmatrix} 7.811 - j8.305 \\ -1.617 - j5.557 \end{bmatrix}$$

The variations caused by a 1% increase in all components are:

$$\delta R_1 = 1.00 \; \Omega; \quad \delta R_2 = 1.50 \; \Omega; \quad \delta C = 0.30 \; \mu f; \quad \delta L = 3.00 \; mh$$

Hence the variation of $\mathbf{V_n}$ caused by these changes is

$$\Delta \mathbf{V_n} = \begin{bmatrix} -0.02186 - j0.00558 \\ -0.06202 + j0.01508 \end{bmatrix} = \begin{bmatrix} 0.0225 \; \underline{/-165.7°} \\ 0.0638 \; \underline{/166.3°} \end{bmatrix}$$

The change in the magnitude of V_1 becomes

$$\Delta |V_1| = \frac{1}{6.219} \{ [(5.782)(-0.02964) + (2.290)(0.00358)](1.00)$$

$$+ [(5.782)(-0.00315) + (2.290)(0.01236)](1.50)$$

$$+ [(5.782)(-36425) + (2.290)(-9277)](3.0 \times 10^{-7})$$

$$+ [(5.782)(7.811) + (2.290)(-8.305)](3.0 \times 10^{-3}) \}$$

$$= -0.0226 \; V$$

Similarly, the change caused by the phase angle of V_1 becomes

$$\Delta\underline{/V_1} = \frac{1}{(6.219)^2}\{[(5.782)(0.00358) - (2.290)(-0.02964)](1.00)$$
$$+ [(5.782)(0.01236) - (2.290)(-0.00315)](1.50)$$
$$+ [(5.782)(-9277) - (2.290)(36425)](3.0 \times 10^{-7})$$
$$+ [(5.782)(-8.305) - (2.290)(7.811)](3.0 \times 10^{-3})\}$$
$$= 0.0004887 \text{ radians} = 0.0280°$$

Using Eqs. (6.52) and (6.53) results in

$$\alpha_1 = 21.6°$$
$$\alpha_2 = 194.3°$$
$$\alpha_2 - \alpha_1 = 172.7°$$
$$\Delta|V_1| = 0.0225 \cos 172.7° = -0.0223$$
$$\Delta\underline{/V_1} = \arctan\frac{0.0225 \sin 7.3°}{6.219 - 0.0223} = \arctan\frac{0.00289}{6.197} = 0.0268°$$

For comparisons, use of Eq. (6.54) gives

$$\Delta\underline{/V_1} = 57.3 \times \frac{0.0225}{6.219} \sin 172.7° = 0.0268°$$

Note again that some of the calculations were done with a slide rule.

An analogous set of relations can be written for the variations of V_2. Note that ω and E are considered fixed; the changes in the node voltages at a fixed frequency are computed in the above. (See Problem 6.6.)

One further set of relationships must be discussed. The parameters I_{g_i} and V_{g_i} occurring in Eqs. (6.31) and (6.32) are complex variables; the derivative relations should be rewritten so that the effects of magnitude change and phase angle change are explicit. The generator current in branch i is

$$I_{g_i} = |I_{g_i}| \exp[\sqrt{-1}\underline{/I_{g_i}}] \quad (6.55)$$

and similarly, V_{g_i} can be written. Direct differentiation of the node-voltage expression for the magnitude and phase angle changes are

$$\frac{\partial \mathbf{V_n}}{\partial|I_{g_i}|} = (\mathbf{AY_bA}^T)^{-1}\mathbf{A}[\delta_i]\exp[\sqrt{-1}\underline{/I_{g_i}}]$$
$$= \mathbf{P}[\delta_i]\exp[\sqrt{-1}\underline{/I_{g_i}}] \quad (6.56)$$

$$\frac{\partial \mathbf{V_n}}{\partial\underline{/I_{g_i}}} = \sqrt{-1}\, I_{g_i}\mathbf{P}[\delta_i] \quad (6.57)$$

$$\frac{\partial \mathbf{V_n}}{\partial|V_{g_i}|} = -\mathbf{PY_b}[\partial_i]\exp[\sqrt{-1}\underline{/I_{g_i}}] \quad (6.58)$$

$$\frac{\partial \mathbf{V_n}}{\partial\underline{/V_{g_i}}} = -\sqrt{-1}\, V_{g_i}\mathbf{PY_b}[\delta_i] \quad (6.59)$$

The effects of these magnitude and phase changes on the magnitudes and phase changes of the node voltages are calculated directly from Eqs. (6.44) through (6.49) or from Eqs. (6.52) through (6.54).

The derivatives calculated above will show the changes in node voltages mathematically. It is, however, difficult to visualize the actual changes in the network as a function of the circuit parameter changes from a knowledge of the derivatives alone. For example, in the last example the numerical values of the derivatives with respect to R_1 and with respect to C differ by about six orders of magnitude; yet the effects of comparable 1% changes in R_1 and C are about the same. Hence the derivatives should be normalized with respect to the parameter values. Let

Δp_k = the percentage change in parameter p_k;
Δp_k^f = the sensitivity of f (either a voltage, a node-voltage magnitude, or a node-voltage phase angle) with respect to parameter p_k.

Furthermore let

$\mathbf{S}_{p_k}^f$ = the vector of all sensitivities Δp_k^f for all node voltages.

Then the node-voltage variation, given in Eq. (6.2), may be written as

$$\Delta \mathbf{V_n} = \sum_{k=1}^{P} \frac{\partial \mathbf{V_n}}{\partial p_k} \cdot \frac{p_k \cdot \Delta p_k}{100}$$
$$= \sum_{k=1}^{P} \mathbf{S}_{p_k}^{\mathbf{V_n}} \Delta p_k \qquad (6.60)$$

and therefore the sensitivity may be defined as

$$\mathbf{S}_{p_k}^{\mathbf{V_n}} = \frac{\partial \mathbf{V_n}}{\partial p_k} \frac{p_k}{100} \qquad (6.61)$$

Note that the above definition is only one of many possible ways of showing the effect of parameter changes. Another procedure is to find the fractional change in the node voltages. For each of the nodes i

$$S_{p_k}^{V_{n_i}} = \left[\frac{\partial V_{n_i}}{\partial p_k} \cdot \frac{p_k}{V_{n_i}} \right] \qquad (6.62)$$

In this latter definition the sensitivity coefficient becomes a dimensionless quantity (as opposed to having the dimension of voltage). The definition adopted for this chapter is Eq. (6.61); in this form the sensitivity is simply the voltage change at the various nodes for a 1% change in the parameter. As long as the parameter is real (passive components, controlled source strengths), the definition may be used directly; for independent generators i_g and v_g, two separate sensitivities may be defined, one for the magnitude changes and one for the phase angle changes of the parameter. Since the phase angle change yields a derivative which is already in volts, it is reasonable to use the following definition for phase angle changes only:

$$S_{\underline{/V_{g_i}}}^{\mathbf{V_n}} = \frac{\partial \mathbf{V_n}}{\partial \underline{/V_{g_i}}} \cdot \frac{\pi}{180} \qquad (6.63)$$

and a similar expression for phase angle changes in I_{g_i}. Eq. (6.63) shows the effect of a one-degree change in the phase angle on the node voltages.

Sec. 6.2   Statistical Variations   267

Thus the node-voltage changes may be calculated from the following:

$$\Delta \mathbf{V_n} = \sum_{k=1}^{P} \mathbf{S}_{p_k}^{V_n} \Delta p_k + \sum_{k=1}^{\#V} \mathbf{S}/\underline{V_{g_k}}^{V_n} \Delta \underline{/V_{g_k}} + \sum_{k=1}^{\#I} \mathbf{S}/\underline{I_{g_k}}^{V_n} \Delta \underline{/I_{g_k}} \qquad (6.64)$$

where

$$\mathbf{S}_{p_k}^{V_n} = \frac{\partial \mathbf{V_n}}{\partial |p_k|} \cdot \frac{|p_k|}{100} \qquad (6.65)$$

with
$\quad \Delta p_k = \%$ change in parameter p_k;
$\quad P = $ number of passive components (G, R, L, C), transfer quantities (M, β, g_m, r_t, μ), and independent generators (i_g and v_g) in the network;
$\quad \#V = $ number of independent voltage sources;
$\quad \#I = $ number of independent current sources;
$\quad \Delta \underline{/V_{g_k}} = $ change in phase angle of V_{g_k}, in degrees;
$\quad \Delta \underline{/I_{g_k}} = $ change in phase angle of I_{g_k}, in degrees;
$\quad \mathbf{S}/\underline{V_{g_k}}^{V_n}$ and $\mathbf{S}/\underline{I_{g_k}}^{V_n}$ are defined by Eq. (6.63).
In DC analysis only the first summation is used.

EXAMPLE 6.4

Calculate the sensitivities for the circuit used in Ex. 6.3.

Solution. By application of Eqs. (6.58) and (6.60) we have

$$\mathbf{S}_{R_1}^{V_n} = \begin{bmatrix} -0.02964 + j0.00358 \\ -0.00608 + j0.01389 \end{bmatrix} \cdot \frac{100}{100} = \begin{bmatrix} -0.02964 + j0.00358 \\ -0.00608 + j0.01389 \end{bmatrix}$$

$$\mathbf{S}_{R_2}^{V_n} = \begin{bmatrix} -0.004725 + j0.01854 \\ -0.02008 + j0.01032 \end{bmatrix}; \quad \mathbf{S}_{C}^{V_n} = \begin{bmatrix} -0.01093 - j0.00278 \\ -0.03101 + j0.00754 \end{bmatrix}$$

$$\mathbf{S}_{L}^{V_n} = \begin{bmatrix} 0.02343 - j0.02492 \\ -0.00485 - j0.01667 \end{bmatrix}; \quad \mathbf{S}_{|V_g|}^{V_n} = \begin{bmatrix} 0.05781 + j0.02292 \\ 0.02494 - j0.01938 \end{bmatrix}$$

$$\mathbf{S}_{\underline{/V_g}}^{V_n} = \begin{bmatrix} -0.248 + j0.626 \\ 0.210 + j0.280 \end{bmatrix}$$

The above sensitivities can be converted to sensitivities involving the magnitude changes and the phase angle changes similar to the procedures outlined in Eqs. (6.46) through (6.54).

6.2 Statistical Variations

The basic derivatives can be applied directly to the calculation of changes in the node voltages if the changes in the components are given. This kind of data is usually not known; instead, only the limits of the permissible variation of the components, the component tolerances, may be known.

For example, resistors specified as 100 $\Omega \pm 10\%$ will lie somewhere between 90 Ω and 110 Ω. If one is interested in an accurate description of the components with which one is building circuits, a large sample (say 5,000) of such resistors can be measured and the measurements displayed in a *histogram*, such as shown by the bars

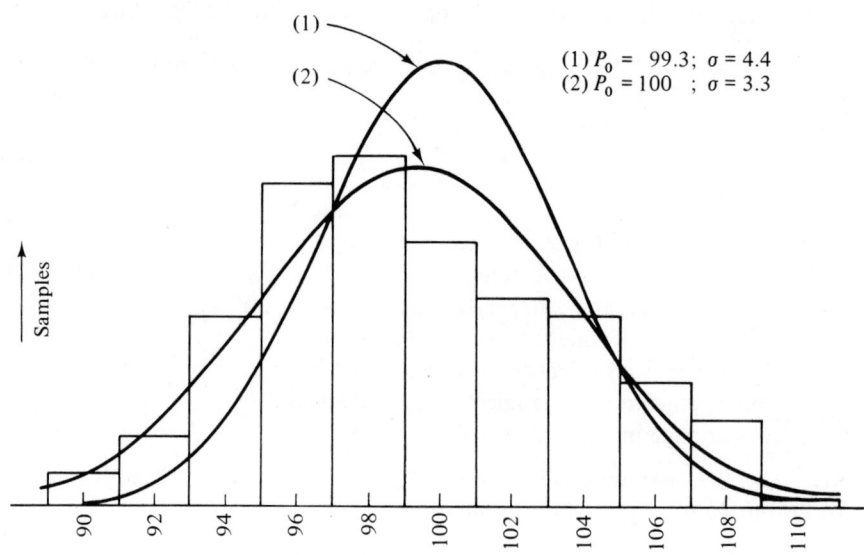

Fig. 6.3 Histogram for Table 6.1

of Fig. 6.3. Note that the accuracy with which the components are measured should be sufficiently fine to define a number of such data points in the range of the allowable variation. The measured data are tabulated in the first three columns of Table 6.1.

Table 6.1 *Measured Values for 5000 Resistors (nominal value 100 Ω)*

Bin # i	Range	Mid-range p_i	Count	Relative Frequency	Cummulative Frequency* f_i
1	89–90.99 ⋯	90	94	0.0188	0.0188
2	91–92.99 ⋯	92	194	0.0388	0.0576
3	93–94.99 ⋯	94	512	0.1024	0.1600
4	95–96.99 ⋯	96	864	0.1728	0.3328
5	97–98.99 ⋯	98	943	0.1886	0.5214
6	99–100.99 ⋯	100	711	0.1422	0.6636
7	101–102.99 ⋯	102	566	0.1132	0.7768
8	103–104.99 ⋯	104	517	0.1034	0.8802
9	105–106.99 ⋯	106	343	0.0686	0.9488
10	107–108.99 ⋯	108	238	0.0476	0.9964
11	109–110.99 ⋯	110	18	0.0036	1.0000
			$\sum = 5000$	$\sum = 1.000$	*(see Sec. 6.6)

$$\text{Average value} = 99.3 \pm 1.0 \ \Omega$$
$$\sigma = 4.4 \pm 1.0 \ \Omega$$
$$3\sigma = 13.0 \pm 1.0 \ \Omega$$

In the example given, an accuracy of 2 Ω was used, resulting in 11 vertical bars for the histogram. The actual counts of all resistors falling between two limits are usually converted to a percentage of the total sample. Such a *normalized histogram* shows the frequency of occurrence vs. the component value. Detailed tolerance analysis requires such data, either actually measured, established from experience, or assumed, as part of the input.

The normalized histogram for real measured data always exhibits a lower and an upper limit. Within these two limits the component values are distributed in some fashion. Analytical treatment for variances can be worked out relatively conveniently only for a few kinds of distributions. Results of such analytic treatment are available in various texts [2, 3]. Two kinds of distribution are easy to consider:

(a) Rectangular distribution, which assumes that between two limits $\underline{P}$ and $\overline{P}$ the probability of occurrence of any component is equal (also called uniform distribution).

(b) Normal distribution, which assumes that the probability of occurrence $f(p)$ of a component value p may be calculated from

$$f(p) = \frac{1}{\sqrt{2\pi}\sigma} \exp\left[-\frac{(p-p_0)^2}{2\sigma^2}\right] \qquad (6.66)$$

where

$\sigma =$ the standard deviation of the component;
$p_0 =$ the mean value of the component.

Note that the normal distribution assigns very small probabilities of occurrence for components that are more than 3σ limits from the mean value. Since in practice all circuit components usually lie within specified limits $\underline{P}$ and $\overline{P}$, the normal distribution at best approximates actual component variations; the uniform distribution on the other hand fails to account for any variation in the actual component values. Thus using either distribution yields only an approximation for the actual circuit component variations.

A normal distribution curve is shown in Fig. 6.4. Note that the curve is symmetric about the peak value (the *mode*) and is characterized only by this value (p_0) and the standard deviation. Several properties of this curve are tabulated below:

(1) The maximum for this curve, the frequency of most likely occurrence, is at p_0 and its value is

$$f_{\max} = \frac{1}{\sqrt{2\pi}} \frac{1}{\sigma} = \frac{0.400}{\sigma} \qquad (6.67)$$

(2) The area under the entire curve is one. This means that the total probability of some value for p is a certainty.

(3) The probability Pr of a component value lying between values p_1 and p_2 is

$$\Pr = \int_{p_1}^{p_2} f(p)\, dp \qquad (6.68)$$

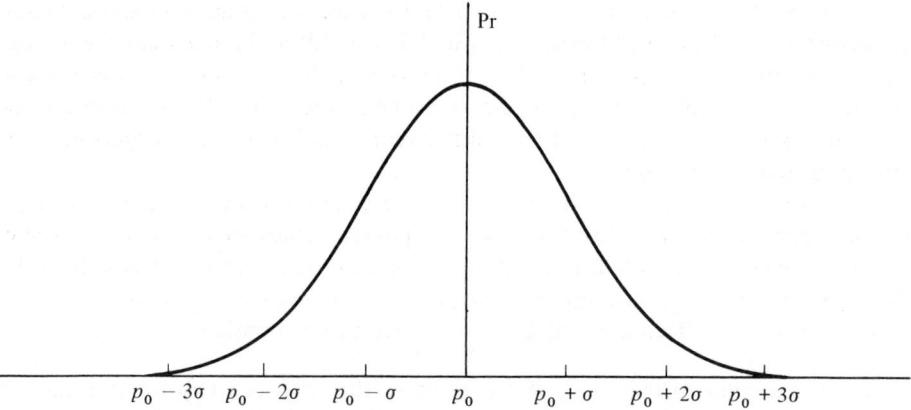

Fig. 6.4 Normal Distribution Curve

(4) The probability of a component being within σ of the average value is

$$\Pr(\sigma) = \int_{p_0-\sigma}^{p_0+\sigma} f(p)\,dp = 0.683 = 68.3\,\% \qquad (6.69)$$

(5) The probability of a component being within 2σ of the average value is

$$\Pr(2\sigma) = \int_{p_0-2\sigma}^{p_0+2\sigma} f(p)\,dp = 0.954 = 95.4\,\% \qquad (6.70)$$

(6) The probability of a component being within 3σ of the average value is

$$\Pr(3\sigma) = \int_{p_0-3\sigma}^{p_0+3\sigma} f(p)\,dp = 0.997 = 99.7\,\% \qquad (6.71)$$

(7) The mode (the value for which the curve has its maximum) is the average value of p.

(8) The second moment about the mean value is

$$M_2 = \int_{-\infty}^{\infty} (p - p_0)^2 f(p)\,dp = \sigma^2 \qquad (6.72)$$

For many applications it will be sufficient to approximate the actual probability distribution by an equivalent normal distribution. The reason that such distributions yield satisfactory predictions for the observed variations of node voltages is because of the central limit theorem of statistics:

If a large number of independent variables are summed, the probability distribution of the sum will tend to be normal no matter what the actual probability distributions of the variables are.

A direct consequence of this statement is that relatively crude approximations to the actual circuit parameter distributions will yield acceptable predictions of the variations in the circuit voltages and currents, provided a large number of parameters vary independently of each other. In practice, variations of four or five parameters can give very acceptable results, but in doubtful cases a more exhaustive analysis, using the actual probability distributions, must be made. Such an analysis procedure is explored in Sec. 6.6.

For finding the parameters of an approximating normal distribution for a set of measured probabilities, one must find the average and the standard deviations for the measured data. The average of data tabulated as in Table 6.1 is

$$p_0 = \frac{1}{L} \sum_{i=1}^{L} p_i' \qquad (6.73)$$

where

p_i' = individual measurement value;
L = number of total measurements.

Normally, measured data are classified in cells, K in number (11 such are used in Table 6.1), with center values p_i; the relative frequency of measurements in each of these cells is

$$f_i = \frac{1}{L} (\text{number of measurements in cell } i) \qquad (6.74)$$

The average value may then be approximated by

$$p_0 = \sum_{i=1}^{K} p_i f_i \qquad (6.75)$$

Similarly, the integral in Eq. (6.72) may be replaced by a summation over the K cells. The result is

$$\sigma^2 = \sum_{i=1}^{K} (p_i - p_0)^2 f_i \qquad (6.76)$$

For the data in Table 6.1 these calculations result in

$$p_0 = 99.2872 \ \Omega$$
$$\sigma = 4.3639 \ \Omega$$

Note that the accuracy of these values cannot exceed the tabular accuracy, $2 \ \Omega$ in this case. Hence

$$p_0 = 99.3 \pm 1.0 \ \Omega$$
$$\sigma = 4.4 \pm 1.0 \ \Omega$$

The resistors measured were nominally 10% resistors. We note that the entire range is contained within $\pm 2.7\sigma$ of the center value. It is customary, and quite accurate, to consider actual distributions to be approximated by the center value and the upper and lower limits as the 3σ values. In this example

$$\tilde{p}_0 = 100.0 \ \Omega$$
$$\tilde{\sigma} = 3.3 \ \Omega$$

These approximations are shown in Fig. 6.3 also. The agreement between the raw data, the calculated p_0, σ and the approximated $\tilde{p}_0$, $\tilde{\sigma}$ is easily discernible.

The components in a network are usually drawn from components which vary within prescribed limits, but are otherwise independent from one another. In order to calculate the expected change in a node-voltage magnitude, Eq. (6.46) must be solved for the expected variation because of the independently varying δx_i. If the variables x_i are independent and have a normal distribution, with a standard deviation of δ_{x_i}, $|V_k|$ will also have a normal distribution with a standard deviation σ_k:

$$\sigma_k = \frac{1}{|V_k|} \sqrt{\sum_{i=1}^{P} \left(A_k \frac{\partial A_k}{\partial x_i} + B_k \frac{\partial B_k}{\partial x_i} \right)^2 \sigma_{x_i}^2} \qquad (6.77)$$

A similar relation will hold for the variation of the angle of V_k; the starting point for this latter expression is Eq. (6.48):

$$\sigma \underline{/k} = \frac{1}{|V_k|^2} \sqrt{\sum_{i=1}^{P} \left(A_k \frac{\partial A_k}{\partial x_i} - B_k \frac{\partial B_k}{\partial x_i} \right)^2 \sigma_{x_i}^2} \qquad (6.78)$$

If the variations in the component values are not independent from one another, the last two formulas must be modified (see Sec. 6.5).

EXAMPLE 6.5

For 10% component values, calculate the voltage magnitude changes in Ex. 6.3.

Solution. Since no data are available on the statistical nature of the component values, the assumption is made that the 3σ limits are at $\pm 10\%$ of the nominal value. Hence for the components shown in Fig. 6.3

$$\sigma_{R_1} = 100 \times \frac{0.10}{3} = 3.33 \, \Omega$$

$$\sigma_{R_2} = 150 \times \frac{0.10}{3} = 5.00 \, \Omega$$

$$\sigma_C = 30 \times 10^{-6} \times \frac{0.10}{3} = 10^{-6} f$$

$$\sigma_L = 0.3 \times \frac{0.10}{3} = 0.01 \, h$$

The node voltages and the derivatives were calculated in Ex. 6.3. The standard deviation of the node voltage 1 is

$$\sigma_{V_1}^2 = \frac{1}{6.219^2} [(-5.782 \times 0.02964 + 2.290 \times 0.00358)^2 (3.33)^2$$
$$+ (-5.782 \times 0.00315 + 2.290 \times 0.01236)^2 (5.00)^2$$
$$+ (-5.782 \times 36425 - 2.290 \times 9277)^2 (10^{-6})^2$$
$$+ (5.782 \times 7.811 + 2.290 \times 8.305)^2 (0.01)^2]$$

$$\sigma_{V_1} = 0.140 \, V$$

The 3σ limits of the voltage are therefore

$$|V_1| = (5.80, 6.64) \, V$$

with the nominal value of 6.22 V (see Ex. 6.3).

The calculation for the standard deviation of V_2 and for the phase angle changes is left as an exercise.

EXAMPLE 6.6

Three resistors taken from the ones tabulated in Table 6.1 are used in the voltage divider circuit shown in Fig. 6.5. Find the expected variation in the output voltage.

Solution.

$$E_{out} = \frac{R_2 // R_3}{R_1 + R_2 // R_3} E_{in} = \frac{R_2 R_3}{R_1 R_2 + R_1 R_3 + R_2 R_3} E_{in}$$

The nominal value of E_{out} is

$$E_{out} = 10 \text{ V}$$

The variation is

$$\sigma_E = \sqrt{\sum_{i=1}^{3} \left(\frac{\partial E_{out}}{\partial R_i} \sigma_{R_i}\right)^2}$$

$$\frac{\partial E_{out}}{\partial R_1} = \frac{-R_2 // R_3}{(R_1 + R_2 // R_3)^2} E_{in}$$

$$= -\frac{49.6}{(148.9)^2} \times 30 = -0.067$$

$$\frac{\partial E_{out}}{\partial R_2} = \frac{\partial E_{out}}{\partial R_3} = 0.100$$

Since $\sigma_{R_i} = 4.36$, the standard deviation of the output voltage is

$$\sigma_E = 4.36 \sqrt{2 \times (0.1)^2 + (0.067)^2} = 0.681 \text{ V}$$

If one uses the rough approximations $R = 100$, $\sigma_R = 3.3$, the standard deviation becomes

$$\tilde{\sigma}_E = 3.3 \sqrt{2(0.1)^2 + (0.067)^2} = 0.516 \text{ V}$$

The formulas outlined in this section and the ones in the previous section form the basis for all small perturbance calculations. All that remains now is to design and implement a program for the indicated computations; this shall be done in Sec. 6.4. We shall now examine the problem of worst-case variations in a network.

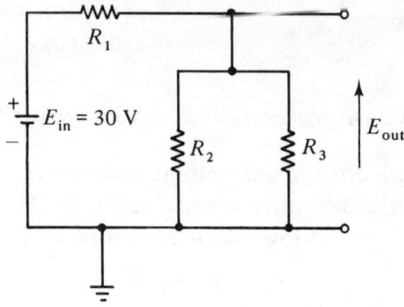

Fig. 6.5 Circuit for Example 6.6

6.3 Worst-Case Analysis

Usually when one tries to estimate the expected variations in the circuit performance parameters (voltages, currents, powers, etc.) data are not available on the distribution of the circuit parameters. Instead, the nominal value p_0 and the nominal percentage tolerance Δp, or the nominal value p_0, the lowest value $\underline{p}$ and the highest value $\bar{p}$ are known. Realistic estimates of the circuit function are hard to make from this scant data. One can, of course, always *assume* that the 3σ limits lie at the extreme values specified by the above number; and a standard deviation calculation may then be made. Such procedure can yield quite usable results, although usually the results are only very approximate.

The simplest tolerance calculation one can make does not involve any statistics on the distribution of the parameter values. One is merely concerned with the following.

If all parameters in the network take on the worst combination of possible values, what is the range of the circuit performance function, that is, the lowest and the highest value the function can take?

We note that in an actual circuit the probability is zero that all parameters are at their worst value. Thus, this kind of analysis gives only the range of possible, but not probable, node voltages (or other circuit functions.)

Conceptually, the problem is simple. For example, Fig. 6.6 shows the variation of V_1 in an arbitrary circuit as a function of two of the circuit parameters R_1 and R_2.

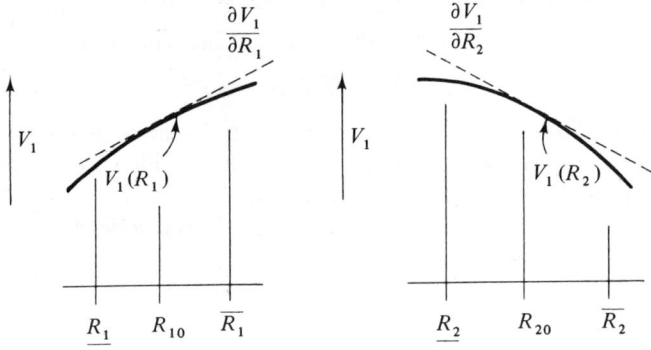

Fig. 6.6 Variation of Circuit Performance

We note that the derivative of the voltage with respect to R_1 is positive at the nominal value of R_1 and the derivative is negative for R_2. Hence to obtain the lowest value for V_1 we decrease R_1 to its lower limit and increase R_2 to its upper limit:

$$\underline{V_1} = V_1(\underline{R_1}, \overline{R_2}) \tag{6.79}$$

and the highest value of V_1 would be obtained from

$$\overline{V_1} = V_1(\overline{R_1}, \underline{R_2}) \tag{6.80}$$

Sec. 6.3   Worst-Case Analysis

In general, the worst lowest value would be obtained for a parameter by taking that parameter as

$$p|_{\text{lowest } f} = \begin{cases} \underline{p} & \text{if } \dfrac{\partial f}{\partial p} > 0 \\ \bar{p} & \text{if } \dfrac{\partial f}{\partial p} < 0 \end{cases} \tag{6.81}$$

The derivative is evaluated at the nominal value for p. For the highest value for f:

$$p|_{\text{highest } f} = \begin{cases} \bar{p} & \text{if } \dfrac{\partial f}{\partial p} > 0 \\ \underline{p} & \text{if } \dfrac{\partial f}{\partial p} < 0 \end{cases} \tag{6.82}$$

Note from the figure that the highest and lowest values of f may be calculated approximately from the slopes. Let

$$\Delta f_j = \begin{cases} \dfrac{\partial f}{\partial p_j}(\bar{p}_j - p_0) & \text{if } \dfrac{\partial f}{\partial p_j} > 0 \\ \dfrac{\partial f}{\partial p_j}(p_0 - \underline{p}_j) & \text{if } \dfrac{\partial f}{\partial p_j} < 0 \end{cases} \tag{6.83}$$

Then the highest value becomes

$$\bar{f} = f_0 + \sum_{j=1}^{P} \Delta f_j \tag{6.84}$$

Similarly,

$$\underline{f} = f_0 + \sum_{j=1}^{P} \nabla f_j \tag{6.85}$$

where

$$\nabla f_j = \begin{cases} \dfrac{\partial f}{\partial p_j}(\underline{p}_j - p_0) & \text{if } \dfrac{\partial f}{\partial p_j} > 0 \\ \dfrac{\partial f}{\partial p_j}(p_0 - \bar{p}_j) & \text{if } \dfrac{\partial f}{\partial p_j} < 0 \end{cases} \tag{6.86}$$

Equations (6.84) and (6.85) can be viewed as a first-order approximation to the true minimum and maximum values of f. As long as the function does not change much in the neighborhood of the nominal values for the parameters, the slope approximations are quite usable. However, consider the possible circuit function variations shown in Fig. 6.7.

In each of these cases the minimum and maximum values of V_1 do not occur for the smallest or the largest values of the variable parameter. Approximating the maximum and minimum values of V_1 by the slope as in Eqs. (6.84) and (6.85) will lead to wholly erroneous numerical values.

Let us examine the curve shown in Fig. 6.7(a). The performance curve changes slope once between the nominal value and either extreme value of the parameter. When an analysis for the variations of circuit functions is carried out, the nominal values are first calculated (by the methods outlined in Chapter 3). Then a sensitivity analysis may be done by using the formulas developed in Sec. 6.1. At this point a

276 Tolerance Analysis

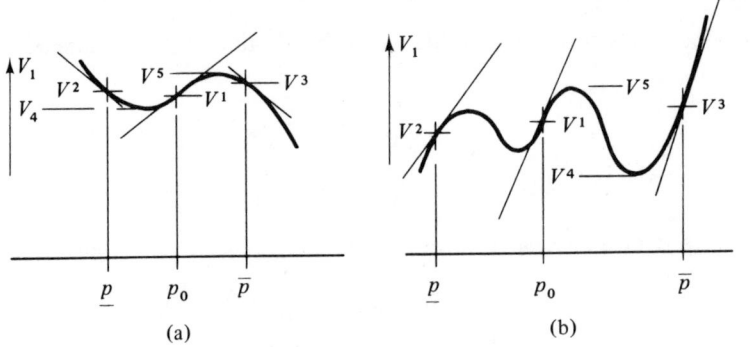

Fig. 6.7 Possible Circuit Performance Variations

slope approximation may be made, but the appropriateness of this approximation is not known until the curve shown in Figs. 6.6 or 6.7 has been calculated. Unfortunately, such luxury in numerical analysis can seldom be justified; it would require recalculation of the node equations many times over for each of the parameters in question.

Comparison of Figs. 6.6 and 6.7 shows that for very slowly varying functions of the independent parameters, the slope will not change radically. In this latter case a fair approximation can be expected by using the slope approximation. As a check one could recalculate the node equations having changed the parameters from their nominal values based on the slopes. That is, if the slope is positive, the parameter is taken at its largest value for the calculation of the highest value of the function. In this manner the sense of the required parameter changes is found; with these changed values the circuit is re-analyzed and the actual values of the function for the extreme parameter changes may be established. A check can then be made to see if the slope has changed. For small parameter variations no change in the sign of the slope can indicate a performance variation akin to what is shown in Fig. 6.6. A change in the sign of the slope will indicate that at least one local extremum of the function lies within the range of the variation of the parameter. Normally, if a sign changes, a message should be printed and the analysis may be abandoned.

Note in Fig. 6.7(b) that no change in the sign of the slope may not indicate that the function does not have a peak or a valley in the range of the parameter variation. We also remark that the sort of performance variation shown in Fig. 6.7 (b) is a very unusual circumstance. Normal responses should be no more complicated than the ones shown in Fig. 6.6 and Fig. 6.7 (a). In this latter case the extreme values may be calculated by estimating the parameter value from a knowledge of the slopes at $\bar{p}$, p_0, and $\underline{p}$.

For example, to estimate the highest value of V_1 for the function in Fig. 6.7 (b), one can plot the slopes as shown in Fig. 6.8.

$$A = \text{slope at } \underline{p}$$
$$B = \text{slope at } p_0$$
$$C = \text{slope at } \bar{p}$$

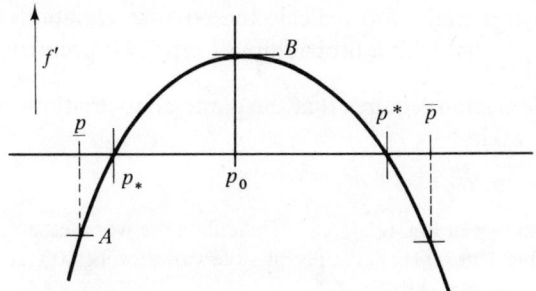

Fig. 6.8 Slope Variation of Fig. 6.7(a)

The value of p^*, the parameter value where the maximum value of V_1 is located may be established graphically, as is shown in Fig. 6.8.

Analytically, we may fit a second-degree equation (a parabolic equation) through the points indicated. Let this equation be

$$f^1 = ap^2 + bp + c \tag{6.87}$$

The values for a, b, and c may be calculated from the following set of simultaneous equations:

$$\begin{bmatrix} \bar{p}^2 & \bar{p} & 1 \\ p_0^2 & p_0 & 1 \\ \underline{p}^2 & \underline{p} & 1 \end{bmatrix} \begin{bmatrix} a \\ b \\ c \end{bmatrix} = \begin{bmatrix} A \\ B \\ C \end{bmatrix} \tag{6.88}$$

and the parameter value for approximating the extremum point is given by the solution of the quadratic equation

$$p^* = \frac{-b \pm \sqrt{b^2 - 4ac}}{2a} \tag{6.89}$$

The value of p^* must be in the range of $(p_0, \bar{p})$. If there are two values of p^* in the range of $(\underline{p}, \bar{p})$, one represents the location of the maximum value of the function, and the other represents the location of the minimum value.

Of course, obvious checks must be satisfied, such as

$$\underline{V_1} < V_1 < \overline{V_1} \tag{6.90}$$

Similarly, when no slope sign changes occur, the minimum and maximum values of the voltage predicted from the slope approximation cannot differ greatly from the actually calculated values using the worst-case mix of parameters. Otherwise, a situation similar to the one shown in Fig. 6.7 (b) may have been encountered.

The following points summarize the above considerations:

(1) Worst-case values for circuit functions are relatively easy to calculate if the variations are smooth and the slopes of the function may be calculated.

(2) Checks on the slope should be made at the indicated extreme combinations; if a change in slope occurs, caution is necessary.

278 Tolerance Analysis

(3) The worst-case limits will indicate the possible variations in the circuit performance (voltage usually) with a probability of zero of it occurring.

Nothing in this section requires that the parameter variations must be symmetric about the nominal value.

EXAMPLE 6.7

For the circuit shown in Fig. 6.2 (Ex. 6.3) calculate the worst-case variation of the magnitude of node voltage 1 at 60 Hz for a permissible variation of 10% in each of the circuit component values. Compare with Ex. 6.5.

Solution. From Ex. 6.3

$$V_1 = A_1 + jB_1 = 5.782 + j2.290 \text{ V}$$

$$|V_1| = 6.219 \text{ V}$$

$$\frac{\partial V_1}{\partial R_1} = \frac{\partial A_1}{\partial R_1} + j\frac{\partial B_1}{\partial R_1} = -0.02964 + j0.00358$$

$$\frac{\partial V_1}{\partial R_2} = \frac{\partial A_1}{\partial R_2} + j\frac{\partial B_1}{\partial R_2} = -0.00315 + j0.01236$$

$$\frac{\partial V_1}{\partial C} = \frac{\partial A_1}{\partial C} + j\frac{\partial B_1}{\partial C} = -36425 - j9277$$

$$\frac{\partial V_1}{\partial L} = \frac{\partial A_1}{\partial L} + j\frac{\partial B_1}{\partial L} = 7.811 - j8.305$$

$$\frac{\partial |V_1|}{\partial R_1} = \frac{1}{|V_1|}\left(A_1\frac{\partial A_1}{\partial R_1} + B_1\frac{\partial B_1}{\partial R_1}\right)$$

$$= \frac{1}{6.219}(-5.782 \times 0.02964 + 2.290 \times 0.00358) = -0.0262$$

Similarly, we calculate

$$\frac{\partial |V_1|}{\partial R_2} = 0.00162$$

$$\frac{\partial |V_1|}{\partial C} = -37300$$

$$\frac{\partial |V_1|}{\partial L} = 4.21$$

Hence for the upper limit on $|V_1|$ we take

$$R_1 = 100 - 10\% (100) = 90 \text{ }\Omega$$
$$R_2 = 150 + 10\% (150) = 165 \text{ }\Omega$$
$$C = 30 \text{ }\mu\text{f} - 10\% (30 \text{ }\mu\text{f}) = 27 \text{ }\mu\text{f}$$
$$L = 0.3 + 10\% (0.3) = 0.33 \text{ h}$$

For the lower limit on $|V_1|$ we take

$$R_1 = 110 \text{ }\Omega$$
$$R_2 = 135 \text{ }\Omega$$
$$C = 33 \text{ }\mu\text{f}$$
$$L = 0.27 \text{ h}$$

With these values the circuit is solved again. The results are
$$\overline{|V_1|} = 6.740 \text{ V}$$
$$\underline{|V_1|} = 5.692 \text{ V}$$

The slopes are recalculated for these two circuit component variations (the program of Sec. 6.4 was used in this part); the slopes are:

$$\begin{array}{ccc} & \text{at } \underline{|V_1|} & \text{at } \overline{|V_1|} \\ \dfrac{\partial |V_1|}{\partial R_1} = & -0.0252; & = -0.0269 \\ \dfrac{\partial |V_1|}{\partial R_2} = & 0.00107; & = 0.00473 \\ \dfrac{\partial |V_1|}{\partial C} = & -35060; & = -59200 \\ \dfrac{\partial |V_1|}{\partial L} = & 5.33; & = 3.20 \end{array}$$

The slope approximation yields

$$\underline{|V_1|} = |V_1| + \frac{\partial |V_1|}{\partial R_1}(10.0) - \frac{\partial |V_1|}{\partial R_2}(15.0)$$
$$+ \frac{\partial |V_1|}{\partial C}(3 \times 10^{-6}) - \frac{\partial |V_1|}{\partial L}(3 \times 10^{-2})$$
$$= 5.695 \text{ V}$$

Similarly, by the slope approximation we get

$$\overline{|V_1|} = 6.743 \text{ V}$$

We conclude from this example that satisfactory answers were obtained for the worst cases and that most likely no slope changes have occurred.

A program for simple tolerance analysis is discussed in the next section.

6.4 A Simple Sensitivity Program

The formulas developed in this chapter may be readily coded into a program for the calculation of sensitivities. Caution is in order, however, because the large amount of results that can be generated by this program may snow the user. We will not attempt here to provide a means of compacting or editing the output; any practical program will have to incorporate such a feature if it is to be accepted by a wide class of potential users.

We note from Eqs. (6.26) through (6.30) that the inverse of the node-admittance matrix Y_n is required in the calculation of the node-voltage derivatives. Also the branch voltages are needed: these are obtained from the node voltages. In Chapter 3 the node voltages were found by solving the node-voltage equations, not by inverting the matrix Y_n. We note that the solution of simultaneous equations is faster than inversion (see Sec. 2.3); thus in Chapter 3, in which only the node voltages were sought, simultaneous equation solution was used. In the sensitivity program only the inversion subroutine need be used; we shall also present here a program for the general case for all types of circuit elements 0 through 10 listed in Table 3.5. In the AC analysis program shown in Listing 3.3 no dependent voltage sources were allowed; the program outlined here starts with the general AC analysis program introduced in Problems 3.11, 3.12, and 3.13. This part of the program is shown in Listing 6.1 in the code from the start of the program to statement number 460 and the statements at the very end of the main program where a check is made on the highest frequency in the analysis (last three FORTRAN statements in the main routine).

The pertinent analysis equations are repeated here for reference.

$$Y_n = AY_bA^T$$
$$V_n = Y_n^{-1}A(I_g - Y_bV_g)$$
$$Y_b = (U + B^* + GZ_e)Z_e^{-1}(U + D^* + R_tZ_e^{-1})^{-1}$$

For the definition of the various matrices, see Sec. 3.2.

The AC analysis program based on the above equations is indicated on the flowchart for the sensitivity program, Fig. 6.9, in the blocks enclosed in the dashed lines. Note from Listing 6.1 that no attempt is made to economize on any storage. For example, the **A** matrix requires 20 by 30 computer words (of which at most 60 are used); the **BSTAR** matrix (as **DSTAR**, **RT**, and **G**) are full 30 by 30 matrices of which again just a few may contain non-zero entries; **ZE** is likewise not economized.

Another drawback of the program listed is its slow execution; the comments in Sec. 3.7 on storage and execution times apply here. This program can, of course, be speeded up considerably; the thrust of this chapter is in exploring the theoretical bases and presenting workable programs rather than on exploring programming details. The interested reader will find Secs. 3.7 through 3.11 profitable in exploring some of the programming details of capable sensitivity programs.

However, as a consequence of these decisions, the program shown here is quite general: Any branch may be coupled to any other branch by any combination of mutual inductances, controlled voltage sources, and controlled current sources.

The flowchart of the program shown in Fig. 6.9 is quite straightforward. The equations used in the calculations are indicated on the flowchart. Three changes, each slight, are made to the AC analysis program:

(1) Inversion is used to get Y_n^{-1} and then the $(N + 1)$st column of Y_n, the I_s vector of Chapter 3, is used to calculate the node voltages. The node voltages are converted to magnitude and phase angle (degrees) for printout.

(2) The matrix **Q** of Eq. (6.27) is produced in the AC analysis during the calculation of Y_b. At statement number 330 this partial result is stored for use later in the program.

(3) Calculation of the derivatives, sensitivities, and voltage changes requires various pointers (branch-to-branch pointers k and j) and values of elements [see Eqs. (6.34), (6.36) and (6.61) as examples]. These numbers occur in the input data and must be used at the end of the analysis to pull out given rows and columns and to

Sec. 6.4 A Simple Sensitivity Program

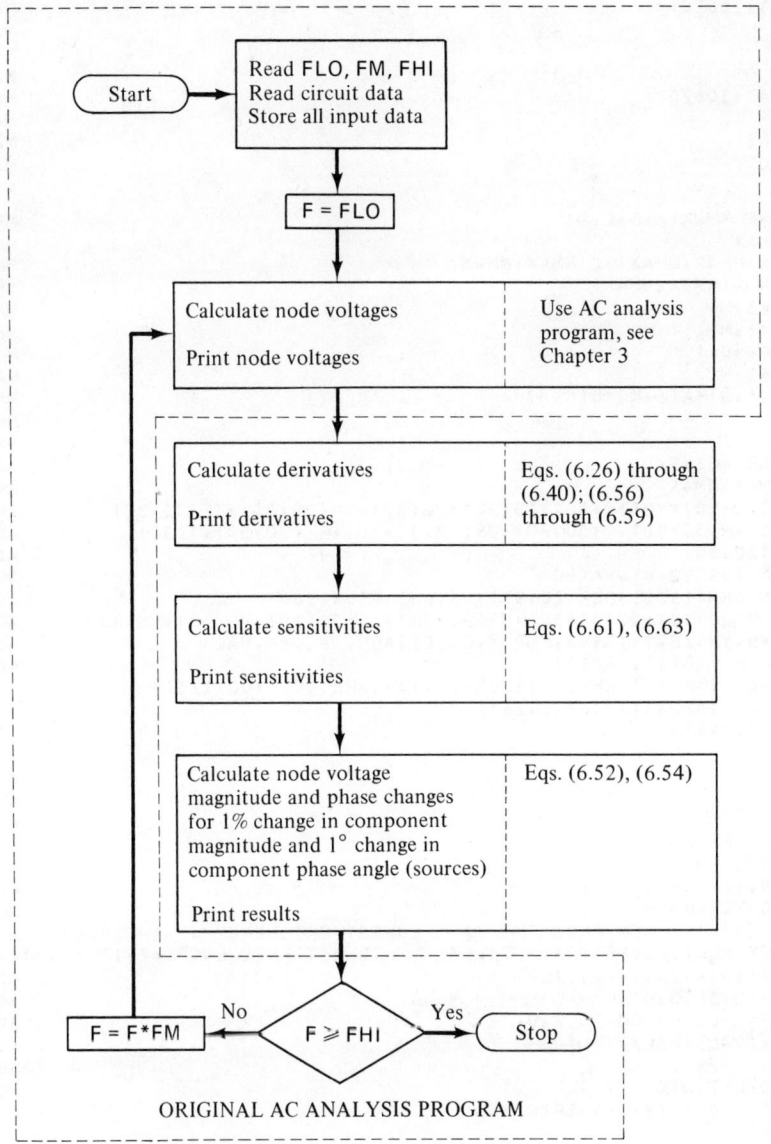

Fig. 6.9 Flowchart of AC Sensitivity Program

calculate sensitivities. Hence the entire input data stream is stored; this latter operation is not required if only the node voltages were sought.

In Listing 6.1 all derivative calculations are done in the DO-loop beginning after statement 510; the necessary matrices **P**, **Q**, **R**, **S**, and **V*** [Eqs. (6.26) through (6.30)] are calculated from statement number 460 through 510. In the program double

```
C           J. STAUDHAMMER NCSU EE503  SENSITIVITY CALCULATIONS         MAIN  10
C           MAIN ROUTINE FOR GENERASENSITIVITY PROGRAM.                 MAIN  20
            COMMON MMAX,NMAX                                            MAIN  30
            COMMON ST(10620)                                            MAIN  40
            NMAX=20                                                     MAIN  50
            MMAX=30                                                     MAIN  60
            MAXM=50                                                     MAIN  70
        100 DO 110 I=1,10620                                            MAIN  80
        110 ST(I)=0.                                                    MAIN  90
            CALL ACSNS                                                  MAIN 100
            GO TO 100                                                   MAIN 110
            END                                                         MAIN 120
            SUBROUTINE MMUX(A,B,C,M)                                    MMUX  10
            COMMON MMAX                                                 MMUX  20
            COMPLEX B(MMAX,MMAX),C(MMAX,MMAX)                           MMUX  30
            DIMENSION A(MMAX,MMAX)                                      MMUX  40
            DO 100 I=1,M                                                MMUX  50
            DO 100 J=1,M                                                MMUX  60
            C(I,J)=(0.,0.)                                              MMUX  70
            DO 100 K=1,M                                                MMUX  80
        100 C(I,J)=C(I,J)+A(I,K)*B(K,J)                                 MMUX  90
            RETURN                                                      MMUX 100
            END                                                         MMUX 110
            SUBROUTINE ACSNS                                            ACSN  10
            COMMON MMAX,NMAX                                            ACSN  20
            COMMON A(20,30),YB(30,30),IG(30),VG(30),YN(20,21),ZE(30,30) ACSN  30
            COMMON BSTAR(30,30),G(30,30),DSTAR(30,30),RT(30,30),T(30)   ACSN  40
            COMMON Q(30,30)                                             ACSN  50
            DIMENSION IJX(90,4),VA(90)                                  ACSN  60
            DIMENSION ARR(1800),DER(20,90),VR(60),AIJX(90)              ACSN  70
            COMPLEX PP(20,30),QQ(30,30),RR(30,30),SS(30,30),VS(30),VB(30) ACSN 80
            COMPLEX YB,YN,ZE,T,IG,VG,CMPLX,Q,DET,ARR,XM,DER,VAL         ACSN  90
            EQUIVALENCE (VB(1),VR(1))                                   ACSN 100
            EQUIVALENCE (ARR(1),RR(1,1),DER(1,1)),(ARR(901),QQ(1,1))    ACSN 110
            EQUIVALENCE (AIJX(1),IJX(1,2))                              ACSN 120
            CON=180./3.141593                                           ACSN 130
            NMM=0                                                       ACSN 140
            L=0                                                         ACSN 150
            N=0                                                         ACSN 160
            M=0                                                         ACSN 170
            READ 95,FL,FM,FH                                            ACSN 180
         95 FORMAT (3E10.0)                                             ACSN 185
            IF (FL.EQ.0.) CALL EXIT                                     ACSN 190
            PRINT 100,FL,FM,FH                                          ACSN 200
        100 FORMAT ('1'//'     NEW PROBLEM'//'   LOWEST FREQUENCY =',1PE18.5/' ACSN 210
           1 FREQUENCY MULTIPLIER =',1PE14.5/'   HIGHEST FREQUENCY =',1PE17.5)ACSN 220
        110 READ 115,I,J,K,IX,VAL1,VAL2                                 ACSN 240
        115 FORMAT (4I3,3E10.0)                                         ACSN 245
            IF ((I+J+K).EQ.0) GO TO 210                                 ACSN 250
            VAL=CMPLX(VAL1,VAL2)                                        ACSN 260
            L=L+1                                                       ACSN 270
            PRINT 120,L,I,J,K,IX,VAL                                    ACSN 280
        120 FORMAT ('    NO.',I3,4I5,1P2E15.5)                          ACSN 290
            IJX(L,1)=I                                                  ACSN 300
            IJX(L,2)=J                                                  ACSN 310
            IJX(L,3)=K                                                  ACSN 320
            IJX(L,4)=IX                                                 ACSN 330
            VA(L)=VAL1                                                  ACSN 340
            IF ((IX.EQ.4).OR.(IX.EQ.5))AIJX(L)=VAL2                     ACSN 350
C           VAL2 IS THE PHASE ANGLE IN DEGREES.                         ACSN 360
            IF (IX.NE.0) GO TO 130                                      ACSN 370
            IX=1                                                        ACSN 380
            VAL=1./VAL                                                  ACSN 390
        130 GO TO (140,140,140,150,150,170,180,190,200,110),IX          ACSN 400
```

Listing 6.1 General Sensitivity Program

```
      140 N=MAXO(N,J,K)                                  ACSN 410
C           PASSIVE COMPONENT.                           ACSN 420
          M=MAXO(M,I)                                    ACSN 430
          IF (J.NE.0)A(J,I)=1.                           ACSN 440
          IF (K.NE.0)A(K,I)=-1.                          ACSN 450
C           CONNECTION DATA                              ACSN 460
          GO TO 110                                      ACSN 470
      150 ANG=VAL2/CON                                   ACSN 480
          RLVAL=VAL1*COS(ANG)                            ACSN 490
          AIVAL=VAL1*SIN(ANG)                            ACSN 500
          IF (IX.EQ.5) GO TO 160                         ACSN 510
          VG(I)=CMPLX(RLVAL,AIVAL)                       ACSN 520
          GO TO 110                                      ACSN 530
      160 IG(I)=CMPLX(RLVAL,AIVAL)                       ACSN 540
          GO TO 110                                      ACSN 550
      170 BSTAR(K,J)=VAL                                 ACSN 560
          GO TO 110                                      ACSN 570
      180 G(K,J)=VAL                                     ACSN 580
          GO TO 110                                      ACSN 590
      190 RT(K,J)=VAL                                    ACSN 600
          GO TO 110                                      ACSN 610
      200 DSTAR(K,J)=VAL                                 ACSN 620
          GO TO 110                                      ACSN 630
      210 IF ((M+N).EQ.0) CALL EXIT                      ACSN 640
          N1=N+1                                         ACSN 650
          LCM=L                                          ACSN 660
          DO 220 I=1,M                                   ACSN 670
          DSTAR(I,I)=1.                                  ACSN 680
      220 BSTAR(I,I)=1.                                  ACSN 690
          F=FL                                           ACSN 700
      230 W=F*6.28318                                    ACSN 710
          DO 240 I=1,M                                   ACSN 720
          DO 240 J=1,M                                   ACSN 730
      240 ZE(I,J)=(0.,0.)                                ACSN 740
          DO 300 I=1,LCM                                 ACSN 750
          IX=IJX(I,4)                                    ACSN 760
          IF (IX.EQ.10) GO TO 290                        ACSN 770
          IF (IX.GT.3) GO TO 300                         ACSN 780
          IX=IX+1                                        ACSN 790
          IB=IJX(I,1)                                    ACSN 800
          GO TO (250,260,270,280),IX                     ACSN 810
      250 ZE(IB,IB)=1./VA(I)                             ACSN 820
          GO TO 300                                      ACSN 830
      260 ZE(IB,IB)=VA(I)                                ACSN 840
          GO TO 300                                      ACSN 850
      270 ZE(IB,IB)=CMPLX(0.,-1./(W*VA(I)))              ACSN 860
          GO TO 300                                      ACSN 870
      280 ZE(IB,IB)=CMPLX(0.,W*VA(I))                    ACSN 880
          GO TO 300                                      ACSN 890
      290 J=IJX(I,2)                                     ACSN 900
          K=IJX(I,3)                                     ACSN 910
          ZE(J,K)=CMPLX(0.,W*VA(I))                      ACSN 920
          ZE(K,J)=ZE(J,K)                                ACSN 930
      300 CONTINUE                                       ACSN 940
          CALL MMUX(DSTAR,ZE,YB,M)                       ACSN 950
          DO 320 I=1,M                                   ACSN 960
          DO 320 J=1,M                                   ACSN 970
      320 YB(I,J)=YB(I,J)+RT(I,J)                        ACSN 980
          CALL INVRX(YB,MMAX,M,DET,T)                    ACSN 990
C           IN-PLACE INVERSION, COMPLEX                  ACSN1000
          DO 330 I=1,M                                   ACSN1010
          DO 330 J=1,M                                   ACSN1020
      330 QQ(I,J)=YB(I,J)                                ACSN1030
          CALL MMUX(G,ZE,Q,M)                            ACSN1040
          DO 340 I=1,M                                   ACSN1050
```

Listing 6.1—*Cont*.

```
      DO 340 J=1,M                                          ACSN1060
  340 Q(I,J)=Q(I,J)+BSTAR(I,J)                              ACSN1070
      DO 360 K=1,M                                          ACSN1080
      DO 350 I=1,M                                          ACSN1090
      T(I)=0.                                               ACSN1100
      DO 350 J=1,M                                          ACSN1110
  350 T(I)=T(I)+YB(J,K)*Q(I,J)                              ACSN1120
      DO 360 J=1,M                                          ACSN1130
  360 YB(J,K)=T(J)                                          ACSN1140
C        YB-MATRIX IS COMPLETE.                             ACSN1150
      DO 390 K=1,N                                          ACSN1160
      DO 370 I=1,M                                          ACSN1170
      T(I)=0.                                               ACSN1180
      DO 370 J=1,M                                          ACSN1190
  370 T(I)=T(I)+YB(J,I)*A(K,J)                              ACSN1200
      YN(K,N1)=(0.,0.)                                      ACSN1210
      DO 380 I=1,M                                          ACSN1220
  380 YN(K,N1)=YN(K,N1)-T(I)*VG(I)                          ACSN1230
      DO 390 L=1,N                                          ACSN1240
      YN(K,L)=(0.,0.)                                       ACSN1250
      DO 390 I=1,M                                          ACSN1260
  390 YN(K,L)=YN(K,L)+T(I)*A(L,I)                           ACSN1270
      DO 400 I=1,N                                          ACSN1280
      DO 400 J=1,M                                          ACSN1290
  400 YN(I,N1)=YN(I,N1)+A(I,J)*IG(J)                        ACSN1300
C        NODE EQUATIONS ARE COMPLETE.                       ACSN1310
      IF (F.GT.FL) GO TO 440                                ACSN1320
      PRINT 410                                             ACSN1330
  410 FORMAT (//'   THE NODE EQUATIONS ARE'/)               ACSN1340
      DO 420 I=1,N                                          ACSN1350
  420 PRINT 430,(YN(I,J),J=1,N1)                            ACSN1360
  430 FORMAT (1P8E15.5)                                     ACSN1370
  440 CALL INVRX(YN,NMAX,N,DET,T)                           ACSN1380
C        YN CONTAINS THE INVERSE NODE ADMITTANCE MATRIX.    ACSN1390
      DO 450 I=1,N                                          ACSN1400
      T(I)=(0.,0.)                                          ACSN1410
      DO 450 K=1,N                                          ACSN1420
  450 T(I)=T(I)+YN(I,K)*YN(K,N1)                            ACSN1430
C        T CONTAINS THE NODE VOLTAGE SOLUTION.              ACSN1440
      PRINT 460,F                                           ACSN1450
  460 FORMAT (//'   FREQUENCY =',1PE12.4/'   NODE VOLTAGES')ACSN1460
      CALL POUT(LCM,N,DER,VB,VR,T,20,VAL1,0.)               ACSN1470
      DO 470 I=1,N                                          ACSN1480
      DO 470 J=1,M                                          ACSN1490
      PP(I,J)=(0.,0.)                                       ACSN1500
      DO 470 K=1,N                                          ACSN1510
  470 PP(I,J)=PP(I,J)+YN(I,K)*A(K,J)                        ACSN1520
      DO 480 I=1,M                                          ACSN1530
      DO 480 J=1,M                                          ACSN1540
      RR(I,J)=G(I,J)                                        ACSN1550
      DO 480 K=1,M                                          ACSN1560
  480 RR(I,J)=RR(I,J)-YB(I,K)*DSTAR(K,J)                    ACSN1570
      DO 490 I=1,N                                          ACSN1580
      DO 490 J=1,M                                          ACSN1590
      SS(I,J)=(0.,0.)                                       ACSN1600
      DO 490 K=1,M                                          ACSN1610
  490 SS(I,J)=SS(I,J)+PP(I,K)*RR(K,J)                       ACSN1620
      DO 500 I=1,M                                          ACSN1630
      VB(I)=VG(I)                                           ACSN1640
      DO 500 J=1,N                                          ACSN1650
  500 VB(I)=VB(I)+A(J,I)*T(J)                               ACSN1660
      DO 510 I=1,N                                          ACSN1670
      VS(I)=(0.,0.)                                         ACSN1680
      DO 510 J=1,M                                          ACSN1690
```

Listing 6.1—*Cont.*

```
  510 VS(I)=VS(I)+QQ(I,J)*VB(J)                                ACSN1700
      KFL=1                                                    ACSN1710
      DO 770 L=1,LCM                                           ACSN1720
      IB=IJX(L,1)                                              ACSN1730
      J=IJX(L,3)                                               ACSN1740
      K=IJX(L,2)                                               ACSN1750
      IX=IJX(L,4)                                              ACSN1760
      VAR=VA(L)                                                ACSN1770
      IF (IX.EQ.0) GO TO 520                                   ACSN1780
      GO TO (530,540,550,710,620,640,660,690,660,550),IX       ACSN1790
  520 XM=1./VAR**2                                             ACSN1800
      GO TO 560                                                ACSN1810
  530 XM=-1.                                                   ACSN1820
      GO TO 560                                                ACSN1830
  540 XM=CMPLX(0.,-1./(W*VAR*VAR))                             ACSN1840
      GO TO 560                                                ACSN1850
  550 XM=CMPLX(0.,-W)                                          ACSN1860
      IF (IX.EQ.10) GO TO 580                                  ACSN1870
  560 DO 570 I=1,N                                             ACSN1880
  570 DER(I,L)=XM*SS(I,IB)*VS(IB)                              ACSN1890
      GO TO 760                                                ACSN1900
  580 DO 590 I=1,N                                             ACSN1910
  590 DER(I,L)=XM*(SS(I,J)*VS(J)+SS(I,K)*VS(K))                ACSN1920
      GO TO 760                                                ACSN1930
  600 DO 610 I=1,N                                             ACSN1940
  610 DER(I,L)=-Q(I,IB)                                        ACSN1950
C         INDEPENDENT VOLTAGES.                                ACSN1960
      GO TO 760                                                ACSN1970
  620 DO 630 I=1,N                                             ACSN1980
  630 DER(I,L)=PP(I,IB)                                        ACSN1990
C         INDEPENDENT CURRENTS                                 ACSN2000
      GO TO 760                                                ACSN2010
  640 DO 650 I=1,N                                             ACSN2020
C         BETA                                                 ACSN2030
  650 DER(I,L)=-PP(I,J)*VS(K)                                  ACSN2040
      GO TO 760                                                ACSN2050
  660 XM=0.                                                    ACSN2060
C         GM VALUES                                            ACSN2070
      DO 670 I=1,M                                             ACSN2080
  670 XM=XM+ZE(K,I)*VS(I)                                      ACSN2090
      IF (IX.EQ.9) GO TO 710                                   ACSN2100
      DO 680 I=1,N                                             ACSN2110
  680 DER(I,L)=-PP(I,J)*XM                                     ACSN2120
      GO TO 760                                                ACSN2130
  690 DO 700 I=1,N                                             ACSN2140
  700 DER(I,L)=Q(I,J)*VS(K)                                    ACSN2150
      GO TO 760                                                ACSN2160
  710 GO TO (720,740),KFL                                      ACSN2170
  720 KFL=2                                                    ACSN2180
C         SET Q=PP*YB  USED WITH VG,RT,AND MU VALUES.          ACSN2190
      DO 730 I=1,N                                             ACSN2200
      DO 730 I1=1,M                                            ACSN2210
      Q(I,I1)=0.                                               ACSN2220
      DO 730 I2=1,M                                            ACSN2230
  730 Q(I,I1)=Q(I,I1)+PP(I,I2)*YB(I2,I1)                       ACSN2240
  740 IF (IX.EQ.4) GO TO 600                                   ACSN2250
      IF (IX.EQ.8) GO TO 690                                   ACSN2260
      DO 750 I=1,N                                             ACSN2270
  750 DER(I,L)=Q(I,J)*XM                                       ACSN2280
  760 VANG=0.                                                  ACSN2290
      IF ((IX.EQ.4).OR.(IX.EQ.5))VANG=AIJX(L)/CON              ACSN2300
      CALL POUT(L,N,DER,VB,VR,T,IX,VAR,VANG)                   ACSN2310
  770 CONTINUE                                                 ACSN2320
      IF (F.GE.FH) RETURN                                      ACSN2330
```

Listing 6.1—*Cont.*

```
            F=F*FM                                              ACSN2340
         GO TO 230                                              ACSN2350
         END                                                    ACSN2360
         SUBROUTINE POUT(LCM,N,DER,VB,VR,VN,IX,VAL,ANGV)        POUT  10
         DIMENSION DER(20,90),HEAD(5,3),VB(30),VR(60),VN(20),ITYP(3)  POUT  20
         COMPLEX DER,CMPLX,VB,T,VN,VM                           POUT  30
         DATA HEAD/'DERI','VATI','VES ',' ',' ',' ','SENS','ITIV','ITIE',POUT  40
        1'S  ',' ',' ','CHAN','GE I',' ','N VO','LTAG','E   '/  POUT  50
         DATA LBL1,LBL2,LBL3,LBL4/'REAL','IMAG','MAG.','PHA.'/  POUT  60
         DATA ITYP/'    ','MAG.','PHA.'/                        POUT  70
         ANG(T)=ATAN2(AIMAG(T),REAL(T))                         POUT  80
         CON=180./3.14159                                       POUT  90
         IT=1                                                   POUT 100
         IF (IX.EQ.20) GO TO 260                                POUT 110
         VM=CMPLX(COS(ANGV),SIN(ANGV))                          POUT 120
         IF ((IX.EQ.4).OR.(IX.EQ.5))IT=2                        POUT 130
     100 DO 110 I=1,N                                           POUT 140
     110 VB(I)=DER(I,LCM)*VM                                    POUT 150
         PRINT 130,LCM,ITYP(IT)                                 POUT 160
     130 FORMAT (/' PART NO.',I3,2X,A4)                         POUT 170
         KIND=1                                                 POUT 180
         LAB1=LBL1                                              POUT 190
         LAB2=LBL2                                              POUT 200
     150 PRINT 160,(HEAD(I,KIND),I=1,5)                         POUT 210
     160 FORMAT (/2X,5A4)                                       POUT 220
     170 ILO=1                                                  POUT 230
     180 IHI=MIN0(N,ILO+5)                                      POUT 240
         PRINT 190,LAB1,(I,VR(2*I-1),I=ILO,IHI)                 POUT 250
     190 FORMAT (/2X,A4,6(I4,' = ',1PE12.4))                    POUT 260
         PRINT 200,LAB2,(VR(2*I),I=ILO,IHI)                     POUT 270
     200 FORMAT (2X,A4,1P6E19.4)                                POUT 280
         ILO=IHI+1                                              POUT 290
         IF (ILO.LE.N) GO TO 180                                POUT 300
         GO TO (210,230,250),KIND                               POUT 310
     210 KIND=2                                                 POUT 320
         LAB1=LBL3                                              POUT 330
         LAB2=LBL4                                              POUT 340
         DO 220 I=1,N                                           POUT 350
         VB(I)=DER(I,LCM)*VAL/100.                              POUT 360
         IF (IT.EQ.3)VB(I)=DER(I,LCM)*VM/CON                    POUT 370
         TTT=CABS(VB(I))                                        POUT 380
         TT=ANG(VB(I))*CON                                      POUT 390
     220 VB(I)=CMPLX(TTT,TT)                                    POUT 400
         GO TO 150                                              POUT 410
     230 KIND=3                                                 POUT 420
         DO 240 I=1,N                                           POUT 430
         ANG1=ANG(VN(I))                                        POUT 440
         ANG2=VR(2*I)/CON                                       POUT 450
         ANG12=ANG2-ANG1                                        POUT 460
         TT=VR(2*I-1)*COS(ANG12)                                POUT 470
         TTT=ATAN2(VR(2*I-1)*SIN(ANG12),TT+CABS(VN(I)))*CON     POUT 480
     240 VB(I)=CMPLX(TT,TTT)                                    POUT 490
         GO TO 150                                              POUT 500
     250 IF (IT.NE.2) RETURN                                    POUT 510
         IT=3                                                   POUT 520
         VM=VAL*VM*(0.,1.)                                      POUT 530
         GO TO 100                                              POUT 540
     260 KIND=3                                                 POUT 550
         LAB1=LBL3                                              POUT 560
         LAB2=LBL4                                              POUT 570
         DO 270 I=1,N                                           POUT 580
     270 VB(I)=CMPLX(CABS(VN(I)),ANG(VN(I))*CON)                POUT 590
         GO TO 170                                              POUT 600
         END                                                    POUT 610
```

Listing 6.1—*Cont.*

letters were used for the matrices (for example, **P** of Eq. (6.26) becomes array **PP** in the program); the vector **V*** is simply called **VS**. The derivative calculations end with statement 760; **SUBROUTINE POUT** is then called to print the derivative, calculate the sensitivity, and calculate the node-voltage changes for a 1% change in component magnitudes and a 1° change in generator phase angles. The derivatives are stored in the matrix **DER**; all calculations and all printouts are done for a component's influence on all the node voltages.

Observe that a very large amount of printout is generated even for modest problems.

The program input is slightly different from the one shown in Listing 3.2 (AC analysis program), but it is not significantly changed. The input data must be structured as follows:

(1) One card containing three floating-point numbers: the lowest frequency, the frequency multiplier, and the highest frequency for which the analysis is to be done. The **FORMAT** is 3E10.0. If the first number is zero, the program **STOP**s.

(2) Each circuit component is then described on a separate card in **FORMAT** (4I3, 2E10.0).

(a) Passive components: branch number, from-node number, to-node number, type number (0, 1, 2, or 3), value.

(b) Sources: branch number, two unused fields of three digits each, type number (4 or 5), magnitude, phase angle in degrees.

(c) Controlled sources and mutual inductances: one unused integer of three digits, from-branch number, to-branch number, type number (6, 7, 8, 9, or 10), value.

(3) One blank card must follow the last circuit component specified. This card initiates execution of the program.

Different circuits may be analyzed by stacking the complete analysis jobs, described in the formats above, one behind another. Note that following the last circuit component specification card will be a single blank card, which initiates execution of the circuit analysis run; jobs are thus separated by a single blank card.

For each frequency the node voltages, then for each component the derivative, the sensitivity, and then the change in each node-voltage magnitude and phase angle for a 1% change in component magnitude and a 1° change in source phase angle are printed. The derivatives are stored.

The listed program can handle circuits containing 90 components, 30 branches, and 20 nodes.

The program described in Fig. 6.9 may be extended relatively easily to the calculation of standard deviations of node voltages and to worst-case analysis. The basic flowchart for such a program is shown in Fig. 6.10.

In addition to the circuit interconnections and the nominal values of the circuit parameters, data must be input for the worst-case variations and/or the standard deviations of the circuit components. Hence the input data requirements are more

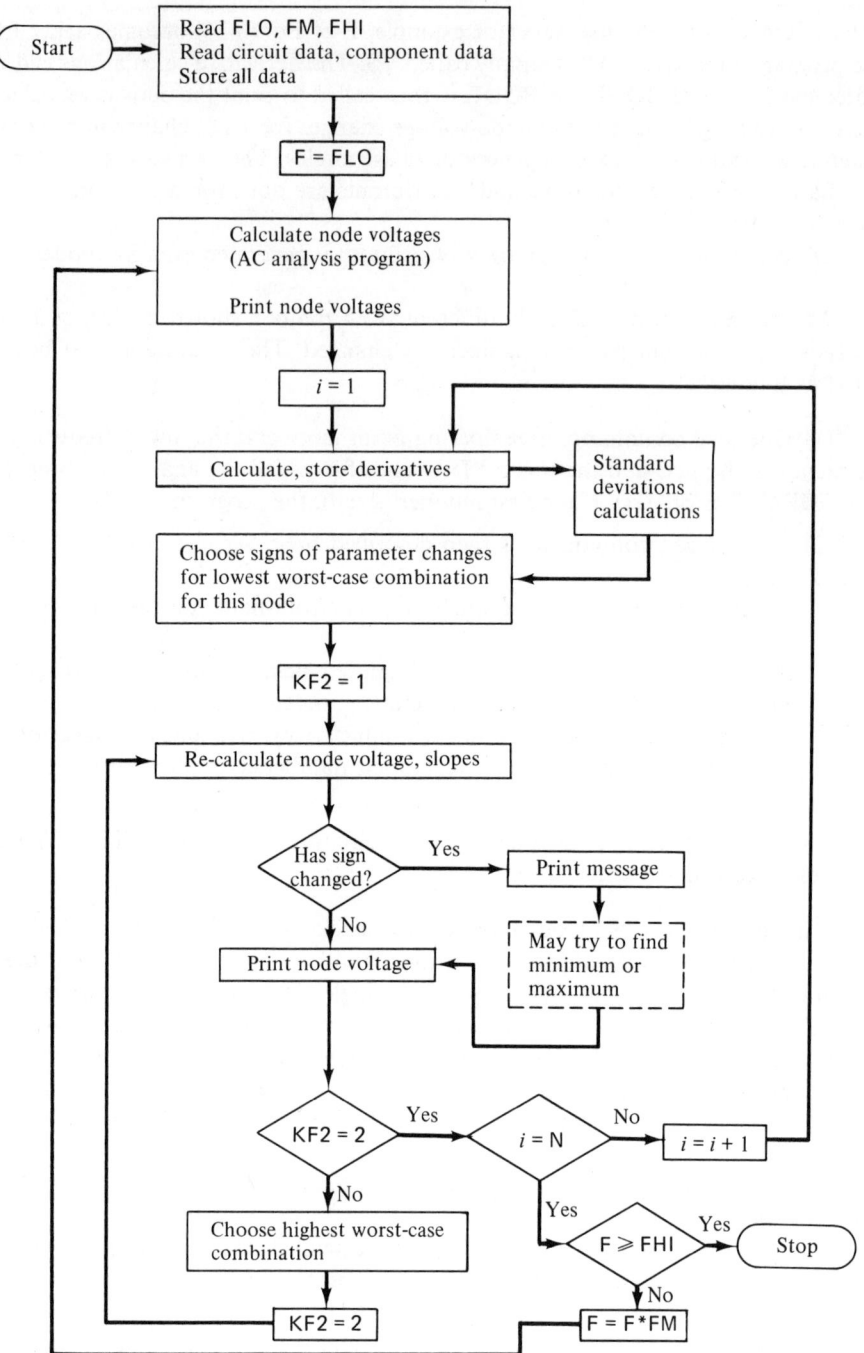

Fig. 6.10 Flowchart for Standard Deviation and Worst-Case Program

extensive than the input provided to the sensitivity calculations. In order to reduce the total input required and the amount of calculations, it is advisable to require only the varying components to be specified in addition to the nominal circuit values. Then the calculations need be done only for those varying components; also only these varying components need be stored.

One can specify the component values, for example, by three numbers: (1) the nominal value p; (2) the lowest worst-case value $\underline{p}$; (3) the highest worst-case value $\bar{p}$. These three numbers can also be used to specify fixed (non-varying) component values by letting p be a blank field on input; a symmetric tolerance can be specified by using a blank field for $\bar{p}$; in this case $\underline{p}$ can denote the nominal tolerance. The tolerance is conventionally taken as the $\pm 3\sigma$ limit of the parameter, i.e., 99.7% of all components will fall within the specified tolerance about the nominal value. In this scheme of input the tolerance t for the case where $\underline{p}$ and $\bar{p}$ are specified can be taken as

$$t = \frac{(\bar{p} - \underline{p})}{2} \tag{6.91}$$

Alternately, one could use a fourth input number to denote a σ value for components whose nominal and worst-case values are specified by the first three input numbers. In any case the input will become a more involved code than the simple READ statement used in Listing 6.1.

Details of coding this program are left as an exercise. Note that it is relatively easy to modify the program given in Listing 6.1 to this task (see Problem 6.15).

6.5 Correlated Parameters

In Sec. 6.2 the standard deviation of a node-voltage magnitude and a node-voltage phase angle were given [Eqs. (6.77) and (6.78), respectively]. The two formulas require that the variations of the various parameters be independent of one another. Sometimes variations of parameters within a circuit are related, at least in a statistical sense. For example, the output capacitance and the output resistance of transistors are somewhat related. (In statistics texts an often quoted pair of related parameters are heights and weights of people; it is obvious that taller people tend to be heavier).

The correlation matrix **R** is used in statistical work to relate the predictability of a variable (j) from a knowledge of another variable (k). For example, let us relate the output capacitance (circuit parameter j) and the output resistance (circuit parameter k) of transistors. Let the output capacitance and the output resistance of N transistors be measured; these measurements will be denoted by v_{ji} and v_{ki}. The nominal (average) values may be obtained by Eq. (6.73):

$$\tilde{v}_j = \frac{1}{N} \sum_{i=1}^{N} v_{ji} \tag{6.92}$$

$$\tilde{v}_k = \frac{1}{N} \sum_{i=1}^{N} v_{ki} \tag{6.93}$$

Similarly, the standard deviations of the two components may be obtained from Eq. (6.76):

$$\sigma_j^2 = \frac{1}{N} \sum_{i=1}^{N} (v_{ji} - \tilde{v}_j)^2 \qquad (6.94)$$

$$\sigma_k^2 = \frac{1}{N} \sum_{i=1}^{N} (v_{ki} - \tilde{v}_k)^2 \qquad (6.95)$$

The correlation coefficient r_{jk} is

$$r_{jk} = \frac{\sum_{i=1}^{N} (v_{ji} - \tilde{v}_j)(v_{ki} - \tilde{v}_k)}{(N-1)\sigma_j \sigma_k} \qquad (6.96)$$

From the above it is obvious that $r_{kj} = r_{jk}$; also

$$\lim_{N \to \infty} r_{jj} = 1 \qquad (6.97)$$

The correlation matrix is simply the matrix of the coefficients r_{jk} shown in Eq. (6.96).

The correlation coefficient must be used with extreme care; it should be computed only when a physical reason exists for two parameters in a network being related. The coefficient must be set to zero otherwise. When a zero coefficient is calculated, that fact alone is not sufficient for the parameters to be uncorrelated. For a more detailed discussion of correlation, the reader is referred to [5].

The changes of voltages (or any other circuit performance functions) in a network are linear combinations of parameter variations, as shown in Eq. (6.2). It can be shown from statistical arguments [6] that the standard deviation of changes in a linear combination, such as

$$\Delta f = \sum_{i=1}^{P} \frac{\partial f}{\partial p_i} \delta p_i \qquad (6.2)$$

is given from

$$\sigma_f^2 = [\mathbf{D}][\mathbf{R}][\mathbf{D}]^T \qquad (6.98)$$

where

$\mathbf{R}$ = correlation matrix, defined in Eq. (6.96);
$\mathbf{D} = [(\partial f / \partial p_i) \sigma p_i]$, an $1 \times p$ vector of derivatives with respect to the parameters multiplied by the standard deviation of the respective parameter;

and T denotes transposition.
This equation holds for real valued derivatives; the above must be modified for AC analysis.

Note that for uncorrelated parameters ($r_{jk} = 0$), the above reduces to

$$\sigma_f^2 = \left(\frac{\partial f}{\partial p_i}\right)^2 \sigma_{p_i}^2 \qquad (6.99)$$

Normally, f would be a node-voltage magnitude or a node-voltage phase angle; the appropriate relationship must be used for f, which must be a real function.

Sec. 6.5   Correlated Parameters   291

EXAMPLE 6.8

Consider a voltage divider made up of two 500-ohm resistors. Let each resistor have a $\sigma = 15\,\Omega$.

(a) Uncorrelated Components

The voltage transfer ratio is

$$\mu = \frac{E_{\text{out}}}{E_{\text{in}}} = \frac{R_2}{R_1 + R_2} = 0.500 \text{ (nominal)}$$

The derivatives are

$$\frac{\partial \mu}{\partial R_1} = \frac{-R_2}{(R_1 + R_2)^2} \qquad \frac{\partial \mu}{\partial R_2} = \frac{R_1}{(R_1 + R_2)^2}$$

The standard deviation of the voltage transfer ratio is

$$\sigma_\mu^2 = \left[\frac{-R_2}{(R_1 + R_2)^2}\right]^2 \sigma_{R_1}^2 + \left[\frac{R_1}{(R_1 + R_2)^2}\right]^2 \sigma_{R_2}^2$$

$$\sigma_\mu = \sqrt{2}\,\frac{500}{(1000)^2}(15) = 0.0106$$

The percent standard deviation is

$$\sigma_\mu(\%) = \frac{0.0106}{0.500} \times 100 = 2.12\%$$

(b) Perfect Correlation

Perfect correlation implies that $R_1 = R_2$ for all combinations.

$$\sigma_\mu^2 = \left[\frac{-500}{1000^2}(15) \quad \frac{500}{1000^2}(15)\right]\begin{bmatrix}1 & 1\\1 & 1\end{bmatrix}\begin{bmatrix}\frac{-500}{1000^2}(15)\\[4pt]\frac{500}{1000^2}(15)\end{bmatrix}$$

$$= 0$$

(c) Complete Mismatch

This condition implies that whenever R_1 is high, R_2 is low by the same amount. From an examination of Eq. (6.96) this condition can be characterized by a coefficient of correlation -1. Hence

$$\sigma_\mu^2 = \left[\frac{-500}{(1000)^2}(15) \quad \frac{500}{(1000)^2}(15)\right]\begin{bmatrix}1 & -1\\-1 & 1\end{bmatrix}\begin{bmatrix}\frac{-500}{(1000)^2}(15)\\[4pt]\frac{500}{(1000)^2}(5)\end{bmatrix}$$

$\sigma_\mu(\%) = 3.0\%$

Although Eq. (6.98) may be applied directly to the calculation of direct-current voltage and current changes, the calculation of the derivative of the magnitude and the derivative of the phase angle must be done akin to Eqs. (6.46) and (6.47). Using the notation of Sec. 6.1, let the voltage of node k be

$$V_k = A_k + jB_k = |V_k|\underline{/V_k} \qquad (6.100)$$

The derivative of the node voltage with respect to the magnitude of the ith independent voltage source is given by Eq. (6.58):

$$\frac{\partial V_k}{\partial |V_{g_i}|} = R_k + jX_k = \frac{\partial A_k}{\partial |V_{g_i}|} + j\frac{\partial B_k}{\partial |V_{g_i}|} \qquad (6.101)$$

Thus the magnitude change in V_k with respect to magnitude change in V_{g_i} is

$$\begin{aligned}\frac{\partial |V_k|}{\partial |V_{g_i}|} &= \frac{1}{|V_k|}\left(A_k\frac{\partial A_k}{\partial |V_{g_i}|} + B_k\frac{\partial B_k}{\partial |V_{g_i}|}\right) \\ &= \frac{1}{|V_k|}(A_k R_k + B_k X_k)\end{aligned} \qquad (6.102)$$

Similarly, the derivatives with respect to phase angle changes can be calculated from Eq. (6.59):

$$\frac{\partial V_k}{\partial \underline{/V_{g_i}}} = C_k + jD_k = \frac{\partial A_k}{\partial \underline{/V_{g_i}}} + j\frac{\partial B_k}{\partial \underline{/V_{g_i}}} \qquad (6.103)$$

The change of magnitude of V_k with respect to phase angle changes in V_{g_i} is

$$\frac{\partial |V_k|}{\partial \underline{/V_{g_i}}} = \frac{1}{|V_k|}(A_k C_k + B_k D_k) \qquad (6.104)$$

Similarly, the changes in the phase angle of the node voltage may be calculated:

$$\frac{\partial \underline{/V_k}}{\partial |V_{g_i}|} = \frac{1}{|V_k|^2}(A_k R_k - B_k X_k) \qquad (6.105)$$

$$\frac{\partial \underline{/V_k}}{\partial \underline{/V_{g_i}}} = \frac{1}{|V_k|^2}(A_k C_k - B_k D_k) \qquad (6.106)$$

A completely analogous set of derivatives for the independent generators may be obtained by substituting I_{g_i} for V_{g_i} and using Eqs. (6.56) and (6.57) in lieu of Eqs. (6.58) and (6.59).

The formulas just derived are used in calculating the standard deviations of variables which are functions of correlated parameters using Eq. (6.98).

For even the most trivial of cases the computational task involved is horrendous; it is left as an exercise to work out some simple cases in the problems. Note that the use of the program discussed in Sec. 6.4 with modifications for standard deviation calculations and correlation may ease any computational difficulties.

6.6 Monte Carlo Analysis

The two tolerance analysis procedures discussed so far, the worst-case and the standard deviation analysis, give limited information about the distribution of circuit performance functions (such as the node voltages) as they are dependent on the variations of the circuit parameters. The worst-case analysis gives the possible extreme values of the functions, each of which has a probability of occurrence of zero. The standard deviation analysis regards all component variations as being Gaussian distributed and thus ignores possible significant information about the actual component variations. Both analysis methods give very useful information. The limits and an approximate distribution of the variations can be calculated with these methods. Moreover these two methods require only relatively limited computer time and storage.

One could also develop the necessary formulas for uniformly distributed component values. Here one can assume that between the extreme limits (the lower worst-case and the upper worst-case values) the component values are distributed with equal probability. This procedure, too, makes assumptions about the actual parameter variations and will yield only an approximate result for any real distribution. The theory relating to uniform distributions leads to fairly complicated convolution formulas; their solution in turn leads to intricate relationships. Such formulas are at best difficult to work with and the amount of labor involved is seldom justifiable; consequently, very little is available in the literature in the way of practical procedures for other than Gaussian distributions when a very large number of parameters must be considered.

The only practical procedure that can account for the actual distributions is to simulate the actual process of constructing the circuit. One simply chooses a set of components at random from distributions that describe the actual expected component variations and then calculates the circuit performance function. Such simulation by repeated analysis using randomly generated data is referred to as Monte Carlo analysis; this mode of analysis is in wide use in system simulation [4].

The analysis procedures to be used have been covered in Chapter 3 (DC and AC analysis); the method is equally applicable to transient analysis (Chapter 4) and state variable analysis (Chapter 5). Commonly, Monte Carlo simulation is used to establish the variation of bias points (DC analysis), to find expected variations in frequency response (AC analysis) and probable changes in pole-zero locations (state-variable analysis). Sometimes the same procedures may be used to calculate other performance function variations, such as rise time, overshoot, etc. In any case, the performance function(s) must be monitored, typically in a subroutine attached to the normal analysis program; it is simply impractical to write a generalized program that will monitor the majority of desired performance functions.

In order for the program to simulate the network performance, data on the circuit parameters must be entered. Conceptually, one could provide a distribution function for each circuit component (and a matrix of correlation coefficients), but clearly the amount of data necessary to describe the components is not normally available and it is not desirable. A compromise between no such data and all components having a distribution is to provide capability to enter histograms and then use pointers to derive actual circuit values from these histograms. Possibly the histograms could be scaled also. Usually, however, only a limited amount of data is available on a circuit component and its variation; commonly, the nominal value and the nominal tolerance may be the only such data items. In these cases the assumption of the 3σ points at the tolerance limits may be the only workable procedure.

At any rate, the input data will be more complicated than it was in the analysis problem alone; the input data will have to contain more information. In order to handle such increased information we may consider the following input forms, assuming that a free-form input, akin to the ones discussed in Chapters 3 and 4, is used:

(1) For components without variation:

$$R = 15.3$$

(2) For components with nominal and stated tolerance values:

$$R = 15.3 \ (0.10)$$

(3) For components with stated nominal, minimum, and maximum values:

$$R = 15.3 \ (14.2, 16.8)$$

(4) For components with stated nominal value and a known (or assumed) histogram input with bin ranges (see Table 6.1) of a given size:

$$R = 15.3 \ (H3, 0.25)$$

(5) For components with stated nominal and stated standard deviation (Gaussian distributed values):

$$R = 15.3 \ (S0.5)$$

Many other forms may also be used (for example, for uniformly distributed values), but, normally, data are not available for classifying component tolerances in any finer detail. Note that forms 2 and 5 will be treated equivalently since the 10% tolerance change in form 2 will be converted to a standard deviation of $(15.3)(0.10)/3 = 0.51$.

The specification of the histogram (histogram #3 is shown in form 4 above) may be done in a standardized form. The nominal value, a stated bin width, an agreed-upon number of bins, and the corresponding relative frequencies are all that is necessary to define a histogram. If 21 relative frequencies are used to specify a histogram (10 values below the nominal and 10 values above the nominal), a very intricate histogram may be described; data are not usually available for justifying any more meaningful input values. The histogram may then be specified by 21 floating-point numbers; normally, such data would be punched free-format on a set of consecutive data cards with blanks or commas separating the floating-point data values. Specification of histogram shapes independent of a component allows a single histogram to be used for several circuit components.

Generation of a circuit component value starts with a uniformly distributed random number in the range (0.0, 1.0). The value of a component is then determined from the cumulative frequency curve of the distribution for that component. This curve is merely the integral under the relative frequency curve. For an arbitrary distribution, specified by a set of relative frequency values as in Table 6.1, the ith point of the curve is obtained by successively summing the relative frequencies up to and including the ith relative frequency, as shown in the last column of Table 6.1. This curve for the data in Table 6.1 is shown in Fig. 6.11. The slope of the curve is non-negative in the entire range of the distribution and is used to generate a random

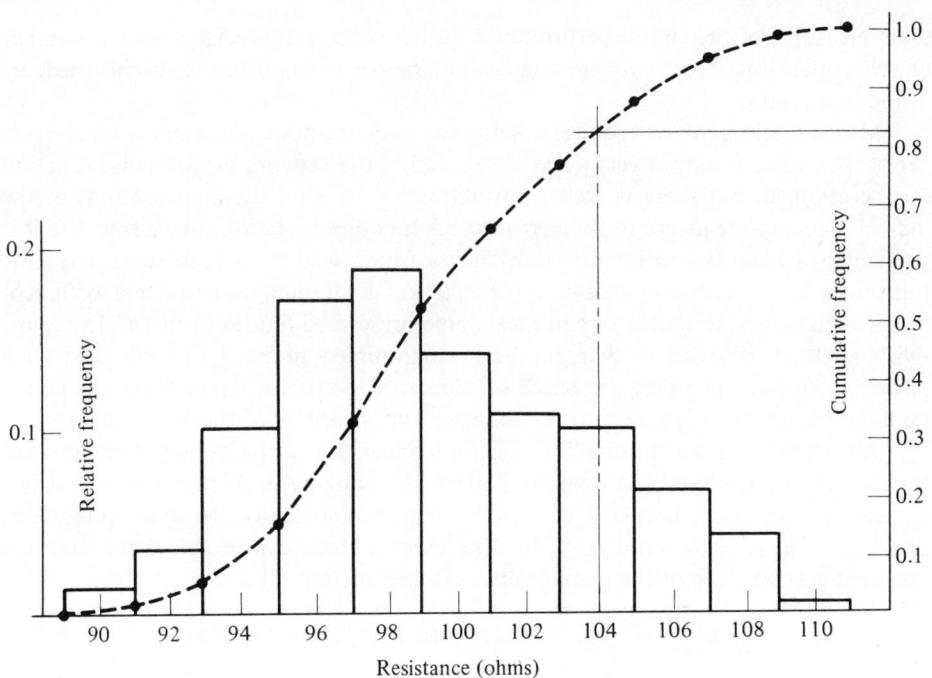

Fig. 6.11 Cumulative Frequency of Resistors Listed in Table 6.1

value having the prescribed distribution from a uniformly distributed number in the range (0., 1.0).

The generation of such a random value is dependent on the particular computer word length. For any machine such a routine is available as part of the scientific subroutine package supplied on the machine. Strictly speaking, all such subroutines generate only an approximately random number; the individual numbers, however, have distributions sufficiently close to truly random numbers and no significant errors are introduced by regarding them truly random. In the data shown in Fig. 6.11 the random value 0.83 yields a component value of 104.1 ohms as indicated on the diagram. All components are assigned a value in this manner. The remainder of the analysis for node voltages or any other circuit performance function proceeds as outlined in Chapter 3. Conceptually, any analysis method may be used, but usually only DC or AC analysis is done in this context.

Next, the performance function must be classified and statistical data must be formed. This is usually done with a subroutine that makes up histograms of the desired performance functions.

The process of generating component values, analyzing the network, and classifying the results is repeated a large number of times in order to build up a performance distribution function that will not change appreciably as additional runs are made. This usually entails keeping a running average and computing the variance each time a new analysis is completed. Program termination may be automatic if these measures of the circuit performance do not change appreciably with a number of additional runs. More conventionally, the number of iterations is specified as part of the input data.

Monte Carlo analysis requires solving the network-node equations a large number of times. Eventually, a very good simulation of the network performance function can be attained, but there is little information about what the contributions of the individual components are to the performance function variation. For design use it is necessary to identify the circuit components which lead to a high sensitivity; the tolerances for these components are then picked small enough, consistent with economic constraints, to let the circuit meet some prescribed figures of merit. The sensitivity relations detailed in Sec. 6.1 and programmed in Sec. 6.4 will yield such information, but since they are based on slopes of the performance functions at the nominal values, they ignore any pronounced nonlinearities of these functions.

An attempt at accounting for these nonlinearities is the "one-parameter-at-a-time" analysis. Circuit equations are solved by changing each parameter to a few equally spaced values between its prescribed outer limits, say at the nominal value p_0, at $p_0 \pm \sigma$, $p_0 \pm 2\sigma$, and $p_0 \pm 3\sigma$. The average slope can be calculated from the resultant seven values of the circuit performance function [7].

$$\frac{\partial f}{\partial p} = \frac{1}{60\sigma}(-f_{-3} + 9f_{-2} - 45f_{-1} + 45f_1 - 9f_2 + f_3) \qquad (6.107)$$

where f_k is the value of f evaluated at $p_0 + k\sigma$. These values are shown in Fig. 6.12; the resultant derivative is sometimes referred to as the central derivative [4].

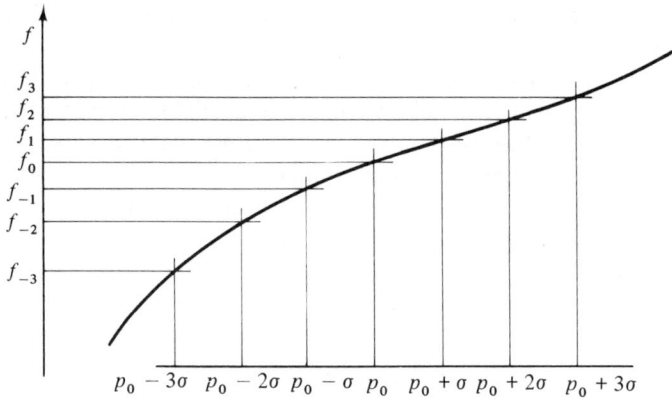

Fig. 6.12 7 Point Central Derivative

Note that the calculation of the central derivative is computationally quite expensive; it involves seven independent circuit analyses for each parameter. Computational costs may be drastically reduced by noting that if all parameters, save one, in a linear network remain at some value, the entire network may be replaced by its Thevenin (or Norton) equivalent circuit as far as the terminals of the changing parameter are concerned. Changes in the circuit can thus be deduced from a reduced equivalent circuit [9, 10].

For the great majority of networks, if the component tolerances are relatively tight (20% or less), the central derivative does not differ markedly from the slope at the nominal point. However, for wide component variations, such as transistor parameter tolerances, there may be more of a difference. Since good design practice dictates that the performance of a circuit should not depend markedly on transistor parameters, well-designed circuits can normally be analyzed using the slope calculations only. Some of these central derivatives were calculated and converted to sensitivities for the example in the next section; they are shown in the last column in Table 6.2. The reader will note only small differences between these values and the corresponding sensitivities shown in column 8. We shall not pursue the central derivative calculations here but will just simply note that if they are desired they can be obtained by repeated analysis of the circuit node equations.

We now have all the necessary formulas and outlines of procedures (algorithms) to assemble a powerful tolerance analysis program. Basically, the program outlined in Fig. 6.10 may be extended by the addition of a Monte Carlo analysis section. Such an extended program is shown in the flowchart in Fig. 6.13. The various tasks within the program are numbered; blocks 1 through 7 represent the worst-case and sensitivity program outlined in Fig. 6.10. One change that must be made is that in block 1 data must be entered on the allowable variations of the performance function and iteration control (the number K) must be provided.

The one-parameter-at-a-time analysis, discussed above, is performed next. The output, block 9, provides detailed data on the contribution of each circuit component to the performance function variation. If the analysis is done by use of a central derivative, such as the seven-point derivative in Eq. (6.107), the results will reflect the influence of curvatures on the variation. Otherwise, the sensitivity results obtained in block 4 will lead to the same results since the contribution of parameter k with a standard deviation of σ_k to the square of the standard deviation of the performance function f is

$$w_k = \left(\frac{\partial f}{\partial p_k}\sigma_k\right)^2 \tag{6.108}$$

The contribution of parameter k to σ_f is

$$\sigma_f|_k = \sigma_f - \sqrt{\sum_{\substack{i=1 \\ i \neq k}}^{P} w_i} \tag{6.109}$$

where P is the total number of parameters in the circuit.

Unless some attempt is made to account for the curvature of the performance function, the necessary derivative for the above equations is obtained in block 4; otherwise, either Eq. (6.107) or an equivalent approximation is used.

The generation of random parameter values was discussed earlier in this section; the entire process of Monte Carlo analysis is contained in blocks 10 through 14 in Fig. 6.13. Note that since many analyses must be done on circuits of slightly differing

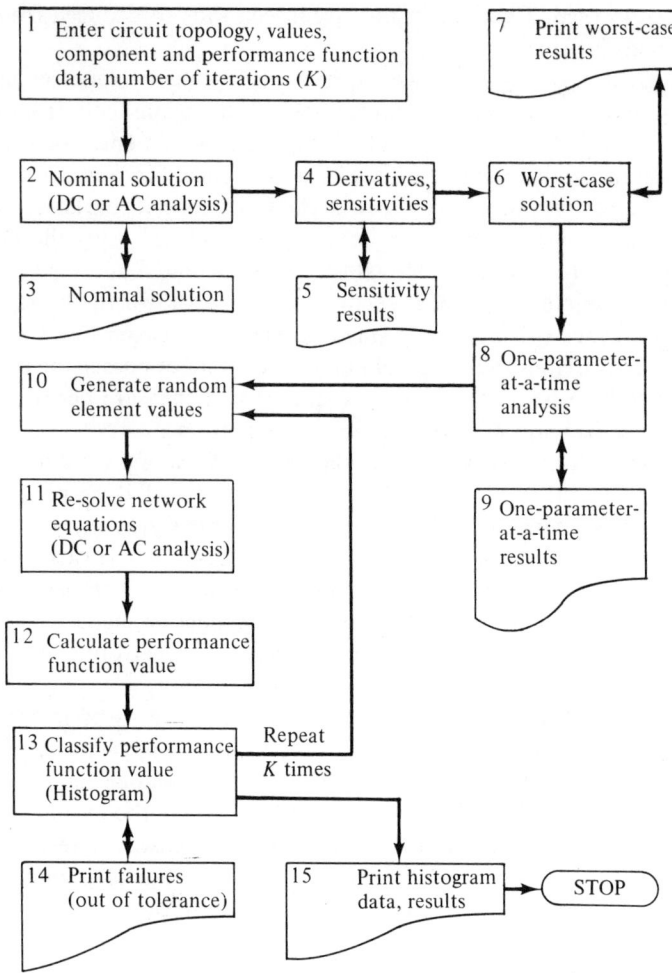

Fig. 6.13 Flowchart of a Tolerance Analysis Program

parameters, iteration schemes become attractive. The methods of Sec. 3.8 should be tried in order to see if they will converge in given situations; normally, passive circuits and some active filters can lead to fast-convergent equations. If the iteration scheme does not work, the fastest solution process should be used: Gauss elimination is in order. Note that in the sensitivity and worst-case analyses inversion was performed (see Secs. 6.1 and 6.3); thus the tolerance program would normally use both an inversion subroutine and a Gauss elimination solution routine. It is possible to combine the two, as may be noted by comparing the Listings in Appendix A, but little is gained by doing so; thus, normally, two separate routines would be used.

As the Monte Carlo analysis proceeds, one might find combinations of parameters that lead to unacceptable performance functions. Whenever such a case is

Sec. 6.7   A Comprehensive Example   **299**

encountered, it should be printed out complete with all circuit parameters. One can then examine such combinations in detail later on. The performance function values may be sorted into histograms as the solutions are being calculated. Usually, 20 bins will suffice for these classifications; the bin limits should be arranged so that they are equally spaced between the lowest and the highest acceptable performance function values. In most instances the worst-case limits obtained in block 7 give far too wide a variation to be usable for establishing the histogram limits unless a very large number of bins are used. In this latter case the majority of the bins will be empty.

Coding of this program is left as an exercise. Note that an input language becomes necessary for handling the large amount of input data conveniently. The example in the following section illustrates some of the possible uses of such a program.

6.7 A Comprehensive Example

To illustrate the material developed in this chapter, consider the circuit shown in Fig. 6.14. The circuit is an audio amplifier; the design specifications were

Input impedance	1000 Ω ± 10%
Power gain	40 db ± 2 db
Output load impedance	1000 Ω
Frequency response	100 Hz to 20 kHz
	±10% at −3 db points
Current drain	10 ma max

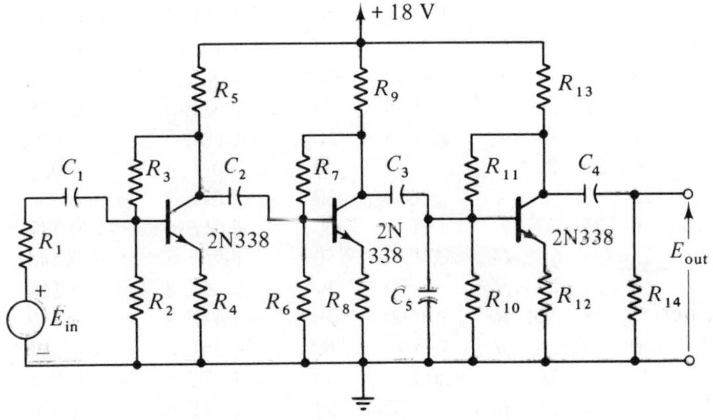

Fig. 6.14 Circuit of Audio Amplifier Example

The circuit component values and component tolerances are listed in Table 6.2. The transistor equivalent circuit for mid-band (1000 Hz) is shown in Fig. 6.15. The values and tolerances were established by measuring 73 transistors at $V_{ec} = 5$ V, $I_b = 1$ ma (Texas Instruments 2N338). The capacitor will be neglected in this section because its effect is very minimal at 1000 Hz.

Table 6.2 *Tolerance Analysis Results*

Component	Name	Branch	From	To	Type	Value	Tolerance %	ΔE_{out} (1% Δp)	ΔE_{out} (mv)	ΔE_{out} (mv) (One parameter variation)
1	R1	1	0	1	R	1000	10	−7.78E-4	−7.780	—
2	E_{in}	1	0	0	E	1	0	1.64E-3	0.	—
3	C1	2	1	2	C	1.5E-6	20	4.57E-6	0.091	—
4	R2	3	2	0	R	3300	5	2.63E-4	1.315	1.49
5	R3	4	3	2	R	22000	5	5.33E-4	2.665	2.99
6	R4	5	4	0	R	100	1	−7.76E-4	−0.776	−0.87
7	R5	6	0	3	R	10000	5	1.93E-4	0.965	0.92
8	C2	7	3	5	C	15.E-6	20	−1.27E-7	−0.003	—
9	R6	8	5	0	R	4300	5	4.49E-4	2.245	1.99
10	R7	9	6	5	R	22000	5	5.18E-4	2.590	2.55
11	R8	10	7	0	R	270	1	−1.04E-3	−1.040	−1.15
12	R9	11	0	6	R	10000	5	1.65E-4	0.825	0.81
13	C3	12	6	8	C	15.E-6	20	1.75E-7	0.004	0.17
14	C5	13	8	0	C	.01E-6	10	−7.03E-6	−0.070	—
15	R10	14	8	0	R	5100	5	3.23E-4	1.615	1.61
16	R11	15	9	8	R	22000	5	8.98E-4	4.490	4.62
17	R12	16	10	0	R	430	1	−8.99E-4	−0.899	−0.92
18	R13	17	0	9	R	10000	5	4.19E-4	2.095	2.02
19	C4	18	9	11	C	3.3E-6	20	6.13E-8	0.001	0.01
20	R14	19	11	0	R	10000	10	4.19E-4	4.190	4.03
21	Q1	20	2	4	R	3600	30	−4.06E-4	−12.180	−7.80
22	Q1	0	21	20	μ	7.0E-4	60	8.81E-6	0.528	—
23	Q1	21	3	4	G	20.E-6	50	−1.60E-5	−0.800	—
24	Q1	0	20	21	β	70	30	5.01E-4	15.030	15.60
25	Q2	22	5	7	R	3600	30	−2.02E-4	−6.060	—
26	Q2	0	23	22	μ	7.0E-4	60	4.05E-6	0.243	—
27	Q2	23	6	7	G	20.E-6	50	−8.96E-6	−0.448	—
28	Q2	0	22	23	β	70	30	3.04E-4	9.120	10.61
29	Q3	24	8	10	R	3600	30	−1.15E-4	−3.450	—
30	Q3	0	25	24	μ	7.0E-4	60	6.36E-6	0.382	—
31	Q3	25	9	10	G	20.E-6	50	−1.47E-5	−0.735	—
32	Q3	0	24	25	β	70	30	1.81E-4	5.430	2.24

The entire set of measurements is far too bulky to be reproduced here, but a sample is shown in Table 6.3, together with three matched 2N930 transistors, which could also be used in the circuit. (The latter three were in fact used in a copy of this circuit for six years; their characteristics have not changed appreciably during this time.)

Sec. 6.7 A Comprehensive Example

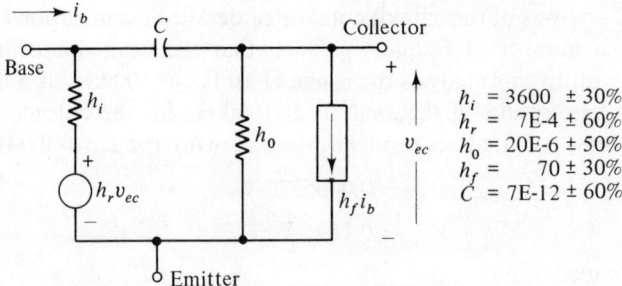

Fig. 6.15 Transistor AC Equivalent Circuit for 2N338

A study of Table 6.3 will reveal that the transistor parameters are correlated; the current gain, for example, and the input resistance are strongly correlated. To begin with, let us assume that these parameters are uncorrelated; we may want to re-examine this premise later on.

Table 6.3 *Sample of Transistor Parameters*

	h_i	$h_r \times 10^4$	$h_o \times 10^6$	h_f
2N338:				
	4500	10.0	26	80
	4700	8.4	20	84
	3400	8.1	20	71
	2600	7.8	9.6	61
	3900	8.5	22	76
	2500	5.1	14	48
	3700	6.1	28	72
2N930:				
	10300	5.8	8.6	220
	10500	4.2	8.5	190
	10200	5.0	9.0	220

The following average values and tolerance limits may be found for the transistor data shown:

$h_i = 3600$ (2500 min, 4700 max) $= 3600 \pm 30\%$

$h_r = 7.0 \times 10^{-4}$ (2.8 min, 10.0 max) $= 7.0 \times 10^{-4} \pm 60\%$

$h_o = 20 \times 10^{-6}$ (9.6 min, 28 max) $= 20 \times 10^{-6} \pm 50\%$

$h_f = 70$ (48 min, 84 max) $= 70 \pm 30\%$

These figures are rough; it is not necessary to expend too much effert in this part of the analysis unless the overall circuit performance is shown to be strongly dependent on these figures.

The overall analysis of this circuit consists of detailed examinations of the circuit performance at a number of frequencies such that the desired operating range is adequately covered. In this analysis the range of 10 Hz to 100 kHz is adequate.

We present here results of the analysis at 1000 Hz for the output voltage $|V_{11}|$. For nominal values, AC analysis results in the following for $E_{in} = 0.001$ V:

$$V_1 = 0.0005276 \text{ V}$$

$$V_{11} = 0.1645 \text{ V} = E_{out}$$

Thus the input impedance is

$$R_{in} = 1118 \ \Omega$$

The input power is

$$P_{in} = \frac{(0.5276 \times 10^{-3})^2}{1118} = 2.49 \times 10^{-10} \text{ watts}$$

The output power is

$$P_{out} = \frac{(0.1645)^2}{10000} = 2.70 \times 10^{-6} \text{ watts}$$

The power gain is

$$G = \frac{2.70 \times 10^{-6}}{2.49 \times 10^{-10}} = 1.086 \times 10^4 = 40.65 \text{ db}$$

The voltage gain is

$$G_V = \frac{0.1645}{0.0005276} = 311$$

The specified allowable power gain variation can be translated to voltage gain variations for constant input and output impedances. The allowable voltage gain variations are $+12.5\%$ and -20%.

Sensitivity analysis by the computer program described in Sec. 6.4 gives the voltage changes for all nodes. For the output node these are given in Table 6.2, under "ΔE_{out}, 1% Δp." The changes in the output voltage are obtained by multiplying the 1% changes with the actual component tolerances. The resultant changes in E_{out} are given in Table 6.2, under "ΔE_{out}, mv."

Since the distributions of the component values are not specified (although some idea of the transistor parameter distributions may be gleaned from Table 6.3), it is reasonable to assume that the tolerance limits specify the $\pm 3\sigma$ limits of a normal distribution. Thus the values in Table 6.2 ("ΔE_{out}, mv") represent the 3σ limits of contribution of the individual component variations. The 3σ value of the total variation is the square root of the sum of the squares with the input generator excluded (its variation is not under the designer's control.)

$$\sigma_f = \frac{1}{3}\sqrt{\sum_{i=3}^{32}(\Delta E_{out}|_i)^2} = 8.2 \times 10^{-3}$$

where σ_f is the standard deviation of the output voltage $|V_{11}|$. The contribution of the individual component to σ_f is obtained from Eq. (6.109).

A careful examination of this column indicates a rather marked influence of the transistor parameters h_i and h_f on σ_f, but the sensitivities have the opposite sign. Thus, if the two parameters have a positive correlation, their combined influence is less than their uncorrelated influences. We also note that the other two transistor parameters h_r and h_o contribute relatively little to the variation of $|E_{out}|$.

The correlation coefficient for h_f and h_i is obtained from Eq. (6.96); accurate values of the averages and the σ must be used. For this case

$$r_{fi} = 0.93$$

A plot of the two parameters will show that the equation $h_f = [(h_i - 3600)/1080] \times 21 + 70$ makes a good prediction between the two parameters.

Thus the contribution of h_f and h_i to the square of the standard deviation for the first stage is (in mv^2).

$$(15.03)^2 + (12.18)^2 - 2(0.93)(15.03)(12.18) = 33$$

as opposed to 374 mv^2 without the correlation. Thus by using the transistor correlation coefficients, the overall standard deviation becomes

$$\sigma_{fc} = 4.4 \text{ mv}$$

The output voltage and the variation limits are

(1) No correlation (normally distributed random variables)

$$E_{out} = 0.1645$$
$$3\sigma = 24.6 \text{ mv } (15\% \text{ of } E_{out})$$
$$\sigma = 8.2 \text{ mv}$$

(2) 0.93 correlation between h_f and h_i (normally distributed random variables)

$$E_{out} = 0.1645$$
$$3\sigma = 13.5 \text{ mv } (8.2\% \text{ of } E_{out})$$
$$\sigma = 4.4 \text{ mv}$$

We note that the 3σ limit for this latter case is within the specified power gain limits.

Since the signs of the changes are known, a worst-case analysis for this performance function ($|V_{11}|$) may be made by setting up the two worst combinations and re-analyzing the networks. The results are

$$\overline{|V_{11}|} = 0.2107 \text{ V (allowable} = 0.185 \text{ V)}$$
$$\underline{|V_{11}|} = 0.1154 \text{ V (allowable} = 0.137 \text{ V)}$$

The input impedance variations may be also calculated from the worst-case results. These are

$$\overline{R_{in}} = 1169 \text{ }\Omega$$
$$\underline{R_{in}} = 1054 \text{ }\Omega$$

We observe from the sensitivity analysis that the voltage gain and its variation are within the allowable limits. The input impedance is a little high; it may be lowered by decreasing the value of the 3.3 kΩ-resistor at the input. A new value of 2.7 kΩ may be tried; we shall, however, continue with our analysis of the circuit without this change.

So far the analysis is confined to blocks 1 through 7 of the program shown in Fig. 6.13. Before the Monte Carlo analyses are done, a check is made by one-parameter-at-a-time variation on the possible curvature of the output voltage magnitude with respect to some of the parameters. For components with stringent tolerance limits (such as the emitter resistors in this example) the average slope over the range of the variation (the central derivative) will differ but slightly from the slope at the nominal point as can be observed in Table 6.2. For parameters with wide variation (such as transistor parameters) curvature can become significant; hence output voltage changes for these parameters may be differing more. This again is demonstrated in Table 6.2.

Monte Carlo analyses were run on this circuit using the program detailed in Fig. 6.13 with the following constraints:

(1) All component values were assumed to be normally distributed random variables with the mean at the given nominal values and the 3σ limits at the given tolerance limits.

(2) Same as in (1) except that the transistor current gains were calculated from the transistor input resistance

$$h_f = \left[\frac{(h_i - 3600)}{(0.30 \times 3600)}\right] \times (0.30 \times 70) + 70$$

This restraint makes the correlation between h_f and h_i equal to one.

(3) Same as in (1) except that the parameter values were assumed to be uniformly distributed between the tolerance limits.

(4) Same as in (3) except that the transistor gains were calculated as in (2).

(5) Same as in (4) except that the transistor gain was generated as a uniformly distributed random variable with a correlation coefficient of 0.93 to the transistor input resistance.

(6) Components were chosen randomly to be at either their highest or lowest values with the transistor gain being calculated as in (2).

Case (6) will give a very pessimistic distribution of circuit performance, but it is still a long way from a worst-case analysis. Such extreme value analysis gives some idea about the performance function limiting distribution including the nonlinear nature of the performance functions for each parameter variation in the neighborhood of the nominal point.

To generate two Gaussian distributed random variables x_1 and x_2, with a prescribed correlation coefficient one might proceed as follows. Let y_1 and y_2 be two

Gaussian distributed random numbers with zero average value and a standard deviation of one.

Such numbers can be generated by computer programs which are included in the mathematical subroutine package of virtually all computers. Let the means and the standard deviations of x_1 and x_2 be $\overline{x_1}, \sigma_1$ and $\overline{x_2}, \sigma_2$ respectively. Then the two numbers

$$x_1 = \overline{x_1} + \sigma_1 y_1$$
$$x_2 = \overline{x_2} + \sigma_2 (r y_1 + \sqrt{1 - r^2} y_2)$$

will have the required correlation coefficient r.

The results of these analyses are summarized in Table 6.4. The histograms obtained in the first three analyses are shown in Fig. 6.16. From this figure it may be noted that good agreement is obtained between the average values shown in the diagrams and the nominal solution value as well as the calculated and computed standard deviations. However, the shape of the distribution curve is not very close to being normal. Good detail can be seen only in case of (1); this curve shows a relatively long tail toward the higher values. On the whole, however, the agreements are very good.

Table 6.4 *Results of the Monte Carlo Analyses*

Case	$\bar{E}_{out}$ (V)	3σ (mv)	σ (mv)	Calculated σ (mv)
(1)	0.1606	24.73	8.24	8.20
(2)	0.1606	11.09	3.70	3.50
(3)	0.1587	38.85	12.96	14.2
(4)	0.1601	17.28	5.76	6.05
(5)	0.1712	29.9	10.0	7.65
(6)	0.1719	54.1	18.0	13.2

The values in the last column of Table 6.4 can be readily obtained from the data in Table 6.2. A coefficient of correlation of 1.00 must be included for entry on row (2) and by replacing a uniform distribution and an extreme-value distribution by equivalent normal ones having the same means and the same standard deviations for rows (3), (4), and (6) respectively. In all cases one must remember that the distribution function is widened away from the center value as nonlinearities in the performance function become more significant and thus calculations based on the slope approximation become less accurate. Nevertheless, very useful results can be obtained from these analyses. Note also that any results from a limited sample are also strongly dependent on the specific random number generator that is used.

To analyze this circuit fully, the steps outlined in this section must be repeated at various frequencies. At least two points should be considered if conformance to the low-frequency specifications is to be verified; similarly, at least two analyses should be run in the neighborhood of the frequency limit. Hence at least five analyses of the

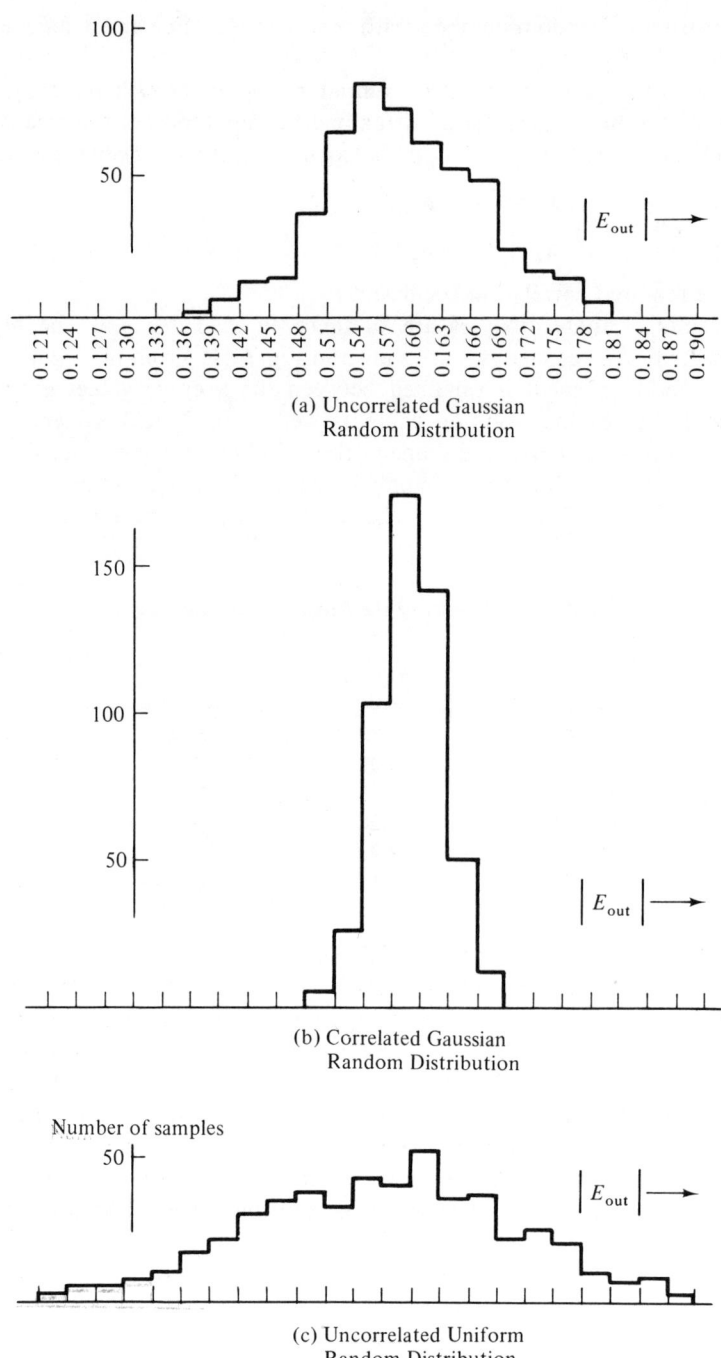

Fig. 6.16 Results of Monte Carlo Analysis

type described here must be made. If any components must be changed (or any tolerances must be modified), additional computations may be required. The computer charges for such an analysis are not negligible; one should always try to change the numerical procedures to the most economical ones in this type of analysis. Often it is cheaper to write special-purpose programs instead of using a general analysis package to obtain the statistical data; however, the first few solution points should be obtained conveniently.

6.8 Summary

The procedures discussed in this chapter form the necessary framework for analysis of network performance variations. Their use is not so much in analyzing networks as in establishing tolerance limits on circuits where only the nominal values are known. As an example, in the circuit shown in Fig. 6.14 the required component tolerances can be established by calculating the sensitivities, as was done in Table 6.2, and then ordering the sensitivities. The most sensitive components are thus identified; their tolerances can be assigned consistent with the state of the art and economic considerations: one could not assign 1% tolerances to transistor parameters, but one could, of course, do so for resistors.

The list of sensitive components will depend on the frequency. Thus at different operating conditions different tolerances may be indicated for the same component or tolerances for different components will be specified from different analyses. The designer's task is to resolve any conflicts that may arise; normally, the tightest tolerance values are chosen with a possible easing of tolerance requirements on other components. When the conflicts cannot be resolved, a circuit redesign is indicated.

In this phase of the design, actual distributions of component values are not known. One can use assumed normal distributions, or, better yet, work with uniform distributions between the tolerance limits. Slope approximation and/or one-parameter-at-a-time analysis is done so that the contribution of each component can be seen directly. Worst-case analysis may be done, but the results are usually too pessimistic. We observe that in the example shown in Sec. 6.7 the worst-case limits were at 5.8σ of the normal distribution, 3.7σ of the uniform distribution, and 2.0σ of the extreme value distribution.

Only when the circuit topology and components are thus established should actual component value distributions be used to analyze the circuit performance more accurately. Usually, accurate data may be needed only about some few highly sensitive components; from the others there may be insignificant contribution to the variance of the circuit performance function and thus a reasonable normal distribution may be usable. One should keep in mind that the central limit theorem will come to one's rescue.

As was demonstrated in the last example, the h-parameter equivalent circuit of transistors is a poor one to be used in tolerance work without an extensive investigation of parameter correlation. The question of properly modeling devices is not unique to this type of analysis and is central in assuring the reliability of the answers.

REFERENCES

1. Parzen, E., *Modern Probability Theory and Its Applications*. John Wiley and Sons, New York, 1960.
2. Feller, W., *An Introduction to Probability Theory and Its Applications*, 2nd ed., Vol. I. John Wiley and Sons, Inc., New York, 1957.
3. Papoulis, A., *Probability, Random Variables and Stochastic Processes*. McGraw-Hill Book Co., New York, 1965.
4. Mark, D. G., and L. H. Stember, Jr., "Variability Analysis," *Electro-Technology*, C-M Technical Publications Corp., New York (July 1965).
5. Davenport, W. B., Jr., *Probability and Random Processes*. McGraw-Hill Book Co., New York, 1970.
6. Mood, A. M., *Introduction to the Theory of Statistics*. McGraw-Hill Book Co., New York, 1950.
7. McCormick, J. M., and M. G. Salvadori, *Numerical Methods in FORTRAN*. Prentice-Hall, Inc., Englewood Cliffs, N.J., 1964.
8. Skilling, H. H., *Electrical Engineering Circuits*. John Wiley and Sons, New York, 1958, Chapter 11.
9. Kron, G., *Tensors for Circuits*. Dover Publications, New York, 1959.
10. Mark, D. G., "Tolerance Analysis of Nonlinear Circuits," *Proc. 1967 National Electronics Conference*, Chicago, p. 204.

PROBLEMS

6.1 Calculate the node-voltage derivatives for the circuit in Fig. 3.1.

6.2 Calculate the node-voltage derivatives for the circuit in Fig. 3.6. What are the node-voltage changes for a 1% increase in passive component values?

6.3 Calculate the node-voltage derivatives for the circuit in Fig. 3.8.
(a) What are the node-voltage changes for a 1% increase in resistances?
(b) What change in the values of μ_1 and μ_2 must occur to compensate a 1% increase in the resistance values?

6.4 In Ex. 6.2 calculate the derivatives associated with V_2.

6.5 Verify the results of Ex. 6.3 by writing the node-voltage expressions, differentiating them directly and calculating the node-voltage changes for the indicated component variations. The chain rule for partial derivatives may be used to advantage. All numerical work must be done with complex variables, hence slide-rule work becomes tedious. A special-purpose computer program may be considered for stepping through the arithmetic details.

6.6 Calculate the variation of V_2 in Ex. 6.3. The considerations of Problem 6.5 apply here also.

6.7 Calculate the sensitivities, Eq. (6.61), for the node voltages with respect to the circuit parameters in Ex. 6.1.

6.8 Calculate the sensitivities for the node voltages in Ex. 6.2.

6.9 Calculate the standard deviation σ_{V_2}, the 3σ limits for $|V_2|$ and the angle changes $\underline{/V_1}$ and $\underline{/V_2}$ in Ex. 6.5.

6.10 Calculate the worst-case voltages for Ex. 6.2 for 20% nominal tolerances on all passive components.

6.11 Calculate the worst-case combinations for the magnitude of node voltage 2 in Ex. 6.7.

6.12 Calculate the worst-case phase angle changes in the node voltages in Ex. 6.7.

6.13 Using the program in Listing 6.1, verify the results given in Exs. 6.1 through 6.7.

6.14 Modify the program in Listing 6.1 for the calculation of standard deviations of node voltage, branch voltages, and branch currents. The input must accommodate two numbers, the nominal value, and a standard deviation (or 3σ limit) for each component that is to be varied.

6.15 Complete the coding of the program for worst-case circuit analysis and standard deviation calculation shown in Fig. 6.10. The input must accommodate three values for each component that is varied (lowest value, nominal value, highest value); alternately, two numbers could describe the nominal value and a 3σ limit on the variations as in Problem 6.14.

6.16 Modify the program given in Listing 6.1 to include magnitude and phase angle changes of node voltages and branch quantities with respect to changes in generator voltage magnitudes and phase angles [see Eqs. (6.99) through (6.105)].

6.17 Simplify the AC sensitivity program to a DC sensitivity program.

6.18 Extend the DC sensitivity program to standard deviation and worst-case analysis similar to the program in Problem 6.15.

6.19 Extend the worst-case analysis program (Problem 6.15) to a Monte Carlo capability. The input should specify the number of trials and the variation of all parameters should be selectable between uniform distribution or 3σ limits at the specified lower and upper component values. A change in the input data format is necessary.

6.20 Modify the sensitivity program of Listing 6.1 for the calculation of central derivatives such as the one shown in Fig. 6.12.

6.21 Code the program shown in the flowchart, Fig. 6.13. Parts of the code have been completed in Problem 6.15.

6.22 Add the central derivative capability to the tolerance analysis program of Problem 6.21. Include in the program a check on changes in slope between adjacent points and a comparison with the derivative at the operating point. Print warnings in case of significant discrepancies.

6.23 Some storage in the tolerance analysis program can be saved by combining the simultaneous solution routine with the matrix inversion program. Code such a combined routine and add to your analysis program.

6.24 Verify the results shown in Table 6.2. Your program must be capable of handling at least 11 nodes and 32 components.

6.25 Repeat the analysis indicated in Table 6.2 at various frequencies in the range 10 to 100,000 Hz. Four steps per decade should be sufficient. Draw the nominal, the mode, and the 3σ limits of the expected response. Use correlated parameters for h.

6.26 What changes are necessary in the circuit of Fig. 6.14 to make the nominal response conform more closely to the design specifications? Re-analyze the modified network to verify your changes.

NONLINEAR ANALYSIS

7

The preceding chapters dealt with circuits that are composed of linear elements. Such elements have voltage-current characteristics that are independent of the magnitude of the applied voltage and of the magnitude of the current through them. Circuits composed of such elements obey the superposition principle: the response of a circuit to a linear combination of inputs may be calculated from the same linear combination of individual responses. That is, if the response (a voltage or a current at a point in the network) to an excitation e_j is r_j and to another input e_k is r_k, then the response (at that same point in the network) to an input combination of $c_j e_j + c_k e_k$ is $c_j r_j + c_k r_k$. The excitations e and the responses r are time functions and the above definition applies equally well to networks whose elements are constants and to networks whose elements are time-variant. Methods for the analysis of constant-element networks were studied in previous chapters. Because of the computational difficulties existing with time-variant networks they are usually analyzed by the methods to be discussed in this chapter, even though they are linear networks.

A brief reflection will show that all electronic circuits violate the linearity and superposition principles. No circuit can produce a proportional response as the constants c are being increased without limit. Proportional response can be achieved only over limited ranges of input conditions; thus the analyses discussed in previous chapters are limited to ranges of input excitation magnitudes and also to ranges of operating speeds and frequencies for which the components used in the mathematical formulation describe the actual behavior with sufficient accuracy. Passive electrical networks usually have a very wide dynamic range, although many mutual inductances normally are produced with saturable cores and many capacitor types exhibit breakdown and polarization properties. The most frequently encountered dependent sources in circuits are produced with transistors; these devices obey linearity only approximately and then only under stringent conditions.

This chapter will deal with some methods of analyzing electronic circuits containing nonlinear elements. Because linearity is violated by these circuits, notions based on linearity have to be abandoned. In particular, the idea of a frequency response must be revised because the various voltages and currents in a network will no longer occur as sinusoids of the same frequency as the input. The notion of input-output transfer matrix, as used in Chapter 5, must be replaced by calculation of a response (or a set of outputs) to a given input set. Thus nonlinear analysis usually becomes the calculation of a transient response to a set of inputs.

In Chapter 4 the transient analysis of linear networks was reduced to a series of DC analyses. There the capacitors and inductors were modeled as constant conductances with variable voltage sources or current sources respectively. In the linear transient analysis calculations the node-admittance matrix was constant, independent of the network voltages and currents. An elementary nonlinear capability was indicated in Sec. 4.7, Problems 4.23 and 4.24. This led to changes in Y_b and Y_n as a function of the voltages and currents in the system.

The basic idea in calculating the response of nonlinear circuits is that each voltage and current change is calculated as if it were occurring in a linear network whose elements are obtained from secants between the old operating point and the new one. At the end of each time step the voltages and currents are calculated as in a DC analysis. The element values are calculated from the differences between the last known solution point and the new one. The new voltages and currents had to be calculated from a set of assumed conductance values. The conductances over the last time step must agree with the assumed values for the solution to be valid. In many analysis programs, several time solutions are stored and slopes are used to predict the differential conductances over the next interval.

This chapter outlines several facets for nonlinear analysis; the emphasis is on transient analysis. Extensions to problems akin to pole-zero analysis and tolerance analysis will be indicated. As in previous chapters, a number of examples will be calculated through, a program will be discussed in some detail, and computational examples will be indicated. Because of the complexity and length of even relatively modest nonlinear capabilities, no extensive listing will be provided, but the program discussed is a widely available one [2, 3].

7.1 Basic Relationships

By way of introduction, let us consider the simple circuit shown in Fig. 7.1. It is desired to find E_{out} and its dependence on R.

From simple voltage summation

$$\begin{aligned} E_{\text{out}} &= I \cdot R \\ V &= E_{\text{in}} - I(R + 50) \end{aligned} \quad (7.1)$$

Sec. 7.1 Basic Relationships

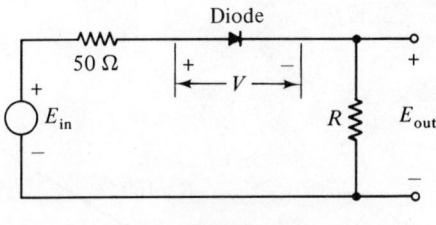

Fig. 7.1

where V is the diode terminal voltage, a function of the current through the diode. Usually, the terminal characteristic is known only as a v-i curve, such as the one shown in Fig. 7.2. Graphical analysis may be used to construct a load line and establish the family of $E_{in} - E_{out}$ characteristics shown in Fig. 7.3. The procedure is tedious, but exact; nevertheless, it is not suited to digital computer use because of its graphical nature. One could use the measurement data shown in Fig. 7.2 and enter a table of values which are then used to calculate E_{out} by iteration of Eq. (7.1). The procedure is to assume a value of V, calculate I from Eq. (7.1), then compare with the value in the stored table. If the two values do not compare favorably, the process is repeated with a new assumed value. The number of steps needed by the process to converge to a solution depends on the choice of the new value (the "prediction" algorithm) and on the nature of the function. We note that as long as the diode conducts the voltage across it will be relatively constant; when it does not conduct, the current will be nearly a constant (namely, zero).

Hence we seek expressions for changes in voltages and currents for circuit elements. These were derived for the linear elements in Chapter 4. Since we write the

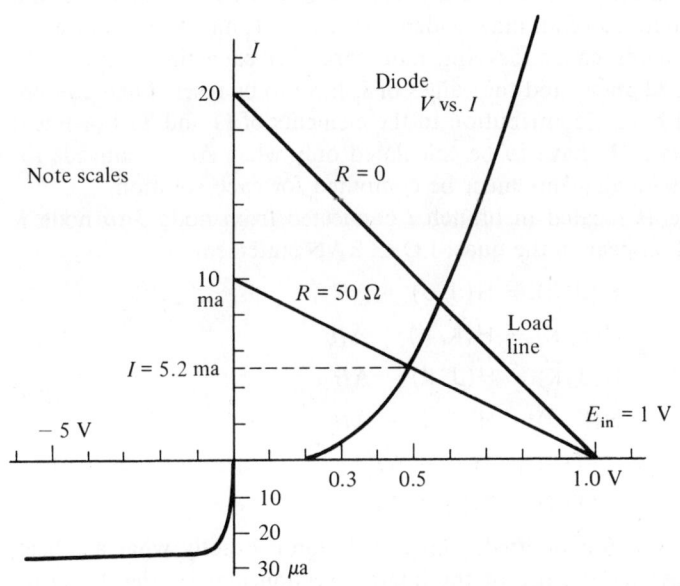

Diode V vs. I	
-5.0	-25.5 μa
-1.0	-25.1 μa
-0.2	-22.9 μa
-0.1	-17.8 μa
0	0
0.2	0.28 ma
0.4	2.2 ma
0.6	9.9 ma
0.8	21.3 ma

Fig. 7.2 Terminal Characteristics of a Diode

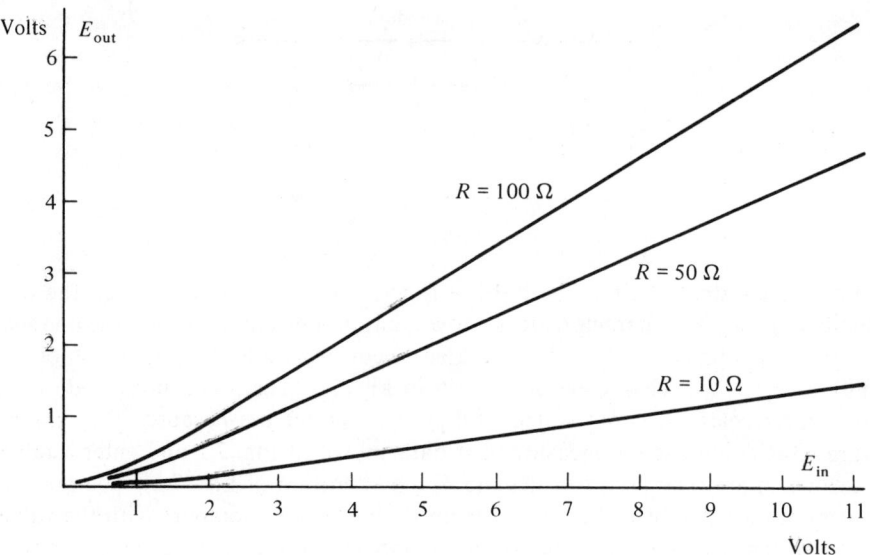

Fig. 7.3 Voltage Input-Output Characteristics for the Circuit of Fig. 7.1

node equations, the contributions to the $\mathbf{Y_n}$ and $\mathbf{I_s}$, the node-admittance matrix, and the vector of node-generator currents will be sought. We adopt the nomenclature of [2, 3]. The node equations at time t will be

$$\mathbf{HV_n} = \mathbf{T} \tag{7.2}$$

where $\mathbf{H}$ is the node-admittance matrix and $\mathbf{T}$ is the vector $\mathbf{I_s}$ used in Chapters 3 and 4. It contains the contributions of all independent generators $\mathbf{I_g}$ and $\mathbf{V_g}$ as well as $\mathbf{I_p}$, the sources used in modeling capacitors and inductors. For each time step the $\mathbf{V_g}$ and $\mathbf{I_g}$ have to be evaluated anew, and the values of $\mathbf{I_p}$ have to be reset. Each component in the network will have a contribution to the elements of $\mathbf{H}$ and $\mathbf{T}$. For linear components the elements of $\mathbf{H}$ have to be calculated only when Δt is changed; for nonlinear components these elements must be computed for each solution.

For linear components located in branch I connected from node J to node K the contribution ΔH will appear in the quasi-FORTRAN statement

$$\begin{aligned}
H(J, J) &= H(J, J) + \Delta H \\
H(K, K) &= H(K, K) + \Delta H \\
H(J, K) &= H(J, K) - \Delta H \\
H(K, J) &= H(K, J) - \Delta H \\
T(J) &= T(J) - \Delta T \\
T(K) &= T(K) + \Delta T
\end{aligned} \tag{7.3}$$

where the term for J or K $= 0$ is omitted. This calculation is exactly what has been done in Sec. 3.7 and obviates the use of the interconnection matrix (the $\mathbf{A}$ matrix of Chapter 3).

Sec. 7.1 Basic Relationships

The contributions of each of the individual components to **T** become the corresponding entries in the transient analysis program. Since the contributions to **H** and **T** are to be calculated at each solution point (time step), there is little to be gained from using ideal components (such as lossless inductors) in the program. We might as well derive the ΔH and ΔT entries for realistic two-terminal devices. It is easily seen that actual circuits will contain resistors, capacitors with losses, inductors with losses, transformers, diodes, and transistors and will not contain controlled sources. On occasion, other devices, such as Zener diodes and magnetic switching cores, will be encountered; these will also have to have the equations for their ΔH and ΔT contribution derived [3].

Although real circuit components do not include controlled sources, we shall include controlled current sources among the "elements" in the nonlinear analysis. From the material presented in Sec. 3.7, Eq. (3.116), it is easily seen that such sources will appear as transconductances in the Y_b matrix and therefore will appear only in the **H** matrix, if the controlling branch (from-branch) does not contain an independent voltage source. Such controlled sources will inject current in the secondary branch (the "controlled branch") proportional to the voltage across the controlling branch (a "g_m source") or proportional to the current in the passive element in the controlling branch (a "β source"). Since we shall derive voltage-current relations in specific forms (contributions to **H** and **T**) of special component assemblies, such as lossy inductors, the general circuit branch is no longer the one shown in Fig. 3.10. Instead, the allowable circuit components will be summarized in this and the next sections.

Voltage controlled sources can lead to many entries in the Y_n (the **H**) matrix, as was demonstrated in Sec. 3.4. Such sources will not be allowed in the program discussed in this chapter. However, a modification of the transient analysis program outlined in Sec. 4.5 may be made to include such sources. Most of the pertinent equations are developed in Sec. 4.3 [Eqs. (4.34) through (4.42)] and may be adopted here; the program will be slowed in execution and will be a complicated one.

Other models of the circuit elements will be similar to the ones used in Chapter 4. Resistors will contain no elements of I_p; inductor and capacitor models will have sources associated with them that give rise to a ΔT term. The nonlinear elements (diodes and transistors) will be treated separately in the next section. There will be terms in **T** resulting only from the independent sources and from the sources in the models of the reactive elements, which in turn will not be injected into nodes other than the ones to which the branch is connected. Hence all basic elements (G, R, L, C, V_g, and I_g) will have at most two terms contributing to **T** and at most four terms contributing to **H**.

7.1a Conductances G_i

These will be treated as simple conductances as shown in Table 3.5. Since the G_i do not contribute to I_p [see Eq. (4.24)], the contributions are

$$\Delta H = G_i$$
$$\Delta T = 0 \qquad (7.4)$$

7.1b Resistances R_i

$$\Delta H = \frac{1}{R_i}$$
$$\Delta T = 0 \qquad (7.5)$$

7.1c Current Source I_{g_i}

If such a current source has a parallel conductance G_i (possibly of zero value), then

$$\Delta H = G_i$$
$$\Delta T = -I_{g_i} \qquad (7.6)$$

where the entry ΔT is added to the node J and subtracted from node K.

7.1d Voltage Source V_{g_i}

If such a voltage source has a series resistance R_s, then a Norton equivalent circuit may be used. The contributions are

$$\Delta H = \frac{1}{R_s}$$
$$\Delta T = \frac{V_{g_i}}{R_s} \qquad (7.7)$$

where ΔT is added to node J and subtracted from node K.

7.1e Capacitor C

A very simple approximation for the capacitor is derived in Sec. 4.1 and shown in Fig. 4.3. The slope of the voltage vs. time function is approximated from the last known voltage and from the voltage being calculated. This approximation gives

$$\Delta H = \frac{C}{\Delta t}$$
$$\Delta T = \frac{v_{-1} C}{\Delta t} \qquad (7.8)$$

where v_{-1} is the voltage across the capacitor branch at the last known solution point.

A more realistic model for the capacitor should account for losses. Such a capacitor model is shown in Fig. 7.4. The internal node voltage V' creates, in effect, an additional node in the calculations, but if the internal equations are solved, only the terminal voltage-current relationships will appear in the calculations. Hence the node-voltage equation, Eq. (7.2), will not contain the node at V' explicitly. The calculation of the terms ΔH and ΔT are simplified by recognizing that ΔH is the conductance and ΔT is the current source strength in a Norton equivalent circuit of the model under consideration.

Sec. 7.1 Basic Relationships

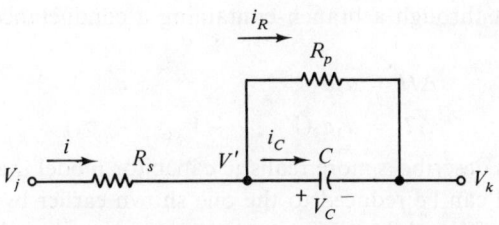

Fig. 7.4 Capacitor Equivalent Circuit

As in Chapter 4, the capacitor voltage must be approximated. If the derivative of the capacitor current is approximated by trapezoidal integration, there results

$$\Delta V_C = V_C - V_{C(-1)} = \frac{\Delta t}{2}(\dot{V}_{C(-1)} + \dot{V}_C) \quad (7.9)$$

(the subscript -1 refers to the previous time solution). From the model shown in Fig. 7.4

$$\dot{V}_C = \frac{1}{C}i_C = \frac{1}{C}(i - i_R) = \frac{1}{C}\left(i - \frac{V_C}{R_p}\right) \quad (7.10)$$

This equation must, of course, also hold at the previous solution time; therefore

$$\dot{V}_{C(-1)} = \frac{1}{C}\left(i_{-1} - \frac{V_{C(-1)}}{R_p}\right) \quad (7.11)$$

Furthermore, the voltage across the capacitor is

$$V_C = V_j - V_k - iR_s \quad (7.12)$$

Similarly,

$$V_{C(-1)} = V_{j(-1)} - V_{k(-1)} - i_{-1}R_s \quad (7.13)$$

Therefore Eq. (7.9) becomes

$$\left(2C + \frac{\Delta t}{R_p}\right)V_C = \left(2C - \frac{\Delta t}{R_p}\right)V_{C(-1)} + \Delta t(i + i_{-1}) \quad (7.14)$$

The capacitor current can be solved from the last three equations. The result is

$$i = a_2 a_4(V_j - V_k) - a_3 a_4(V_{j(-1)} - V_{k(-1)} + a_5) \quad (7.15)$$

where

$$a_1 = \frac{\Delta t}{R_p}$$

$$a_2 = 2C + a_1$$

$$a_3 = 2C - a_1$$

$$a_4 = \frac{1}{\Delta t + R_s a_2}$$

$$a_5 = i_{(-1)}\left(\frac{\Delta t}{a_3} - R_s\right)$$

The last equation shows ΔH and ΔT immediately since it is in the form

$$i = \Delta H(V_j - V_k) - \Delta T \quad (7.16)$$

which is the current through a branch containing a conductance ΔH and a source ΔT. Therefore

$$\Delta H = a_2 a_4$$
$$\Delta T = a_3 a_4 (V_{j(-1)} - V_{k(-1)} + a_5) \quad (7.17)$$

The last expressions describe a more realistic capacitor model than the one shown in Sec. 4.1; this model can be reduced to the one shown earlier by letting $R_s = 0$ and $R_p \to \infty$. Because of the different integration approximation the models will take different parameter values.

We note that the ΔH and ΔT terms require only the terminal voltages V_j and V_k at the previous solution point and the terminal current at that point.

7.1f Inductor L

Again, the model shown in Sec. 4.1 can be used. The results are

$$\Delta H = \frac{\Delta t}{2L}$$
$$\Delta T = i_{-1} + \frac{\Delta t}{2L}(V_{j(-1)} - V_{k(-1)}) \quad (7.18)$$

where i_{-1} is the current through the inductor branch at the previous time solution point. The trapezoidal integration approximation was used in Sec. 4.1 and the model is as accurate as the capacitor model; however, inductor losses were neglected.

A more realistic model for the inductor is shown in Fig. 7.5. The resistances are

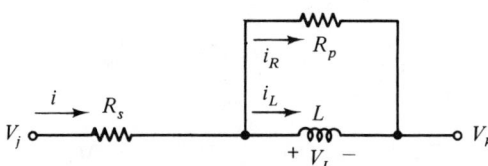

Fig. 7.5 Inductor Equivalent Circuit

similar to the ones used for the capacitor model. The same methods of solving for the terminal characteristics are employed to calculate ΔH and ΔT:

$$\Delta H = a_2(a_1 + \Delta t)$$
$$\Delta T = a_3(V_{j(-1)} - V_{k(-1)} + a_4) \quad (7.19)$$

where

$$a_1 = \frac{2L}{R_p}$$

$$a_2 = \frac{1}{2L + R_s(\Delta t + a_1)}$$

$$a_3 = a_2(a_1 - \Delta t)$$

$$a_4 = \left(\frac{2L}{\Delta t - a_1} - R_s\right) i_{-1}$$

Comparing with the inductor model shown in Fig. 4.2 shows complete agreement between the two for $R_s = 0$ and $R_p \to \infty$.

In neither the capacitor nor the inductor model derived here is there any requirement that the circuit elements be constants. Each component value can be dependent on the voltages and currents; one must only calculate the proper circuit element values for each solution time interval.

7.1g Mutual Inductance

The general case of mutual inductance coupling without losses in the coils is covered in Chapter 4. The case of arbitrary mutual inductance with losses in each of the coils may be modeled by moving the losses into external resistances and then using the formulas developed in Sec. 4.3b. This will entail nodes in the circuit which are not found in the circuit itself. We develop the formulas here only for pairwise mutual coupling.

Let the mutual inductance voltages, currents, and element values be given as shown in Fig. 7.6.

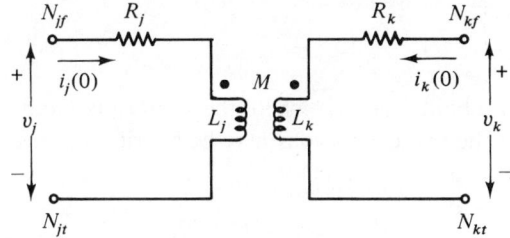

Fig. 7.6 Pairwise Mutual Inductance

A single series resistance has been included in series with each of the physical coils to account for the normal coil losses. Although these linear resistances normally cannot faithfully account for the coil losses at all frequencies, they do provide an acceptably accurate description of the coil losses in many applications.

The pertinent equations are

$$i_j(t) = \frac{1}{L_j} \int_0^t \left(v_j - i_j R_j - M \frac{di_k}{dt} \right) dt + i_j(0)$$

$$i_k(t) = \frac{1}{L_k} \int_0^t \left(v_k - i_k R_k - M \frac{di_j}{dt} \right) dt + i_k(0)$$

(7.20)

In order to recast the above into a form usable in the node-voltage equations, the integrals and derivatives occurring above must be approximated by some differences. The admittances of the four nodes involved will be interrelated by these difference equations and there will be current contributions to ΔT for each of the nodes. Hence there will be 16 ΔH terms and four ΔT terms; because a transformer is a passive device, the ΔH and ΔT terms will be symmetric:

$$\Delta H_{\alpha\beta} = \Delta H_{\beta\alpha} \qquad (7.21)$$

where α, β = nodes $jf, jt, kf,$ or kt and

$$\Delta T_{jf} = -\Delta T_{jt}$$
$$\Delta T_{kf} = -\Delta T_{kt} \qquad (7.22)$$

The simplest approximation for the derivative is taking the secant between the last known current and the present current:

$$M\frac{di_k}{dt} \cong \frac{M}{\Delta t}(i_k - i_{k(-1)}) \qquad (7.23)$$

and for the integral, the trapezoidal approximation:

$$I(t) = \int_0^t F(x)\,dx = \frac{1}{2}\Delta t[F + F_{(-1)}] + I_{(-1)} \qquad (7.24)$$

where $F = F(t)$, $F_{(-1)} = F(t - \Delta t)$, and $I_{(-1)} = \int_0^{t-\Delta t} F(x)\,dx$.
Consequently,

$$i_j(t) = \frac{\Delta t}{2L_j}[v_j - i_j R_j + v_{j(-1)} - i_{j(-1)} R_j]$$
$$+ \frac{M}{2L_j}[i_k - i_{k(-1)}] + i_{j(-1)} \qquad (7.25)$$

A similar equation holds for $i_k(t)$ where the subscripts j and k occurring in Eq. (7.25) are exchanged. The two expressions may be rewritten in the form

$$ai_j + ci_k = A + Cv_j$$
$$di_j + bi_k = B + Dv_k \qquad (7.26)$$

with

$$a = 1 + \frac{\Delta t\, R_j}{2L_j}$$

$$b = 1 + \frac{\Delta t\, R_k}{2L_k}$$

$$c = \frac{M}{L_j}$$

$$d = \frac{M}{L_k}$$

$$A = \left(1 - \frac{\Delta t\, R_j}{2L_j}\right)i_{j(-1)} + \frac{M}{L_j}i_{k(-1)} + \frac{\Delta t}{2L_j}v_{j(-1)}$$

$$B = \left(1 - \frac{\Delta t\, R_k}{2L_k}\right)i_{k(-1)} + \frac{M}{L_k}i_{j(-1)} + \frac{\Delta t}{2L_k}v_{k(-1)}$$

$$C = \frac{\Delta t}{2L_j}$$

$$D = \frac{\Delta t}{2L_k}$$

Sec. 7.1   Basic Relationships   **321**

The equations can be solved for i_j and i_k directly

$$i_j = \frac{bC}{\Delta}v_j + -\frac{cD}{\Delta}v_k + \frac{bA - cB}{\Delta}$$

$$i_k = \frac{aD}{\Delta}v_k + -\frac{dC}{\Delta}v_j + \frac{aB - dA}{\Delta} \quad (7.27)$$

$$\Delta = ab - cd$$

These expressions show the components of the currents in the primary and secondary of the transformer each as a self-conductance, a mutual conductance, and as a current source in parallel. These are entered in the **H** and the **T** matrices:

$$\Delta H(JT, JT) = \Delta H(JF, JF) = \frac{bC}{\Delta}$$

$$\Delta H(JT, JF) = \Delta H(JF, JT) = -\frac{bC}{\Delta}$$

$$\Delta H(KT, KT) = \Delta H(KF, KF) = \frac{aD}{\Delta}$$

$$\Delta H(KT, KF) = \Delta H(KF, KT) = -\frac{aD}{\Delta}$$

$$\Delta H(JT, KF) = \Delta H(KF, JT) = \frac{cD}{\Delta}$$

$$\Delta H(JF, KF) = \Delta H(KF, JF) = -\frac{cD}{\Delta} \quad (7.28)$$

$$\Delta H(JT, KT) = \Delta H(KT, JT) = -\frac{cD}{\Delta}$$

$$\Delta H(JF, KT) = \Delta H(KT, JF) = \frac{cD}{\Delta}$$

$$\Delta T(KF) = -\Delta T(KT) = \frac{aB - dA}{\Delta}$$

$$\Delta T(JF) = -\Delta T(JT) = \frac{bA - cB}{\Delta}$$

The reader is urged to verify these results; the development given here can be used as a guide for the derivation of computational models for other devices.

7.1h Voltage Controlled Current Source

This device will inject current proportional to the voltage across branch j into branch k. This element is Type Number 7 in Table 3.5. The contributions are

$$\Delta T = 0$$
$$\Delta H = g_m \quad (7.29)$$

and appear in the FORTRAN statements

$$
\begin{aligned}
H(KF, JF) &= GM \\
H(KF, JT) &= -GM \\
H(KT, JF) &= -GM \\
H(KT, JT) &= GM
\end{aligned}
\quad (7.30)
$$

where GM is the source strength, KF, KT, JF, JT are the from-to node numbers of the controlled and the controlling branch respectively.

7.1i Current Controlled Current Source

This device is element Type Number 6 in Fig. 3.5; the current injected in branch k will be β times the current in branch j. If the controlling branch is a resistor, then

$$\Delta T = 0$$
$$\Delta H = \frac{\beta}{R_j} \quad (7.31)$$

and the FORTRAN statements will be as Eq. (7.30) with β/R_j replacing GM.

If the controlling branch is not a resistor, the current in the controlling branch must be established first. For capacitors and inductors there will be contributions to ΔT, as can be seen from Eqs. (7.17) and (7.19). Should the controlling inductor branch be also used in a mutual inductance, additional complications arise, as can be seen from Eq. (7.28). As a consequence of these coding difficulties, we shall restrict our allowable element to a resistor (or a conductance) controlling branch. Should it become necessary to obtain currents through inductors and capacitors, an appropriate model will have to be used in the analysis.

The models listed in this section differ very slightly from the element models given in Chapter 4. The major difference is the use of more realistic capacitor and inductor models. Every model derived here assumes that the various element values (R, C, L, g_m, β, M, etc.) are constants during the time step Δt. The values can be recalculated for other time steps; thus saturable inductors could be analyzed in this way. However, the equations are meant to be used on essentially constant linear circuit elements; the essentially nonlinear diodes and transistors will be analyzed separately in the next sections.

7.2 Diode Equivalent Circuits

The terminal steady-state voltage-current relationship of a diode can be given easily and conveniently by a curve as in Fig. 7.2. For computational use, however, equations have to be written which can be solved for the instantaneous voltage and current. One would also like to be able to specify merely the fact that a specific diode is to be used in a branch: the branch number (or a pair of node numbers) and the diode type (i.e., 1N4727) should suffice. This convenient way of specifying a branch should, of course, allow for all operating possibilities of the branch element; hence a very good description of the voltage-current relations is needed. The derivation of an accurate circuit model is an art, but what is needed is a sufficiently accurate model for any given application. The question of sufficiency is very difficult to assess for any problem until a complete solution has been calculated; yet the model must be known before a solution can be computed. Although there is no generally "best" model, engineering judgment will eventually indicate "sufficient models." Such a model will be the least complicated consistent with the results sought in a given problem.

For general computer analysis the action of the diode must be described in fine enough detail to be usable for all "normal" applications. Virtually all diodes in use in modern circuits are semiconductor diodes, and the circuit models derived here are for such diodes. Semiconductor device theory is used to derive the terminal characteristics of diodes here and transistors in the next section. For more details of these derivations the reader is referred to [4, 5, and 6].

The voltage-current relationship of a semiconductor diode junction is at thermal equilibrium [4]

$$I = I_s \left[\exp\left(\frac{qV}{kT}\right) - 1 \right] \qquad (7.32)$$

where
- I_s = saturation current, a function of geometry and carrier concentration and hence a fixed value for any given junction; given in units of I;
- q = electronic charge = 1.602×10^{-19} coulombs;
- k = Boltzmann's constant = 1.380×10^{-23} joule/°K;
- T = absolute temperature (°K);
- I = junction current (amps);
- V = junction voltage (volts).

As an aside, the value of I_s is given by theoretical considerations alone as [4]:

$$I_s = qA\left(\frac{D_p p_0}{L_p} + \frac{D_n n_0}{L_n}\right) \qquad (7.33)$$

where
- A = cross-sectional area of the junction region (m^2);
- D_p = diffusion constant for holes in the N region;
- p_0 = normal hole density in the N region;
- L_p = diffusion length of holes in the N region;
- D_n, n_0, L_n = corresponding constants for electrons in the P region.

The last two equations are valid for abrupt semiconductor junctions and neglect carrier recombinations in the diode depletion region as well as surface effects on the outer edges of the diode junction. Since diodes are produced by the diffusion of dopants and are not abrupt junctions, modifications must be introduced in the above to get a more accurate description of the diode terminal characteristics. The recombination effects alone indicate that the following is a more accurate equation [9]:

$$I = I_s\left[\exp\left(\frac{qV}{mKT}\right) - 1\right] = I_s\left[\exp\left(\frac{V}{\theta}\right) - 1\right] \qquad (7.34)$$

with m, the "curvature constant," being a factor that depends on current density in the junction. However, a reasonably good description occurs for m a constant, determined from the average terminal characteristics. For silicon diodes, m is in the range of 1.0 to 2.0. This model does not give a good description of the reverse current characteristics of the diode.

The terminal characteristics are strongly dependent on the temperature T occurring in Eq. (7.34). This temperature is not the ambient temperature but rather the instantaneous temperature of the semiconductor junction. This temperature can be obtained approximately from a thermal equivalent circuit, but for the majority of applications it is assumed to be a constant value for an entire problem run.

Mathematically, the junction saturation current I_s is the current which would flow for $V \rightarrow -\infty$. Actually, it is impossible to increase the reverse voltage without limit and observed data indicate rather an increase in the current proportional to the junction voltage to a power $1/n$ with n in the range 2 to 3. Such an effect can crudely be modeled by a resistance in parallel with the diode junction, termed the "leakage resistance." For convenience in calculations, this resistance (R_s) is taken as a linear resistance whose value will account with sufficient accuracy over the usual back-voltage range. From a set of measured terminal voltage-current data, the value of R_s can be found from the slope of the characteristic for large negative diode voltages.

Since the observed forward diode characteristic fairly well approximates Eq. (7.34), values for m are obtained from moderate forward current-voltage data. For heavy currents, the diode bulk resistance must be considered (see below).

For large negative voltages the junction exhibits reverse voltage breakdown (Zener diode action). The voltage-current relations are essentially that of a constant voltage generator; many diodes cannot maintain sustained operation in this region. For analysis purposes the diode reverse voltage is limited by this voltage V_r, but it is usually assumed that operation in this region does not destroy the diode.

Hence the diode is now modeled with four parameters: (1) I_s and θ which occur in the nonlinear I vs. V relation, Eq. (7.34); (2) the junction leakage resistance R_s; (3) the reverse voltage rating V_r.

Some reflection reveals that the junction voltage-current characteristic does not fully describe the diode characteristic: One of the most obvious sources of discrepancy is the bulk resistance of the diode, R_b. This resistance then appears in series with the junction. Two such resistances should be considered: (1) the anode bulk resistance and (2) the cathode bulk resistance; but since they are in series with one another, their combined effect may be represented by a single resistance, R_b.

Thus the static characteristic of the diode may be represented by the five parameters shown in Fig. 7.7. Note that the diode voltage V_d will differ from the junction voltage V by the ohmic drop across the series bulk resistance of the diode.

The equivalent circuit should provide some indication when the maximum allowable current I_m in the device is exceeded. Hence the DC characteristic of a given diode is specified by:

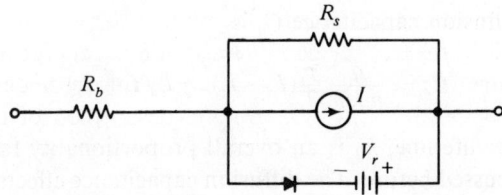

Fig. 7.7 Static Large Signal Equivalent Circuit for a Diode Junction

(1) I vs. V, Eq. (7.34).
(2) Temperature T.
(3) I_s, the saturation current.
(4) m, the curvature constant.
(5) V_r, the maximum reverse voltage (Zener voltage).
(6) I_m, the maximum allowable forward current.

In the calculation of the transient behavior of diodes the lead inductances and the diode capacitances must be considered. The lead inductances can normally be added externally to the diode model; also, they usually represent but minor effects in the overall behavior. For these reasons they will not be considered as part of the diode model developed here. For all but extremely high-frequency applications, the diode can be represented quite adequately with two capacitances: (1) the junction diffusion capacitance C_d and (2) the junction transition capacitance C_t. Another capacitance that influences the high-frequency behavior of the diode is the header capacitance; this, however, can be added externally to the diode model. Also, the header capacitance does not vary with applied voltage and/or with current flow; thus its effect is relatively easy to account for. The junction transition capacitance is voltage sensitive; a good approximation for it is [4, 7]:

$$C_t = C_T\left(1 - \frac{V}{V_Z}\right)^{-N}; \qquad V \leq 0.9 V_Z \qquad (7.35)$$
$$= C_T(0.1)^{-N}; \qquad V > 0.9 V_Z$$

where
C_t = transition capacitance at $V = 0$;
V_Z = junction contact potential;
N = junction grading constant.

The junction transition capacitance is the predominant capacitance for negatively biased diodes. The contact potential V_Z as well as C_t are normally determined from three V, C_t measurements. Some manufacturers' data sheets will list C_t at $V = 0$ and give N as well. For example, the 1N914 diode has $N = 0.5$ and $V_Z = 0.5$ V; V_Z is approximately 0.6 V for silicon and 0.3 V for germanium.

The junction diffusion capacitance C_d is

$$C_d = \frac{q}{mkT} \frac{\tau}{2} (I + I_s) = k_d (I + I_s) \tag{7.36}$$

where τ is the carrier lifetime, k_d is an overall proportionality factor, and all other parameters were discussed before. The diffusion capacitance effects become noticeable at higher forward currents. The various constants are normally lumped into k_d, which becomes a function of temperature and is determined experimentally. With these elements the complete large-signal equivalent circuit for the diode will contain equivalent parameters (I_s, m, T, R, V_z, N, k_d, R_s, R_b, V_r, I_m). All but T must be determined for the diode from measurements. The diode equivalent circuit becomes the one shown in Fig. 7.8.

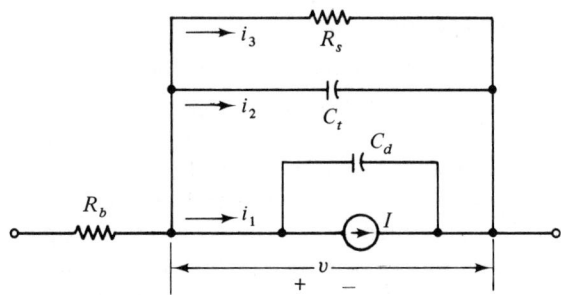

Fig. 7.8 Complete Large Signal Equivalent Circuit for a Diode Junction

In assessing the action of a diode, one more parameter must be observed, the maximum allowable power dissipation P of the device. Should analysis indicate that this value is exceeded at any time, appropriate action must be taken in the analysis program; at the very least a message should be printed.

For a particular diode the above parameters must be determined experimentally and/or be read from the device specification sheets. Since all measurements are subject to inaccuracies, a set of data points must be taken and a "best fit" must be obtained to these points. Normally, a computer program would be used for such fitting [8]. Results for a 1N914 diode are summarized in Table 7.1 and apply to the parameters of the diode model shown in Fig. 7.8.

The model accommodates virtually all possible uses of the junction diode. For most analyses the data required by an elaborate model are simply not available, and often the amount of calculation demanded by the circuit model is not justified because of limitations in observable accuracy. Also, often it is known in what region the diode operates for a given input; in these instances a simple small circuit model might suffice. Crude models of DC characteristics are shown in Figs. 7.9 and 7.10. All models shown require a capability of setting switches as voltages reach certain values (zero, V_1, V_2, etc.). Such capability is available in piecewise linear analysis programs, such

Table 7.1 *Model Data and Specification Sheet Information 1N914—Diffused Silicon Mesa Computer Diode*

			Model	Spec. Sheet
R_b	Bulk resistance	ohms	8.0	—
R_s	Leakage resistance	ohms	3.5×10^6	—
I_s	Saturation current	μA	.0021	—
	current at -20 V bias	μA		.025
T	Temperature	°C	25	25
P_m	Power dissipation	mW	250	250
V_r	Max reverse voltage	V	75	75
I_m	Max forward current	μA	110	75
	Recurrent peak forward current	μA		225
V_z	Junction contact potential	V	.5	—
N	Grading constant	—	.5	—
C_t	Capacitance at 0 V	pf	4.0	4
k_d	Proportionality factor	$\frac{\text{pf}}{\text{A}}$	1.81×10^7	—
m	Curvature coefficient	—	1.72	—

as the one discussed in Sec. 4.5. As the accuracy of representation of the diode function increases, the computational involvement increases disproportionately faster.

A reduction in the accuracy of the diode model is often attainable by carefully considering the circuit parameters external to the diode. For instance, if a relatively large resistance is in series with the diode, it is often possible to represent the diode-resistor combination as an ideal diode and a series resistance. Such insight by the user is the best insurance against excessive accuracy in modeling. Unfortunately, such procedure is very much dependent on the circuit configuration and circuit action and is best described as engineering judgment. For automated analysis the model equations, Eqs. (7.34) through (7.36), and the circuit shown in Fig. 7.7 are used. Implicit in this model is the neglect of Zener action and maximum power checks; the user has to check on these modes of operation outside the automated analysis program.

In order to set up the node-voltage equations for diode elements, their contributions to the **H** and **T** matrices, Eq. (7.2), must be calculated. Since the diode model contains an internal node between R_s and R_b (Fig. 7.8), and capacitors, recurrence relations between previous iterates and the present voltages will be sought in the form of Eqs. (7.15) and (7.17).

Let the previously calculated ideal diode current be I_{-1} and the corresponding voltage be V_{-1}. The current through the diode junction I and the current through the transition capacitance C_t are both exponential functions of the junction voltage.

The current through the diffusion capacitance can be considered as a current through an equivalent current source:

$$i_d = C_d \frac{dV}{dt} = T_d \frac{dI}{dV} \cdot \frac{dV}{dt} = T_d \frac{dI}{dt} \qquad (7.37)$$

328 Nonlinear Analysis   Ch. 7

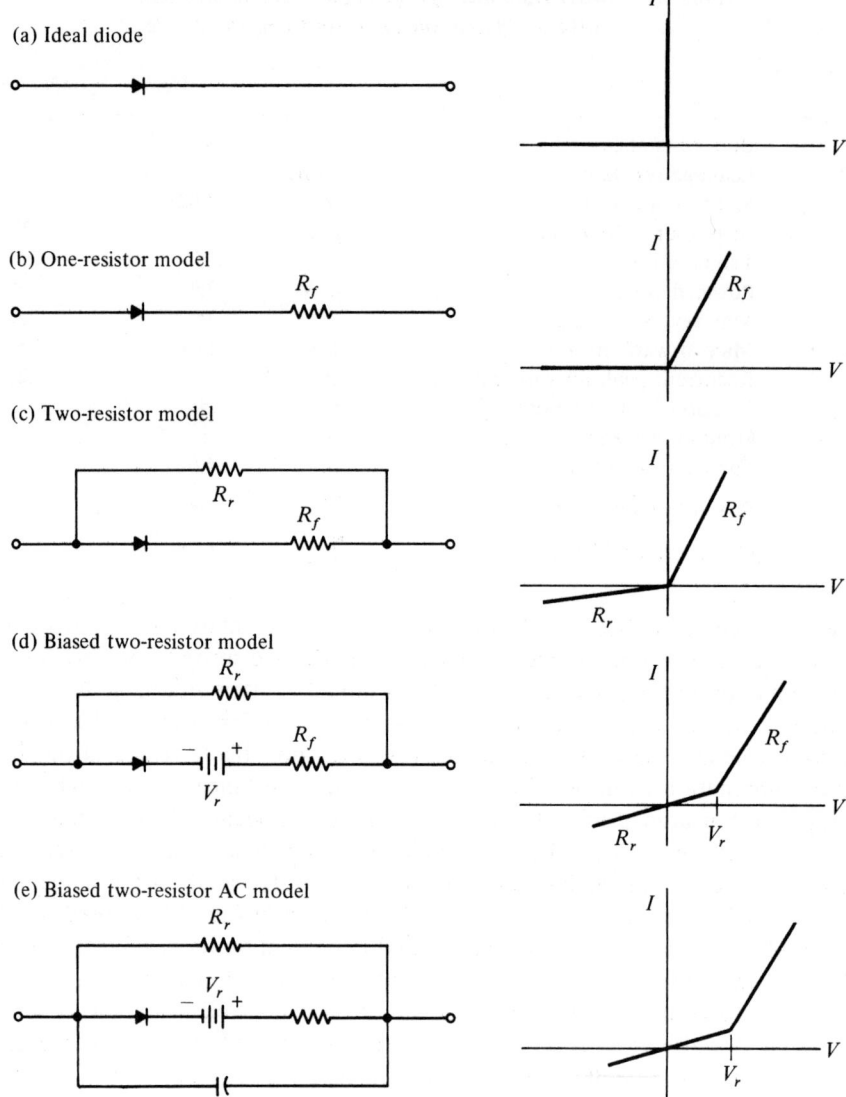

Fig. 7.9 One-Switch Diode Models

where

i_d = current through the diffusion capacitance;
T_d = diffusion capacitance time constant.

From Eq. (7.34):

$$V = \frac{mkT}{q} \ln\left(1 + \frac{I}{I_s}\right) \qquad (7.38)$$

Sec. 7.2   Diode Equivalent Circuits   **329**

(a) Two-switch model

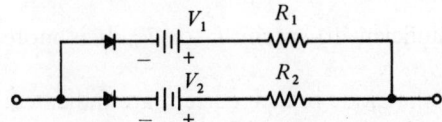
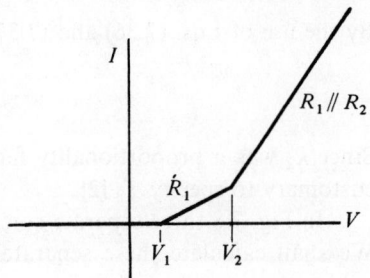

(b) Two-switch model with leakage

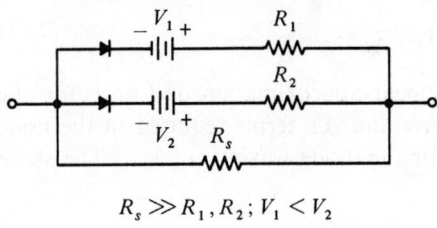

$R_s \gg R_1, R_2; V_1 < V_2$

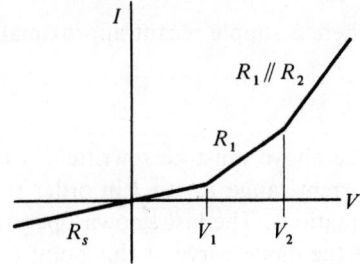

(c) Multiple-switch AC model with leakage

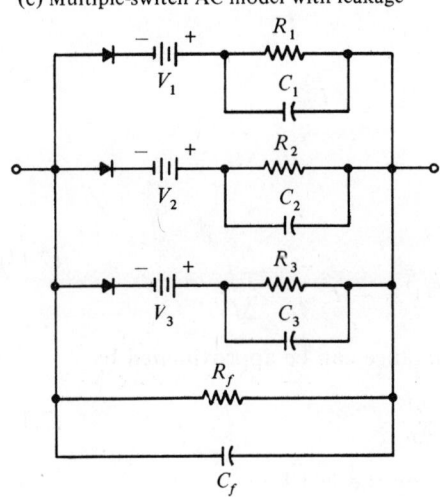
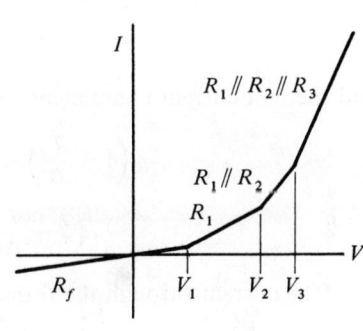

Fig. 7.10 Multiple-Switch Diode Models (Piecewise Linear)

By the use of Eqs. (7.36) and (7.37) there results

$$T_d = k_d \frac{mkT}{q} \tag{7.39}$$

Since k_d was a proportionality factor, it is sufficient to specify k_d or T_d. It is more customary to specify T_d [2].

In Fig. 7.8 three separate components of the diode branch current are indicated. We shall calculate these separately and combine them to get the complete diode voltage-current expression.

$$i_1 = I + T_d \frac{dI}{dt} = I + T_d \frac{I - I_{-1}}{\Delta t} \tag{7.40}$$

where a simple secant approximation for the diode current was used. Thus

$$i_1 = I\left(1 + \frac{T_d}{\Delta t}\right) - \frac{T_d}{\Delta t} I_{-1} \tag{7.41}$$

The above must be rewritten in terms of a linear approximation of I or V for the current range I_{-1} to I in order to produce ΔH and ΔT terms required in the node equations. The last known operating point for the diode was at I_{-1}, V_{-1}. The slope of the diode curve at this point is

$$s = \frac{dI}{dV}\bigg|_{V=V_{-1}} = \frac{I_{-1} + I_s}{\frac{mkT}{q}} \tag{7.42}$$

Using this slope and $\theta = mkT/q$

$$I = s(V - V_{-1}) + I_{-1}$$
$$= \frac{I_{-1} + I_s}{\theta}(V - V_{-1}) + I_{-1} \tag{7.43}$$

and the first current component becomes

$$i_1 = \left(1 + \frac{T_d}{\Delta t}\right) \frac{I_{-1} + I_s}{\theta} V$$
$$- \left[\left(1 + \frac{T_d}{\Delta t}\right)\left(\frac{I_{-1} + I_s}{\theta} V_{-1} - I_{-1}\right) + \frac{T_d}{\Delta t} I_{-1}\right] \tag{7.44}$$

The current through the transition capacitance can be approximated by

$$i_2 = C_t \frac{V - V_{-1}}{\Delta t} \tag{7.45}$$

where C_t is calculated from Eq. (7.35) by using the last known voltage across the junction, V_{-1}.

The current through R_s is

$$i_3 = \frac{1}{R_s} V \tag{7.46}$$

Sec. 7.2   Diode Equivalent Circuits

The total current can be written as

$$i = i_1 + i_2 + i_3 = GV - F \qquad (7.47)$$

with

$$G = \left(1 + \frac{T_d}{\Delta t}\right)\frac{I_{-1} + I_s}{\theta} + \frac{C_t}{\Delta t} + \frac{1}{R_s}$$

and

$$F = \left(1 + \frac{T_d}{\Delta t}\right)\left(\frac{I_{-1} + I_s}{\theta}V_{-1} - I_{-1}\right) + \frac{T_d}{\Delta t}I_{-1} + \frac{C_t}{\Delta t}V_{-1}$$

The ΔH and ΔT terms are obtained from

$$V = v_b - iR_b \qquad (7.48)$$

where v_b is the voltage applied across the diode. Using Eq. (7.47), we see that the above may be rewritten in the desired form:

$$i = \frac{G}{1 + GR_b}v_b - \frac{F}{1 + GR_b} \qquad (7.49)$$

$$= \Delta H\, v_b - \Delta T$$

Thus the components needed for the node equations are

$$\Delta H = \frac{G}{1 + GR_b} \qquad (7.50)$$

$$\Delta T = \frac{F}{1 + GR_b} \qquad (7.51)$$

An approximation that leads to very small errors in circuits containing appreciable resistances external to the diode is $R_b = 0$. The last two equations simplify somewhat, but not very much is gained since the bulk of the computations is in calculating G and F.

The calculation of G and F, Eq. (7.47), involves terms that are dependent on the previous solution point (I_{-1}, V_{-1}) and terms that depend on Δt alone. By calculating the latter terms just once each time a new Δt is chosen and lumping the other terms into multiplicative constants of I_{-1} and V_{-1} a significant speed up in the calculations may be achieved.

The Δt-dependent terms are

$$d_1 = \frac{T_d}{\Delta t}$$

$$d_2 = 1 + \frac{T_d}{\Delta t} = 1 + d_1 \qquad (7.52)$$

$$d_3 = \frac{C_t}{\Delta t}$$

The others can be grouped by using Eqs. (7.47), (7.45), and (7.35) as follows:

$$c_1 = 0.9 V_z$$
$$c_2 = \frac{1}{\theta} = \frac{q}{mkT}$$
$$c_3 = \frac{1}{R_s} \tag{7.53}$$
$$c_4 = \frac{1}{V_z}$$

Then the ΔH and ΔT terms may be rewritten as

$$\Delta H = \frac{a}{1 + aR_b}$$
$$\Delta T = \frac{b}{1 + aR_b} \tag{7.54}$$

with

$$a = a_1 + a_2$$
$$a_1 = d_1(I_{-1} + I_s)c_2 + c_3$$
$$a_2 = \frac{d_3}{\sqrt{1 - c_4 \min(V_{-1}, c_1)}}$$
$$b = b_1 + a_2 V_{-1}$$
$$b_1 = d_2[(I_{-1} + I_s)c_2 V_{-1} - I_{-1}] + d_1 I_{-1}$$

The reader is urged to verify these terms.

For a diode extending from node J to node K, ΔH is added at H(K, K) and H(J, J), ΔH is subtracted from H(K, J) and H(J, K); ΔT is added to T(K) and subtracted from T(J) as shown in Eq. (7.3).

The above equations represent an approximation of the diode characteristics based on the slopes of the I vs. V curve and the C_t vs. V curve at V_{-1}, the last known diode voltage. In order to be able to use these approximations for realistic solutions of the nonlinear circuit's behavior, the equations above must be iterated in such a way that the errors introduced are kept below an acceptable limit. In using these equations, an iterative scheme will be used, the Newton-Raphson method [10], and the diode equations will be solved in conjunction with the node equations until two successive iterations of I and V will result in values that differ from each other by less than a predetermined amount. Hence I_{-1} and V_{-1} will refer to previous solutions within the iterative scheme, not necessarily to iterations in the time domain. Details of this scheme of calculation will be discussed a little later.

The diode model was discussed at length so that the derivation can serve as an example of similar models for other nonlinear devices. The transistor model, to be discussed next, is based entirely on the diode model presented in this section.

7.3 Large-Signal Model for Transistors

In construction a transistor resembles two closely spaced diode junctions, the normally forward-biased emitter-base junction and the normally reverse-biased base-collector junction. The static (DC) characteristics of an NPN transistor can thus be approximated by the two-diode model (shown in Fig. 7.11) originally introduced by Ebers and Moll [5]. As with the diode model, bulk resistances must be used in the transistor leads to properly account for the difference between terminal voltages on the device and the junction and the collector diode equations. Such a modified model is shown in Fig. 7.12.

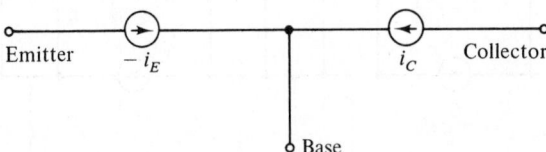

Fig. 7.11 Basic Ebers-Moll Model for the Junction Transistor

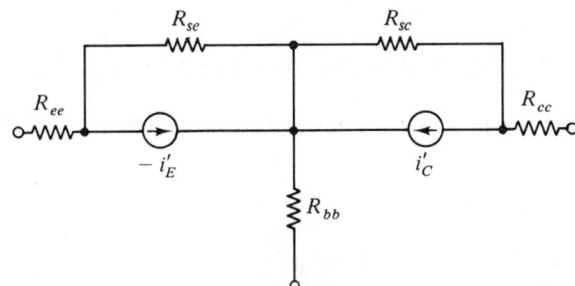

Fig. 7.12 Modified Ebers-Moll Static Model

Following the arguments in the previous section two capacitors for each junction, C_{te} and C_{de} for the emitter and C_{tc} and C_{dc} for the collector, will account for the storage effects of these junctions. As with the diode model, the diffusion capacitance will be described with a time constant T_d similar to Eq. (7.37).

The "transistor action" consists of an injected current from the emitter junction to the collector junction, also the collector current will have a similar transistor action in the emitter junction. These can be represented as controlled current sources, as indicated in Fig. 7.13. Note the polarities of the various applied voltages, the various current components, and their directions.

The common base current gains α_n and α_i are difficult to measure accurately;

334 Nonlinear Analysis Ch. 7

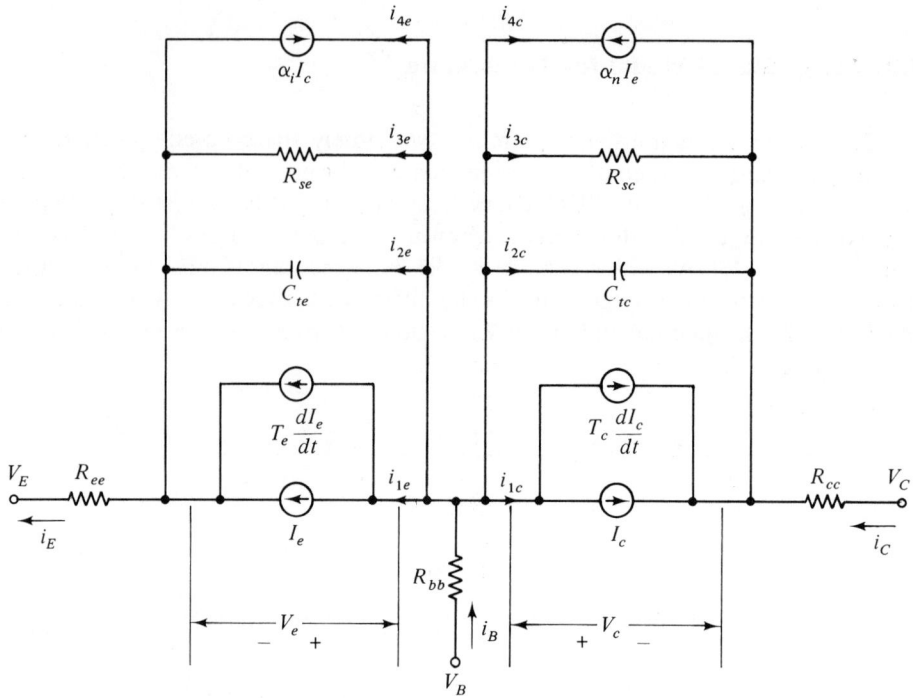

Fig. 7.13 Modified Ebers-Moll for NPN Transistor

normally, they are derived from the common emitter current gains β_n and β_i:

$$\alpha_n = \frac{\beta_n}{\beta_n + 1}$$
$$\alpha_i = \frac{\beta_i}{\beta_i + 1}$$
(7.55)

The time constants T_c and T_e are, to a first approximation, independent of the temperature T as a comparison of Eq (7.36) and (7.39) shows.

The base-collector current $I_{bc} = -i_C$ has the three components

$$I_{bc} = i_{1c} + i_{2c} + i_{3c} + i_{4c}$$
(7.56)

The first three components are described by the diode model equations, Eq. (7.47), with V_c used in place of V. The fourth component is

$$i_{4c} = -\alpha_n I_e$$
(7.57)

I_e in turn is given from the voltage V_e; its value may be calculated directly from Eq. (7.34) by using V_e in place of V and I_e in place of I:

$$i_{4c} = -\frac{\alpha_n (I_{e_{-1}} + I_{se})(V_e - V_{e_{-1}})}{\theta_e} + I_{e_{-1}}$$
(7.58)

where the subscript e refers to the emitter and the subscript -1 indicates the previous operating point.

From Eqs. (7.47) and (7.58) the base-collector current becomes

$$I_{bc} = G_c V_c + P_c V_e - F_c \qquad (7.59)$$

with

$$G_c = \left(1 + \frac{T_c}{\Delta t}\right)\frac{I_{c-1} + I_{sc}}{\theta_c} + \frac{C_{tc}}{\Delta t} + \frac{1}{R_{sc}}$$

$$P_c = \alpha_n \frac{I_{e-1} + I_{se}}{\theta_e}$$

$$F_c = \left(1 + \frac{T_c}{\Delta t}\right)\left(\frac{I_{c-1} + I_{sc}}{\theta_c}V_{c-1} - I_{c-1}\right) + \frac{T_c}{\Delta t}I_{c-1}$$

$$+ \frac{C_{tc}}{\Delta t}V_{c-1} + \alpha_n\left(I_{e-1} - \frac{I_{e-1} + I_{se}}{\theta_e}V_{e-1}\right)$$

The base-emitter current $I_{be} = i_E$ also has four components, each of which has a counterpart in the current components above. A similar derivation will yield

$$I_{be} = G_e V_e + P_e V_c - F_e \qquad (7.60)$$

where

$$G_e = \left(1 + \frac{T_e}{\Delta t}\right)\frac{I_{e-1} + I_{se}}{\theta_e} + \frac{C_{te}}{\Delta t} + \frac{1}{R_{se}}$$

$$P_e = -\alpha_i \frac{I_{c-1} + I_{sc}}{\theta_c}$$

$$F_e = \left(1 + \frac{T_e}{\Delta t}\right)\left[\frac{I_{e-1} + I_{se}}{\theta_e}V_{e-1} - I_{e-1}\right] + \frac{T_e}{\Delta t}I_{e-1}$$

$$+ \frac{C_{te}}{\Delta t}V_{e-1} + \alpha_i\left(I_{c-1} - \frac{I_{c-1} + I_{se}}{\theta_c}V_{c-1}\right)$$

The junction voltages V_e and V_c can be calculated from the terminal voltages and the terminal currents. The base to collector voltage rise is

$$V_{BC} = V_c + I_{bc}R_{cc} + (I_{bc} + I_{be})R_{bb} \qquad (7.61)$$

Similarly, the base to emitter voltage rise is

$$V_{BE} = V_e + I_{be}R_{ee} + (I_{bc} + I_{be})R_{bb} \qquad (7.62)$$

From Eqs. (7.60) and (7.59) the junction voltages are expressed in terms of the terminal currents:

$$V_e = \frac{(I_{be} + F_e)G_c - (I_{bc} + F_c)P_e}{G_e G_c - P_e P_c}$$
$$V_c = \frac{(I_{be} + F_e)P_c - (I_{bc} + F_c)G_e}{G_e G_c - P_e P_c} \qquad (7.63)$$

The last two equations are used to eliminate V_e and V_c from Eqs. (7.61) and (7.62);

the terminal currents are then calculated:
$$I_{bc} = K_1 V_{BC} + K_2 V_{BE} - K_3$$
$$I_{be} = K_4 V_{BC} + K_5 V_{BE} - K_6 \qquad (7.64)$$
with
$$D_1 = G_e G_c - P_e P_c$$
$$A_1 = \frac{P_c}{D_1}$$
$$A_2 = -\frac{G_e}{D_1}$$
$$A_3 = \frac{F_e P_c - F_c G_e}{D_1}$$
$$B_1 = \frac{G_c}{D_1}$$
$$B_2 = \frac{-P_e}{D_1}$$
$$B_3 = \frac{F_e G_c - F_c P_e}{D_1}$$
$$D = (R_{bb} + R_{cc} + A_2)(R_{bb} + R_{ee} + B_1) - (R_{bb} + A_1)(R_{bb} + B_2)$$
$$K_1 = \frac{R_{bb} + R_{ee} + B_1}{D}$$
$$K_2 = -\frac{R_{bb} + A_1}{D}$$
$$K_3 = \frac{A_3(R_{bb} + R_{ee} + B_1) - B_3(R_{bb} + A_1)}{D}$$
$$K_4 = \frac{R_{bb} + B_2}{D}$$
$$K_5 = -\frac{R_{bb} + R_{cc} + A_2}{D}$$
$$K_6 = \frac{A_3(R_{bb} + B_2) - B_3(R_{bb} + R_{cc} + A_2)}{D}$$

and the constants G_e, G_c, P_e, P_c, F_e, and F_c are given in Eqs. (7.59) and (7.60).

From the above the ΔH and ΔT terms can be found. The transistor is a three-terminal device. ΔH and ΔT terms will appear at the summation of currents at each of the three nodes. There will be three ΔT terms are nine ΔH terms. Let the voltages at the three terminals be V_B, V_C, and V_E and let the currents leaving each node be I_B, I_C, and I_E as indicated in Fig. (7.14).

The base current of the transistor is obtained from Eq. (7.64) and Fig. 7.14:
$$\begin{aligned} I_B &= I_{bc} + I_{be} \\ &= (K_1 + K_4)V_{BC} + (K_2 + K_5)V_{BE} - (K_3 + K_6) \end{aligned} \qquad (7.65)$$

Sec. 7.3 Large-Signal Model for Transistors 337

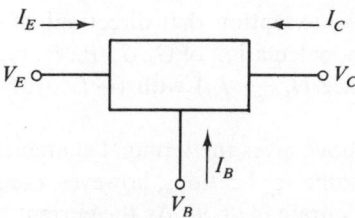

Fig. 7.14 Transistor Voltages and Currents

Since $V_{BC} = V_B - V_C$ and $V_{BE} = V_B - V_E$, the last expression becomes

$$I_B = \Delta H_{bb}V_B + \Delta H_{bc}V_C + \Delta H_{be}V_E - \Delta T_b \tag{7.66}$$

with

$$\Delta H_{bb} = K_1 + K_2 + K_4 + K_5$$
$$\Delta H_{bc} = -(K_1 + K_4)$$
$$\Delta H_{be} = -(K_2 + K_5)$$
$$\Delta T_b = K_3 + K_6$$

For the collector current

$$I_C = -I_{bc}$$
$$= \Delta H_{cb}V_B + \Delta H_{cc}V_C + \Delta H_{ce}V_E - \Delta T_c \tag{7.67}$$

with

$$\Delta H_{cb} = -(K_1 + K_2)$$
$$\Delta H_{cc} = K_1$$
$$\Delta H_{ce} = K_2$$
$$\Delta T_c = -K_3$$

The emitter current becomes

$$I_E = -I_{be}$$
$$= \Delta H_{eb}V_B + \Delta H_{ec}V_C + \Delta H_{ee}V_E - \Delta T_e \tag{7.68}$$

with

$$\Delta H_{eb} = -(K_4 + K_5)$$
$$\Delta H_{ec} = K_4$$
$$\Delta H_{ee} = K_5$$
$$\Delta T_e = -K_6$$

In establishing the instantaneous node equations, currents leaving the node are considered positive. Hence the 15 terms (nine ΔH and six ΔT terms) appearing in Eqs. (7.66) through (7.68) will be positive contributions to the **H** and **T** matrices, Eq. (7.2), at the rows, and column positions corresponding to the node number for the base, collector, and the emitter for the NPN transistor.

The modeling of PNP transistors proceeds exactly the same way as the derivation given in this section with the exception that directions of all voltages and currents must be reversed. Thus in the calculation of G_e, G_c, P_e, P_c, F_e, and F_c, replace $(I_{c_{-1}} + I_{sc})$ with $(-I_{c_{-1}} + I_{sc})$ and replace $(I_{e_{-1}} + I_{se})$ with $(-I_{e_{-1}} + I_{se})$. No other changes are necessary.

The model described above gives the terminal characteristics of the transistor to a reasonable accuracy; for some applications, however, the assumption of a constant gain α_n and α_i is not very accurate [4, 8, 9]. As the current levels increase, i.e., as the carrier injection level increases, the gain of the device changes. For most devices the change in gain may be calculated from the approximation

$$\beta_n = B_N(a_1 + a_2 V_e + a_3 V_e^2 + a_4 V_e^3) \tag{7.69}$$

where V_e is the emitter junction voltage and with B_N and the a_i found experimentally by a best fit to a set of observed data taken in the range of current levels which the transistor encounters [8]. Similarly,

$$\beta_i = B_I(b_1 + b_2 V_c + b_3 V_c^2 + b_4 V_c^3) \tag{7.70}$$

with V_c the collector junction voltage and with B_I and b_i determined from a "best-fit" approximation. The α_n and α_i are then determined from Eq. (7.55).

We remark here that although Eqs. (7.69) and (7.70) are in wide use, a more appropriate relation which gives the variation of β for various operating conditions involves the junction current:

$$\beta_n = \frac{k_1}{1 + \dfrac{k_2}{\sqrt{I_c}} + k_3 I_c}$$

This equation reflects the observed behavior in that $\beta_n \to K_1/I_c$ for large values of I_c and $\beta_n \to K_2\sqrt{I_c}$ for small values of I_c. Eqs. (7.69) and (7.70), however, are entrenched in practice.

We should remark that the model has now become more complicated than our ability to obtain meaningful data for it. Nevertheless, the capability has been provided to model the transistor behavior in considerable detail. Not all the elaborations given above are needed in most cases since typical transistor parameters are subject to a very wide spread of values. Particularly the α_n and α_i in Eqs. (7.69) and (7.70) are usually considered constants, but some programs have provisions for the indicated variations [7, 11].

Thus the Ebers-Moll model presented here has the following parameters:

R_{bb} = base bulk resistance (Fig. 7.13)
R_{cc} = collector bulk resistance (Fig. 7.13)
R_{ee} = emitter bulk resistance (Fig. 7.13)
R_{sc} = collector leakage resistance (Fig. 7.13)
R_{se} = emitter leakage resistance (Fig. 7.13)
I_{sc} = reverse saturation current, collector [Eqs. (7.34), (7.59)]
I_{se} = reverse saturation current, emitter [Eqs. (7.34), (7.59)]
m_e = recombination constant, emitter [Eqs. (7.34), (7.59)]
m_c = recombination constant, collector [Eqs. (7.34), (7.59)]

B_N = normal "beta" [Eqs. (7.69), (7.55), (7.57), (7.59)]
a_1, a_2, a_3, a_4 = B_N fitting constants [Eq. (7.69)]
B_I = inverted beta [Eqs. (7.70), (7.55), (7.60)]
b_1, b_2, b_3, b_4 = B_I fitting constants [Eq. (7.70)]
N_c = collector junction grading constant [Eqs. (7.35), (7.59)]
N_e = emitter junction grading constant [Eqs. (7.35), (7.60)]
V_{Zc} = collector junction potential [Eq. (7.35)]
V_{Ze} = emitter junction potential [Eq. (7.35)]
C_{tc} = collector transition capacitance [Eq. (7.35)]
C_{te} = emitter transition capacitance [Eq. (7.35)]
T_c = collector diffusion capacitance constant [Eqs. (7.37), (7.59)]
T_e = emitter diffusion capacitance constant [Eqs. (7.37), (7.60)]
$I_{c(\max)}$ = maximum collector current: specification sheet
$V_{BC(\max)}$ = maximum base-collector voltage: specification sheet
$V_{BE(\max)}$ = maximum base-emitter voltage: specification sheet
$V_{CE(\max)}$ = maximum collector-emitter voltage: specification sheet
$P_{C(\max)}$ = maximum collector power dissipation: specification sheet
T = Temperature: operating conditions

Hence some 33 parameters are required to describe the transistor model. The various maximum voltage, current, and power ratings are not modeled in the transistor. Zener action has been neglected: a maximum voltage value can be specified for the constants in Eqs. (7.69) and (7.70); if the voltage across either the emitter or collector junction exceeds this value, abnormal (and usually irreversible) action occurs and an appropriate warning message could be printed. Most of the parameters are very difficult to obtain from the manufacturer's data and recourse is taken to a multi-parameter fit to measured data [8].

For comparison, Table 7.2 lists values for the 2N2222 Silicon Epitaxial Planar Transistor, a high-speed computer switching transistor. The model library data for the NET-1 computer program and manufacturer's data sheet were used. We note that the transistor gains are modeled as constants.

The model presented here is in wide use. Part of its appeal is that model data are stored on a mass-access device and called in whenever they are needed in a particular circuit. Thus to use the data in Table 7.2 one merely specifies the base, collector and emitter node numbers, and the transistor type.

There are several shortcomings with this model, aside from its sometimes unnecessary elaborateness. The most obvious source of inaccuracy is the constancy of the base spreading resistance R_{bb}. In practice it is found that R_{bb} varies with base current and V_{BE} voltage. Although this "base resistance modulation" is neglected in virtually all analysis programs, it is possible to model such a nonlinear resistance in the base lead at some cost of computational complexity [11]. One might also use a separate equation or a table of R_{bb} vs. I_B values to effect such model refinement.

In order to speed up the solution of the transistor model equations, the various terms appearing in Eqs. (7.59), (7.60), (7.64), (7.66), (7.67), and (7.68) can be separated

Table 7.2 *Comparison of NET-1 Data and Specification Sheet Information 2N2222 Silicon Epitaxial Planar Transistor (High-speed Switching)*

			NET-1	*Spec. Sheet*
R_{bb}	Base spreading resistance	ohms	50	130
R_{sc}	Collector leakage resistance	ohms	1×10^9	—
R_{cc}	Collector bulk resistance	ohms	2.8	1.6
R_{ee}	Emitter bulk resistance	ohms	.5	.4–2.0
R_{se}	Emitter leakage resistance	ohms	1×10^9	—
a_1	Current gain curve shape	—	1	—
a_2	$S_n = \sum_{i=1}^{4} a_i(V_e)^{i-1}$	—	0	—
a_3			0	—
a_4	$0 \leq V_e \leq A$	—	0	—
A	Max V_{BE} for the a_i and b_i	V	1	—
b_1	Inverted current gain curve shape	—	1	—
b_2	$S_i = \sum_{i=1}^{4} b_i(V_c)^{i-1}$	—	0	—
b_3			0	—
b_4	$0 \leq V_c \leq A$	—	0	—
B_N	Current gain amplitude	—	136	100 (min) 300 (max)
B_I	Inverted current gain amplitude		2.3	—
I_{sc}	Collector-base saturation current	A	3.7×10^{-11}	$I_{cBo} = 10^{-6}$
P	Max power dissipation	mW	500	500
T	Junction temperature	°C	25	25 (ambient)
V_{CE}	Max collector-emitter voltage	V	30	30 (min)
V_{BE}	Max base-emitter voltage	V	5	5 (min)
V_{CB}	Max collector-base voltage	V	60	60 (min)
I_C	Max collector current rating	ma	500	530
V_{ze}	Collector-base contact potential	V	.9	1.
N_c	Collector grading constant	—	.35	.4
C_{tc}	Collector transition capacitance	pf	12	4
V_{Ze}	Emitter-base contact potential	V	0.9	1
N_e	Emitter grading constant	—	0.4	0.53
C_{te}	Emitter transition capacitance	pf	20	20
I_{se}	Emitter-base saturation current	A	9.3×10^{-12}	—
m_c	Collector curvature coefficient	—	1.02	—
m_e	Emitter curvature coefficient	—	0.97	—
T_c	Collector diffusion constant	μsec	0.62	—
T_e	Emitter diffusion constant	nsec	1.58	—

into components dependent on Δt only (which are calculated just once for each run where Δt remains unchanged) and terms dependent on the previous iteration. The reader is asked to derive the two sets of components as was done with Eqs. (7.52) through (7.54) for the diode model.

In the analysis of critical very high-speed switching circuits the transistor header capacitance and the socket capacitance (if any) may be included in the analysis as external capacitors. This refinement is typically unneeded, however.

The reader might wonder where he might obtain the necessary data for using the model shown here. Fortunately, there are libraries that contain data on many transistor types [1, 7, 11]. Although the specific models used there may be slightly different from the one derived here, the data can be readily used to set parameters in our derived model. The library listed in [1] is especially useful and easily available. Although the particular transistor type used in a specific example is not likely to be found in the library, similar transistor types are usually in the library and a few parameter changes can be made to yield a usable model.

It should also be pointed out that few applications require all the parameters given here. The program to be described in subsequent sections does in fact use only about half as many parameters by letting α_n and α_i be constants and dispensing with R_{bb} as well as with the maximum ratings. A brief list of data on some transistor types is given in Appendix C; the list is suitable for Program TRAC, the transient analysis program to be discussed in this chapter.

Transistor and diode data have been measured for many devices and libraries are available for programs CIRCUS and SCEPTRE. These programs model transistors with all the parameters discussed here; from their libraries simpler models can be derived which give sufficiently good answers to many problems. Computer routines exist that allow the data from one program to be used with other programs. Later we shall use a somewhat simplified model derived from the model in this section; data for the simplified model can be obtained from the more extensive CIRCUS library by a routine [13]. The data included in Appendix C were obtained partially through this routine and partially from the SCEPTRE library.

Only as the very last resort should measurement of parameters be attempted. The required data are difficult to obtain and individual parameters are difficult to isolate. A guide to obtaining such measurements is contained in [13]. A thorough understanding of transistor device theory is a must in deriving useful models.

Many applications do not need the nonlinear aspects of the transistor characteristics to be very accurate. For example, a piecewise linear model similar to the piecewise linear diode models shown in Figs. 7.9 and 7.10 is possible. Such a model is discussed in the next section.

Similarly, for high-frequency small-signal applications the capacitive effects and time-delay effects become dominant with the transistor operating essentially as a linear device. The transistor is then best represented by a hybrid-Π model and treated as a linear device. Although many references exist for this model, the interested reader will find a good introduction in [4] and further work in [9].

Finally, the methods developed here can also be applied to vacuum tube technology. Basically, the vacuum tube space-charge equations will play the role of the exponential diode equation. Models for large-signal equivalent circuits of vacuum tubes are not widely available, but they can be derived starting with the analytical work summarized in [12].

7.4 Piecewise Linear Transistor Model*

In this section we present a simplified circuit model for transistors using the switching capabilities discussed in Sec. 4.7. Thus the model here will be a piecewise linear network model whose component values can be altered by the direction of current flow of some controlling branches.

The usefulness of any general circuit analysis program depends critically on the appropriateness of the models chosen to represent active devices. All the ingenuity and optimization which can be applied to improve a program are of no avail if the models used by the computer to represent the component devices in the system are inadequate. The optimum model to be used for a given device depends strongly on the nature of the results expected from the computation. There is no perfect model that completely describes the behavior of an electronic device.

Design computations done by hand are restricted to simple models or to small subsystems. Computers are not so restricted and can accommodate complicated models for a single transistor. Such a complex model presents severe problems in parameter evaluation, however. For some purposes a much simpler model with more easily measured parameters may provide satisfactory results. This section describes an effort to develop a reasonably simple piecewise linear (PWL) model useful in the analysis of large-signal switching transistor operation.

A simple saturating inverter circuit shown in Fig. 7.15 is used to establish the usefulness of the chosen PWL transistor model for large-signal circuit analysis. The transistor is driven from cutoff into saturation and back to cutoff. Representative waveforms are as shown in Fig. 7.16.

The transistor model to be used is shown in Fig. 7.17. It is a PWL common-base model with a capacitance added across each "junction." The current controlled switching capability described in Sec. 4.7 is used here to change resistance and capacitance values in approximation to the changes in the static and dynamic characteristics of

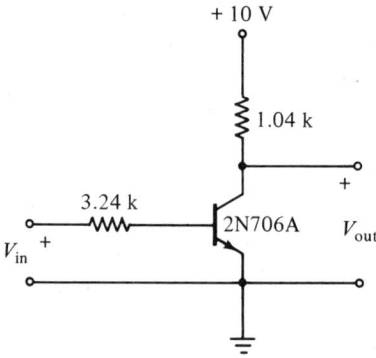

Fig. 7.15 Transistor Inverter Circuit

*This material was contributed by Dr. C. O. Harbourt, Department of Electrical Engineering, University of Missouri, Columbia, Mo.

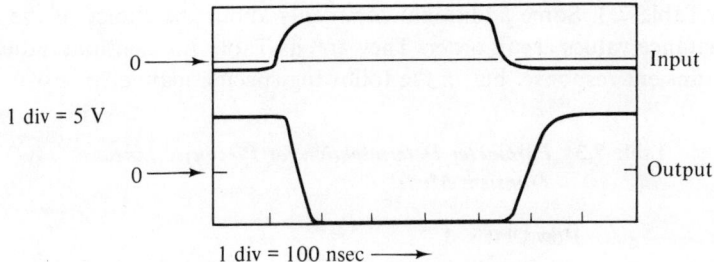

Fig. 7.16 Inverter Circuit Waveforms

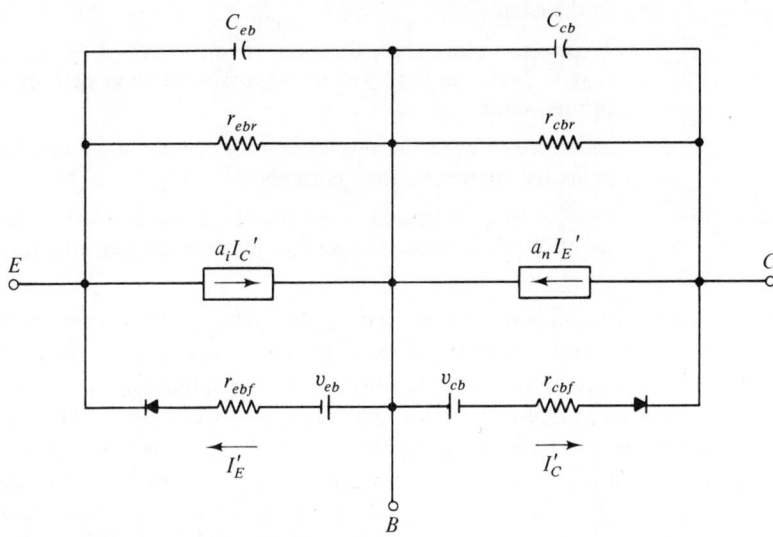

Fig. 7.17 Piecewise Linear Transistor Model

on- and *off-biased* junctions. The model uses only two values for each changing parameter. It could be made much more elaborate if desired, but we wish to see how well a simple piecewise linear model can be made to perform.

Only minimal efforts are made to identify internal transistor processes with the model. We model the individual diode resistive nonlinearities, the current transfer effects of normal and inverse transistor action, and an equivalent lumped capacitive representation of depletion-region capacitances and the dynamics of the diffusion process in the base region. The base-emitter and the base-collector diodes are thus modeled by one-switch diode models each as shown in Fig. 7.9(c). The capacitor effects will also be switched between a forward conducting junction where the C_d capacitance is preponderant to a reverse-biased junction where the C_t capacitance has more effect. The two capacitance values will be held constant, however, and thus some of the computational difficulties of Sec. 7.3 will be avoided.

A summary of the means used to evaluate the parameters of the transistor model is given in Table 7.3. Some additional comments about the choice of the higher or "on" capacitance values are in order. They are, as Table 7.3 mentions, adjusted to fit observed transient response, but in the following specific manner.

Table 7.3 *Parameter Determination for Piecewise Linear Transient Model*

Parameter	How Obtained
r_{ebf}, r_{ebr}, v_{eb}	Piecewise linear approximation to measured $i_E - v_{EB}$ curve ($v_{CB} = 0$) chosen to fit well for approximate range of i_E expected in the circuit
r_{cbf}, r_{cbr}, v_{cb}	Piecewise linear approximation to measured $i_c - v_{CB}$ curve ($v_{EB} = 0$) chosen to fit well for approximate range of i_C expected in the circuit
α_n, α_i	Measured forward and reverse short circuit common-base current gains for "representative" currents
C_{eb}, C_{cb}	*Low values*, measured capacitance, reverse-biased junctions (at 10 V); *high values*, adjusted to fit observed transient response

The charge control time constants τ_F and τ_R associated with forward and inverse transistor action, respectively, were measured using the compensated pulse attenuator method [14]. This measurement technique consists of using a pulse input to completely switch the transistor from ON to OFF and back to the ON condition. The input and output waveforms are observed on an oscilloscope and are shown on Fig. 7.16. The rise and fall time constants can be measured directly from the oscilloscope traces. In this example $t_d + t_r = 95$ nsec and $t_s + t_f = 110$ nsec. We note that both these times are measured to 90% of the final values of the output voltage. More detail for the output wave is shown in Fig. 7.18.

These measured values were divided by the "on" (low) resistance values, r_{ebf} and r_{cbf}, respectively, to obtain first-guess "on" values for C_{eb} and C_{cb}. Computer runs made using these values showed transient times ($t_r =$ rise time; $t_f =$ fall time; $t_d =$ delay time; $t_s =$ storage time; see Fig. 7.18) to be much longer than experimentally observed times. At this point a trial-and-error procedure of halving the "on" capacitance values was repeated three times, the final result being shown in Fig. 7.18. Thus, the "on" capacitance values used in the model are:

$$C_{eb(\text{on})} = \frac{\tau_F}{8 r_{ebf}}$$

$$C_{cb(\text{on})} = \frac{\tau_R}{8 r_{cbf}}$$

No effort has been made to justify these choices by subsequent analysis.

The *off*-values of the capacitors were chosen to be representative of the back-biased transistor junctions.

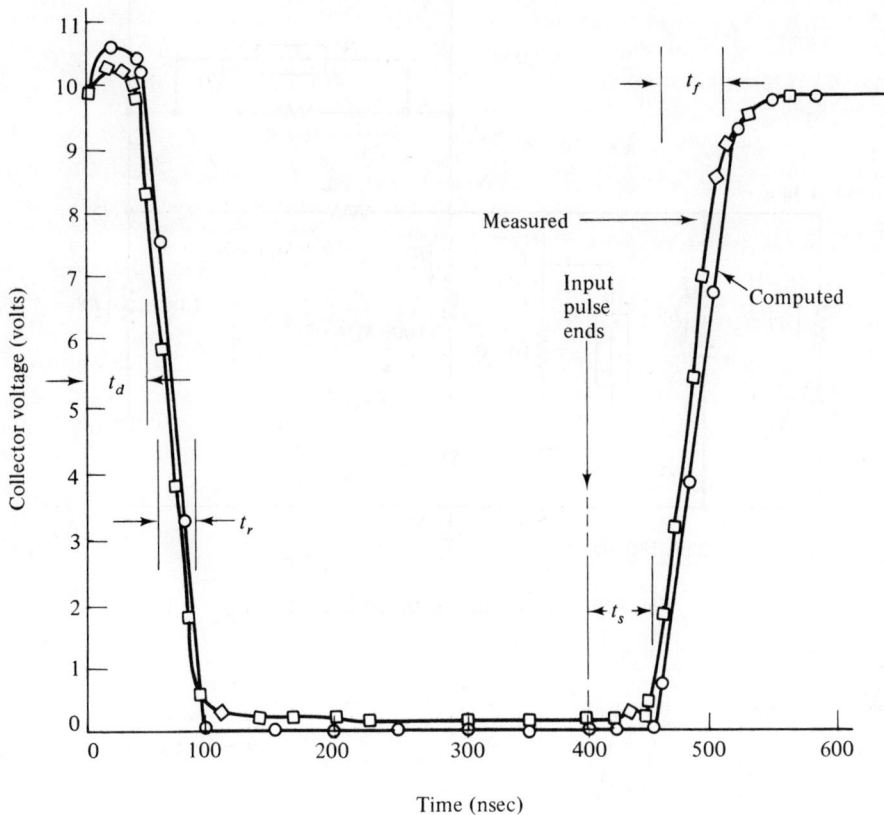

Fig. 7.18 Measured and Computed Waveforms

It is one thing to "fudge in" the best parameter values for a particular set of circuit conditions, but it may be quite another to devise a model that is also useful under varying external conditions. As an alternative to attempting to justify the final (capacitance) parameter choices by analysis, measured and computed waveforms are now compared for varying *on* and *off* drive conditions in the inverter circuit.

For convenience, *on* and *off* drive are given in volts applied to the series combination of the 3.24K resistor and the base-to-emitter terminals of the transistor. When the *on* drive is varied, the *off* drive is held constant at -1 volts. When the *off* drive is varied, the *on* drive is held constant at $+3$ volts, which is more than sufficient to saturate the transistor. Figures 7.20 through 7.23 present the comparison of measured and computed transient times. The measured quantities were obtained by scaling distances on oscilloscope photographs of the input and output waveforms. The calculated quantities were obtained by interpolation between points on the output waveforms computed using the PWL model of Fig. 7.19 and a forty-point approximation to the actual input voltage waveform. The overall imprecision of data obtained in this manner complicates any attempt to discuss deviations between calculated and

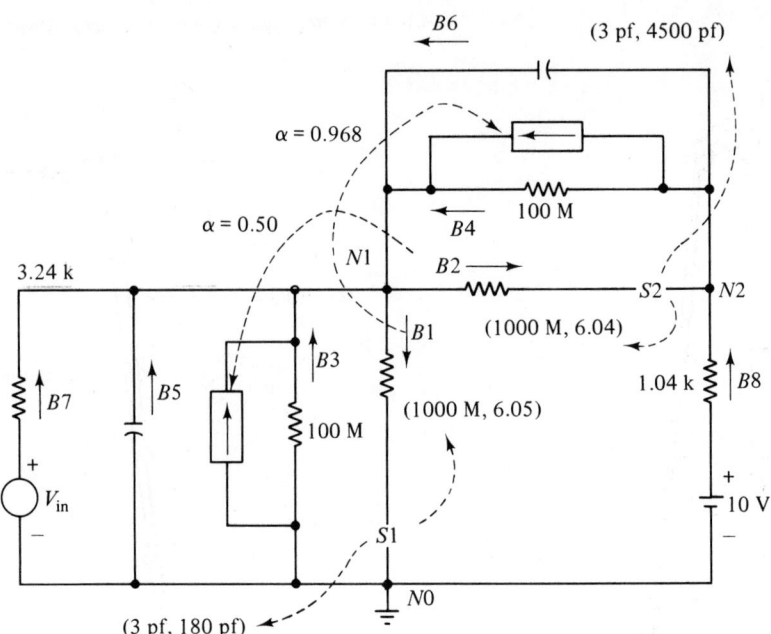

Fig. 7.19 Computational Model of the Circuit

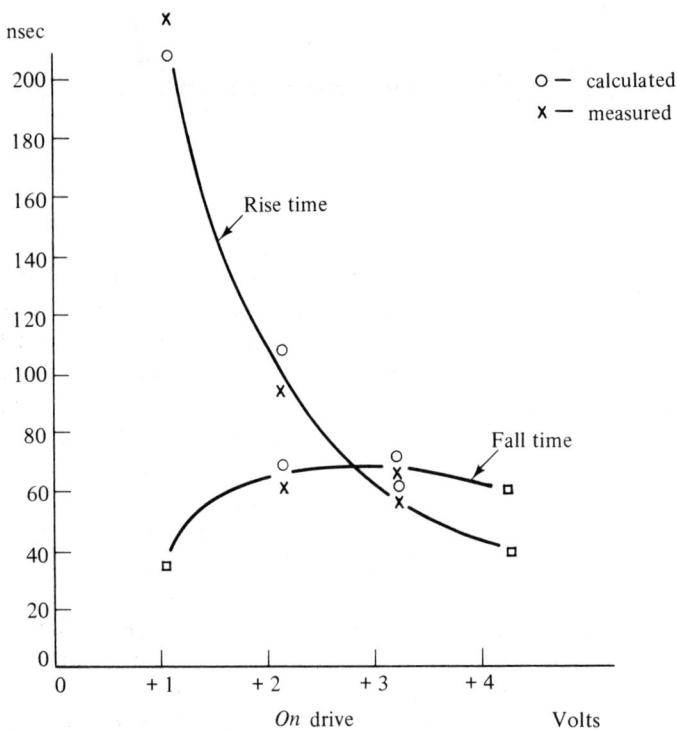

Fig. 7.20 Rise and Fall Times vs. *on* Drive

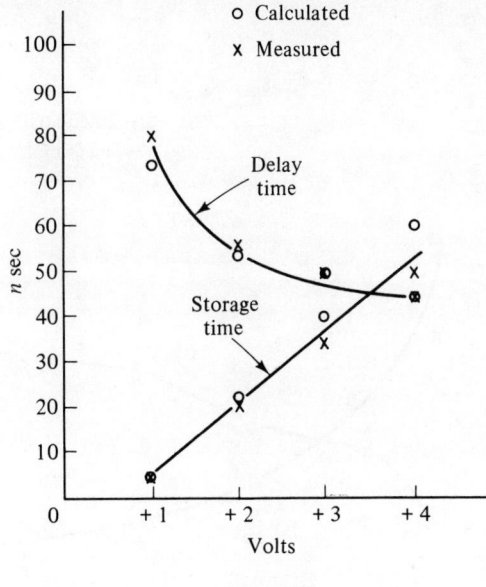

Fig. 7.21 Delay and Storage Times vs. *on* Drive

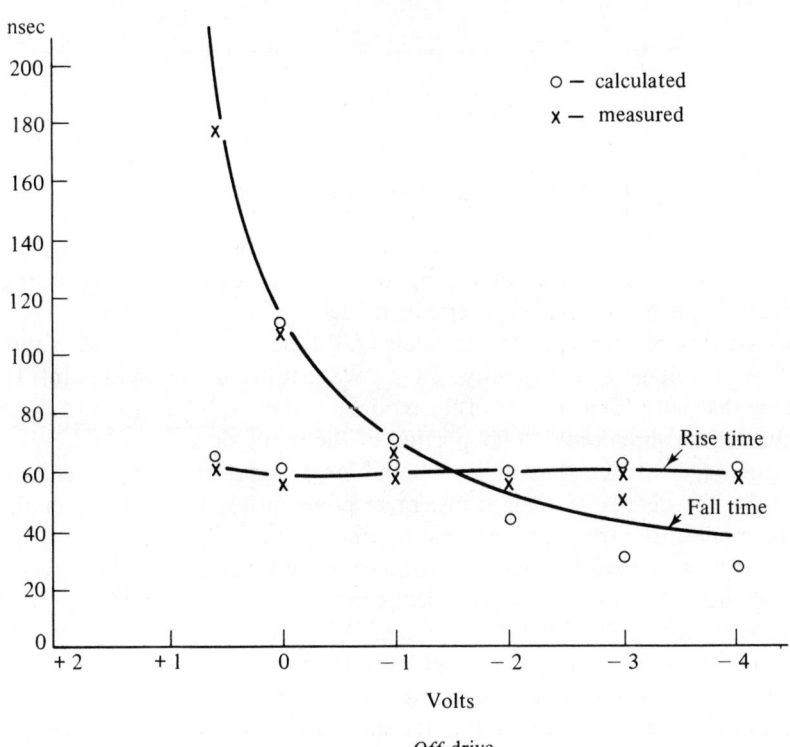

Fig. 7.22 Rise and Fall Times vs. *off* Drive

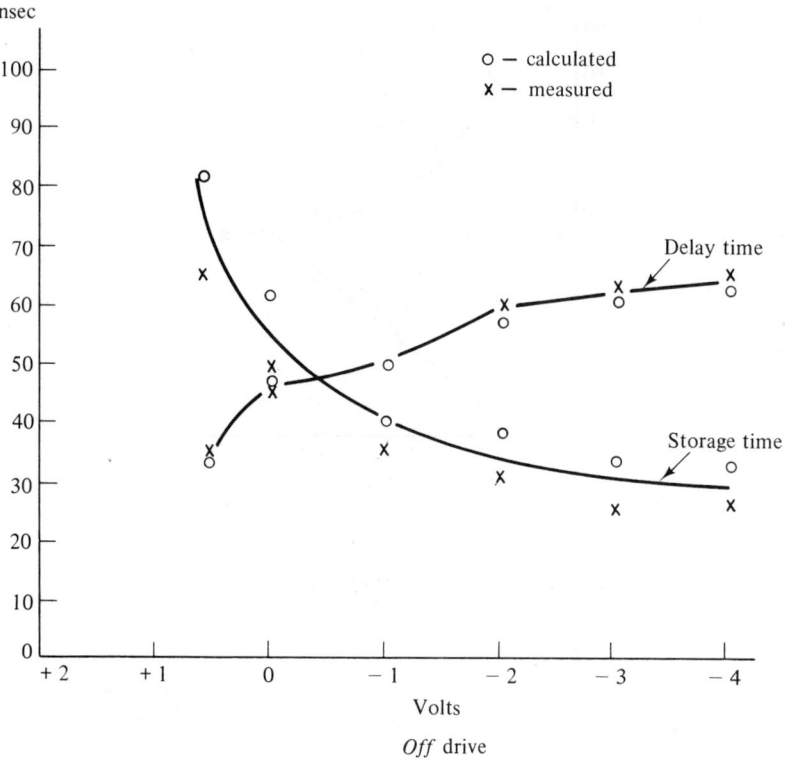

Fig. 7.23 Delay and Storage Times vs. *off* Drive

measured points. The qualitative nature of the model's behavior is obviously correct for all four transient times. It is interesting to note the success with which storage time is predicted by such a simple PWL circuit model.

The one very bad point on the calculated fall-time curve of Fig. 7.22 is the use of an "off" drive voltage in the vicinity of the PWL emitter diode breakpoint. It seems most likely that both capacitances of the model failed entirely to switch to their lower values during the supposedly "off" portion of the input signal.

In summary, we see that at least some large-signal switching circuits can be analyzed for the details of their transient response using a relatively simple PWL transistor model. In particular, reasonably accurate predictions of storage time can be obtained from a model formed of ordinary circuit elements without recourse to overly complicated models. Other transient times can also be satisfactorily predicted, as might be expected. The PWL transistor model shown here can be of considerable practical use in computer-aided analysis of transistor switching circuits, of which the pulse inverter is the simplest representative.

It should be stressed, however, that the success of this model and of similar simple models hinges on a knowledge of the expected ranges of collector and emitter currents expected in the circuit. If no such *a priori* knowledge exists, an iterative approach may be usable: one guesses the range of currents, obtains a set of component values for this range (the diode elements must be fitted), does an analysis, and based on the result of the analysis refines the current-range guesses. This process may be repeated until satisfactory results are reached using the computational model only. Of course, the true success of the process can be judged only by comparison with experimental results.

7.5 Equation Iterations

In this book we study the solution of network node-voltage equations based on summing currents. At each node the currents are summed; by Kirchhoff's current law these sums must all be zero. The currents leaving each node are expressed as linear combinations of the node voltages. Solving the resultant set of simultaneous linear algebraic equations gives the node voltages in the network.

For linear networks the node voltages that result from the process just outlined will be the actual observed voltages since the currents in the network obey the linearity conditions necessary for the voltage-current relationship (Ohm's law) for each branch. In a nonlinear network the voltages and currents obey this law only as an approximation in the neighborhood of an operating point. In the developments of Secs. 7.1 through 7.3 the voltage-current expressions will quite accurately describe the actual observed behavior of the components only for very small changes in the voltages and currents. In order to accurately model the circuit behavior, the linearization of the terminal characteristics must be obtained for an operating point, typically the solution at the previous time step. The network equations are then solved and the newly calculated voltages and currents in the nonlinear elements are checked against the starting conditions. If the difference is small enough, the solution is accepted; otherwise, a new linearization is performed about the just-obtained "operating point." All nonlinear elements must satisfy this condition before a new time step may be taken.

Fortunately, the nonlinear components in transistor and diode circuits are subject to rather large component-to-component variations and good design practice will result in circuits that are not critically dependent on the parameters of these components. Hence the agreement between assumed operating conditions and calculated values need not be too close; 10% differences are typically acceptable.

Since the solutions from the node equations are the node voltages from which the branch voltages are easily computed, the iterative process for the circuit voltage-current conditions will be started with assumed branch voltages for each nonlinear element. These voltages will be the accepted solutions at the previous time step. For each element a linear approximation at this assumed voltage is made. The slopes of the linear approximations are the conductance values used in setting up the set of network node equations. These equations are then solved and the element voltages based on the linear approximation are calculated. If these voltages are within acceptably close values to the voltages assumed in obtaining the linear approximation, the solution is accepted. Otherwise, these voltages are used for finding a new operating point on the nonlinear characteristic and the entire procedure must satisfy the convergence criterion; if any one violates it, the entire iteration is repeated for all circuit components.

The procedure is illustrated for a single diode without capacitive effects in Fig. 7.24. The steps of the algorithm are as follows:

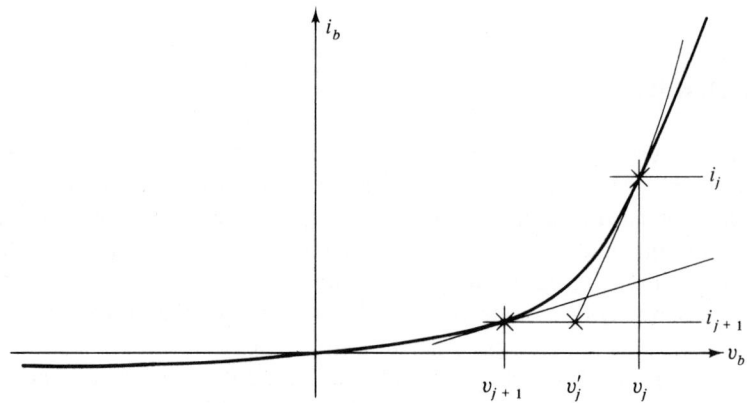

Fig. 7.24 Iterations for a Simple Diode Circuit

(1) $j = 0$; let V_0 denote the solution at the last time step.

(2) Compute $di/dv|_{v=v_j}$ for all nonlinear elements.

(3) Using the slopes from Step 2, set up the node equations: $\mathbf{HV_n = T}$.

(4) Solve for the node voltages $\mathbf{V_n}$.

(5) Calculate the branch voltages v'_j for the nonlinear elements and the branch currents i_{j+1}; use the approximation of Step 2.

(6) If $|v'_j - v_j| \leq \epsilon_{max}$ for all elements, accept v'_j as the branch voltage. ϵ_{max} is an error limit that can depend on the absolute value of v'_j as well as some constant. For example,

$$\epsilon_{max} = \max(0.1v_j, 10^{-6})$$

would establish a tolerance limit of at least 1 μV. If all errors are within the allowable error limit, stop the iteration and accept this iteration result as the solution for this time value.

(7) Use i_{j+1} of Step 5 to solve for a new v_{j+1} as indicated in Fig. 7.24.

(8) $j \leftarrow j + 1$; go to Step 2.

Fig. 7.24 clearly shows that at each approximation the nonlinear function is approximated by the slope of the function. Approximation of this nature is known as the Newton-Raphson approximation. Its convergence properties are well known and the reader is referred to other sources [10] for a discussion of these properties.

We illustrate the above algorithm by a very simple circuit.

Sec. 7.5   Equation Iterations   **351**

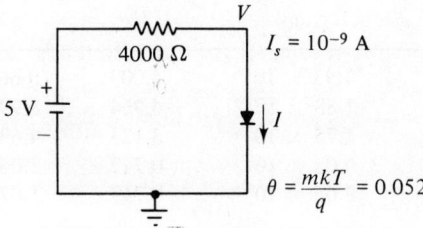

Fig. 7.25   Diode Example Circuit

EXAMPLE

Find the voltage across the diode in the circuit shown in Fig. 7.25.

Solution. The diode equation is

$$I = I_s(e^{V/\theta} - 1) = 10^{-9}(e^{19.3V} - 1)$$

The slope of the diode equation is

$$\frac{dI}{dV} = y = \frac{I_s(e^{V/\theta})}{\theta} = 19.3 \times 10^{-9} e^{19.3V}$$

The diode is modeled as in Eq. (7.47) with $T_d = 0$, $C_t = 0$, $R_s \to \infty$;

$$I_d = GV - F$$

with $G = \dfrac{I + I_s}{\theta}$

$$F = \frac{(I + I_s)V}{\theta} - I$$

The calculated node voltage is obtained from the expression

$$\frac{V_c - 5}{4000} + GV_c - F = 0$$

Thus the calculated node voltage becomes

$$V_c = \frac{F + \frac{5}{4000}}{G + \frac{1}{4000}}$$

The current flowing in the linear circuit is

$$I' = I - y(V - V_c)$$

The diode voltage for the current I' is

$$V_d = \theta \ln\left[\frac{I'}{I_s} + 1\right]$$

The procedure will be stopped if V_d and V agree within 5%, the chosen convergence criterion. The calculations are summarized in the table below with an arbitrary starting value of $V = 0$ V.

V	I	y	V_c	I'	V_d
0.	0.	1.93×10^{-8}	5.009	9.66×10^{-8}	0.237
0.237	9.68×10^{-8}	1.88×10^{-6}	4.964	8.99×10^{-6}	0.473
0.473	8.99×10^{-6}	1.75×10^{-4}	3.123	4.69×10^{-4}	0.678
0.678	4.69×10^{-4}	9.05×10^{-3}	0.742	1.06×10^{-3}	0.719
0.719	1.06×10^{-3}	2.05×10^{-2}	0.719	1.07×10^{-3}	0.719

The diode voltage is therefore 0.72 V. The process converges for arbitrary starting value V and requires at most seven steps for a 5% convergence (see Problem 7.8).

7.6 Organization of a Nonlinear Program

The nonlinear circuit analysis program can be viewed as an extension of the linear transient analysis program discussed in Sec. 4.5. The essential difference is in the calculation of the node voltages since iterative procedures must be used. However, this difference dictates that the Y_n matrix must be set up at each iteration, at least for the nonlinear elements. Thus it becomes uneconomical to invert the Y_n matrix as was done in Sec. 4.5. Instead, a simultaneous equation solution routine will be called for each solution of the node voltages.

This difference makes extensive changes to the transient analysis program shown in Fig. 4.13. Basically, there will be no difference between the initial time solution (the right-side loop in Fig. 4.13) and the subsequent time-step solutions (the left-side loop). As in the linear transient analysis case, much output will be generated; it is convenient to use plotted outputs rather than printed numbers. The plotting routines of Appendix B will be useful in this. The input to the routine will consist of two kinds of data: (1) solution controls and (2) circuit element descriptors. As in the linear analysis case it is desirable to input such data in virtually any order using a free-form input language. The input language of Sec. 3.11 and the subroutine of Fig. 4.14 can be expanded to handle the input for this program; provisions must be made to include transistors and diodes as permissible "elements."

A flowchart for the overall organization of the nonlinear transient analysis program is shown in Fig. 7.26. The flowchart gives the flow of the program for iterative solution of the node equations, but it must be modified for initial condition solutions. As in Chapter 4, this capability may be achieved with a very small DT value at the beginning of the analysis. Program logic for the initial condition solution has been omitted from Fig. 7.26.

The various subroutines of Sec. 4.5 have been retained in this program whenever convenient. Thus Subroutine INPUT obtains the input data, both control statements and circuit descriptors, from a free-format input data stream. Specific details of the valid input forms are given in the next section; however, they are similar to the format details discussed in Sec. 4.5.

Similarly, Subroutine SETEI evaluates the time variable voltage and current sources. Subroutine SETUP calculates the entries for the **H** and **T** matrices; however,

Sec. 7.6   Organization of a Nonlinear Program

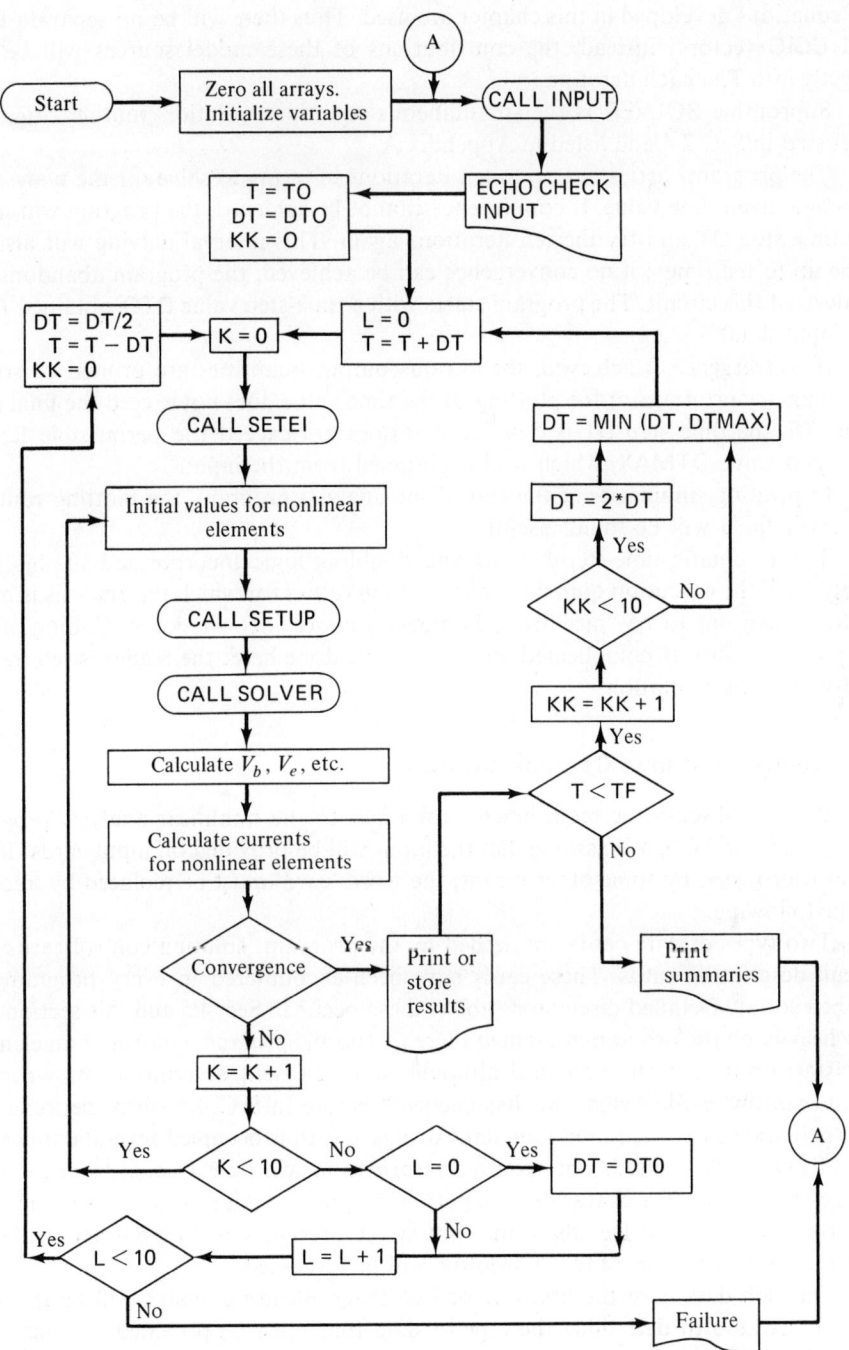

Fig. 7.26   Flowchart of Nonlinear Program

the equations developed in this chapter are used. Thus there will be no separate **LLIC** and **CCIC** vectors; instead, the contributions of these model sources will be put directly into **T** at each iteration.

Subroutine **SOLVER** is the simultaneous equations solution routine originally discussed in Sec. 2.4 and listed in Appendix A.

The program carries out up to ten iterations in trying to solve for the node voltages at a given time value. If convergence cannot be achieved, the program will halve the time step **DT** and try the ten iterations again. This interval halving will also be done up to ten times; if no convergence can be achieved, the program abandons the analysis of this circuit. The program starts with a time-step value **DT0** obtained from the input data.

If convergence is achieved, the various output quantities are printed or stored for summary printouts or for plotting. If the time value does not exceed the final time value **TF**, the time step **DT** is doubled if it does not exceed the permissible largest time-step value **DTMAX**, which is also obtained from the input.

In printing summaries at the end of the analysis program, the plotting routines of Appendix B will be found useful.

The automatic time-step halving and doubling logic incorporated in this program make the execution quite fast, but the time values for which the analysis is made will not turn out to be "nice" round values, a minor inconvenience. Coding of the program is relatively complicated and will not be done here; the reader is referred to [2] for an implementation.

7.7 Input Language Considerations

We now discuss the requirements for input to the nonlinear analysis program described in Sec. 7.6. We assume that the input will be provided on input cards; if the input is provided by some other means, the word 'card' must be replaced by 'record' in the following.

Two types of data cards are needed by the program: solution control cards and circuit description cards. These cards have been encountered in every program described so far. Detailed discussions about these occur in Sec. 4.5 and this section will rely heavily on the program presented there. A specific program control or an element will start on a separate data card although some element descriptions may occupy more than one card. Again, this has occurred before in Sec. 4.4 where periodic and non-periodic sources required long data strings and thus occupied several data cards. Free format will be used. The required information will be extracted from the card images by means of a character recognition program similar to Subroutine **INPUT** described in Fig. 4.14. Details of the routine required here will be omitted; only the structure and functions of the subroutine will be discussed.

On each data card the first one or two letters (leading blanks will be ignored) will act as keys to determine the type of data that must be provided on that card. As in Sec. 4.5, one can write a program that ignores all characters until the next key character is encountered, but it is probably better to provide for a simple capability

Sec. 7.7 Input Language Considerations

only by requiring blanks (or other specific delimiters) to separate information. The details of the free-format routine are left to the reader; the implementation in [2] allows only a limited number of delimiters.

I. Control Cards

(a) PRINT J[, A]

PR must be the first two letters. J will denote the ratio of calculated to printed results. Since an automatic step-size adjustment is used, a great many solutions may be calculated and only a few selected points would be printed. J can be omitted; a default value will then be used (J = 10). 'A' should be a flag indicating if branch voltages and/or currents are to be printed in addition to node voltages. The print control of Table 4.3 could also be implemented. ", A" is optional.

(b) PLOT V

As data values are printed, they should be stored for selected variables V and plotted, as a function of time at the end of the calculations.

(c) PLOT X/Y

This form can indicate plotting variables X vs. variable Y. The delimiter '/' is important.

(d) TIME $d_1, t_1, d_2, t_2, \ldots$

The key letters TI must be the first two characters, followed by pairs of time-increment and time-value pairs. The iteration of the equations would start with a step of d_1 up to time t_1, followed by step-sizes d_2 up to time t_2, etc.

(e) EXECUTE

The key letters EX will initiate execution of the program.

(f) END

The key letters EN will terminate the program.

(g) MODIFY

The key letters MO will cause the re-use of the previous input and add to it whatever data cards follow. This necessitates storing all input data, preferably on disc storage.

(h) CONTINUE

This control card (CO) enables the previously calculated voltage and current values to be used in a network where some component values may have been changed. Time-variant elements can be modeled with this control statement.

(i) NODE VOLTAGES

The key letters NO will indicate that initial voltages at the various nodes are to follow. These values can be used to start the iteration process. The node voltages should follow in free-format on subsequent data cards.

(j) BRANCH CURRENTS

The key letters BR will indicate that initial branch currents are to follow. These values can be used to start the iterative solution procedure. Such currents need be given only for inductors; the data cards must identify the inductor and its current as pairs of input data groups.

(k) TEMPERATURE = V or TE = V

(l) FINAL TIME = V or FI = V

(m) INITIAL TIME = V or IN = V

The above all set variables in the program using the first two letters as keys.

II. *Component Specification*

The program generates the **H** and **T** entries directly; the reader will note that no branch number has been used in the derivations. Hence no branch numbers need be given with each element; if dependent current sources (β's and g_m's) are not allowed, no difficulties will be encountered in setting up the node equations. Real circuits do not have such elements anyway; one needs these controlled sources to model transistors. The program described here uses transistors directly.

However, in order to be able to modify components in the network, some labeling of the components is necessary. Although some programs will not allow a MODIFY or a CONTINUE control card, we describe a program that allows these. Since branch numbers cannot be used for identification, let us use sequence numbers. For example,

R 2 (0, 5) = 500

will refer to resistor number 2, extending from ground (i.e., node number 0) to node number 5; the value of the resistor is 500 ohms. Then the subsequent statements

MODIFY
R 2 = 300
EXECUTE

will change the value of R 2 to 300 ohms and will re-run the analysis program.

The MODIFY and CONTINUE controls require that the input data cards be stored (and probably be sorted) so that subsequent calculations could use them. However, only those elements that will be referred to later on need be identified. Consequently, the identifier will be optional. In coding the free-form input routine, one might decide to use a very high sequence number (say 500) and then sort the element by this number. If the element sequence numbers are required to be serial without duplication, then the elements can be stored in sequence by types, i.e., first all resistors, sorted by their sequence numbers, followed by all capacitors, sorted, etc. If these passive linear components are put into one array, a five-column matrix will hold information about all of them:

(1) From-node numbers.

(2) To-node numbers.

(3) Element value.

(4) Series resistance (ignored with resistor elements).

(5) Parallel resistance (ignored with resistor elements).

In addition, the count of resistors, capacitors, and inductors is needed. Also independent constant sources can be put in the same array.

Diodes and transistors must have more data associated with them. With most diodes the bulk internal resistance is very small compared with the resistances of the circuit in which the diode is used. As a result, usually little error is introduced by ignoring the diode bulk resistance (R_b in Fig. 7.7); the diode equations simplify in this case. Hence diodes will have the following data:

(1) From-node number (anode).

(2) To-node number (cathode).

(3) I_s [see Eq. (7.34)].

(4) m [see Eq. (7.34)].

(5) R_s (see Fig. 7.7).

(6) C_t [see Eq. (7.35)].

(7) V_z [see Eq. (7.35)].

(8) T_d [see Eq. (7.37)].

The junction grading constant N can be set to 0.5 without introducing serious errors.

Hence six parameters will be used for each diode. A library can be assembled (usually on a mass-storage device, for example, a tape) and the library entry can be read to obtain pre-stored data. Some such data are shown in Appendix C. Provision must be made to read the library when diode data are not supplied in the input stream.

With transistors the collector and emitter bulk resistances are not negligible. Thus each junction in the transistor may be modeled with seven parameters: the six used above for each of the two diodes and the bulk resistances of the emitter and the collector respectively. Therefore transistor data will have seventeen entries:

(1) Base node number.

(2) Collector node number.

(3) Emitter node number.

(4) Nornal beta [β_n, see Eq. (7.55)].

(5) Inverted beta [β_i, see Eq. (7.55)].

(6) Emitter time constant [T_e, see Fig. 7.13].

(7) Collector time constant [T_c, see Fig. 7.13].

(8) I_{sc} [see Eqs. (7.34) and (7.59)].

(9) m_c [see Eqs. (7.34) and (7.59)].

(10) C_{tc} [see Eq. (7.35)].

(11) V_{zc}, collector junction capacitance [see Eq. (7.35)].

(12) R_{sc}, collector leakage resistance (see Fig. 7.13).

(13) I_{se}
(14) m_c
(15) C_{te}
(16) V_{ze}
(17) R_{se}
} Emitter junction model data, similar to collector junction data described in (8) through (12) [see Fig. (7.13)].

As in the diode, the junction grading constants can be set to 0.5 without introducing significant errors. Since the function SQRT executes far faster than any exponentiation ('**') operation in virtually all FORTRAN systems, its use is preferred.

One must also have a way of flagging NPN as distinct from PNP transistors.

The various component specification cards that need be recognized are as follows.

(a) Resistor
R#(J, K) = v
The letter R must be first, the other symbols denote:
# (optional) = sequence number for resistor; if omitted, it will be added to the end of the resistor list.
J = from-node number.
K = to-node number.
v = any floating-point or integer value; the value of the resistor.

(b) Capacitor
C# (J, K) = v, r_s, r_p
The letter C must be first; the other symbols are as with resistors except
r_s = value of the series resistor R_s (see Fig. 7.4).
r_p = value of parallel resistor R_p (see Fig. 7.4).
r_s and r_p are optional. If omitted, $r_s = 0$ and $r_p^{-1} = 0$

(c) Inductor
L# (J, K) = v, r_s, r_p
The letter L must be first; the other symbols are similar to capacitor and resistor data, except
r_s = value of series resistor R_s (see Fig. 7.5);
r_p = value of parallel resistor R_p (see Fig. 7.5).
r_s and r_p are optional. If omitted, $r_s = 0$ and $r_p^{-1} = 0$.

(d) Voltage Source (Constant)
V# (J, K) = v, r_s
The letter V must be first; all symbols are as with capacitors; r_s denotes the series resistance with the voltage source [see Eq. (7.7)]. If r_s is omitted, a non-zero default value must be used; 0.001 Ω is suggested.

(e) Current Source (Constant)
I# (J, K) = v, r_p
The letter I must be first, the other symbols are as with capacitors. If r_p is omitted, $r_p^{-1} = 0$ is used.

Sec. 7.7 Input Language Considerations 359

(f) Time-variant Sources
These will be of the form shown in Sec. 4.4, with the important change that the branch number must be replaced by an element identification (i.e., R 2). Otherwise, the same input formats can be used.

(g) Diode
D# (J, K) = *
where
J denotes the anode node number;
K denotes the cathode node number;
is optional as before;
* indicates that a stream of diode model constants are to follow.

On the next card the six required diode model constants are to be listed in proper order, but in free format.

(h) Diode
D# (J, K) = 1NXYZ
where 1NXYZ is a library entry; the other symbols are as before.

(i) Transistor

$$\begin{array}{rl} T\#(I,J,K) &= NPN, *\\ T\#(I,J,K) &= PNP, *\\ T\#(I,J,K) &= NPN, 2NXYZ\\ T\#(I,J,K) &= PNP, 2NXYZ \end{array}$$

These component specifications are similar to the two diode specifications cards [items (g) and (h)]; the numbers I, J, K denote;
I = base node number.
J = collector number.
K = emitter node number.

It is, of course, possible to implement other input card forms. For example, mutual inductance and controlled current sources may be added to the above list; the corresponding entries in the **H** and **T** matrices were derived in Sec. 7.1.

Various numerical values must be resolved on each of the data cards. Probably the easiest way to handle this task is to require specific delimiters between key letters. Such delimiters would be opening and closing parentheses, equal signs, commas, and stars. These are easy to search for and will still allow nearly arbitrary use of blanks and other alpha characters in the cards to improve readability.

A good way to handle the input sorting is to read all data into temporary arrays that utilize the H matrix at input time. These arrays can be EQUIVALENCEd to the H array, sorting can be done as required, and the component data can be entered in the data arrays after sorting. In this manner the data preparation routines do not have to have extensive working arrays and storage requirements do not become too severe.

The details of the input decoding are not presented here because they are very tedious. The free-form input routine discussed in Fig. 4.14 and Sec. 3.11 may be used as a guide in writing such programs. Alternately, there are available several completed

360 Nonlinear Analysis Ch. 7

programs that can be used for analyzing nonlinear networks. Each of these programs has its own input format, similar to the one described here and described in detail in the respective user manual [1, 2, 11].

7.8 Illustrative Examples

First Example

In order to use the nonlinear program for analyzing the circuit of Fig. 7.1, a diode model must be derived. The diode data shown on Fig. 7.2 are used to find the required diode parameters.

$$R_s = \frac{-5.0 - 1.0}{(-25.5 - 25.1) \times 10^{-6}} = 10 \text{ meg}$$

The parameters R_b, I_s, and θ must be solved from

$$V_t = V + IR_b$$
$$I = I_s(e^{V/\theta} - 1)$$

where V_t is the diode terminal voltage and I is the current through the diode. The effect of R_s can be neglected if the (V_t, I) pairs of data are taken at higher forward-biased voltage points. Capacitive effects are neglected. These equations may be solved by trial and error at $V_t = 0.4, 0.6, 0.8$ volts. The results are

$$R_b = 12\Omega$$
$$I_s = 0.025\text{ma}$$
$$\theta = 0.08V$$
$$m = 3.116$$

The input data for the analysis becomes

```
R1 (0,1) = 50.
R2 (2,0) = 10.
D1 (1,2) = *
0.025 E-3, 3.116, 1.E + 7, 10.E - 12, 0.5, 10.E - 12
E R1 N 0.,0.,10.,10.
TIME 0.1, 10.
FINAL TIME = 10.
EXECUTE
MODIFY
R2 = 20
EXECUTE
MODIFY
. . . . . .
```

The computed results match the graphically obtained results shown in Fig. 7.3. Note that the capacitive terms were set very small. Ideally, they should be set as high as possible without affecting the results sought. Very small capacitors lead to instability in the model currents and often result in very small time steps used in the program. In this case, the capacitive effects become negligible since the input voltage rise is very slow.

Second Example [15]

We now discuss the details of obtaining a realistic computer model for an integrated circuit. The model parameters will be derived and performance of the model will be checked against measured data, much as was done in Sec. 7.4.

A particular diode-transistor logic (DTL) gate was selected. The DTL gate to be studied is a NAND gate whose basic circuit is shown in Fig. 7.27. The manufac-

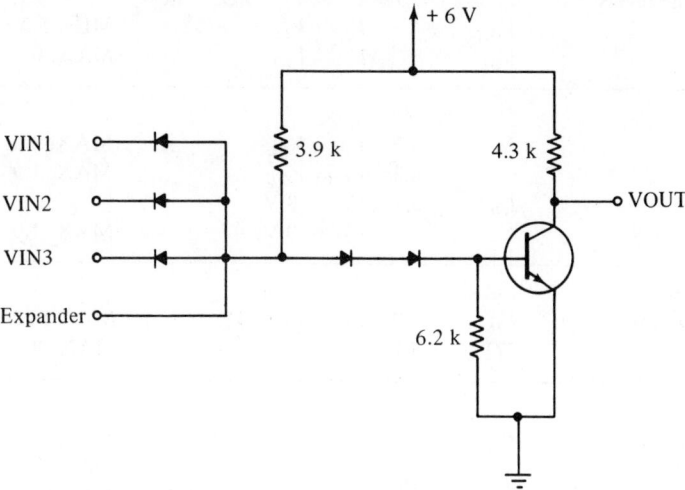

Fig. 7.27 Basic DTL NAND Gate

turer's data sheet is slightly paraphrased in Table 7.4. The associated loading circuit and switching waveforms for which the manufacturer's data are given are shown in Figs. 7.28 and 7.29. The exact internal construction of the integrated circuit is not known; thus the circuit shown in Fig. 7.27 can not be verified directly. Since it is extremely difficult to probe an integrated circuit without destroying it after it has been assembled, the measurement of the characteristics of each separate component is not practical. Therefore, in order to determine these characteristics, it is necessary to go through a pseudo-design of the DTL gate. During this process the automated nonlinear analysis program was used as a design tool. The characteristics that must be determined are the transistor parameters, the diode parameters, and the capacitances that are present.

In the design of an integrated circuit, the resistivities and profiles of the various layers are usually chosen to provide optimum characteristics for the transistors [16]. Unless additional diffusions are economically feasible, one or more of the layers provided for transistors have to be used for the resistors and diodes. In order to

Table 7.4 *Electrical Characteristics for DTL Gate with Collector Resistor (temperature = 25°C)*

Parameter	Symbol	Test Conditions			Test Limits	Units
		VCC	Input	Output		
STATIC						
Input threshold voltages		6				
Input low	V_{IL}				MIN. 1.1	volts
Input high	V_{IH}				MAX. 2.1	volts
Output voltage levels		5.7		$FO = \text{max}$		
High state	V_{OH}		$V_I = V_{IL}$		MIN. 5.3	volts
Low state	V_{OL}		$V_I = V_{IH}$		MAX. 0.4	volts
Input current						
Input low	I_{IL}	5.7	$V_I = V_{OL}$		MAX. 1.6	milliamps
		6.3	$V_I = V_{OL}$		MAX. 1.9	milliamps
Input high	I_{IH}	5.7	$V_{I1} = 8\text{ V}$			
			$V_{I2} = 0\text{ V}$		MAX. 1.0	microamps
Output current						
Output low, input high	I_{OL}	5.7	$V_I = V_{IH}$	$V_O = V_{OL}$	MIN. 13.2	milliamps
Fan-out	FO	5.7			MAX. 9	
DYNAMIC						
Fall propagation delay time	t_{PHL}	6		$FO = 1$	Typ. 26 / Max. 65	nanosec / nanosec
Rise propagation delay time	t_{PLH}	6		$FO = 1$	Typ. 50 / Max. 85	nanosec / nanosec

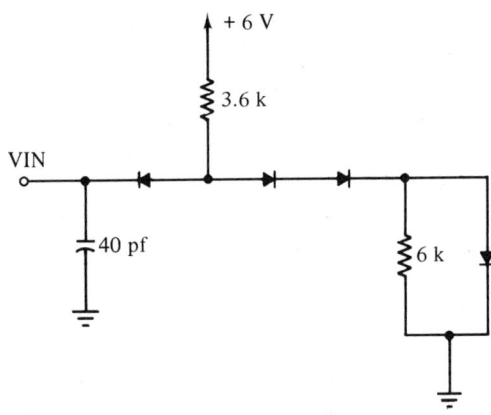

Fig. 7.28 Gate Dynamic Loading Circuit, $FO = 1$

Sec. 7.8   Illustrative Examples   363

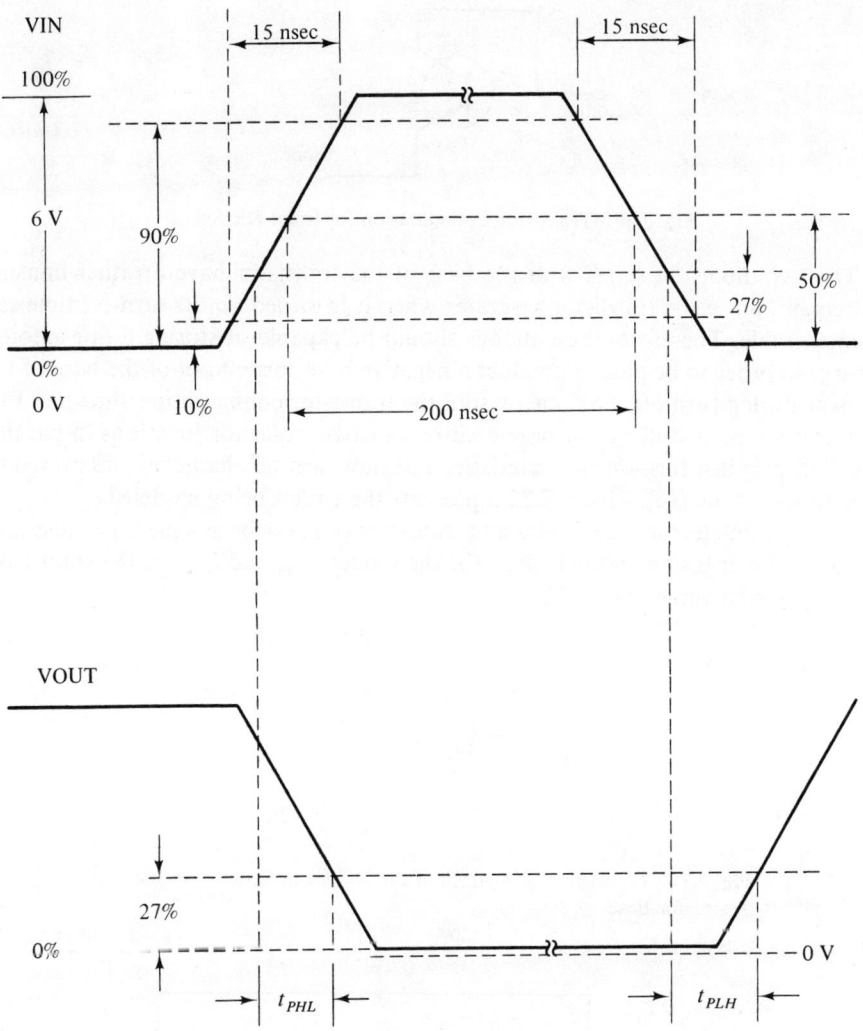

Fig. 7.29 Gate Switching Waveforms

establish a model that describes the terminal characteristics of the gate adequately, the diode and transistor parameters should be looked at concurrently.

The diodes may be assumed to be formed from transistors with the terminals connected in order to achieve the desired diode characteristics. The diodes in the "AND" gate should be fast in both turning on and in turning off. For this reason, the transistor configuration for these diodes is that shown in Fig. 7.30. It utilizes the base-emitter junction of the transistor with the base-collector junction shorted, and the resultant diode is one of the fastest that can be formed from the transistor [16].

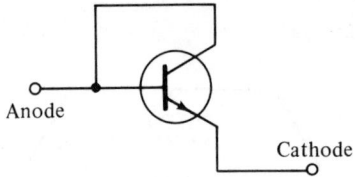

Fig. 7.30 Transistor Configuration for Input Diodes

The two diodes in series with the base of the transistor have a rather unusual requirement. Since the transistor saturates when it is turned on, its turn-off time can be considerable. Therefore, these diodes should be capable of storing a fair amount of charge in order to be able to conduct a negative base current out of the base of the transistor during turn off. For this reason, the transistor configuration shown in Fig. 7.31 was chosen. It utilizes the base-emitter and base-collector junctions in parallel and will display fast turn-on characteristics and slow turn-off characteristics caused by a large storage time [16]. Figure 7.32 represents the circuit being modeled.

Some parameters for the diodes and transistor can now be assigned. Because they are silicon, the diffusion voltages (V_{zD} for the diodes, V_{ze} and V_{zc} for the transistor) can be assigned a value of 0.75 V.

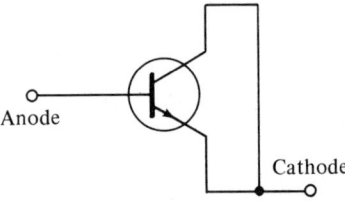

Fig. 7.31 Transistor Configuration for Diodes in Series with the Transistor Base

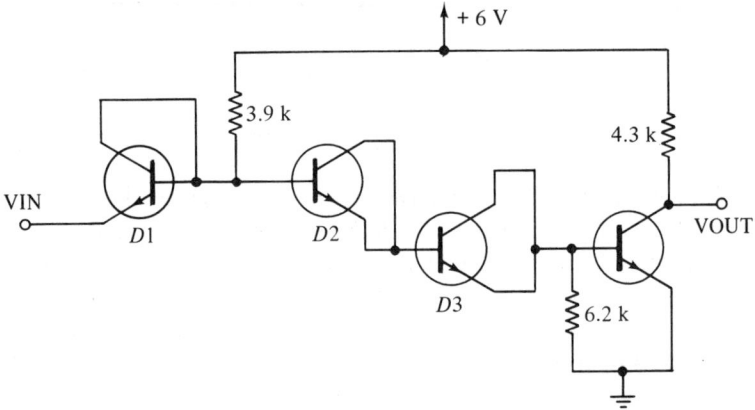

Fig. 7.32 Circuit Being Modeled

In order to select a starting point for the values for the other major transistor and diode parameters, the 2N2369A high-speed switching transistor is investigated. This device is a silicon NPN transistor of planar epitaxial construction (the same construction technique used in the manufacture of the DTL gate) [16]. The characteristics of this device must be studied in relation to the characteristics of integrated diodes of the types shown in Figs. 7.30 and 7.31.

Based on a combination of the information gleaned from these various inputs, the transistor and diode models may be developed. During the model development the nonlinear analysis program is used as a design tool. Various calculations are made for varying parameters in order to observe their effects on the model's response. The resultant diode and transistor characteristics are shown in Tables 7.5 and 7.6 respectively. In order to demonstrate how some of these parameters can be developed, a couple of examples will be given.

Table 7.5 *Diode Parameters*

	D1	D2 and D3	
IS	3×10^{-12}	7×10^{-12}	amperes
MD	1.5	1.5	—
RDL	1×10^9	1×10^9	ohms
CDO	3×10^{-12}	6×10^{-12}	farads
VDBI	0.75	0.75	volts
TD	1×10^{-9}	20×10^{-9}	seconds

Table 7.6 *Transistor Parameters*

HFEN	100	VCBI	0.75 V
HFEI	1	RCL	5×10^9 ohms
TN	1×10^{-9} sec	IES	3×10^{-12} amperes
TI	3×10^{-9} sec	ME	1.5
ICS	7×10^{-9} amp	CEO	3×10^{-12} farads
MC	1.3	VEBI	0.75 volts
CCO	2×10^{-12} farads	REL	1×10^9 ohms

One of the parameters whose derivation is interesting is the storage time constant (TD) for D2 and D3. Note that in order to keep the various T_d and other parameters distinct for the diodes and the transistors, various non-subscripted groups of letters are used in this section to denote the different device parameters, as is evident in Tables 7.5 and 7.6.

This time constant must be selected, as explained previously, in such a way as to ensure proper turn off of the transistor. Figures 7.33 and 7.34 are from the three final computer runs used to determine this parameter. Figure 7.33 shows the waveform at the transistor base with the storage time constant too short (15 nsec), and with the time constant a little too long (25 nsec). Figure 7.34 illustrates the base waveform with

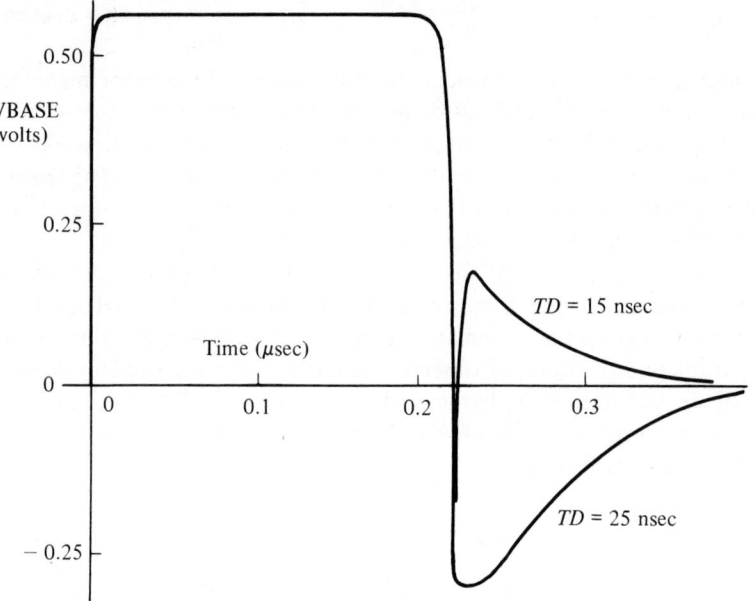

Fig. 7.33 Waveforms at the Transistor Base

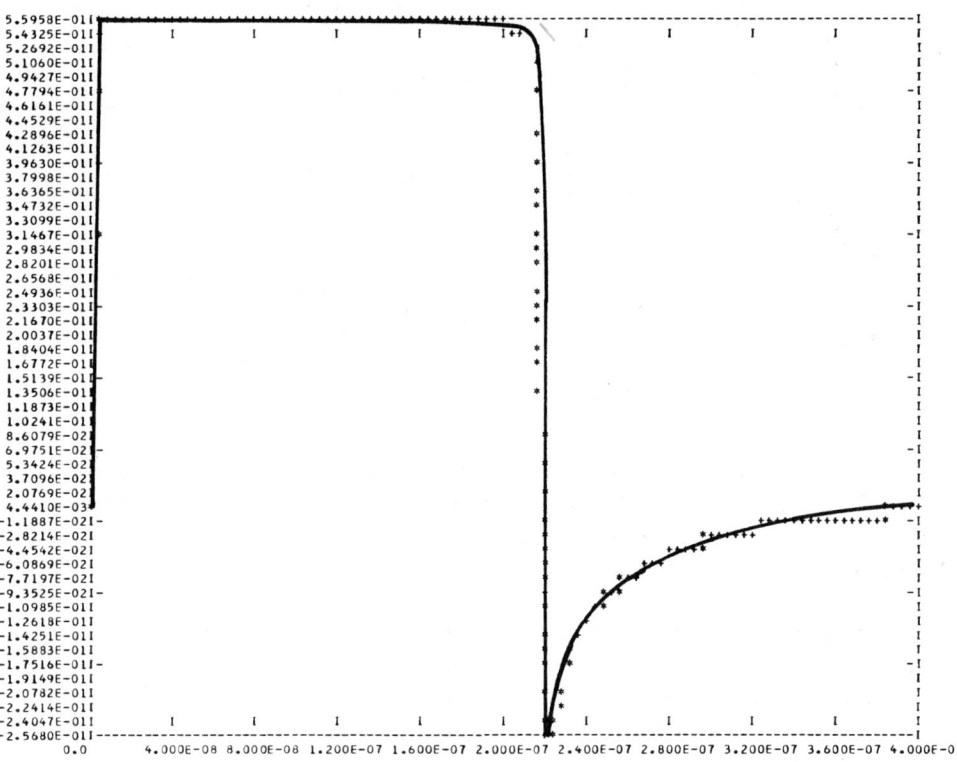

Fig. 7.34 Waveform at the Transistor Base, TD = 20 ns

the storage time constant finally used for the diodes (20 nsec). Figure 7.33 was drawn from the computer-produced printer plot, whose appearance is as shown in Fig. 7.34.

Other parameters whose derivation proves to be of interest are the proportionality constants. MD can vary from about 0.8 to 2.0. A value of 1.0 is assumed for MD, ME, and MC for the initial runs. After the models are pretty well developed, these constants are determined more accurately.

From the specifications for the DTL gate, the two major breakpoints for the transfer function occur at VIN = 1.2 and 1.6 V as illustrated in Fig. 7.35. In order for this to happen a forward voltage drop of 0.75 V is required of the diodes.

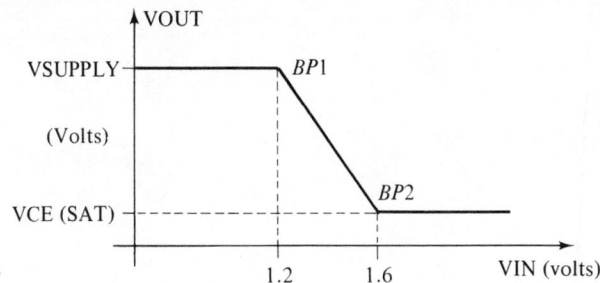

Fig. 7.35 Transfer Characteristic for DTL Gate

A calculation using MD = 1.0 yields the following values for D1: ID (forward current) = 1.4×10^{-3} amperes; VD (forward voltage) = 0.52 volts; IS (diffusion current) = 3×10^{-12} amperes. Since the required diode forward voltage drop is 0.75 V, these parameters must be adjusted appropriately. The relevant equation follows:

$$\mathrm{ID} = \mathrm{IS}(e^{(q\mathrm{VD}/\mathrm{MD}kT)} - 1).$$

Solving this equation for VD, we get

$$\mathrm{VD} = \mathrm{MD}\left(\frac{kT}{q}\right)\ln\left(\frac{\mathrm{ID}}{\mathrm{IS}} + 1\right)$$

At room temperature (27°C)

$$\frac{kT}{q} = 0.026 \text{ V}$$

Therefore, for D1

$$\mathrm{VD} = \mathrm{MD}\,(0.026)\ln\left(\frac{1.4 \times 10^{-3}}{3 \times 10^{-12}} + 1\right)$$

$$= \mathrm{MD}\,(0.520).$$

It is quite obvious that either ID or IS will have to change by an order of magnitude to have much effect on changing the value for VD. Inserting a value of 0.75 for VD in the last equation yields a value of 1.44 for MD.

Going through the same procedure for the other diodes and the base-emitter and base-collector junctions for the transistor results in the following approximate values: MD = 1.4 ME = 1.4, and MC = 1.2. To illustrate the above process, Fig. 7.36 shows the transfer function with all of the proportionality constants equal to 1.0.

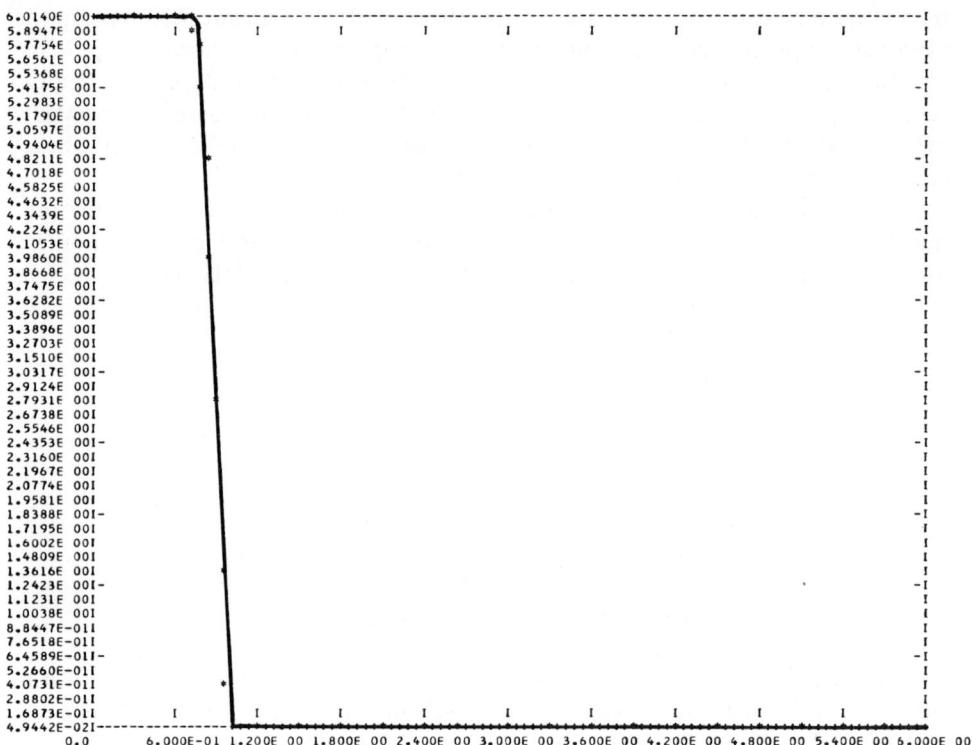

Fig. 7.36 Transfer Function with Proportionality Constants Equal to 1

Here the EIN values for BPI and BP2 are 0.6 V and 0.9 V respectively, and VCE(SAT) = 0.04 V. The input waveform is shown in Fig. 7.37. Figure 7.38 illustrates the transfer function using the final proportionality constants of MD1 = 1.5, MD2 = 1.5, MD3 = 1.5, ME = 1.5, MC = 1.3. The same input waveform is used with the breakpoints occurring approximately at 1.2 and 1.6 V as desired, and VCE(SAT) = 0.16 V.

During device optimization, resistances can be added to the circuit for the diodes and transistor. A value of 10 ohms may be assumed for the base spreading resistance of the transistor. Since the transistor is of integrated planar construction (the collector contact is made from the same side of the substrate as the emitter and base contacts), a finite collector resistance of 10 ohms is reasonable.

The diodes, when conducting in the forward direction have a finite resistive impedance which is small and strongly current dependent. If the forward resistance is defined as $d\text{VD}/d\text{ID}$, then

$$\frac{d\text{ID}}{d\text{VD}} = \frac{q\text{IS}}{\text{MD}kT} e^{(q/\text{MD}kT)} \cong \frac{q\text{ID}}{\text{MD}kT}$$

$$r_e = \frac{\text{MD}kT}{q\text{ID}} = \frac{\text{MD}}{\text{ID}}(0.026)$$

Sec. 7.8 Illustrative Examples

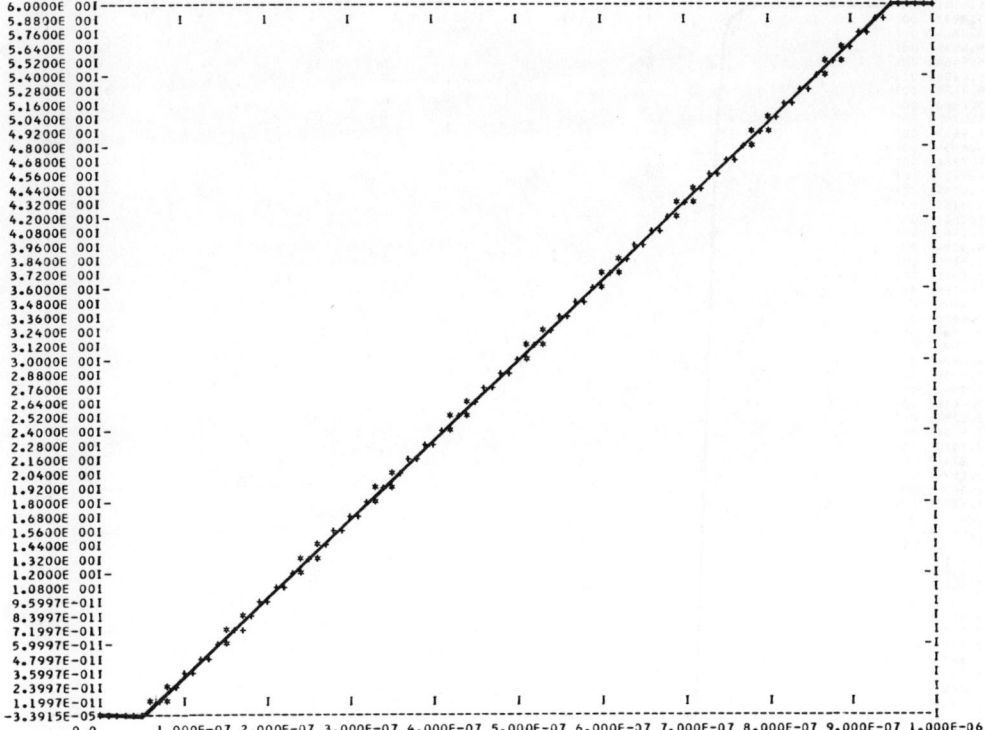

Fig. 7.37 Input Waveform

A value for r_e of 20 ohms is used for all three diodes throughout the computer simulation. These resistances can be calculated more accurately, but because their value is low, they have little effect on the circuit's operation at the speeds being considered.

The final circuit characteristics to be determined are the input and output capacitances. Figure 7.39 schematically illustrates the location of the parasitic capacitances for the integrated DTL circuit.

The values of these capacitances are dependent on the geometries of the devices and the impurity concentrations of the different diffusions. The values cannot be calculated since none of the manufacturing process information is available for the integrated circuit. Instead, typical values for these capacitances can be selected from the literature [17]. Table 7.7 contains the values used for this analysis.

Table 7.7 *Values for Parasitic Capacitances in Picofarads*

$CR1 = 0.7$	$CD1 = 1.4$	$Ce = 0.7$
$CR2 = 1.2$	$CD2 = 1.0$	$Cc = 0.9$
$CR3 = 0.7$	$CD3 = 1.1$	$Cs = 2.4$

370 Nonlinear Analysis Ch. 7

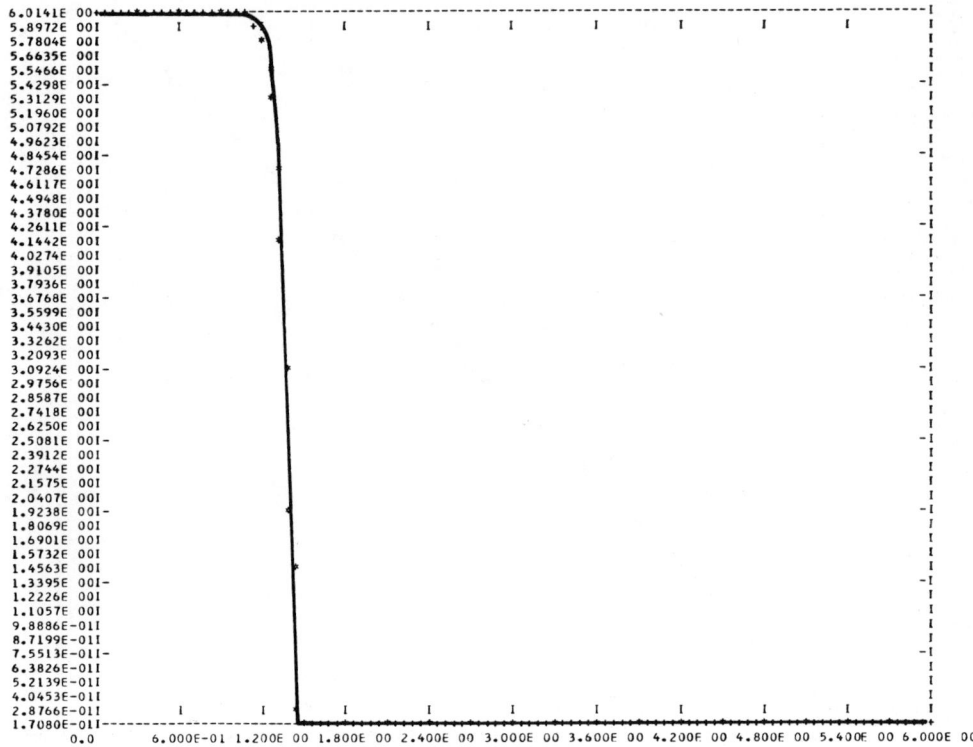

Fig. 7.38 Transfer Function with the Final Proportionality Constants

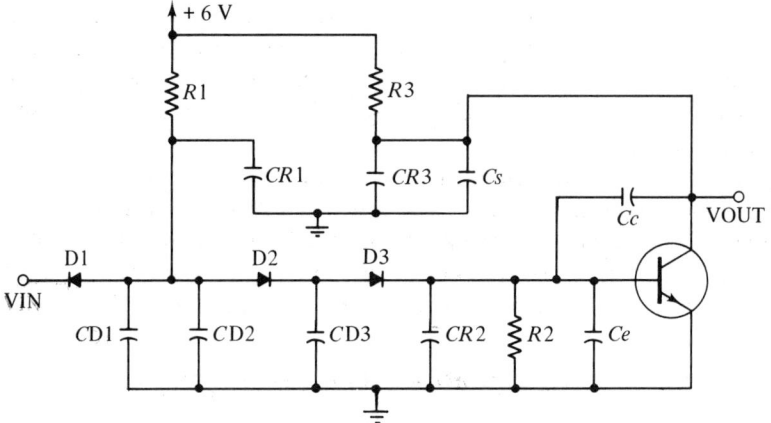

Fig. 7.39 DTL Circuit Showing Capacitive Parasitics

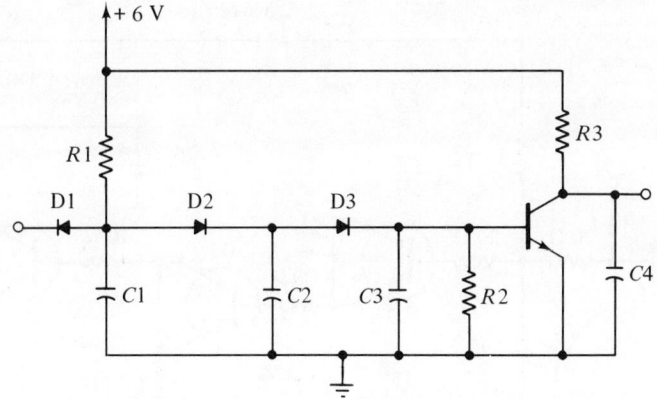

Fig. 7.40 DTL Gate with Total Node Capacitances

Some of the capacitances shown in Fig. 7.39 can be lumped together as shown in Fig. 7.40.

In this figure

$$C1 = CR1 + CD1 + CD2$$
$$C2 = CD3$$
$$C3 = CR2 + Ce + Cc$$
$$C4 = CR3 + Cs + Cc$$

In order to simplify the circuit further, the charging rates of $C1$, $C2$, and $C3$ during turn on are examined. When $D1$ is on (input low) and gets turned off (input goes high), $C1$ charges from $+6$ V through $R1$, $C2$ charges from $V1$ through the resistance of $D2$, and $C3$ charges from $V1$ through the resistances of $D2$ and $D3$. If the time constants involved are examined, it turns out that a good approximation is to lump $C1$, $C2$, and $C3$ at the location of $C1$.

If the values shown in Table 7.7 are used, the approximate capacitances shown in Figure 7.41 are determined. This figure shows the final discrete component model.

The response of the circuit without load is shown in Fig. 7.42. The wave of Fig. 7.29 is the input.

In order to check the validity of this model, the dynamic propagation delay times may be determined. The dynamic loading circuit for a fanout of 1 is shown in Fig. 7.43.

The diode models used for D4 and D7 are the same as that derived for D1 in the DTL gate. The models for D5 and D6 are equivalent to D2 in the DTL gate. In addition, the diode resistances are the same as those for the gate.

The input waveform is the same as that shown in Fig. 7.29. The output waveform is shown in Fig. 7.44. From this computer run the dynamic propagation delay times are determined. The model displays a fall propagation delay time (t_{PHL}) of approxi-

372 Nonlinear Analysis Ch. 7

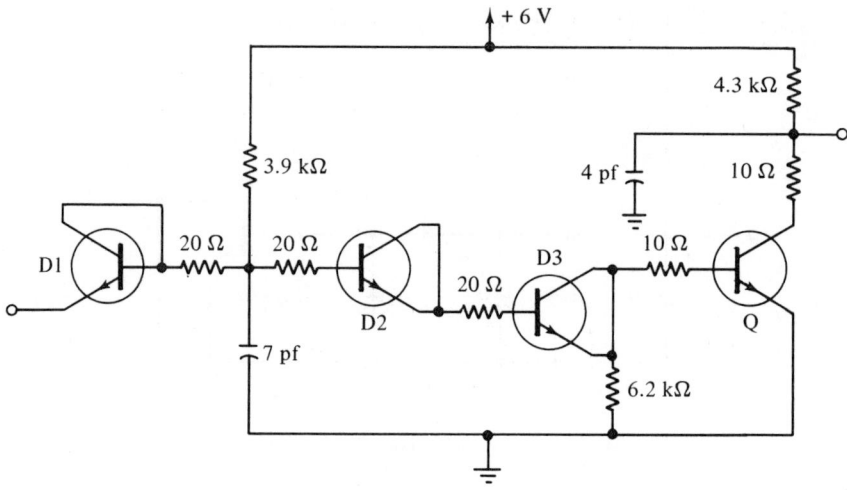

Fig. 7.41 The Final Discrete Component Model

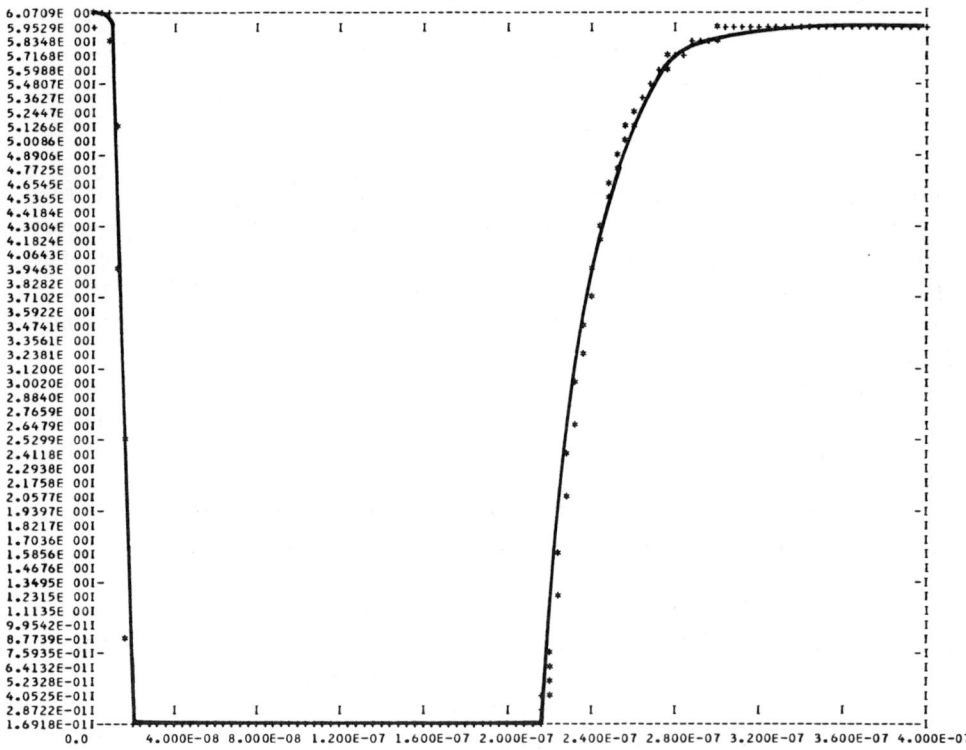

Fig. 7.42 Output Waveform for the DTL Gate with No Load

Sec. 7.8   Illustrative Examples   373

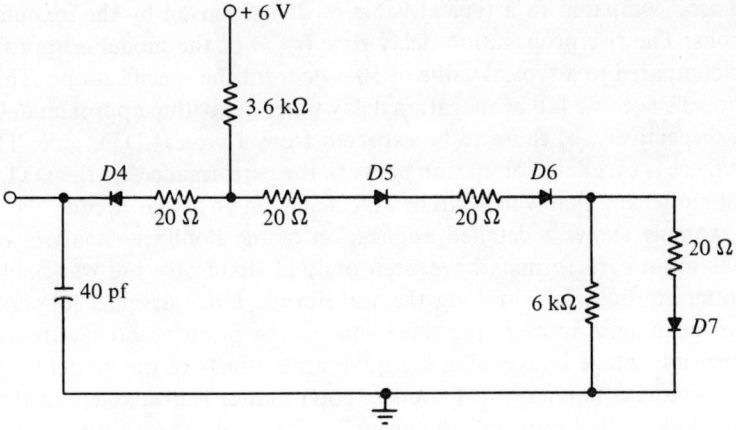

Fig. 7.43 Dynamic Loading Circuit, $FO = 1$

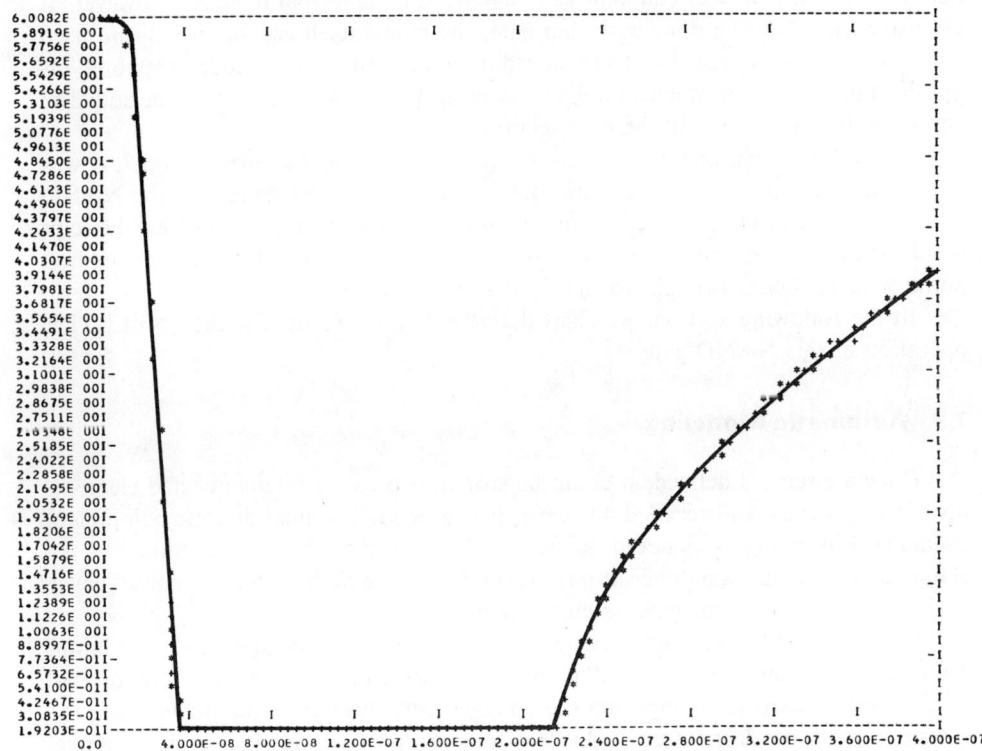

Fig. 7.44 Output Waveform for DTL gate, $FO = 1$

mately 23 nsec compared to a typical value of 26 nsec given by the manufacturer's specifications. The rise propagation delay time (t_{PLH}) of the model is approximately 44 nsec as compared to a typical value of 50 nsec from the specifications. This means that the model's rise and fall propagation delay times are within approximately 11.5% and 12%, respectively, of those to be expected from a typical DTL gate. This illustrates that there is excellent correlation between the performances of the DTL discrete component model and that which can be expected from an actual circuit.

This example shows a detailed application of the nonlinear analysis program. The various input card formats have been omitted. Extensive use was made of the printer-plotter routines for checking the waveforms. Since detailed numerical data are far less meaningful in assessing wave shapes, the printer-plotter outputs can be used to great advantage in assessing the qualitative effects of the model parameters. Of course, in most computer installations a good plotter is available, but the cost of the printer plots is far lower and the accuracy required is compatible with printer plots.

Once the parameters of the equivalent circuit have been determined, each time that gate is used in a circuit one can re-enter all model parameters. However, if extensive work is to be done with that gate, the model itself can be stored and used much like the library-stored transistor data. Such automatic model capability is provided in the more advanced nonlinear analysis programs [1, 11]. It can be added to any program as outlined in the next section.

The development of the final discrete component model shown in Fig. 7.41 was done with one input only. Thus, strictly speaking, this model represents the NAND gate for a single input, i.e., used as an inverter. The total model of the original circuit will have three D1 diodes at the input all tied to the 3.9-kΩ resistor and the common point of these diodes brought to the EXPANDER point.

In the following section, we shall derive a simpler model for the INVERTER operation of this NAND gate.

7.9 Automatic Modeling

Once a circuit is defined, it could be stored as a list of interconnected elements on a library device and recalled whenever it is needed. The final discrete component model shown in Fig. 7.41 could be stored. We note that this circuit, model of the actual DTL gate, has a number of internal nodes. These nodes are of no consequence to the circuit in which the gate might be used.

When the model is stored, its external connections must be identified. The input, ground, output, and the +6V points must be labeled as external nodes. The internal nodes must be used for computing the voltages and currents in the model from the external voltages and/or currents. One way to accomplish this is to insert in the model node numbers following the node numbers used in the rest of the circuit. A similar mechanism was used in Sec. 3.10 in dealing with the conversion of dependent voltage sources to dependent current sources. One must know the highest node number used in the network (at the end of the input, indicated by the EXECUTE statement) and then insert transistor models between the defined external node numbers and the required internal nodes. There are eight internal nodes in the model:

(1) Emitter and base of D1.

(2) Junction of the 7-pf capacitor and the 20-Ω, 20-Ω, and 3.9-kΩ resistors.

(3) Base of D2.

(4) Emitter and collector of D2.

(5) Base of D3.

(6) Emitter and collector of D3.

(7) Base of Q.

(8) Collector of Q.

Hence each DTL NAND model made up as shown in Fig. 7.41 will add eight nodes to the circuit. Although such increase in node numbers may be acceptable, often this increase will raise excessively the execution time of the program. A similar model of the circuit would be desirable where only the input and the output of the gate are modeled in enough detail to mimic the performance of the gate acceptably. Here we develop such a simpler model useful for a supply voltage of 6 V.

The model is such that it can be easily inserted in different kinds of networks. It is expected that the networks in which the model will be used are ones in which the rise and fall times are at least an order of magnitude lower than the capabilities of the DTL gate. Hence the model stresses accurate steady-state simulation at the expense of rougher transient prediction. After the equivalent circuit model is developed, its performance is compared to the performance of the discrete component model.

The model developed here is to predict input and output terminal behavior of the DTL gate. The input circuit is investigated first and then the output circuit, for use of the NAND gate as an INVERTER.

Figure 7.45 is an approximate straight-line representation of the input resistance as a function of the input voltage. The breakpoints occur at input voltages of 1.2 and 1.6 V as determined from the manufacturer's data.

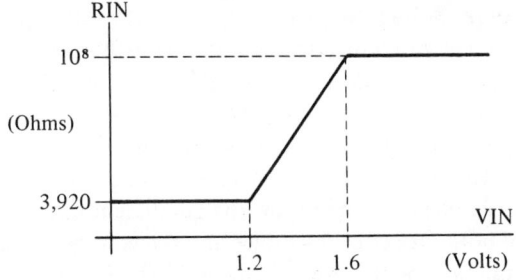

Fig. 7.45 Input Resistance as a Function of Input Voltage

Figure 7.46 is a schematic representation of the input circuit. This circuit can be reduced to the circuit in Fig. 7.47. It consists of two floating voltage sources, two diodes, a resistor, and a capacitor. D1 has the characteristics of an input diode and D2 that of a diode in series with the transistor base.

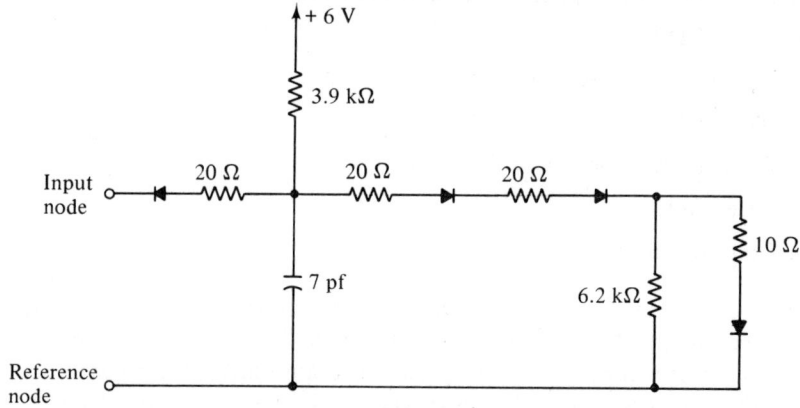

Fig. 7.46 Schematic Diagram of the Input Circuit

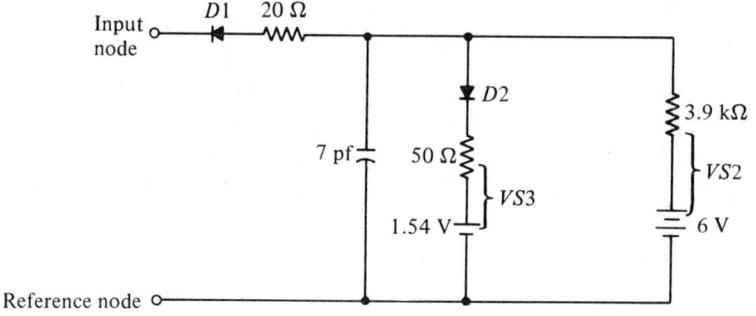

Fig. 7.47 Schematic Diagram of Equivalent Input Circuit

In order to make the model as nearly like the original circuit as possible, the voltage source for VS3 was calculated by using the results from a discrete component model run. The resultant voltage is 1.54 V.

The output circuit is treated in somewhat the same fashion. Figures 7.48 and 7.49 show straight line approximations of the output voltage and the output resistance as a function of the input voltage. Again, the two breakpoints appear. In Fig. 7.48 the high and low levels for the output voltage are the supply voltage and the collector-to-emitter transistor saturation voltage, respectively. Since the output of the DTL gate is a function of what happens at the input, the equivalent circuit shown in Fig. 7.50 is used. The values of both the series resistance and the voltage source are programmed so that they are a function of the input voltage. If the input voltage is less than or equal to 1.2 V, RS1 is set equal to 4.3 k ohms and VS1 equal to 6 V. If the input

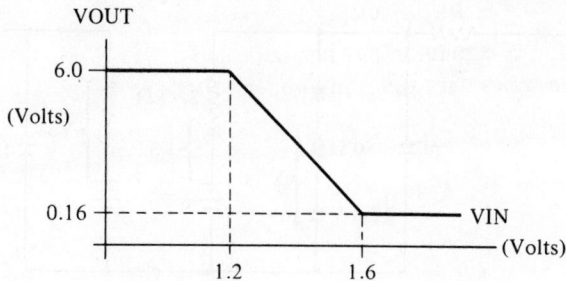

Fig. 7.48 Output Voltage as a Function of the Input Voltage (transfer function)

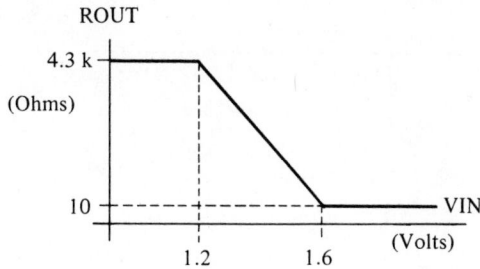

Fig. 7.49 Output Resistance as a Function of Input Voltage

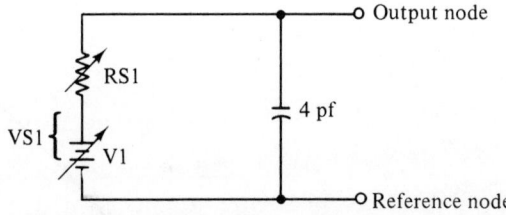

Fig. 7.50 Schematic Diagram of the Equivalent Output Circuit

voltage is greater than or equal to 1.6 V, RS1 and VS1 are set equal to 10 ohms and 0.16 V respectively. If the input voltage is between 1.2 and 1.6 V, then an equation for a straight line between the two breakpoints is used to calculate the values for RS1 and VS1.

Figure 7.51 shows the final equivalent circuit model. From this point on the equivalent circuit model will be referred to as DTLM and the discrete component model as DTL.

The subsequent figures show comparisons of the operation of the DTL model with the DTLM model. Figure 7.52 shows the output waveform of the DTLM model with a fanout of one (loaded by the circuit shown in Fig. 7.43) with the input wave shown in Fig. 7.29. This output, when compared to the computed values shown in Fig. 7.44, makes it evident that making the output voltage source a linear function of

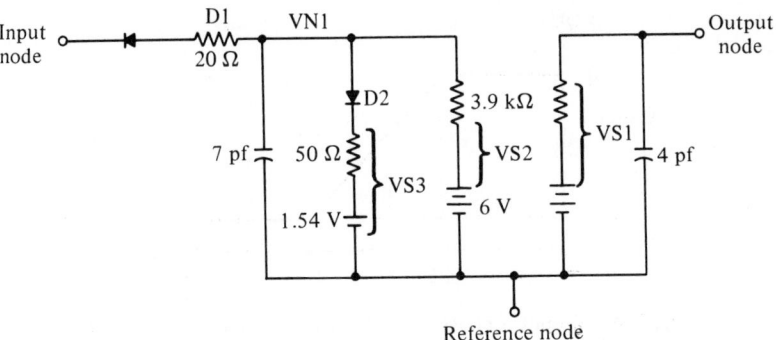

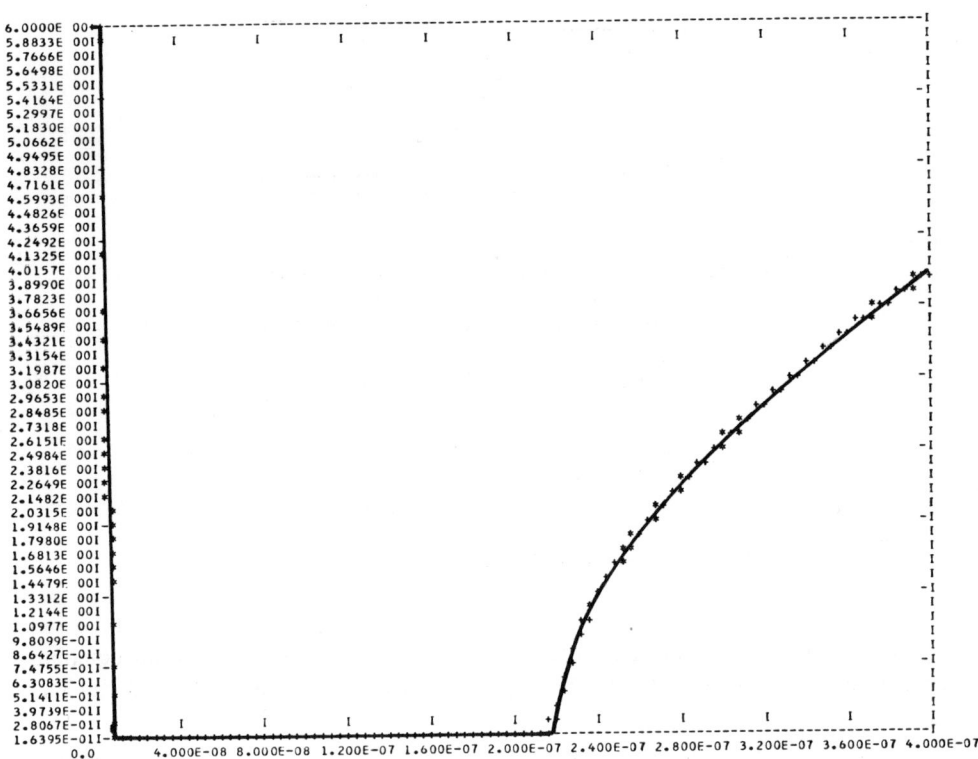

Fig. 7.51 Equivalent Circuit Model

Fig. 7.52 Output Waveform for the DTLM model, $FO = 1$, VS1 is a function of EIN.

the input waveform does not yield a very satisfactory model. Therefore, in order to improve on the model, VS1 can be made a function of the voltage at the node labeled VN1 in Fig. 7.51. The breakpoints for VN1 are determined by computer runs to be 1.95 and 2.25 V.

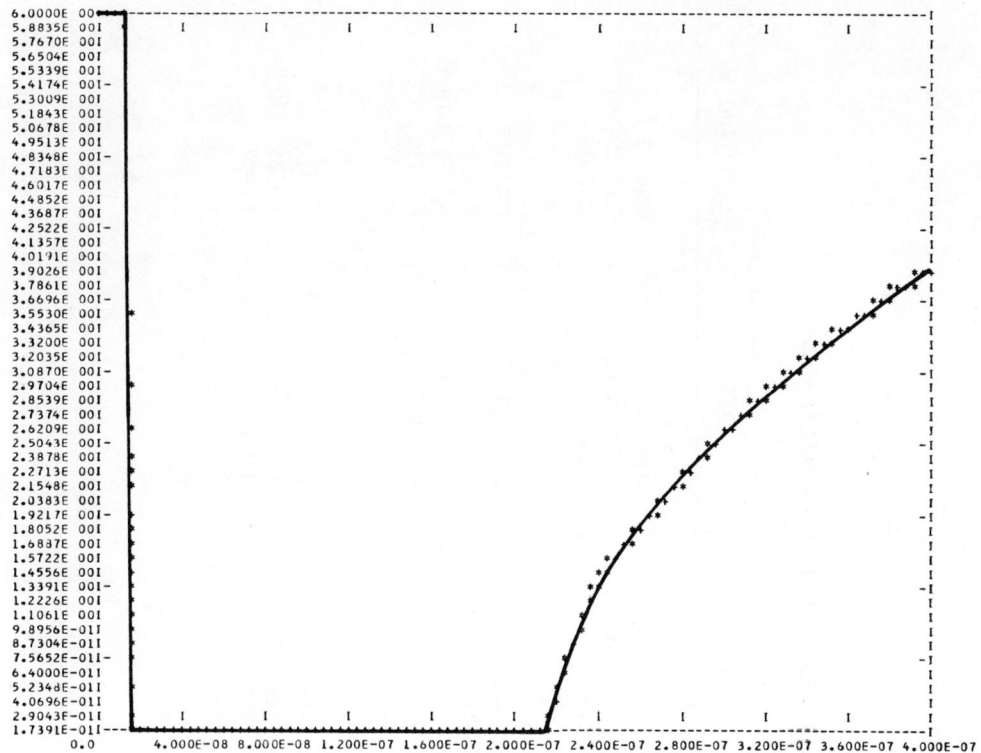

Fig. 7.53 Output Waveform for the DTLM model, $FO = 1$, VS1 is a function of VN1.

Figure 7.53 is the output waveform for the DTLM model with fanout of one and VN1 controlling VS1. As can be seen by comparing this figure with Fig. 7.44, the fall time of the DTLM model more nearly approximates that of the DTL model. The output falls to its lower level (100% to 0%) in approximately 18.3 nsec for the DTL model and in 15.9 nsec for the DTLM model. There is some ringing present, although it is not large.

A caution is in order at this point: The plots cannot be compared point-for-point because of the manner in which the plotting subroutine determines the points for the abscissa. A point that falls between two adjacent points on the abscissa is plotted on the nearest point. Precise information on the performance of the models is obtained from the computer printouts.

At this point a number of computer simulations of the DTL and the DTLM models are made to obtain qualitative assessments of their performances. Table 7.8 shows comparisons of voltage at node one, the current through the input diode, and the output voltage. The comparisons are shown at time zero, at 123 nsec from time zero (output at low level), and at the end time (400 nsec). As can be seen from the table, the input and output voltages and input currents of the two models are very nearly

Table 7.8 Comparison at Selected Points of the DTL and DTLM Models

	Fanout = 1		No Load	
	DTL	DTLM	DTL	DTLM
Initial Conditions				
VN1 (volts)	0.80314	0.80315	0.80314	0.80315
ID1 (amperes)	1.3323×10^{-3}	1.3325×10^{-3}	1.3323×10^{-3}	1.3325×10^{-3}
VOUT (volts)	6.0000	6.0000	6.0000	6.0000
ID LOAD (amperes)	3.7155×10^{-9}	3.7155×10^{-9}	—	—
TIME (seconds)	0.11925×10^{-3}	0.124625×10^{-3}	0.122625×10^{-3}	0.124625×10^{-3}
VN1 (volts)	2.3277	2.3271	2.3123	2.3171
ID1 (amperes)	7.9889×10^{-9}	6.4913×10^{-9}	8.9223×10^{-9}	6.1583×10^{-9}
VOUT (volts)	0.20082	0.17403	0.17125	0.16136
ID LOAD (amperes)	1.3975×10^{-3}	1.4024×10^{-3}	—	—
TIME = 0.4×10^{-3} (sec)				
VN1 (volts)	0.80314	0.80315	0.80313	0.80315
ID1 (amperes)	1.3324×10^{-3}	1.3325×10^{-3}	1.3321×10^{-3}	1.3325×10^{-3}
VOUT (volts)	3.8964	3.9392	5.9991	5.9998
ID LOAD (amperes)	1.8011×10^{-5}	1.7760×10^{-5}	—	—

the same. The largest difference occurs when the output is low (the output voltages differ by approximately 10 millivolts and the input currents differ by approximately 2.8 nanoamps).

Figure 7.53 is the output waveform from the DTLM model with a fanout of one. The dynamic propagation delays calculated from this run yield a t_{PHL} of 9.4 nsec and a t_{PLH} of 42.9 nsec. When these results are compared with the DTL run with a fanout of one, it is seen that the fall propagation delay time for the DTLM model is less than one-half of the value for the DTL model. This is partly because the DTLM model includes no provisions for the delay in turn on of the transistor. The rise propagation delay time of the DTLM model is very close to that of the DTL model (within approximately 3%). The main reason for this agreement is that this delay is almost solely due to the R-C time constant of the DTL load resistor (4.3 k) and the capacitance at the output (44 pf).

When the load on the circuit is a fanout of one (load shown in Fig. 7.43), there is acceptable agreement between the DTL and DTLM models also. The difference of approximately 27 millivolts for VOUT when the output is low is because the current through the saturated transistor in the DTL model is a summation of the current through the 4.3K DTL load resistor and that from the dynamic load circuit (ID LOAD).

Table 7.8 shows that for the stable conditions the two models agree quite closely. An examination of the diode currents will show a larger disagreement. The computed peak currents are compared in Table 7.9 for the input wave shown in Fig. 7.29. The peak current for the DTLM model is greater than three times that for the DTL model.

Table 7.9 *Comparison of the Peak Currents for the Fall and Rise of the Output Waveform*

	FALL		RISE	
	DTL	DTLM	DTL	DTLM
ID1 (mA)	−1.6368	−1.7655	6.5437	8.7442
ID LOAD (mA)	2.0627	6.8302	−0.09997	−0.10176

This is to be expected since the output voltage waveform of the DTLM model falls much faster. Therefore, in order to discharge the 44-pf capacitor in less time, larger currents are required, as observed.

As can be seen from the computed results, the equivalent circuit is a fairly good overall approximation of the discrete component model. The area of largest disagreement is the transient response of the DTLM model during the fall of the output voltage waveform.

The supply voltage may be changed by changing the 6V source now in the model to whatever the supply voltage might be.

The simplified model, Fig. 7.51, contains only three internal nodes:

(1) Anode of D1.
(2) Anode of D2.
(3) Cathode of D2.

The model does require the computation of a dependent voltage, VS1, which is a function of another node voltage. This dependent voltage source can easily be converted to a single dependent current source through a Norton transformation. The number of nodes required by the model of the DTL gate has been decreased drastically. The new model can be stored and recalled whenever necessary.

In using such models in analysis programs, provisions must be made to recognize some set of key letters in the input, such as MODEL = ABC (J, K, L) with ABC denoting a model name on the library; J, K, L denoting the ordered external node numbers.

Storing the model data is probably best done with a separate routine that reads input data and adds it at the end of the device library.

The use of the model may require recalculation of circuit parameters during the solution, as is the case with the output resistance in the model developed here. Such calculations can easily be performed in Subroutine SETUP; appropriate flags must be set for that subroutine for such calculations.

The automatic model library, as opposed to a device library, allows very complex subsystems to be included as circuit "elements." This aids greatly in the convenience of use of automated circuit analysis programs for the anslysis of electronic systems.

Automatic model capability in extensive analysis programs merely consists of retrieving from a mass-storage device all parameters needed for a subcircuit and letting the main program assign required additional (internal) nodes in the model. If several models are used, a great amount of additional computational complexity can be created. An attempt at establishing simplified models should always be made, much as was shown in this section.

We derived here a simple model for an INVERTER made of the four-input NAND gate. To establish a model for the general NAND gate, the equivalent circuit of Fig. 7.41 must be modified for the inclusion of three input diodes. Similar changes must be made for all equivalent circuits in this section. The total circuit will change considerably as far as the inputs are concerned, but the output will still be a variable voltage source controlled from the internal VN1 voltage.

The final equivalent circuit will have four input nodes (VIN1, VIN2, VIN3, EXPANDER of Fig. 7.27), one output node (VOUT) and one common node (REFERENCE). The EXPANDER voltage will be VN1 of Fig. 7.51. The completed total model can be stored and recalled whenever necessary.

The method discussed in the last two sections can, of course, be applied directly to any integrated circuit where an equivalent discrete component circuit can be established.

7.10 Summary

The nonlinear program discussed is a relatively modest one. Far more sophisticated programs exist [1, 11], but the program here is reasonably accurate, execution is fast (because of its simplicity), and the program yields very acceptable solutions for networks of around 60-node complexity [18].

Nonlinear circuits must also be analyzed for component variations. Tolerance analysis of nonlinear networks is typically much more difficult than that of linear networks. Basically, the derivatives used so successfully in Chapter 6 cannot be formed easily and the simple equations cannot be obtained. Although special methods exist for special networks, the general case is best treated by making changes in the circuit element values and recalculating the entire analysis.

State analysis of nonlinear networks consists of collecting all the first derivatives together and keeping the remaining circuit equations as an algebraic set. The resultant differential-algebraic set must be integrated. Integration is very time consuming for other than linear differential equations, but it can be accomplished [11]. Procedures exist for assembling the nonlinear differential equations for integration [1, 11], but the required code is vastly more intricate than what was discussed here and the calculations can take a very long time. The solution of such equations involves a study of various integration schemes. A good exposition of these is given in the references of Chapter 5 [7, 9].

In this chapter we have discussed a technique for reducing the time-domain analysis of nonlinear networks to a series of DC analyses at each time point. The resultant program is quite similar to the transient analysis program discussed in Chapter 4 with the vital difference that normally at each time point a series of analyses have to be made in order to solve a set of DC analysis problems. Each DC analysis starts by assuming voltages and/or currents for the nonlinear elements, linearizing the element characteristics about the assumed operating point, and solving the resultant linear network. The solutions are then compared with the assumed operating point. The solution is accepted as the solution for the network if the assumed and computed values are close enough.

The procedure is critically dependent on realistic linear equivalent circuits valid for limited ranges of voltages and currents. Models of diodes and transistors were derived which can be used in linearizations of the terminal characteristics. Equivalent circuits for computation may be derived for other elements by following the procedures discussed in this chapter.

Finally, a detailed example was shown which indicated how a simplified equivalent circuit can be obtained for a device in which most of the constituent elements are not available for measurements. During the development of that model we became deeply embroiled in solid-state device technology: this is typical of any attempt at modeling any integrated semiconductor gate or gates. The cost of computation for a circuit analysis can depend heavily on the care expended on obtaining an appropriate computational model.

In earlier chapters all developments were made toward writing good programs in the particular topic studied. The procedures in this chapter can also be implemented relatively easily. However, before any reader would implement his own program containing the details of this chapter, he should carefully study the nonlinear electronic circuit analysis program(s) which are available in his own computer center. Great economies can be effected by using existing programs that have been checked out. The procedures presented here can, however, be used as a guide for modifications and improvements to existing programs.

REFERENCES

1. Bowers, J. C., and S. R. Sedore, *SCEPTRE: A Computer Program for Circuit and System Analysis*. Prentice-Hall, Inc., Englewood Cliffs, N.J., 1971.

2. Johnson, E. D., et al., *Transient Radiation Analysis by Computer Program (TRAC)*. Harry Diamond Laboratories, Washington, D.C., Contract No. DAA 639-68-C-0041, 1968.

3. Nitzan, D., "MTRAC: Computer Program for Transient Analysis of Circuits Including Magnetic Cores," *IEEE Transactions on Magnetics*, MAG-5, No. 3 (September 1969).

4. Gray, P. E., et al., *Physical Electronics and Circuit Models of Transistors* (SEEC, Vol. 2). John Wiley and Sons, New York, 1964.

5. Ebers, J. J., and J. L. Moll, "Large-Signal Behavior of Junction Transistors," *Proc. IRE*, Vol. 42, No. 12 (December 1954).

6. Millman, J., and H. Taub, *Pulse, Digital and Switching Waveforms*. McGraw-Hill, Inc., New York, 1967.

7. Malmberg, A. F., et al., *NET-1 Network Analysis Program—7090/94 Version*. Los Alamos Scientific Laboratories, LA-3119, August 1964.

8. Purdue, C. H., *Computer Programs for Obtaining the Modified Ebers-Moll Transistor Parameters*. Sandia Corp., Albuquerque, N.M., SC-DR-66-2613, 1966.

9. Burger, R. M., and R. P. Donovan, eds., *Fundamentals of Silicon Integrated Device Technology*, Vol. II. Prentice-Hall, Inc., Englewood Cliffs, N.J., 1968.

10. Ralston, A., *A First Course in Numerical Analysis*. McGraw-Hill, Inc., New York, 1965.

11. Dembart, B., and L. Milliman, *CIRCUS—2, A Digital Computer Program for Transient Analysis of Electronic Circuits, User's Guide*. The Boeing Co., Report 0070-1, Seattle, Wash., July 1971.

12. Spangenberg, K. R., *Vacuum Tubes*. McGraw-Hill Book Company, New York, 1948.

13. Puttcamp, R. R., *TRACLIB, A Computer Program to Calculate the TRAC Transistor and Diode Parameters from the CIRCUS Library*. Harry Diamond Laboratories, Washington, D.C., HDL-TR-1532, September 1970.

14. Searle, C. L., et al., *Elementary Circuit Properties of Transistors*, (SEEC, Vol. 3). John Wiley and Sons, New York, 1964.

15. BOLASH, L. J., "Modeling Diode-Transistor Logic Integrated Circuits for Automated Network Analysis." Master's thesis, Department of Electrical Engineering, North Carolina State University, Raleigh, 1972.

16. CAMENZIND, H. R., *Circuit Design for Integrated Electronics*. Addison-Wesley Publishing Company, Reading, Mass., 1968.

17. LYNN, D. K., C. S. MEYER, and D. J. HAMILTON, eds., *Analysis and Design of Integrated Circuits*. McGraw-Hill Book Company, New York, 1967.

18. SCHNEIDER, W. J., "Operating Characteristics of Computer Programs for Nonlinear Transient Analysis," Technical Report 32-1429, Jet Propulsion Laboratory, California Institute of Technology, Pasadena, Calif., June 1970.

PROBLEMS

7.1 Compute output voltage $e_{(out)}(t)$ for the circuit and excitation shown in Fig. P7.1. Use the diode in Fig. 7.2. The initial capacitor voltages are 0 V.

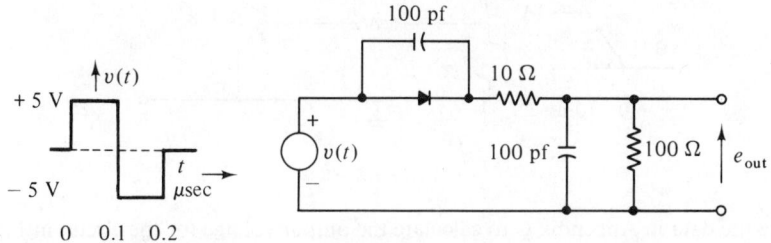

Fig. P7.1

7.2 Derive the ΔH and ΔT equations for two coupled coils if each coil is modeled with a parallel resistance as in Fig. 7.5.

7.3 Describe a procedure for obtaining the ΔH and ΔT terms for arbitrarily coupled coils (more than two) each with a series resistance.

7.4 Repeat Problem 7.2 if each coil is modeled as in Fig. 7.5.

7.5 A crude but often adequate model of a diode is a current sensitive switch in series with a resistor [see Fig. 7.9(b)]. Derive the ΔH and ΔT terms for such a device.

7.6 Verify the terms derived for the transistor model, particularly Eq. (7.64), and the ΔH and ΔT terms in Eq. (7.68).

7.7 Derive a PNP transistor model along the lines of Sec. 7.3.

7.8 In the diode circuit shown in Fig. 7.25, the voltage across the diode can be at most 5 V, or at least 0 V.
   (a) Show that the iteration process will converge for any starting values between these two limits.
   (b) How many steps does it take to obtain an accuracy of p%?

386 Nonlinear Analysis Ch. 7

7.9 Replace the voltage source in Fig. 7.25 by a ± 5-V square wave generator of 1 μsec duration and put a 100-pf capacitor across the diode. Use the equation iteration procedure of Sec. 7.5 to calculate the current (as a function of time) for two cycles of the input wave. How do you decide what value of Δt is reasonable?

7.10 Design the code details for changing your transient analysis program (see Problem 4.17 and Fig. 4.13) into a nonlinear transient analysis program (see Fig. 7.26). Implementation is complicated and time consuming.

7.11 Most large computer installations as well as time-sharing services have at least one nonlinear transient analysis program (TRAC, CIRCUS, SCEPTRE, or others). Use the most convenient such program to verify the results given in Sec. 7.8. The input data may have to be reformulated.

7.12 Use the data in Appendix C to calculate the output voltage of the circuit in Fig. P7.2 for the input waveform shown.

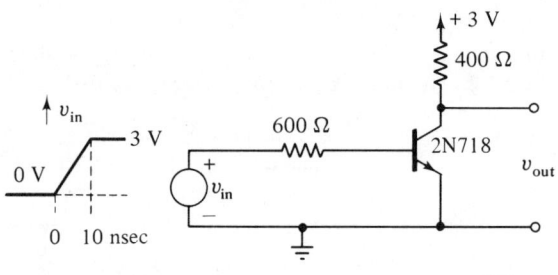

Fig. P7.2

7.13 Use the data in Appendix C to calculate the output voltage for the circuit in Fig. P7.3 for the input wave shown.

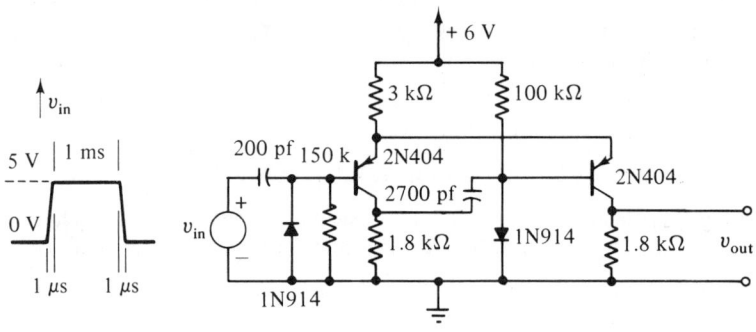

Fig. P7.3

7.14 Verify the results of the previous two problems using a nonlinear analysis program.

COMPUTER PROGRAM LISTINGS

Routines for the Solution of Simultaneous Algebraic Equations and for Matrix Inversion

APPENDIX

A

A.1 Subroutine SOLVER

```
      SUBROUTINE SOLVER (A,NA,N,DET)
C         LIBRARY ROUTINE FOR THE SOLUTION OF A SET OF SIMULTANEOUS
C         ROUTINE FOR THE SOLUTION OF A SET OF SIMULTANEOUS
C         EQUATIONS.    FORTRAN IV VERSION -- NO PRINTING.
      DIMENSION A(NA,NA)
C         GAUSS ELIMINATION WITH PIVOT SEARCHING BELOW MAIN DIAGONAL ONLY.
C         A       AUGMENTED ARRAY OF COEFFICIENTS
C         NA      COLUMN SIZE OF ARRAY IN CALLING ROUTINE
C         DET     VALUE OF THE DETERMINANT OF COEFFICIENTS
C         N       ACTUAL SIZE OF ARRAY (NUMBER OF EQUATIONS)
C      THE SOLUTION VECTOR IS PRODUCED IN COLUMN N+1
C      THE INPUT ARRAY IS DESTROYED
      N1=N+1
      NN=N-1
      DET=1.
      DO 100 I=1,N
      II=I+1
      IF (I.EQ.N) GO TO 27
      AMAX=ABS(A(I,I))
      JM=I
      DO 20 J=II,N
      IF (AMAX.GT.ABS(A(J,I))) GO TO 20
      AMAX=ABS(A(J,I))
      JM=J
   20 CONTINUE
C         PIVOT ELEMENT AT ROW JM
C         START EXCHANGE PROCEDURE IF NECESSARY
      IF (JM.EQ.I) GO TO 27
```

```
C         EXCHANGE ROWS
          DO 23 J=I,N1
          T=A(JM,J)
          A(JM,J)=A(I,J)
   23     A(I,J)=T
          DET=-DET
C         END OF EXCHANGE SEQUENCE
   27     IF (ABS(A(I,I)).LE.1.0E-30) GO TO 500
          DET=DET*A(I,I)
          DO 30 J=II,N1
   30     A(I,J)=A(I,J)/A(I,I)
          IF (I.EQ.N) GO TO 100
          DO 50 J=II,N
          DO 50 K=II,N1
   50     A(J,K)=A(J,K)-A(J,I)*A(I,K)
  100     CONTINUE
C         FORWARD COURSE COMPLETE, START RETURN COURSE.
          DO 300 I=1,NN
          II=N-I
          III=II+1
          DO 300 K=III,N
  300     A(II,N1)=A(II,N1)-A(II,K)*A(K,N1)
C         END OF RETURN COURSE
          RETURN
  500     DET=0.
          RETURN
          END
```

A.2 Subroutine INVRT

```
          SUBROUTINE INVRT(A,NA,N,DET,IX)
          DIMENSION A(NA,1),IX(NA)
          DET=1.
          NN=N+N
          DO 20 I=1,N
          IX(I)=I
          NI=N+I
          DO 20 J=1,N
          NJ=N+J
   10     A(I,NJ)=0.
   20     A(I,NI)=1.
C         UNIT MATRIX IS SET IN COLUMNS N+1 THROUGH N*2
C         PERFORM GAUSS-JORDAN FEDUCTION ON INPUT ARRAY AND THE
C         APPENDED UNIT MATRIX, KEEPING ROW-WISE EQUALITY.
          DO 90 I=1,N
          II=I+1
          IF (I.EQ.N) GO TO 50
          AMAX=ABS(A(I,I))
          KJ=I
          DO 30 K=II,N
          IF (AMAX.GT.ABS(A(K,I))) GO TO 30
          AMAX=ABS(A(K,I))
          KJ=K
   30     CONTINUE
          IF (KJ.EQ.I) GO TO 50
C         NEED TO EXCHANGE ROWS I AND KJ
          DO 40 K=I,NN
          T=A(I,K)
          A(I,K)=A(KJ,K)
   40     A(KJ,K)=T
          IT=IX(I)
          IX(I)=IX(KJ)
          IX(KJ)=IT
C         EXCHANGE OF ROWS COMPLETE.
   50     IF(ABS(A(I,I)).LT.1.E-30) GO TO 500
          DET=DET*A(I,I)
```

```
C           DIVIDE BY MAIN DIAGONAL ELEMENT.
      DO 60 J=II,NN
   60 A(I,J) = A(I,J)/A(I,I)
      IF (I.EQ.N) GO TO 90
      DO 80 J=1,N
      IF (I.EQ.J) GO TO 80
      DO 70 K=II,NN
   70 A(J,K) = A(J,K) -A(J,I)*A(I,K)
   80 CONTINUE
C           REDUCTION COMPLETE.
   90 CONTINUE
C           PUT COLUMNS OF INVERSE IN ORDER
      DO 150 I=1,N
      DO 140 J=1,N
      IF (IX(J).NE.I) GO TO 140
      IJ=N+J
      DO 130 K=1,N
  130 A(K,I) = A(K,IJ)
  140 CONTINUE
  150 CONTINUE
      RETURN
  500 DET=0.
      RETURN
      END
```

A.3 Code Changes for Subroutine SOLVX

```
C  THESE CHANGES WILL CONVERT SOLVER, LISTED IN APPENDIX A, TO
C  HANDLE COMPLEX COEFFICIENTS.  THE SUBROUTINE IS RENAMED SOLVX
C  BY REPLACING THE SUBROUTINE SOLVER... CARD BY THE FOLLOWING
      SUBROUTINE SOLVX (A,NA,N,DET)
      COMPLEX A,DET,T
C  * * * *
C  THE TWO STATEMENTS AFTER   DO 20 J=I,N   ARE REPLACED
C  BY THE FOLLOWING TWO STATEMENTS.
      IF (CABS(A(J,I)).LE.AMAX) GO TO 20
      AMAX=CABS(A(J,I))
C  * * * *
C  THE NEXT CARD IS THE NEW STATEMENT 27
   27 IF (CABS(A,I,I).LE.1.E-30) GO TO 500
C  * * * *
C  NO FURTHER CHANGES ARE REQUIRED TO CONVERT SOLVER TO SOLVX,
C  A ROUTINE TO SOLVE A SET OF N LINEAR ALGEBRAIC EQUATIONS WITH
C  COMPLEX COEFFICIENTS.
C  ON SOME COMPILERS IT IS ADVISABLE TO REPLACE THE NAME AMAX WITH
C  SOME OTHER NAME, SUCH AS  ABIG, TO AVOID CONFUSION WITH
C  THE MAXIMUM FUNCTION AMAX1.
```

A.4 Code Changes for Subroutine INVRX

```
C   COMPLEX MATRIX INVERSION (IN-PLACE).
C   THESE CHANGES WILL CONVERT SUBROUTINE INVERT, LISTED IN
C   APPENDIX A, TO HANDLE COMPLEX COEFFICIENTS IN THE INPUT ARRAY.
C   THE NEXT TWO STATEMENTS REPLACE THE SUBROUTINE INVERT ..
C   STATEMENT.
      SUBROUTINE INVRX(A,NA,N,DET,IR)
      COMPLEX A,DET,T
C   * * * *
C   THE TWO STATEMENTS AFTER DO 20 J=I,N  ARE REPLACED
C   BY THE FOLLOWING TWO STATEMENTS.
      IF (CABS(A(J,I)).LE.AMX) GO TO 20
      AMAX=CABS(A(J,I))
C   * * * *
C   STATEMENT NUMBER 50 IS REPLACED BY THE FOLLOWING.
   50 IF (CABS(A(I,I)).LE.1.E-30) GO TO 500
C   * * * *
C   NO OTHER CHANGES ARE NECESSARY TO HANDLE COMPLEX
C   COEFFICIENTS.
C   ON SOME COMPILERS IT IS ADVISABLE TO REPLACE THE NAME AMAX WITH
C   SOME OTHER NAME, SUCH AS  ABIG, TO AVOID CONFUSION WITH
C   THE MAXIMUM FUNCTION AMAX1.
```

PLOTTING SUBROUTINES

APPENDIX **B**

B.1 Instructions for Use of Plotting Routines

The three plotting subroutines listed here are intended to be used in conjunction with various programs discussed in this book. Each subroutine produces one page of output consisting of various labels and a line printer plot of 51 lines and 101 columns (50 line spacings by 100 print-position spacings). The plots are complete with borders and x/y axis labels.

These programs are self-scaling, i.e., they find the range of the input variables and adjust the scaling on the x and the y axes so that the plots extend over the full page. The programs detect constant (unchanging) input data and will omit printing in this case; for small changes input values, such as would be produced by round-off errors, however, the programs will attempt to plot margin to margin.

The programs are designed for byte-oriented machines, such as the IBM 360-370 systems, but they can be converted to other machines without much effort since care was taken not to use statements that are strongly machine-organization dependent. The one exception is the LOGICAL*1 statement.

(A) Subroutine PLOTXY allows the plotting of any two variables. The data must be supplied to the subroutine as follows:

(1) X, the array of **x**-data points, containing in a vector N data points.

(2) the integer N (the number of data points in the arrays X as well as in Y).

(3) Y, the array of **y**-data points: the arrays X and Y must be paired, i.e., X(I) and Y(I) (I = 1, ..., N) will contain one x/y data pair.

(4) XNAM, a location containing four alphanumeric characters, to be used for labeling the x axis.

(5) NUMX, an integer used also in labeling the x axis (for example, "node voltage two" may be generated as

DATA XNAM/'VN-'/, NUMX/2/

and will be printed by this routine as VN-2.

(6) YNAM, a label similar to XNAM, but used for labeling the y axis.

(7) NUMY, an integer similar to NUMX, but used for the y axis.

(8) PR, an array used temporarily in this subroutine, of at least 5151 bytes long. This array is used to generate the plot image and may be used for other purposes elsewhere in the calling program(s).

The values for every print line (y axis) are labeled and eleven x values (ten increments) are printed also.

(B) Subroutine PLT2F is designed to plot decibel magnitude and phase angle of an input variable vs. logarithmic frequency. Hence this program will generate Bode plots for use with AC analysis programs. The required inputs are:

(1) FREQ, an array of floating-point numbers, the frequencies for which data are supplied.

(2) VAL, a doubly dimensioned array /DIMENSION VAL (NS, 2)/ which contains magnitude and phase angle (in degrees) of the variable to be plotted. The triplet of numbers FREQ(I), VAL(I, 1), VAL(I, 2) contains the data for the Ith point to be plotted (I = 1, ..., N); the array VAL is stored NS numbers per column, with NS $\geq$ N; the magnitude data must be positive numbers.

(3) N, the number of data points in FREQ as well as in VAL (N $\geq$ 2).

(4) NS, the storage size of VAL in the calling routine.

(5) PR, a temporary array of at least 5151 bytes, usable elsewhere in the calling program for other storage.

(6) NAM, an alphameric label of up to four characters, similar to XNAM used in subroutine PLOTXY.

(7) NUM, an integer number similar to XNUM in subroutine PLOTXY.

The program converts all frequencies to log frequencies in the array FREQ, all values VAL(I, 1), I = 1, ..., N are converted to decibel values and stored in place of the original data. The program finds the lowest whole decade of frequencies and the highest whole decade plus one for scaling the frequency axis. Thus, frequency data between 300 and 4500 Hz will be plotted in the range of two decades, from 100 to 10,000 Hz. The lowest and highest decades are labeled and all intervening decade points are indicated by a vertical bar around the top and the bottom of the plots.

The program sets all magnitude data less than 1.E-30 to 1.E-30; otherwise the magnitude data is self-scaling. The phase angle data are plotted in 49 lines (two less than the magnitude data, from lines 2 to 50 of the plot); the phase angle is always plotted in the range of $+180.00°$ to $-180.00°$. The choice of 48 line spacings allows a step size of $7.5°$ per line, a relatively round figure to work with.

Every line for the y data (51 magnitudes and 48 angles) is labeled. Only the lowest decade and highest decade in frequency are printed as labels.

(C) Subroutine PLT4XY is intended to plot up to four variables (x/y plots). The data pairs X(I, J), Y(I, J) represent the x-y data pair for the Ith point of the Jth plot (I = 1, ..., N and J = 1, ..., NP with $1 \leq NP \leq 4$). The data to be plotted need not be on any common base; the routine will find the minimum and maximum values of all x data and for all y data; in each array, however, the same number of data points (N) will be examined.

The input data must be supplied as follows:

(1) X, a doubly dimensioned array /DIMENSION X(NS, NP)/ containing the x data values for the Jth plot $1 \leq J \leq NP$, with $NP \leq 4$ and the Ith point $1 \leq I \leq N$, with $2 \leq N \leq NS$, and NS = storage size for the columns of **x** and of **y** in the calling routine.

(2) N, the number of data pairs in X and Y.

(3) Y, a doubly dimensioned array, similar to the array X, containing the **y** data corresponding to the **x** data.

(4) NS, the storage size for X and Y in the calling routine.

(5) PR, an array of at least 5151 bytes, used for temporary storage, similar to the array PR in subroutine PLOTXY.

(6) NP, the number of plots for which data are supplied in X and Y. $1 \leq NP \leq 4$;

(7) NAMX, an array of NP alphameric labels used to label the x values for the plots, similar to XNAM used in subroutine PLOTXY.

(8) NAMY, an array of NP labels used to label the y values, similar to NAMX.

(9) NUMX, an array of NP integers used in conjunction with NAMX to complete the x-value labels; similar to NUMX used in Subroutine PLOTXY.

(10) NUMY, an array of NP integers used in conjunction with NAMY, similar to NUMX.

The listings of these subroutines follow.

B.2 Subroutine PLOTXY

```
      SUBROUTINE PLOTXY(X,N,Y,XNAM,NUMX,YNAM,NUMY,PR)
      DIMENSION X(N),Y(N),PR(51,101),XL(11)
      LOGICAL*1 PR,BLANK,DASH,BAR,STAR
      DATA BLANK/' '/,DASH/'-'/,BAR/' '/,STAR/'*'/
      DO 14 I=1,51
      DO 10 J=1,101
   10 PR(I,J)=BLANK
      PR(I,1)=BAR
   14 PR(I,101)=BAR
      DO 18 I=1,101
      PR(1,I)=DASH
   18 PR(51,I)=DASH
      XLO=X(1)
      YLO=Y(1)
      XHI=XLO
      YHI=YLO
      DO 20 I=1,N
      XLO=AMIN1(XLO,X(I))
      XHI=AMAX1(XHI,X(I))
      YLO=AMIN1(YLO,Y(I))
   20 YHI=AMAX1(YHI,Y(I))
      YM=50./(YHI-YLO)
      XM=100./(XHI-XLO)
      YINC=1./YM
      XINC=10./XM
      XL(1)=XLO
      DO 25 I=2,11
   25 XL(I)=XL(I-1)+XINC
      DO 30 I=1,N
      K=(YHI-Y(I))*YM+1.49
      L=(X(I)-XLO)*XM+1.49
   30 PR(K,L)=STAR
      PRINT 40,YNAM,NUMY,XNAM,NUMX,YNAM,NUMY,(XL(I),I=1,11,2),
     1(XL(I),I=2,10,2),(BAR,I=1,11)
   40 FORMAT ('1'/   2X,A4,I2,52X,A4,I2,54X,A4,I2,/9X,1PE10.3,1P5E20.3/
     119X,1PE10.3,1P4E20.3/13X,A1,10A10)
      YL=YHI+YINC
      DO 50 I=1,51
      YL=YL-YINC
   50 PRINT 60,YL,DASH,(PR(I,J),J=1,101),DASH,YL
   60 FORMAT (1X,1PE10.3,1X,A1,101A1,A1,1PE12.3)
      PRINT 70,(BAR,I=1,11),(XL(I),I=1,11,2),(XL(I),I=2,10,2),
     1YNAM,NUMY,XNAM,NUMX,YNAM,NUMY
   70 FORMAT (13X,A1,10A10/9X,1PE10.3,1P5E20.3/9X,1P5E20.3/
     12X,A4,I2,52X,A4,I2,54X,A4,I2)
      RETURN
      END
```

B.3 Subroutine PLT2F

```
      SUBROUTINE PLT2F (FREQ,VAL,N,NS,PR,NAM,NUM)
      DIMENSION FREQ(N),VAL(NS,2),PR(51,101),LINE(101)
      LOGICAL*1 PR,BAR,DASH,STAR,PLUS,EKS,BLANK,LINE
      DATA BAR,DASH,STAR,PLUS,EKS,BLANK/1H ,1H-,1H*,1H+,1HX,1H /
      IF (N.LE.1) RETURN
      FMIN=FREQ(1)
      FMAX=FREQ(N)
      FL=ALOG10(FMIN) - 0.01
      FH=ALOG10(FMAX) + 0.99
      JFMN=FL
      JFMX=FH
      IF (JFMN.GE.JFMX) JFMN=JFMN-1
      FRMN=10.**JFMN
      FRMX=10.**JFMX
C        CONVERT TO LOG-FREQUENCY.
      DO 10 I=1,N
   10 FREQ(I)=ALOG10(FREQ(I))
C        CONVERT TO DB VALUES
      DO 40 I=1,N
      IF (VAL(I,1).LT.1.E-30) VAL(I,1)=1.E-30
      VAL(I,1)=20.*ALOG10(VAL(I,1))
      IF (I.EQ.1) VMN=VAL(1,1)
      IF (I.EQ.1) VMX=VMN
      VMN=AMIN1(VMN,VAL(I,1))
      VMX=AMAX1(VMX,VAL(I,1))
   40 CONTINUE
      IF (VMX.LE.VMN) RETURN
      AL=187.5
      AINC=7.5
C        USE 48 LINES FOR ANGLE DATA.
      VINC=(VMX-VMN)/50.
      VM=1./VINC
      JFD=JFMX-JFMN
      FINC=FLOAT(JFD)/100.
      FM=1./FINC
      DO 60 I=1,51
      DO 50 J=1,101
   50 PR(I,J)=BLANK
      PR(I,1)=BAR
   60 PR(I,101)=BAR
      DO 70 I=1,101
      LINE(I)=BLANK
      PR(1,I)=DASH
   70 PR(51,I)=DASH
      PR(1,1)=PLUS
      PR(1,101)=PLUS
      PR(51,1)=PLUS
      PR(51,101)=PLUS
      EK=1.
      LINE(1)=BAR
      DO 80 I=1,JFD
      EK=EK+FM
      K=EK
   80 LINE(K)=BAR
      LINE(101)=BAR
      FRR=JFMN
      DO 100 I=1,N
      K=(VMX-VAL(I,1))*VM+1.49
      L=(FREQ(I)-FRR ) *FM + 1.49
      M=(AL-VAL(I,2))/AINC + 1.49
      PR(K,L)=EKS
      PR(M,L)=STAR
```

```
      100 CONTINUE
          PRINT 130,FRMN,FRMX,LINE
      130 FORMAT('1'/ 55X,'FREQUENCY   (LOGARITHMIC SCALE)'/
         19X,1PE10.3,90X,1PE13.3/13X,101A1)
          YL=VMX
          PRINT 140,YL,(PR(1,J),J=1,101)
      140 FORMAT (1X,1PE10.3,' -',101A1)
          DO 150 I=2,50
          YL=YL-VINC
          AL=AL-AINC
      150 PRINT 160,YL,(PR(I,J),J=1,101),AL
      160 FORMAT(1X,1PE10.3,' -',101A1,'- ',0PF6.1)
          YL=YL-VINC
          PRINT 140,YL,(PR(51,J),J=1,101)
          PRINT 170,LINE,FRMN,FRMX
      170 FORMAT(13X,101A1/9X,1PE10.3,90X,1PE13.3/45X,
         1'FREQUENCY   (LOGARITHMIC SCALE)')
          PRINT 180, NAM,NUM,NAM,NUM
      180 FORMAT (/10X,'X =',A4,I2,' MAGNITUDE (DB.)',20X,'* =',
         1A4,I2,' PHASE ANGLE')
          RETURN
          END
```

B.4 Subroutine PLT4XY

```
      SUBROUTINE PLTX4Y (X,N,Y,NS,PR,NP,NAMX,NAMY,NUMX,NUMY)
      DIMENSION X(NS,NP),Y(NS,NP),PR(51,101),N(NP),NAMX(NP),NAMY(NP),
     1NUMX(NP),NUMY(NP),XL(11)
      LOGICAL*1 PR,BLANK,DASH,BAR,STAR,OH,PLUS,EKS
      LOGICAL*1 CHR(4)
      EQUIVALENCE (CHR(1),STAR),(CHR(2),PLUS),(CHR(3),EKS),(CHR(4),OH)
      DATA BLANK,DASH,BAR,STAR,OH,PLUS/1H ,1H-,1H ,1H*,1HO,1H+/
      DATA EKS/1HX/
      DO 14 I=1,51
      DO 10 J=1,101
10    PR(I,J)=BLANK
      PR(I,1)=BAR
14    PR(I,101)=BAR
      DO 18 I=1,101
      PR(1,I)=DASH
18    PR(51,I)=DASH
      XLO=X(1,1)
      XHI=X(1,1)
      YLO=Y(1,1)
      YHI=Y(1,1)
      DO 22 J=1,NP
      NN=N(J)
      IF (NN.LE.1) GO TO 22
      DO 20 I=1,NN
      XLO=AMIN1(XLO,X(I,J))
      XHI=AMAX1(XHI,X(I,J))
      YLO=AMIN1(YLO,Y(I,J))
20    YHI=AMAX1(YHI,Y(I,J))
22    CONTINUE
      IF (XHI.EQ.XLO) RETURN
      IF (YHI.EQ.YLO) RETURN
      XM=100./(XHI-XLO)
      YM=50./(YHI-YLO)
      YINC=1./YM
      XINC=10./XM
      XL(1)=XLO
      DO 21 I=2,11
21    XL(I)=XL(I-1)+XINC
      DO 40 J=1,NP
      DO 30 I=1,NN
      K=(YHI-Y(I,J))*YM+1.49
      L=(X(I,J)-XLO)*XM+1.49
      GO TO (23,24,26,28),J
23    PR(K,L)=STAR
      GO TO 30
24    PR(K,L)=PLUS
      GO TO 30
26    PR(K,L)=EKS
      GO TO 30
28    PR(K,L)=OH
30    CONTINUE
40    CONTINUE
      PRINT 50,(XL(I),I=1,11,2),(XL(I),I=2,10,2),(BAR,I=1,11)
50    FORMAT('1'/1PE19.3,1P5E20.3/9X,1P5E20.3/A14,10A10)
      YL=YHI+YINC
      DO 60 I=1,51
      YL=YL-YINC
60    PRINT 70,YL,DASH,(PR(I,J),J=1,101),DASH,YL
70    FORMAT (1X,1PE10.3,A2,102A1,1X,1PE10.3)
      PRINT 80,(BAR,I=1,11),(XL(I),BAR,I=1,9,2),XL(11),(XL(I),I=2,10,2),
     1(CHR(I),NAMY(I),NUMY(I),NAMX(I),NUMX(I),I=1,NP)
80    FORMAT (13X,A1,10A10/1PE19.3,5(4X,A1,1PE15.3)/1PE29.3,1P4E20.3//
     11X,4(A1,' = ',2X,A4,I2,' VS. ',A4,I2,8X))
      RETURN
      END
```

SEMICONDUCTOR MODEL LIBRARY

APPENDIX C

C.1 Diode Parameters

This section lists data for Ebers-Moll models of some diodes. The diode model is a simplified version of the one used in Sec. 7.2 and is defined in Fig. C.1. Note that the diode bulk resistance is neglected. The various model parameters are listed here for reference (see Table 7.1 and Fig. 7.8 for further details).

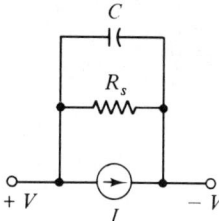

Fig. C.1 Diode Model

$$I = I_s[\exp(V \cdot \Theta) - 1]$$
I_s = reverse saturation current
Θ = constant (measured at 25°C)
R_s = leakage resistance
$C = C_T + C_d$
$$C_T = \frac{C_t}{(V_Z - V)^N}$$
C_t = transition capacitance
V_Z = contact potential
N = grading constant
$C_d = k_d(I + I_s)$
k_d = diffusion capacitance constant

Note that Θ is the reciprocal of θ used in Chapter 7.

The data for Table C.1 (p. 400) was obtained from various sources. Because of the large variations among the individual diodes of the same type designation, the characteristics were approximated for a representative diode. Large variations, typically 50%, can be expected for diodes of the same type number. The model parameters interact greatly; the numbers given here have been obtained for representative normal uses of the devices.

In many applications, one may use $V_Z = 0.75$ V and $N = 0.5$ V without any great discrepancies.

Table C.1 Diode Parameters

Diode	I_s (μA)	Θ (V^{-1})	R_s (meg)	C_t (pf)	V_Z (V)	N	k_d (pf/μA)
1N100	2.5	14.7	0.01	0.354	0.50	0.50	0.023
1N191	1.25E-6	38.9	0.40	0.353	0.50	0.50	0.620
1N279	1.24	23.1	3.80	24.5	0.50	0.50	0.730
1N645	2.51E-2	26.1	1.2E6	10.8	0.86	0.58	125
1N647	1.60E-3	22.0	4.2E4	7.70	1.00	0.51	19
1N649	2.51E-2	26.1	1.2E6	10.8	0.86	0.58	—
1N659	1.40E-5	33.7	560	19.0	0.80	0.40	3.0
1N695	0.780	25.0	0.32	0.70	0.60	0.17	0.36
1N746	2.69E-4	35.3	0.10	463	0.75	0.50	4.09
1N748	1.14E-4	33.6	0.10	414	0.75	0.50	3.88
1N750	4.84E-5	32.7	0.50	368	0.75	0.50	3.45
1N752	1.25E-5	30.8	1.00	331	0.75	0.50	3.10
1N754	4.80E-6	29.7	10.0	285	0.75	0.50	2.67
1N756	7.24E-6	30.3	10.0	238	0.75	0.50	2.23
1N758	3.40E-5	32.2	10.0	199	0.75	0.50	1.86
1N827	1.04E-5	30.6	10.0	219	0.75	0.50	2.05
1N914	2.90E-3	21.5	1.10	24	0.90	0.50	18.1
1N961	7.15E-4	31.0	7.51	217	0.95	0.45	—
1N963	4.25E-6	33.3	1.82	332	0.75	0.50	4.34
1N965	1.37E-5	34.1	2.28	2.90	0.75	0.50	3.81
1N967	4.35E-5	35.7	2.75	290	0.75	0.50	3.81
1N969	2.53E-4	38.1	3.34	265	0.75	0.50	3.48
1N971	5.60E-3	42.3	4.14	199	0.75	0.50	2.61
1N973	2.41E-4	38.2	5.02	185	0.75	0.50	2.44
1N3016	2.53E-4	27.8	0.07	595	0.75	0.50	7.85
1N3018	2.81E-4	27.8	0.25	5.95	0.75	0.50	7.85
1N3020	4.55E-4	28.2	7.60	496	0.75	0.50	6.52
1N3022	1.77E-4	28.6	1.82	4.96	0.75	0.50	6.52
1N3024	2.84E-4	29.7	2.28	477	0.75	0.50	6.28
1N3026	4.64E-4	30.6	2.74	450	0.75	0.50	5.92
1N3028	4.55E-4	31.5	3.34	425	0.75	0.50	5.57
1N3070	4.23E-3	21.8	8.64	1.76	1.00	0.17	.628
1N3600	5.06E-3	20.9	1.4E3	0.90	0.85	0.50	.00398
1N3669	1.30E-4	28.0	1.10	23	0.80	0.46	13
1N4003	4.20E-3	22.0	2.0E4	230	1.00	0.48	150
1N4572	7.42E-6	28.5	2.4E3	5.6	0.85	0.61	13.5

C.2 Transistor Parameters

This section lists data for Ebers-Moll models of some transistors. The transistor model is a simplified version of the one used in Sec. 7.3 and is defined in Fig. C.2. For PNP transistors, the substitutions indicated following Eq. (7.68) must be used.

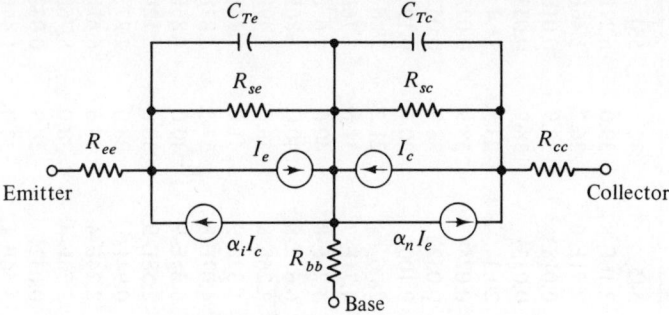

Fig. C.2 NPN Transistor Model

The definitions of the transistor model parameters are given in Sec. 7.3. Compare also to the diode model equations.

The base bulk resistance R_{bb} is included in some models. When the equations in Sec. 7.3 are used, R_{bb} must be included as an external branch separately input to the program.

The parameters C_{Te} and C_{Tc} are defined similar to C_T of the diode model.

As with the diode data, the numerical values for the transistor models come from several sources. The models were fitted to measured data on a representative transistor of a given type. Considerable variation can be expected between individual copies of a given transistor type.

Since the transistor model contains many parameters, considerable interplay between these numbers may be allowed in order to fit observed data. Depending on the closeness of fit desired over the normal operating regions, quite different numbers will result in the models. The individual numerical entries may be quite at variance with other similar entries in other similar transistor models; however, the active set of numbers for a given transistor describes the overall transistor behavior adequately.

REFERENCES

1. Bowers, J. C., and S. R. Sedore, *SCEPTRE: A Computer Program for Circuit and System Analysis*, Prentice-Hall, Inc., Englewood Cliffs, N.J., 1971.
2. SCEPTRE Model Library, AFWL (EST), Kirtland AFB, New Mexico. Report AFWL-TR-69-44.
3. Measured data.

Table C.2 Transistor Parameters

Transistor	α_n	α_i	R_{ee} (Ω)	R_{se} (meg)	R_{cc} (Ω)	R_{sc} (meg)	R_{bb} (Ω)	I_{se} (μA)	Θ_e (V^{-1})	I_{sc} (μA)	Θ_c (V^{-1})
*2N174	0.982	0.833	0	0.40	0.02	0.16	3.0	545	32.1	551	32.4
*2N329	0.972	0.866	2.5	50.	32.	50.	0	2.16E-8	39.0	6.22E-5	29.0
2N336	0.991	0.584	2.5	50.	60.	50.	1000.	2.91E-6	36.7	1.46E-5	28.2
2N356	0.937	0.833	0	0.8	1.	0.8	20.	0.0183	38.9	0.0183	38.9
*2N384	0.909	0.900	0	0.04	20.	1.0	100.	0.033	38.9	0.033	38.9
*2N404	0.984	0.900	0	6.47	0.1	7.46	1.9	20.3	23.3	21.4	21.9
*2N457	0.968	0.833	0	0.06	0.1	0.06	10.0	0.0258	38.9	0.0258	38.9
*2N597	0.986	0.900	0	3000	2500	0.30	6.0	0.0522	38.9	0.0522	38.9
2N657	0.981	0.50	4.5	50.	6.1	50.	18.0	0.100	20.7	0.196	20.7
*2N705	0.976	0.667	0	1.0	12.0	1.0	50.	28.6	21.2	28.6	21.2
2N706	0.968	0.50	0	10.5	0.1	3.6	1.3	4.71E-7	31.6	7.76E-4	23.8
2N718	0.991	0.50	0.10	50.	7.4	50.	30.	3.3 E-8	40.0	2.6 E-7	38.0
2N720	0.991	0.50	1.0	50.	8.7	50.	50.	6.6 E-8	40.0	1.86E-7	39.0
2N743	0.98	0.80	0	∞	8.3	∞	200	3.26E-8	36.0	1.49E-8	38.0
2N834	0.984	0.878	0	∞	11.0	∞	10.0	1.32E-9	41.0	1.14E-8	38.0
2N910	0.985	0.285	0.3	50	8.0	50	13.0	4.07E-8	37.5	4.87E-8	37.5
2N915	0.992	0.50	1.0	50	17.0	50	15.0	6.65E-9	40.0	1.86E-7	27.0
2N918	0.977	0.492	8.5	50	11.5	50	12.0	2.58E-9	38.0	2.12E-8	36.3
*2N964	0.980	0.909	0	1.83	0.04	72.9	1.49	0.941	28.0	2.27	28.9
*2N995	0.986	0.667	0	1.3 E5	2.0	1.3 E5	25.0	2.74E-4	26.6	6.61E-4	26.1
2N1016B	0.990	0.833	0	∞	0.6	∞	40.0	2.73E-4	38.0	8.76E-4	35.0
*2N1039	0.976	0.833	0	0.027	0.25	0.240	10.0	0.0321	38.9	0.0321	38.9
*2N1131	0.952	0.667	0	1990	0.1	1890	1.82	3.56E-12	33.0	1.44E-9	23.9
*2N1184	0.986	0.882	0	∞	1.2	∞	20	8.12	47.0	11.3	40.0
*2N1228	0.941	0.667	0	1.3 E4	0.1	8.1 E4	2.0	1.75E-4	23.0	5.33E-5	25.0

*PNP transistor. All unmarked ones are NPN transistors.

	C_{ie} (pf)	V_{ze} (V)	N_e	K_{de} (pf/µA)	C_{ic} (pf)	V_{zc} (V)	N_c	K_{dc} (pf/µA)	Remarks†
*2N174	1410	1.00	0.50	51.1	1410	1.00	0.50	516	Ge, 100 W, 80 V, 12 A, 50, SP
*2N329	28.0	0.80	0.47	3.43	96.0	0.80	0.50	435	0.35 W, 30 V, 36, A
2N336	57.0	1.00	0.43	0.160	10.0	0.80	0.32	18.0	0.15 W, 45 V, 150, AH
2N356	235	0.50	0.50	2.06	328	0.50	0.50	7.74	Ge, 0.1 W, 20 V, 40, S
*2N384	10.9	0.50	0.33	0.0619	6.96	0.50	0.33	0.310	Ge, 0.12 W, 40 V, 50, AH
*2N404	15.6	1.18	0.59	0.286	56.6	3.13	0.83	3.48	Ge, 0.15 W, 25 V, S
*2N457	122	0.50	0.50	14.4	122	0.50	0.50	61.9	Ge, 50 W, 60 V, 5 A, AP
*2N597	20.8	0.50	0.50	2.06	70.4	0.50	0.50	12.4	Ge, 0.25 W, 45 V, 40, FB = 3 MHZ, A
2N657	80.0	0.90	0.41	.373	47.0	0.90	0.39	9.11	4.0 W, 100 V, 50, A
*2N705	3.5	0.30	0.70	0.0225	11.9	0.30	0.25	1.28	Ge, 0.3 W, 15 V, 30, SH
2N706	8.41	0.50	0.33	0.0251	6.96	0.50	0.33	3.78	0.3 W, 25 V, 20, SH
2N718	37.0	0.90	0.41	0.0308	25.0	0.80	0.38	95.0	0.4 W, 60 V, 60, 50 MHZ, AH
2N720	78.0	1.00	0.41	0.060	32.0	0.90	0.39	164.0	0.4 W, 120 V, 60, 50 MHZ, AH
2N743	5.3	0.80	0.25	0.0159	4.6	0.90	0.39	1.642	0.3 W, 20 V, 30, 200 MHZ, SH
2N834	9.0	0.80	0.50	0.0186	7.3	0.90	0.31	1.868	0.3 W, 40 V, 30, 350 MHZ, SH
2N910	50.0	0.90	0.20	0.102	22.0	0.90	0.31	3.47	0.5 W, 100 V, 75, 60 MHZ, A
2N915	7.1	0.90	0.31	0.0208	4.60	0.80	0.41	48.6	0.36 W, 70 V, 75, 200 MHZ, AH
2N918	2.0	0.50	0.15	0.0083	1.7	0.80	0.12	0.457	0.2 W, 30 V, 30, 600 MHZ, AH
*2N964	2.29	0.50	0.33	0.0097	4.81	0.50	0.33	0.092	Ge, 0.15 W, 15 V, 40, 300 MHZ, SH
*2N995	12.7	1.08	0.32	0.0423	15.2	1.01	0.18	0.415	0.36 W, 20 V, 40, 100 MHZ, AH
2N1016	400	1.0	0.20	7.90	1500.	0.70	0.51	58.3	150 W, 30 V, 5 A, 10, SP
*2N1039	1220	0.5	0.50	774	255	0.50	0.50	6190	Ge, 20 W, 60 V, 1 A, 30, FE = 8 KHZ, AP
*2N1131	83.8	1.6	0.50	0.105	72.7	0.94	0.36	0.475	0.6 W, 50 V, 25, 50 MHZ, S
*2N1184	200	0.4	0.50	3.61	240	0.4	0.48	6.00	Ge, 7.5 W, 45 V, 0.4 A, 50, FB = 500 KHZ, SP
*2N1228	178	1.08	0.32	3.05	137	1.01	0.18	33.2	0.4 W, 15 V, S

†See legend on p. 407.

Transistor	α_n	α_i	R_{ee} (Ω)	R_{se} (meg)	R_{cc} (Ω)	R_{sc} (meg)	R_{bb} (Ω)	I_{se} (μA)	Θ_e (V^{-1})	I_{sc} (μA)	Θ_c (V^{-1})
*2N1301	0.986	0.833	0	0.04	5.0	0.65	50	5.72E-4	38.9	5.72E-4	38.9
2N1306	0.990	0.833	0	15.0	1.0	4.0	40	1.96	25.7	2.00	25.9
2N1308	0.993	0.833	0	15.0	1.0	4.0	40	1.99	25.7	2.03	25.9
2N1483	0.952	0.833	0	0.8	2.7	2.0	12	3.38E-4	38.9	3.38E-4	38.9
*2N1499A	0.968	0.667	0	0.5	10	1.6	50	2.82	29.3	2.82	28.2
2N1613	0.988	0.833	0	500	1.0	6000	10	2.19E-2	32.0	2.26E-2	32.4
2N1724	0.976	0.800	0	3.9 E4	100	9.5 E5	100	2.10E-5	27.7	2.08E-5	27.8
2N1900	0.993	0.578	0	∞	0.6	∞	5	1.60E-5	37.8	2.31E-2	37.8
2N2060	0.992	0.286	6.0	50	2.0	50	94.0	4.61E-9	35.8	2.24E-8	35.8
*2N2187	0.971	0.379	0	∞	15	∞	100	5.67E-8	38.9	5.67E-7	36.1
2N2192	0.995	0.875	0	∞	3.5	∞	40	1.31E-7	42.0	9.49E-7	38.0
2N2223	0.987	0.167	0	∞	9.0	∞	32	1.08E-7	38.4	1.78E-6	35.7
*2N2258	0.980	0.500	0	6000	0.1	5000	1.5	9.81E-6	35.4	9.81E-5	27.8
2N2369	0.978	0.800	0	500	2.0	500	20	2.30E-8	37.3	1.04E-3	23.5
2N2453	0.995	0.500	10	50	20	50	100	2.85E-9	46.1	1.23E-8	36.0
2N2484	0.996	0.627	0	6.0	0.186	4.0	2.0	1.41E-3	21.5	1.04E-2	19.2
2N2656	0.990	0.500	0	∞	19	∞	35	4.85E-6	40.0	3.00E-5	38.0
2N2708	0.991	0.834	1.0	50	15	50	80	1.61E-10	39.0	7.20E-9	34.0
2N2808	0.958	0.565	0	∞	36	∞	80	4.17E-11	41.0	9.02E-9	35.0
2N2887	0.980	0.242	0	∞	0.3	∞	13	2.85E-7	37.3	1.16E-3	37.3
*2N2905	0.993	0.583	0	1000	2.0	1000	22	2.32E-8	40.6	6.13E-8	38.7
2N3019	0.980	0.833	0	500	2.0	500	20	2.73E-7	37.3	1.23E-3	23.5
*2N3026	0.992	0.666	0.2	50	0.4	50	7.8	95.2	46.0	214	46.0
2N3108	0.976	0.800	0	12800	0.1	23800	1.44	1.77E-5	30.2	1.63E-5	30.3
2N3119	0.976	0.500	0	4000	0.01	4000	0.6	1.56E-6	31.4	9.76E-6	28.8
*2N3244	0.991	0.667	0	∞	1.4	∞	20	1.71E-7	38.0	1.02E-7	39.0

	C_{te} (pf)	V_{ze} (V)	N_e	K_{de} (pf/μA)	C_{tc} (pf)	V_{zc} (V)	N_c	K_{dc} (pf/μA)	Remarks
*2N1301	13.6	0.50	0.33	0.177	22.4	0.50	0.33	0.619	Ge, 0.15 W, 13 V, 30, 35 MHZ, SH
2N1306	21.0	1.40	0.40	0.340	42.8	1.40	0.41	2.06	Ge, 0.15 W, 25 V, 60, FB = 10 MHZ, S
2N1308	21.0	1.40	0.40	0.204	42.8	1.40	0.41	2.06	Ge, 0.15 W, 25 V, 80, FB = 10 MHZ, S
2N1483	309	0.50	0.33	4.96	600	0.50	0.33	61.9	25 W, 60 V, 0.75 A, 20, SP
*2N1499A	5.72	0.50	0.33	0.233	5.60	0.50	0.33	0.449	Ge, 0.025 W, 20 V, 20, SH
2N1613	54.6	0.71	0.32	0.0565	2690	1.68	4.46	0.258	0.8 W, 75 V, 50, 60 MHZ, S
2N1724	2320	0.92	0.40	0.350	968	0.92	0.40	3.51	50 W, 120 V, 2 A, 30, 10 MHZ, AP
2N1900	7000	0.80	0.38	0.064	1800	0.70	0.47	40.8	125 W, 140 V, 10 A, 8, 50 MHZ, AHP
2N2060	71.5	0.98	0.45	0.0018	34.5	0.98	0.45	0.0277	0.5 W, 100 V, 50, 60 MHZ, AM
*2N2187	5.40	0.80	0.36	0.0755	7.8	0.70	0.41	0.246	0.15 W, 30 V, 6.5 MHZ, SC
2N2192	39.0	0.90	0.45	0.0343	21.0	0.80	0.41	3.99	0.8 W, 60 V, 100, SH
2N2223	59.0	0.80	0.41	0.0606	33.0	0.90	0.41	22.0	0.5 W, 100 V, 50, 50 MHZ, AM
*2N2258	39.4	0.80	0.30	0.113	12.4	0.20	0.20	7.37	Ge, 0.15 W, 7 V, 17, SH
2N2369	3.31	1.06	0.31	0.0119	2.93	1.10	0.08	0.0747	0.36 W, 40 V, 40, 500 MHZ, SH
2N2453	4.5	0.70	0.41	0.129	5.6	0.70	0.23	1.91	0.5 W, 60 V, 150, 60 MHZ, AM
2N2484	6.0	0.50	0.33	6.83	10.5	0.50	0.33	30.6	0.36 W, 60 V, 100, 15 MHZ, A
2N2656	4.3	0.90	0.37	0.586	5.8	0.80	0.31	388.7	0.36 W, 25 V, 50, 260 MHZ, AH
2N2708	1.2	0.90	0.14	0.0062	1.5	0.90	0.15	1.97	0.2 W, 35 V, 40, AH
2N2808	0.9	1.00	0.08	0.0055	1.3	1.00	0.23	0.87	0.3 W, 30 V, 30, 1 GHZ, AH
2N2887	440	0.90	0.39	0.0106	103	0.80	0.49	18.05	25 W, 100 V, 1 A, 20, 140 MHZ, AHP
*2N2905	18.0	1.00	0.42	1.96	13.0	0.80	0.38	3.08	3 W, 60 V, 0.15 A, 100, 200 MHZ, SH
2N3019	3.31	1.05	0.31	0.0119	2.93	1.10	0.08	0.0747	0.8 W, 140 V, 100, 100 MHZ, AH
*2N3026	715	1.22	0.48	0.0046	332	0.61	0.43	7.64	25 W, 60 V, 3 A, 50, 60 MHZ, SHP
2N3108	80	0.50	0.33	0.0501	43.5	0.50	0.33	0.502	0.8 W, 100 V, 40, 60 MHZ, A
2N3119	130	0.80	0.29	17.2	65.5	0.23	0.11	91.8	1 W, 100 V, 50, 250 MHZ, SH
*2N3244	55	1.00	0.45	0.0188	43	0.80	0.41	1.953	1 W, 40 V, 50, 175 MHZ, SH

Transistor	α_n	α_i	R_{ee} (Ω)	R_{se} (meg)	R_{cc} (Ω)	R_{sc} (meg)	R_{bb} (Ω)	I_{se} (μA)	Θ_e (V^{-1})	I_{sc} (μA)	Θ_c (V^{-1})
2N3252	0.976	0.500	0.10	50	1.80	50	18.0	5.22E-2	40.0	4.20E-1	37.0
2N3309	0.976	0.500	0.10	50	1.60	50	3.00	8.40E-3	40.0	8.4 E-2	37.0
2N3498	0.988	0.666	1.3	50	1.66	50	12.7	6.98E-3	43.7	1.49E-2	43.0
2N3501	0.993	0.909	0	128	0.10	238	1.44	3.99E-5	30.2	3.66E-5	30.3
*2N3503	0.995	0.833	0	∞	0	3000	6.5	2.32E-5	27.2	2.34E-5	27.8
2N3600	0.977	0.492	8.5	50	11.5	50	12.0	9.18E-4	30.4	2.12E-3	28.4
2N3737	0.986	0.333	0.42	50	0.10	50	20.0	3.96E-9	38.0	7.89E-9	38.0
2N3766	0.990	0.834	0.41	50	0.32	50	23.0	7.29E-5	31.3	8.75E-4	32.0
2N3828	0.960	0.677	0	∞	17.0	∞	1.20	3.10E-8	38.0	1.40E-7	38.0
2N3904	0.990	0.500	0	5000	1.0	10000	2.0	7.92E-8	33.0	1.19E-5	25.9
*2N3913	0.990	0.910	5.8	50	34	50	127	8.24E-10	43.5	8.96E-10	43.5
*2N3915	0.997	0.969	2.1	50	18	50	131	2.08E-10	43.5	2.14E-10	43.5
2N3960	0.982	0.834	2.0	50	1.5	50	39	3.51E-10	38.1	1.22E-9	38.0

	C_{te} (pf)	V_{ze} (V)	N_e	K_{de} (pf/μA)	C_{tc} (pf)	V_{zc} (V)	N_c	K_{dc} (pf/μA)	Remarks
2N3252	55	0.90	0.40	0.0200	14	0.80	0.23	1.85	1 W, 60 V, 0.5 A, 40, 200 MHZ, SH
2N3309	24	1.00	0.33	0.0164	16	1.00	0.25	0.481	3.5 W, 50 V, 0.25 A, 10, 300 MHZ, AHP
2N3498	74.4	1.28	0.52	0.0874	19	0.90	0.38	27.5	1 W, 100 V, 40, 150 MHZ, AH
2N3501	80.0	0.50	0.33	0.0688	54.3	0.50	0.33	0.241	1 W, 150 V, 100, 150 MHZ, AH
*2N3503	21.8	1.63	0.50	0.217	10.7	0.94	0.36	0.885	0.7 W, 60 V, 130, 200 MHZ, SH
2N3600	2.0	0.50	0.15	0.0064	17	0.80	0.12	0.352	0.2 W, 30 V, 25, 850 MHZ, AH
2N3737	71.0	0.75	0.37	0.0012	9.1	0.75	0.27	2.69	0.5 W, 75 V, 20, 250 MHZ, SH
2N3766	224	0.94	0.43	0.332	122	1.15	0.31	11.0	20 W, 80 V, 1 A, 40, 15 MHZ, AP
2N3828	18	0.80	0.31	1.54	6.8	0.75	0.23	1.09	0.3 W, 40 V, 30, 360 MHZ, AH
2N3904	22.7	0.90	0.20	1.31	15.3	0.20	0.70	2.75	0.31 W, 60 V, 100, 300 MHZ, SH
*2N3913	21	0.80	0.50	0.220	21	0.80	0.50	2.48	0.4 W, 60 V, 40, 4 MHZ, SC
*2N3915	21	0.80	0.50	0.059	21	0.80	0.50	1.10	0.4 W, 60 V, 90, 10 MHZ, SC
2N3960	2.8	0.75	0.25	0.0037	3.75	0.75	0.26	0.0031	0.4 W, 20 V, 40, 1.6 GHZ, SH

Remarks:
Device dissipation (W); V_{CB} (V); $I_{C\,max}$ (A); Typical h_{FE} (dimensionless); f_T (KHZ, MHZ, or GHZ), FE = f_E and FB = f_B; Code. Some data omitted.

Code:

A = amplifier
AM = amplifier, multiple device
S = switch
SHP = switch, high speed, power

AH = amplifier, high frequency
AP = amplifier, power
SC = switch, chopper

AHP = amplifier, high frequency, power
SH = switch, high speed
SP = switch, power

INDEX

A

AC analysis, 50
    general case, 124
AC analysis program:
    flowchart, 87
    listing, 88
Addition, matrix, 19
Analog, 2
Analysis tasks, 2
Analysis tools, 5
Angle variations, 263
Automatic modeling, 374
Auxiliary circuit, 181
Average, statistical, 271

B

Branch, 54
    Network, 65
Branch-admittance matrix, 66
Branch current, 1

C

Capacitance, 73
Capacitor, transient model, 131
Central derivative, 296
Central limit theorem, 270
Circuit:
    definition, 1

Circuit (*cont.*)
    design task, 3
    diagram, 53
    graph, 53
Coefficient of coupling, 85
Compensated pulse attenuator method, 344
Complex eigenvalues, 231
Component specification, 356
Component value, random, 294
Computational considerations, 7
Conductance, 72
Conformable matrices, 20
Contact potential, 325
Controlled current source, 73
Controlled voltage source, 73
Correlated parameters, 289
Correlation coefficient, 290
Convolution, 241
Cramer's rule, 25
Cumulative frequency of distribution, 268
Current controlled current source, 73
Current controlled voltage source, 73
Curvature constant, 324

D

DC analysis, 50
DC analysis program 72
    flowchart, 79
    listing, 80

Dependent voltage source, 108
    replacement, 75
Derivatives, 254
Determinant, 25
Diagonal matrix, 22
Diffusion capacitance, 325
Diode, 322
    data, 357
    equation, 323
    models, 360
    model, piece-wise linear, 328
Direct notation, matrix, 19
Discrete component model, 372
Distinct eigenvalues, 230
Distribution of components, 268
    relative frequency, 268
    gaussian, 269
    rectangular, 269
DTL gate, example, 361
DTLM model, 377
Dynamic loading circuit, 373

E

Ebers-Moll equations, 333
Eigenvalues, 228
    complex, 231
    distinct, 230
Eigenvectors, 232
Elastance, 145
Elements, 74
Element value iteration, 170
Equilibrium condition, 135
Exponential matrix, 191
Extreme value analysis, 304

F

Final value solution, 165
Forced response, 190
Free-form input, 113
    flowchart, 115
Frequency response, 238

G

Gauss Elimination, 25
    Program, 30
Gauss-Jordan elimination, 29
Gauss-Seidel procedure, 32
Grading constant, 325

H

Header capacitance, 325
Histogram, 267

I

Ideal transformer coupling, 128
Identity matrix, 22
Incidence matrix, 51
Independent:
    current source, 73
    voltage source, 73
    loops 50
    nodes, 51
Inductance, 73
Inductor, transient model, 131
Initial condition, 176
Initial value solution, 152
Inner product, 20
In-place inversion, 43
Interval halving, 168
Inverse inductance matrix, 180
Inversion, matrix, 35
Iterative solutions
    Gauss-Seidel method, 32
    Newton-Raphson, 349
    simple iteration, 33

J

Jordan canonical form, 229

K

Kernel index notation, 19
Kirchhoff's laws, 50

L

Leakage resistance, 324
Linear algebra, 18
   equations, 24
Link, 50
Load line, 313
Loop, 50

M

Magnitude variations, 262
Main diagonal, matrix, 22
Matrix, 19
   addition, 19
   conformable, 20
   direct notation, 19
   inner product, 20
   inversion, 35
   kernel index notation, 19
   main diagonal, 22
   multiplication, 20
   norm, 193
   null, 22
   products, 23
   scalar multiplication, 20
   skew-symmetric, 22
   symmetric, 22
   transpose, 21
   whole matrix notation, 19
Mean value, 269
Mode, 269
Modeling, 10
Monte Carlo analysis, 292
Multiplication, matrix, 20
Mutual inductance, 73
   Model, 96

N

NAND gate example, 361
Network graph, 53
   vectors, 65
Newton-Raphson iteration, 349

Node, 1
   network, 50
Node admittance, 53
   direct calculation, 97
Node voltage derivatives, 257
   equations, 67
   variations, 254
Nonlinear analysis, 311
   input language, 354
Nonlinear elements, 315
Nonlinear program organization, 382
Non-periodic sources, 150
Norm, matrix, 193
Normal distribution, 269
NPN transistor model, 334
Null matrix, 22

O

Objective description, 16
Ohm's law, 52
One-parameter-at-a-time analysis, 296
Optimization, 16

P

Part, 50
Path, 50
Periodic sources, 148
Piecewise linear networks, 166
PNP transistor model, 337
Pointer vector, 103
Poles and zeros, 220
Program features, 14

R

Rectangular distribution, 269
Reference node, 50
Resistance, 72
Round-off errors, 37
   in addition/subtraction, 38
   in multiplication/division, 42
   minimization rules, 42

Round-off errors (*cont.*)
    quadratic equation, 48

S

Schur's theorem, 244
Sensitivity, 266
Sensitivity analysis program, 279
    flowchart, 281
    input, 287
    listing, 282
    results, 287
Simple iteration, 33
Simultaneous equations, 24
Sinusoidal sources, 148
Skew-symmetric matrix, 22
Small parameter changes:
    angles, 262
    formulas, 260
    magnitude, 262
Souriau-Frame algorithm, 222
Sparse matrix:
    AC analysis, 100
    AC analysis flowchart, 104
    transient analysis, 170
Standard deviation, 269
    calculation, 271
State, 177
    equation, 176
        program, 200
        recursive solution, 192
State analysis program, 203
State space analysis, 176
Statistical average, 271
    formulas, 269
    variations, 267
Stiff differential equations, 240
Superposition principle, 311
Surplus variables, 185
Symmetric matrix, 22

T

Tabular calculations:
    Gauss elimination, 27

Tabular calculations (*cont.*)
    Gauss-Jordan, 29
    matrix inversion, 44
    matrix multiplication, 22
Time-domain solution, 190
Time step selection, 244
Time variant sources, 147
Tolerance analysis, 252
    example, 299
Transfer function, 223
Transfer matrix, 223
Transient analysis, 130
    basic program, 150
    flowchart, 153
    general equations, 141
    mutual inductance, 142
    rudimentary program, 138
Transient solution, 190
Transistor data, 357
    gain variation, 338
Transistor models:
    Ebers-Moll, 333
    piecewise linear, 342
Transition capacitance, 325
Transpose, matrix, 21

U

Unaccompanied energy sources, 105
Uniform distribution, 269
Unit matrix, 22
Unit source method, 179

V

Vanishing derivative, 242
Vector, 22
Voltage controlled current source, 73
Voltage controlled voltage source, 73

W

Worst-case analysis, 274
    extreme values, 275
    formulas, 275

Worst-case analysis (*cont.*)
    slope approximation, 276
    slope changes, 277

Z

Zener voltage, 324